THE INFLUENCE OF BINARIES ON STELLAR POPULATION STUDIES

ASTROPHYSICS AND SPACE SCIENCE LIBRARY

VOLUME 264

THE INFLUENCE OF BINARIES ON STELLAR POPULATION STUDIES

Edited by

DANY VANBEVEREN

Astrophysical Institute,
Vrije Universiteit Brussel, Belgium &
Technological Institute Group T,
Leuven, Belgium

KLUWER ACADEMIC PUBLISHERS

DORDRECHT / BOSTON / LONDON

A C.I.P. Catalogue record for this book is available from the Library of Congress.

ISBN 0-7923-7104-6

Published by Kluwer Academic Publishers,
P.O. Box 17, 3300 AA Dordrecht, The Netherlands.

Sold and distributed in North, Central and South America
by Kluwer Academic Publishers,
101 Philip Drive, Norwell, MA 02061, U.S.A.

In all other countries, sold and distributed
by Kluwer Academic Publishers,
P.O. Box 322, 3300 AH Dordrecht, The Netherlands.

Coverpicture:
Near-infrared image of young binary stars with a faint companion (a planet?).
(Produced with the Near Infrared Camera and Multi-Object Spectrometer (NICMOS),
Hubble Space Telescope.) Image credit: Susan Terebey and NASA.

Printed on acid-free paper

Printed in the Netherlands.

Contents

Part VI CHEMICAL EVOLUTION OF GALAXIES

Preface

A star in a binary may interact with its companion and this complicates significantly its structure and evolution. During several decades in the past many astronomers hoped that the number of binaries is small. This is sometimes reflected in theoretical studies dealing with large populations of stars where it is ASSUMED (without any justification) that binaries do not alter the results.

The number of confirmed binaries increases as observations improve and we therefore thought that the time was right to organize a meeting where the main topic would be *the effect of binaries on stellar population studies.* The SOC proposed a program with extended general overviews and highlights of non-evolved and evolved binaries, with low mass, intermediate mass and massive components, the link with supernova explosions, pulsars and gamma-ray bursts, the effect of binaries on the chemical evolution of galaxies. Specialists in the different research fields were invited and have given very interesting and highly educational talks. As can be noticed by reading the proceedings, the authors have put a lot of effort in preparing their contribution.

Brussels, one of the 9 cultural cities of Europe in the year 2000, was choosen as stellar-population-meeting city. The 5-day-meeting started on the 21th of August and...in less than 2 days Andre Maeder felt in love with binaries, Tom Marsh had his first *Clochard* experience on Sunday night, Debra Wallace noticed that the difference between 120 V and 220 V cannot be solved by a simple wall-plug-adaptor, Brian Mason saw stars and stripes while trying to follow Walter van Rensbergen joging in the woods arround Brussels, about 20 participants became used to cold showers, all participants saw Brussels with clear sky during the whole meeting-week, Walter van Rensbergen and his son Hans organised a very nice excursion to Gent on Wednesday, Simon Portegies Zwart broke his record and lost the group in Gent already 5 min after arrival, and I...I hope that everyone tasted a significant number of *Great Beers of Belgium.*

DANY VANBEVEREN

I

OBSERVATIONS OF NON-EVOLVED BINARIES

BINARY STATISTICS IN CLUSTERS AND THE FIELD

Jean-Claude Mermilliod

Institut d'Astronomie, Université de Lausanne, CH-1290 Chavannes-des-Bois, Switzerland

Jean-Claude.Mermilliod@obs.unige.ch

Keywords: Binary statistics, cluster stars, halo field stars.

Abstract The available evidences for the frequency of binaries in clusters and for nearby and halo field stars are reviewed. More emphasis is given to the star-forming regions and star clusters, which have been surveyed in the past years with various techniques. In spite of the large efforts undertaken, the available material is still incomplete at several levels: (1) the survey of cluster members along the main sequence is insufficient, this is especially true for radial velocities, (2) the coverage of separation or periods is incomplete, especially in the period domain from 10^3 to 10^4 days for main-sequence stars in open clusters and for periods shorter than 10^3 days for pre-main-sequence objects, (3) evident biases exist in the samples due to the change of the detection capabilities as a function of luminosity ratios, and separation or periods. All these reasons lead to a complicated situation and one should be very careful when comparing the binary properties over different period domains and primary masses.

1. INTRODUCTION

It is not an easy task to describe the present status of binary statistics, because numerous biases still exist in the samples of detected binary systems. The situation is complicated because the various techniques used in binary surveys do not cover similar separation- or period domains, and, therefore, comparisons of binary rates over different mass ranges, obtained with several instruments with quite different detection limits, may have little meaning.

To illustrate these difficulties, the distribution of the known binaries in the Pleiades published by Bouvier et al. (1997) is a good example. This figure has two well separated peaks: the one on the left hand reflects the

D. Vanbeveren (ed.), The Influence of Binaries on Stellar Population Studies, 3–19.

distribution of periods for spectroscopic binaries, with periods less than 1000 days, and the second one results from the adaptive-optics binary survey. The limited information available, due to the small number of binaries, and the large gap in the period domain, especially between 10^3 and 10^4 days, are readily apparent.

Not only do the various techniques employed result in gaps over the period range, but they also have their own biaises. For example, Bouvier et al. (1997) show very clearly how the binary detection depends on the separation and the K-magnitude difference. If, for stars with similar magnitudes, i.e. $\Delta K \leq 1$ mag. binary stars can be resolved down to 0.1 arcsec, when $\Delta K \sim 6$, a secondary component is detected only if the separation is larger than 1 arcsec. Therefore, at large ΔK close companions are missed. This trend in function of the separation and magnitude difference is different for speckle interferometry surveys.

In a similar way, radial-velocity surveys favour the detection of systems with mass ratios larger than $q \sim 0.4$. Even with as much as 15 observations, the detection rate is less than 50% for periods of a few months (Melo & Mayor 2000). Due to the difficulties of pursuing long term radial-velocity observing programs, we find strong biaises against short-period spectroscopic binaries.

CCD photometric surveys, like those published by the Warsaw groups (Kaluzny 2000), discover mostly short periods eclipsing- and contact binaries, with periods usually shorter than 1 day. As the detection probability declines with period length, the information brought by this technique is limited to short periods.

Finally, analysis of binarity in photometric colour-magnitude diagrams gives also biaised information. Figure 1 of Hurley & Tout (1998), which presents, at various masses along the main sequence, binary loops on which the position of the mass ratio $q = 0.5$ is indicated, shows that the detection efficiency varies along the main sequence as a function of q. For primary masses less than 2 $M_{\odot}$, only binary systems with $q > 0.5$ can be detected, while for more massive stars, i.e. primaries with masses larger than 2.5 $M_{\odot}$, the detection can be extended to $q \sim 0.3$. Extensive radial-velocity surveys conducted in the Pleiades and Praesepe clusters (Mermilliod et al. 1992, Mermilliod & Mayor 1999) have shown that half of the spectroscopic binaries may be missed by the photometric analysis of $(V, B-V)$ colour-magnitude diagrams. The systems are not detected when the light contribution of the secondary is smaller than 0.02 - 0.03 mag. In diagrams involving redder colour indices, the contributions of the fainter and redder companions are more important and the binary effects more apparent.

The above limitations and biaises imply that one should be very careful in building an overall picture of duplicity along the main sequence. In no case have we reached complete binary detection in terms of mass ratio and separation. Therefore, we are allowed to perform comparisons, for a given primary mass, over limited mass-ratio values and within given period- or separation ranges. However, it is very encouraging that binary-detection surveys have been made in various clusters, by complementary techniques and we can hope that the present situation will improve drastically in a near future.

2. FIELD STARS

Nearby F7 - K stars. Binarity in nearby field stars later than F7 has mainly been studied by the Geneva CORAVEL group. Duquennoy & Mayor (1991) have obtained a period distribution which became a reference. Results for the G, K and M dwarfs have been presented by Mayor et al. (2000) at the IAU Symposium 200 based on a systematic observing program run over nearly 20 years with the northern and southern CORAVEL spectrometers (Baranne et al. 1979). The global percentage of binaries (or multiple stars) in the solar vicinity is close to 2/3. The frequency of F7-K dwarfs with periods less than 10 years is equal to 11% and reaches 9±2% for systems with periods less than 1000 days. The frequency is the same for the *unbiaised* sample (249 stars, 26 SBs) as for the *extended* sample (570 stars, 57 SBs). Properties of the combined G+K sample have been presented by Halbwachs et al. (2000).

Nearby M stars. Results of a long-term radial-velocity search for companions of nearby M-type dwarf stars have been presented by Udry et al. (2000). The total number of M dwarfs observed with the two CORAVEL scanners over both hemispheres is 450. Orbital elements were obtained for 31 binaries among which 23 were previously unknown. A high-precision radial-velocity survey have been initiated from the Haute-Provence Observatory with the ELODIE spectrograph (Baranne et al. 1996) to observe all northern M dwarfs closer than 10pc. A similar survey has been undertaken in the southern sky with FEROS on the 1.5m telescope at La Silla. The percentage of spectroscopic binaries detected among nearby M dwarfs is equal to 8±3% for systems with $P < 1000$ days.

Proper-motion stars. Orbital elements for 80 halo stars have been published by Latham et al. (1988, 1992), and Latham et al. (1997) report 201 orbits for spectroscopic binaries belonging to their extensive sample of 1447 proper-motion stars. But no binary frequency is given.

Early F-type stars. Nordström et al. (1997) have investigated the binarity of early F-type stars. Among a sample of 595 stars, the binary frequency is comprised between 22% and 29%, depending on the corrections made to transform the magnitude-limited sample into a volume-limited sample.

Am stars. A large number of radial-velocity observations have been obtained by Debernardi (2000) to extend the sample of orbital elements for Am stars. Seventy new orbits have been determined. The (eccentricity, period) diagram for the double-lined systems (SB2) shows a well defined period cut-off around 7 days, while the transition from circular to elliptic orbits is quite fuzzy for the single-lined systems, resulting probably from various evolutionary histories of the binaries. The binary frequency has not been computed yet because of the biaises introduced by the observability of Am stars with the CORAVEL.

3. GLOBULAR CLUSTERS

Binaries in globular clusters have been thoroughly discussed by Hut et al. (1992). The overall percentage of binaries is small: < 2 - 3 %. Mateo (1996) lists the eclisping binaries identified in CCD surveys in 11 globular clusters. The number ranges from 1 to 10 EBs per cluster. Extensive on-going observations described by Kaluzny et al. (1996) identified 12 EW and EA binaries in ω Cen. Extrapolation to the number of non-eclipsing binaries is still hazardous. Radial-velocity observations of 121 red giants in M71 (Barden & Armandroff 1996) result in binary-rate estimates which have still large uncertainties.

4. PRE-MAIN-SEQUENCE STARS

The binarity of pre-main-sequence stars has been summerized by Mathieu (1994) and Duchêne (1999). In this stellar domain, most binaries known result from speckle interferometry or adaptive optics surveys. Therefore the periods are in the domain of 10^4 to 10^7 days. There seems to be a tendency to find a higher binary frequency in looser aggregates, like Taurus-Auriga, than in Orion, with respect to the mean value defined by the nearby open clusters.

Recently, Levine et al. (2000) and Barsony & Ressler (2000) have surveyed the ρ Oph star-forming region for binaries with K- and N band imaging and found many double stars. A near-infrared speckle-interferometry survey yielded 67 binaries in Taurus-Auriga, Upper Scorpius, Chamaeleon and Lupus (Woitas & Leinert 2000). The resulting mass-ratio distribution is flat.

Radial-velocity surveys searching for spectroscopic binaries are not advanced enough to produce results comparable to those available for the lower main-sequences of the nearby open clusters. Melo & Mayor (2000) describe a high-resolution survey to study the binarity of T Tauri stars. So far, only 1 to 3 observations have been obtained. New multifiber spectrographs will permit, in a near future, to study in more detail the incidence of spectroscopic binaries in PMS stars. One interesting point will be the frequency of triple systems in Taurus-Auriga, due to the large number of "wide" binaries.

5. NGC 6231 AND O-TYPE STARS

Open clusters, and especially the young ones, are the direct results of star formation processes. Their characteristics are important to describe the products of star formation and the incidence of binarity or the star multiplicity. Among the characteristics that are important, we can mention the number of stars of various masses formed (initial mass function), the percentage of binaries, triples, .., formed along the main sequence and, of course, the distribution of their parameters (P, e, q).

NGC 6231 seems to be one of the more promising object to determine this information, although its distance is about 1600 pc. It contains 14 O-type stars, and presents a high binary rate among the most massive stars. Recent observations, combined with available data, have permitted to determine new orbital elements. Among the 14 O-type stars, 11 are binaries and all but one periods are shorter than 8 days. Preliminary results on NGC 6231 have been published by Mermilliod & García (2000) and the whole study has been submitted to A&A (García & Mermilliod 2000). Raboud (1996) also obtained a high binary frequency for B-type stars in this cluster, although the results are based on only 2 radial velocities per star.

To compare with NGC 6231 result, the WEBDA open cluster database (*http://obswww.unige.ch/webda/*) has been searched for literature data and information on detected binaries in other very young open clusters containing O-type stars. Data were found for eight clusters and a clear variation of the binary frequency was obtained from rates as high as 80% in IC 1805 (8/10) and NGC 6231 (11/14) down to values as low as 14% in Trumpler 14 (1/7). This is the first well documented case of variation of the binary frequency among open clusters. The orbital periods, available mainly for NGC 6231 (García & Mermilliod 2000), Tr 16 (Levato et al. 1991) and Tr 14 (Levato et al. 1991), are mostly shorter than 10 days.

In poorer clusters, typically those with only one apparent O-type member, like NGC 2264, 2362, 6383, 6604 and others, Mermilliod & García (2000) have found that all except one O-type stars are in fact close binaries, sometimes eclipsing, with a short period around 3 days, and that all present a multiplicity equal to or larger than 3, with a few examples of trapezium systems. Several, but not all, massive multiple systems are located close to the cluster centers producing a strong mass segregation.

Therefore, although binary detection of low-mass stars has received more attention in the recent years, the duplicity of massive stars also contains much information on the star- and binary formation in dense surroundings like open clusters. A list of 30 open clusters with 1 to 7 O-type members has been compiled by Mermilliod & García (2000) to serve as a basis for further observations.

6. G-TYPE STARS IN OPEN CLUSTERS

Extensive radial-velocity obervations have been conducted in several nearby open clusters (Duquennoy et al. 1992) covering the spectral domain F5 - K0. Results have been published so far for the Pleiades and Praesepe clusters (Mermilliod et al. 1992; Mermilliod & Mayor 1999), 12 and 20 orbits have been respectively determined. But all orbits have not been completed due to the length of some periods and to the fact that both CORAVEL spectrographs are out of commissionning since 1998. Latham and collaborators are working on extensive surveys in the Hyades and M67 (Stefanik & Latham 1992; Latham et al. 1997) which will provide a wealth of information.

The results proved the validity of the interpretation of the main sequence spread in terms of duplicity. However, about half of the binaries cannot be detected by photometric means alone, while a number of obvious photometric binaries are not detected from radial-velocity observations. Part of the solution has been brought by adaptive-optics observations. A large part of these latter stars are in fact seen as double with separations too wide for spectroscopic detection and to small for visual detection by classical means (Bouvier et al. 1997). Extensive observations of the Praesepe cluster gave the same results (Bouvier et al. 2000).

Altough we begin to have more precise ideas of the binary characteristics of lower main sequence stars, i.e. with spectral type later than F5, a large fraction of the binaries with mass ratios $q < 0.3$ have certainly not been detected, and the domain of types later than K0 remains to be investigated, except in the Hyades cluster. Very few radial-velocity observations have been obtained so far because the magnitude of these stars require larger telescopes or longer exposure times.

Photometric CCD surveys have discovered numerous contact or eclipsing binaries in old open clusters for which no spectroscopic survey have been made so far, like for exemple in Cr 261 (Mazur 1995). As only one such system is known in Praesepe (TX Cnc), these findings are a strong argument for the evolution of binary characteristics with time.

The (eccentricity, period) diagram with 78 binaries from the Hyades, Praesepe, Coma Ber, Pleiades and α Persei clusters shows a well defined cut-off period at 8 days and similar behaviour for all clusters. The similarity of the transition between circular and elliptic orbits is predicted by the theory of Zahn & Bouchet (1989).

For the Hyades cluster, Perryman et al. (1998) derived an overall binary frequency of 40%, but a higher rate is found in the central 2pc: 61%, which is mostly due to spectroscopic binaries. On the opposite, Patience et al. (1998) found that the binary frequencies inside and outside 3pc are nearly identical: 14/33 within 3pc and 52/129 outside 3pc.

In the Pleiades as well as in Praesepe, Raboud & Mermilliod (1998) have found that the mass function of the primaries of spectroscopic binaries and of single stars are not the same. This results needs to be confirmed for more clusters, but can be so far explained as resulting either from the formation, or, from dynamical interactions with component exchange.

7. CLUSTER RED GIANTS

The study of spectroscopic binaries with red-giant primaries is important for two main reasons:

1 to investigate the empirical tracks in the colour-magnitude diagram, one need first to detect the binaries, which can be displaced to the blue due to the effect of the companion on the combined colours, and would alter the definition of the evolutionary path,

2 to investigate the binary frequency and the binary properties.

So far, slightly more than one hundred orbits have been determined for spectroscopic binaries with a red-giant primary in open clusters. Among them 90% have been obtained from CORAVEL observations.

The overall binary frequency is 23%. The only exception seems to be NGC 6705 (M11) for which 3 binaries have been discovered among 31 red giants (Mathieu et al. 1986). Further observations should be done to check for the presence of additional binaries. Old data on NGC 188 lead to 6 binaries among 55 red giants, but the long term program on this cluster done at Victoria has probably changed this value.

Most information on binary properties comes from the "classical" (eccentricity, period) diagram, in which two significant periods have been found. The first one is related to the shortest period that a binary red giant can have without entering in contact. This happens for a 2 $M_{\odot}$ stars at about 40 days (Figure 1). So we do not observe red-giant binaries with periods smaller than 40 days. It also implies that the numerous binaries on the main sequence which have periods shorter than this value do follow other evolutionary path because they got into contact before reaching the bottom of the ascending red-giant branch.

The second period, like for dwarfs, corresponds to the transition between circular and elliptical orbits. This transition occurs for periods between 120 and 150 days depending on the mass of the primary, as shown by Mermilliod & Mayor (1996). This phenomenon is related to the primary mass and the maximum radius that the red giants reach at the top of the red giant branch.

Figure 1 shows the lines of limiting periods at which the evolving stars fill their Roche lobes at various ages. The solid lines are the isochrones computed from the models of Schaller et al. (1992) for $\log t = 8.0$, 8,2, 8.6, 8.8 and 9.0. The dashed lines are the lines of equal Roche-lobe filling periods. Typically, for binaries with periods of 1 day (mass ratio $q = 0.5$), the Roche lobe is filled before the primary reaches the end of the core hydrogen-burning phase. The mass of a red giant at the red giant tip for $\log t = 9.0$ is close to 2 $M_{\odot}$. From figure 1, it can be seen that the limiting period should be close to 40 days, which corresponds to the observed limit. We can therefore consider that the main features of the (eccentricity, period) diagram for red giants are rather well understood.

However, it is difficult to correct the population of red giants for missing objects in the computation of the ratio of the number of red giants to that of main-sequence stars, because such a correction depends both on the binary frequency and on the distribution of periods. Indeed it remains to be proved that the overall properties of binaries in open clusters (frequency, period- and mass-ratio distributions) are similar.

Integration of the equations of Zahn & Bouchet (1989) have permitted to follow the evolution of the orbital parameters as a result of the stellar evolution in the primary. Figure 2 displays this evolution for a binary with a primary mass of 2 $M_{\odot}$ and an initial orbital period of 100 days, and a rotational period of 1 days.

The upper panel displays the evolution of the primary radius during its evolution from the upper main sequence to red giant tip and core

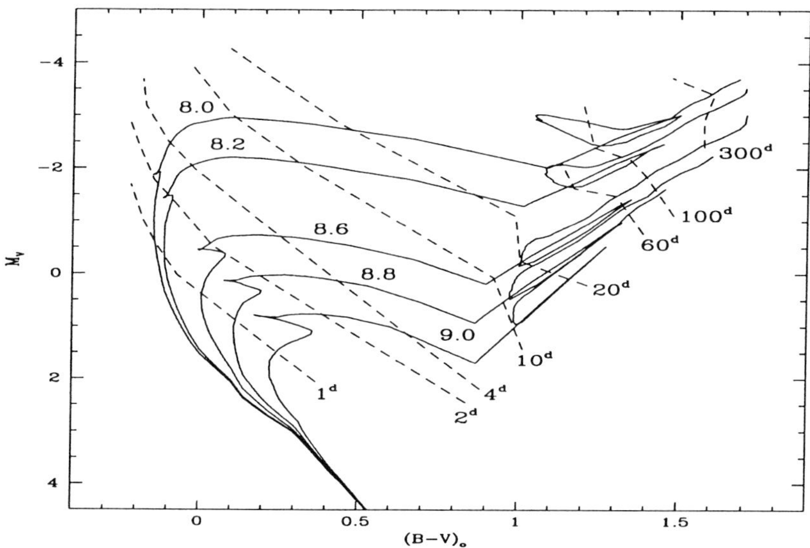

Figure 1 Limiting periods for filling the Roche lobe across the colour-magnitude diagram in the age interval $8.0 < \log t < 9.0$, for a mass ratio $q = 0.5$.

helium burning phase (lower line). The upper line show the limiting radius for Roche lobe overflow.

The middle panel shows the evolution of eccentricity, which abruptly decreases to zero when orbital and rotational periods are synchronized.

The bottom panel represents the evolution of the orbital period (upper curve) and the rotational period (lower curve). We notice the small diminution of the orbital period which appears simultaneously with the circularization and the rapid increase of the rotational period. Synchronism is temporarily lost when the primary radius decreases due to the settlement on the core-helium-burning phase (clump).

For an initial orbital period of 150 days, the eccentricity decreases to 0.05, not to zero, and synchronism is therefore not achieved during the core-helium-burning phase. The eccentricity finally reaches zero when the star moves on the asymptotic giant branch, slightly before the red giant fills its Roche lobe (Baumgartner 1993).

Computations for initial rotation periods of 1 and 2 days lead to the same final result, because the final rotational period is imposed by the synchronization with the orbital period. Binaries with initial mass ratios

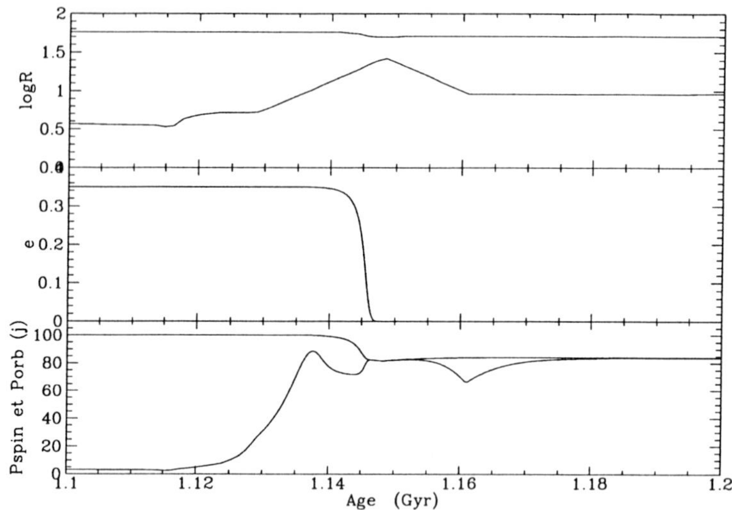

Figure 2 Evolution of the radius, eccentricity, and orbital and rotational periods during the circularization process for a binary with an initial period of 100 days and a primary mass of 2 $M_\odot$.

$q = 0.5$ and $q = 0.96$ also reach the same final state, although the individual evolutionary tracks of the rotation periods are different. However, differing initial eccentricities, computations have been performed with $e = 0.2$ and $e = 0.4$, produce different results. Although the synchronisation occurs at the same evolutionary phase, the final orbital periods are not identical. The period is longer for the binary with $e = 0.2$.

8. RECENT RESULTS ON MS BINARY FREQUENCY

Janes & Kassis (1997) have investigated the photometric binary frequency in the colour-magnitude diagrams of 11 open clusters covering a large age range, and on the basis of various colour-indices ($B-V$, $B-I$, $V-I$). They obtained binary rates ranging from 14% to 36%, with no obvious trends with ages, concentration or richness, and conclude to the probable variation of the binary frequency among open clusters.

The indices used to describe the frequency of companions are : **MSF** = (b + t) / (s + b + t), **CSF** = (b + 2t) / (s + b + t) as used by Patience et al. (1998), and **BF** = (b + t + q) / (s + b + t + q) (Levine et al. 2000), where *s* stands for "single", *b* for binary, *t* for triples and *q* for quadruples. The binary percentage (col. "Rate" in Table 1) is usually computed by simply dividing the number of binaries by the total number of stars in the sample.

Table 1 Recent results on binary frequency of F - M dwarfs in various environnements

	Number of			Rate or		Domain of		
Sample	Stars	Bin	Orb	BF	CSF	P or Sep.	Technique	Reference
Early-F	595	172	20	29%		no limit	RV	Nordström et al. (1997)
F7 - K	249	26	26	11%		< 10 yr	RV	Mayor et al. (2000)
F7 - K	249	26	26	9%		< 1000 d	RV	Mayor et al. (2000)
M dwarfs				8%		< 1000 d	RV	Udry et al. (2000)
T Tauri	195	31		0.14		120 - 1800 AU	Imaging	Brandner et al. (1996)
M67	406	96	53	13.3%		< 1000 d	RV	Latham et al. (1997)
Hyades	167	33			0.30	0.”044 - 1.”34	Speckle	Patience et al. (1998)
Praesepe	103	24	20	20%		< 1000 d	RV	Mermilliod & Mayor (1999)
Pleiades	100	13	12	12.5%		< 1000 d	RV	Mermilliod et al. (1992)
Orion	42	4		9.5%		0.”14 - 0.”5	Imaging	Petr et al. (1998)
Taurus	70	27		0.43		0.”13 - 13.”	Speckle	Köhler & Leinert (1998)
IC 348	66	12		19%		0.”1 - 8.”	AO imaging	Duchêne et al. (1999)
NGC 2024	495	159		0.55		> 0.”3	HST	Levine et al. (2000)
ρ Oph	47	14		30%		0.”25 - 3.”5	Imaging	Costa et al. (2000)

9. DISCUSSION

During a large fraction of the twentieth century, binary surveys by classical techniques, namely observations of visual binaries and radial-velocity determination from spectra taken on glass plates brought a lot of information, but have finally reached their limits in terms of separation, periods and limiting magnitudes. The new instruments and recent techniques have given an important new impetus to the search and study of binary stars.

Two-dimensional CCD photometry has permitted to efficiently search for variable stars over the whole cluster area (Kaluzny 2000), which was not possible when stars were observed one by one with photoelectric photometers. These observations cover mainly the domain of periods shorter than 1 day. Radial-velocity scanners, like the CORAVEL (Baranne et al. 1979) or the CfA Digital Speedometers (Latham 1992) have permitted to investigate the duplicity of solar-type stars in the field and halo, and in nearby open clusters, and to cover the period domain from 1 day to several thousand days. In addition to these modern techniques, near-infrared imaging with adaptive optics and speckle interferometry have provided a completely new approach to binary detection and have filled gaps between the spectroscopic- and visual observations, i.e. covering the domain of separation between 0.05 and 1.0 arcsec.

However, because each technique is dedicated to specific targets, we have not been able to completely survey any open cluster and cover all domains of periods, separations and mass ratio. The Hyades is probably the only exception thanks to the systematic observing programs undertaken by Mason et al. (1993) and Patience et al. (1998) (speckle interferometry) and Stefanik & Latham (1992) (radial velocity). The example of the Pleiades mentionned in the introduction (Bouvier et al. 1997) illustrates very clearly the point. If star-forming region, like Taurus-Auriga, have been extensively surveyed for "wide" binaries, not enough work has been devoted to radial-velocity observations. As a consequence we do not know how many spectroscopic binaries are hidden both among "single" stars or the components of the detected double stars.

Basic questions to which we would like to have an answer concern the proportions of double, triple and quadruple systems, the possibility of variations of the binary frequency along the main sequence, the existence of changes in the orbital-period distributions with the primary masses and whether the mass-ratio distribution varies along the main sequence. Simulations and synthetic colour-magnitude diagrams (Carraro et al. 1993) and mathematical analysis of the main-sequence characteristics in the Pleiades (Kähler 1999) and Praesepe (Kroupa & Tout 1992) cannot really solve the question. One can add as many binaries as seem necessary to fit the observed distributions as long as they remain undetectable due to the faintness of the secondary (white and brown dwarfs, planets). The discovery of the double-lined brown dwarf PPl 15 in the Pleiades (Basri & Martín 1999) proves that duplicity is still present among very low-mass substellar objects.

Clearly, larger telescopes and new instruments, like the multi-fiber spectrographs in use or in construction or the new interferometers, will provide fundamental insights in these questions thanks to their capabilities to resolve very close binaries and even spectroscopic binaries, and the possibility multi-fiber spectrographs offer to get spectra of 100 stars at the time. In a near future, large and systematic surveys, like those undertaken by the WYIN open-cluster study (Mathieu 2000) or by Latham et al. (1997) on M67 and the Hyades, will provide unrivalled binary statistics down the main sequence for several open clusters, which is badly needed.

After individual clusters will be known in detail, the next questions will concern the existence of significant cluster to cluster differences in the above mentionned properties. The results based on the O-type stars, with a variation of the binary frequency between 14% and 80% (Mermilliod & García 2000), and on the analysis of high-quality colour-magnitude diagrams (Janes & Kassis 1997) provide first evidences for the existence of such variations.

We would also like to know also how the stellar and dynamical evolutions alter these distributions, because many clusters observed so far have been submitted to such evolutions. The physical condition inside a cluster, namely its stellar density, seems to play a role, because of the lack of large separation binaries found in the Orion Trapezium cluster which possesses a very dense core. The underlying question is whether these differences are due to formation conditions or to dynamical evolution of the stellar systems. Accordingly, the whole research field on double and multiple stars has strong implications for the scenarios of stellar formation. Clear information on binary frequency, multiplicity rate, period distributions and mass ratios should, in principle, severly constrain theoretical models of cloud fragmentation and star formation. The rich variety of binary star phenomena in star clusters makes them natural laboratories for studying the formation and evolution of binary stars (Janes & Kassis 1997).

A whole new era has opened for double- and multiple star research and for star-cluster investigations.

References

Baranne A., Mayor M. & Poncet J.-L. 1979, Vistas in Astron. **23**, 279

Baranne A., Queloz D. & Mayor M. 1996, A&AS **119**, 373

Barden S.C. & Armandroff T.E. 1996, in "The Origins, Evolutions, and Destinies of Binary Stars in Clusters", Eds. E.F. Milone & J.-C. Mermilliod, ASPC **90**, 89

Barsony M. & Ressler M.E. 2000, in "Birth and Evolution of Binary Stars", poster proceedings of IAU Symp. 200, Eds. B. Reipurth & H. Zinnecker, p. 47

Basri G. & Martín E.L. 1999, AJ **18**, 2460

Baumgartner G. 1993, Diploma thesis, (Ecole Polytechnique Fédérale de Lausanne)

Bouvier J., Duchêne G., Mermilliod J.-C. & Simon T. 2000, in preparation

Bouvier J., Rigaut F. & Nadeau D. 1997, A&A **323**, 139

Brandner W., Alcalá J.M., Kunkel M., Moneti A. & Zinnecker H. 1996, A&A **307**, 121

Carraro G., Bertelli G., Bressan A. & Chiosi C. 1993, A&AS **101**, 381

Costa A., Jessop N.E., Yun J.L. & Santos C.A. 2000, in "Birth and Evolution of Binary Stars", poster proceedings of IAU Symp. 200, Eds. B. Reipurth & H. Zinnecker, p. 49

Debernardi Y. 2000, in "Birth and Evolution of Binary Stars", poster proceedings of IAU Symp. 200, Eds. B. Reipurth & H. Zinnecker, p. 161

Duchêne G. 1999, A&A **341**, 547

Duchêne G., Bouvier J. & Simon T. 1999, A&A 343, 831

Duquennoy A. & Mayor M. 1991, A&A **248**, 485

Duquennoy A., Mayor M. & Mermilliod J.-C. 1992, in "Binaries as Tracers of Stellar Formation", Eds A. Duquennoy & M. Mayor, (Cambridge Univ. Press), p. 52

García B. & Mermilliod J.-C. 2000, A&A (submitted)

Hut P., McMillan S., Goodman J., Mateo M., Phinney E.S., Pryor C., Richer H.B., Verbunt F. & Weinberg M. 1992, PASP **104**, 981

Janes K.A. & Kassis M. 1997, in "Third Pacific Rim Conference on Recent Development of Binary Star Research", Ed. K.-C. Leung, ASPC **130**, 107

Halbwachs J.-L., Arenou F., Mayor M. & Udry S. 2000, in "Birth and Evolution of Binary Stars", poster proceedings of IAU Symp. 200, Eds. B. Reipurth & H. Zinnecker, p. 132

Hurley J. & Tout C.A. 1998, MNRAS **300**, 977

Kähler H. 2000, A&A **346**, 67

Kaluzny J., 2000, in "The Impact of Large-Scale Surveys on Pulsating Star Research", IAU Coll. No 176, Eds. L. Szabados & D. Kurtz, ASPC **203**, 19

Kaluzny J., Kubiak M., Szymański M. & Udalski A. 1996, in "The Origins, Evolutions, and Destinies of Binary Stars in Clusters", Eds. E.F. Milone & J.-C. Mermilliod, ASPC **90**, 38

Köhler R. & Leinert C. 1998, A&A **331**, 977

Kroupa P. & Tout C.A. 1992, MNRAS **259**, 223

Latham D.W. 1992, in "Complementary Approaches to Double and Multiple Star Research". IAU Coll. No 135, Eds. H. McAlister & W. Hartkopf, ASPC **32**, 110

Latham D.W., Mathieu R.D. & Milone A.A.E. 1997, in "Third Pacific Rim Conference on Recent Development of Binary Star Research", Ed. K.-C. Leung, ASPC **130**, 113

Latham D.W., Mazeh T., Carney B.W., McCrosky R.E., Stefanik R.P. & Davis R.J. 1988, AJ **96**, 567

Latham D.W., Mazeh T., Stefanik R.P., Davis R.J., Carney B.W., Krymolowski Y., Laird J.B., Torres G. & Morse J.A. 1992, AJ **104**, 774

Latham D.W., Stefanik R.P., Mazeh T., Torres G. & Carney B.W. 1997, in "Cool Stars, Stellar System and the Sun", ASP Conf. Ser. **154**, 2129

Levato H., Malaroda S., Morrell N., García B. & Hernández C. 1991, ApJS **75**, 869

Levato H., Malaroda S., Morrell N., García B. & Grosso M. 2000, PASP **112**, 359

Levine J.L., Lada E.A. & Elston R.J. 2000, in *Birth and Evolution of Binary Stars*, poster proceedings of IAU Symp. 200, Eds. B. Reipurth & H. Zinnecker, p. 69

Mason B.D., McAlister H.A., Hartkopf W.I. & Bagnuolo Jr. W.G. 1993, AJ **105**, 220

Mathieu R.D., Latham D.W., Griffin R.F. & Gunn J.E. 1986, AJ **92**, 1100

Mathieu R.D., 1994, ARA&A **32**, 465

Mathieu R.D., 2000, in "Stellar Clusters and Associations: Convection, Rotation, and Dynamos", Eds R. Pallavicini, G. Micela & S. Sciortino, ASPC **198**, 517

Mateo M. 1996, in "The Origins, Evolutions, and Destinies of Binary Stars in Clusters", Eds. E.F. Milone & J.-C. Mermilliod, ASPC **90**, 21

Mayor M., Udry S., Halbwachs J.-L. & Arenou F. 2000, IAU Symp. No 200, *Birth and Evolution of Binary Stars*, Eds. B. Reipurth & H. Zinnecker, ASPC (in press)

Mazur B., Krzeminski W. & Kaluzny J. 1995, MNRAS **273**, 59

Melo C. & Mayor M. 2000, in *Birth and Evolution of Binary Stars*, poster proceedings of IAU Symp. 200, Eds. B. Reipurth & H. Zinnecker, p. 13

Mermilliod J.-C. & García B. 2000, IAU Symp. No 200, *Birth and Evolution of Binary Stars*, Eds. B. Reipurth & H. Zinnecker, ASP Conf. Ser. (in press)

Mermilliod J.-C. & Mayor M. 1996, in "Cool Stars, Stellar Systems and the Sun", Eds. R. Pallavicini & A.K. Dupree, ASPC **109**, 373

Mermilliod J.-C. & Mayor M. 1999, A&A **352**, 479

Mermilliod J.-C., Rosvick J.M., Duquennoy A. & Mayor M. 1992, A&A **265**, 513

Nordström B., Stefanik R.P., Latham D.W. & Andersen J. 1997, A&AS **126**, 21

Patience J., Ghez A.M., Reid I.N., Weinberg A.J. & Matthews K. 1998, AJ **115**, 1972

Perryman M.A.C., Brown A.G.A., Lebreton Y., Gómez A., Turon C., Cayrel de Strobel G., Mermilliod J.-C., Robichon N., Kovalevski J. & Crifo F. 1998, A&A **331**, 81

Petr M.G., Coudé du Foresto V., Beckwith V.V.W., Richichi A. & McCaughrean M.J. 1998, ApJ **500**, 825

Raboud D. & Mermilliod J.-C. 1998, A&A **333**, 897

Raboud D. 1996, A&A **315**, 384

Schaller G., Schaerer D., Meynet G. & Maeder A. 1992, A&AS **96**, 269

Stefanik R.P. & Latham D.W. 1992, in "Complementary Approaches to Double and Multiple Star Research". IAU Coll. No 135, Eds. H. McAlister & W. Hartkopf, ASPC **32**, 173

Udry S., Mayor M., Halbwachs J.-L. & Arenou F. 2000, in "Microlensing 2000: A New Era of Microlensing Astrophysics", Eds. J.W. Menzies & P.D. Sackett, ASPC (in press)

Woitas J. & Leinert Ch.. 2000, in "Birth and Evolution of Binary Stars", poster proceedings of IAU Symp. 200, Eds. B. Reipurth & H. Zinnecker, p. 57

Zahn J.-P. & Bouchet L. 1989, A&A **223**, 112

STATISTICS OF NON EVOLVED BINARIES WITH A B TYPE PRIMARY

Walter van Rensbergen

Astrophysical Institute, Vrije Universiteit Brussel, Pleinlaan 2, 1050 Brussels, Belgium

wvanrens@vub.ac.be

Keywords: Non-evolved binaries, mass-ratio distribution, B primaries

Abstract Assuming that binaries originate in star forming clouds with a Salpeter-type IMF an analytical expression was derived for the mass-ratio distribution for non evolved binaries. Selection effects governing the observations were taken into account in order to compare theory with observations. Theory was optimized so as to fit best with the observed q-distribution of SB1s and SB2s with a non-evolved B-type primary. Many real binaries will not be observed as SB1s, so that the number of real SB1s could be more than twice as high as observed.

1. INTRODUCTION

The mass-ratio distribution of binary stars was treated extensively by Hogeveen (1991). He showed how an observed q-distribution can be transformed into a real one. In the present contribution the Hogeveen method was critically revisited and applied onto the population of non evolved spectroscopic binaries (SBs) with a B type primary.

2. THE OBSERVED MASS RATIO DISTRIBUTION

From the catalogue of Batten et al. (1989) we selected 113 SBs with a non evolved early B type [B0-B3] and 119 SBs with a non evolved late B type primary [B4-B9.5].

The mass-ratio for SB2s was determined from observations for the primary (subscript 1) and the companion (subscript 2):

$$q = \frac{M_2}{M_1} = \frac{K_1}{K_2} = \frac{a_1}{a_2} \tag{2.1}$$

D. Vanbeveren (ed.), The Influence of Binaries on Stellar Population Studies, 21–35.

q being the mass-ratio, M the mass, K the maximum radial orbital velocity and a the semi-major axis of the orbital ellipse.

The results are included in table 2.

The mass ratio for SB1s was determined from quantities which are observed for the primary only:

$$f(m) = \frac{M_2^3 sin^3 i}{(M_1 + M_2)^2} = (1.0385 \times 10^7) K_1^3 (1 - e^2)^{3/2} P \qquad (2.2)$$

f(m) being the mass function, with K in km/s and the orbital period P in days. The calibration of the mass of the primary was taken from Vanbeveren et al. (1998) for early B stars. For late B stars we used the evolutionary tracks from Schaller et al. (1992). In that case, the mass of a star on a track in the middle between the ZAMS and the TAMS has been taken as representative for a particular spectral type.

Abt and Levy (1985) proposed to evaluate the mass ratio using relation (2.2) sampled at $i = 15^o$, 45° and 75°, representing respectively the intervals $[0^o - 30^o]$, $[30^o - 60^o]$ and $[60^o - 90^o]$. Assuming a random distribution of the inclination angle i, one writes the fractional number of systems in the range $[i, i + di]$ as: df = $\sin i \, di$. The probability for the inclination angle to occur between α and β is thus f$[\alpha$ - $\beta]$ = cosα - cosβ. This determines f(15°) = 0.134, f(45°) = 0.366 and f(75°) = 0.5.

A more refined method is given by Vanbeveren et al. (1998), evaluating relation (2.2) between $[i_{min} - 90^o]$. The value of i_{min} is the smallest value of the inclination angle for which expression (2.2) is meaningful. The range $[i_{min} - 90^o]$ determines for each binary a range of possible mass-ratio's between q_{min} $(i = 90^o)$ and q_{max} $(i = i_{min})$. The probability for a mass-ratio to occur has been calculated with this method for the five q-bins used in table 2.

Figure 1 considers a typical case HD 19820: a well known SB2 with q $\approx$ 0.486. For such a system the mass function f(m) can also be calculated with expression (2.2). Hence the probability for a certain q-value to occur can be calculated with the method of Abt and Levy and with the Vanbeveren et al. method. Figure 1 compares both methods mutually. Their comparison with the real value is shown for completeness. For individual SB1s, significant differences occur between the methods of Abt and Levy, and Vanbeveren et al. for the determination of q. In order to obtain an idea of the reliability of both methods, table 1 shows the distribution over five q-bins of the 95 SB2s with a non evolved O or B type primary, which can be found in the catalogue of Batten et al. (1989). Table 1 shows that individual differences between the two methods tend to cancel when summed over a large number of stars. Since

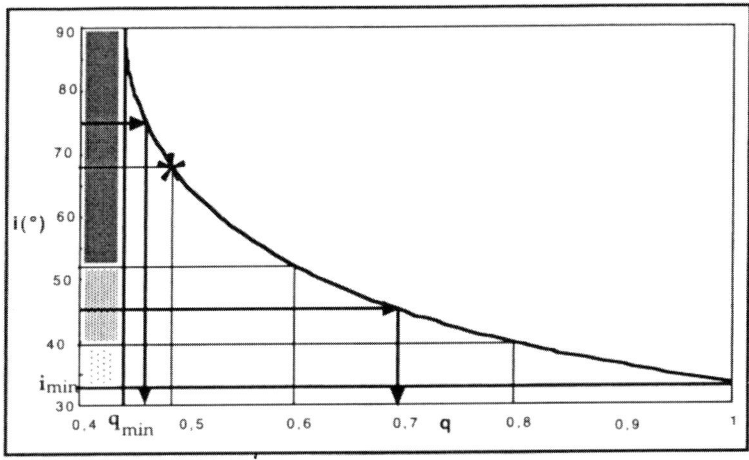

Figure 1 Typical relation between q and i, imposed by relation (2.2) for a SB1. Here HD 19820 is shown. The mass-ratio varies in the interval $[1 - q_{min}]$ with i in $[i_{min} - 90^o]$. Since 15° is smaller than i_{min} in this case, Abt and Levy find only two possible q-values, corresponding to $i = 75^o$ and 45°, respectively. Vanbeveren et al. evaluate the probabilities for every possible q-value. The q-bins $[0.4 - 0.6]$, $[0.6 - 0.8]$ and $[0.8 - 1]$ are shown using gradually weaker speckling. Since HD 19820 is in fact a SB2, the real value of q (and hence i) is indicated with a star-symbol.

the real q-distribution is known for SB2s, table 1 also shows that the two methods represent reality rather well. Only the highest q-bin [0.8-1] is poorly reproduced. The population of this bin is however generated by systems with relatively small inclination angles, which are in practice more difficult to be observed when dealing with SB1s. As a consequence we adopt the Vanbeveren et al. method to establish the observed q-distribution of SB1s. It is possible that a few cases in the highest q-bin are missed by this method. These cases correspond however to low inclination angles and are therefore more likely to be missed by observation. The method gives thus some correction for this observational bias.

Table 2 gives the summary of the data; i.e. the q-distribution of 113 SBs with a non evolved early B type [B0-B3] in the first two lines of the table and 119 SBs with a non evolved late B type primary [B4-B9.5] in the last two lines of the table.

Table 1 95 SB2s treated as if they were SB1s. Method 1 is from Vanbeveren et al. (1998) and Method 2 from Abt and Levy (1985). Both methods give comparable results and differ from reality, especially in the highest q-bin.

q-bin	*[0-0.2]*	*[0.2-0.4]*	*[0.4-0.6]*	*[0.6-0.8]*	*[0.8-1]*
95 SB2s	1	13	14	22	45
Method 1	2.2552	14.9844	20.7158	23.0779	33.9687
Method 2	1.8660	15.2583	20.7321	22.2472	34.8964

Table 2 q-distribution of 232 SBs with a non evolved B type primary. The data are from Batten et al. (1989). The mass-ratio for the SB2s is determined with relation (2.1), whereas for the SB1s relation (2.2) was used with the method of Vanbeveren et al. (1998).

q-bin	*[0-0.2]*	*[0.2-0.4]*	*[0.4-0.6]*	*[0.6-0.8]*	*[0.8-1]*
39 SB2s	0	3	9	12	15
74 SB1s	26.9302	23.9957	12.9764	6.2714	3.8263
40 SB2s	1	7	4	7	21
79 SB1s	22.8451	28.5728	14.0751	7.8322	5.6748

3. THE THEORETICAL MASS-RATIO DISTRIBUTION

The observed mass-ratio distribution $\Phi(q)$ for our sample of binaries with a non-evolved B type primary is given by table 2.

A Monte Carlo procedure with 50.000 binaries simulates a theoretical distribution $\Psi'(q)$. This distribution is transformed into an observed distribution $\Phi'(q)$ due to the impact of observational selection effects. The choice of a theoretical distribution $\Psi'(q)$ and of selection effects is altered until the function $\Phi'(q)$ coincides sufficiently well with $\Phi(q)$.

Modeling the mass of the primary and the mass-ratio of the binary. Consider a cloud fragment with mass M out of which a binary has been formed with primary mass M_1. This mass is subject to the following restrictions:

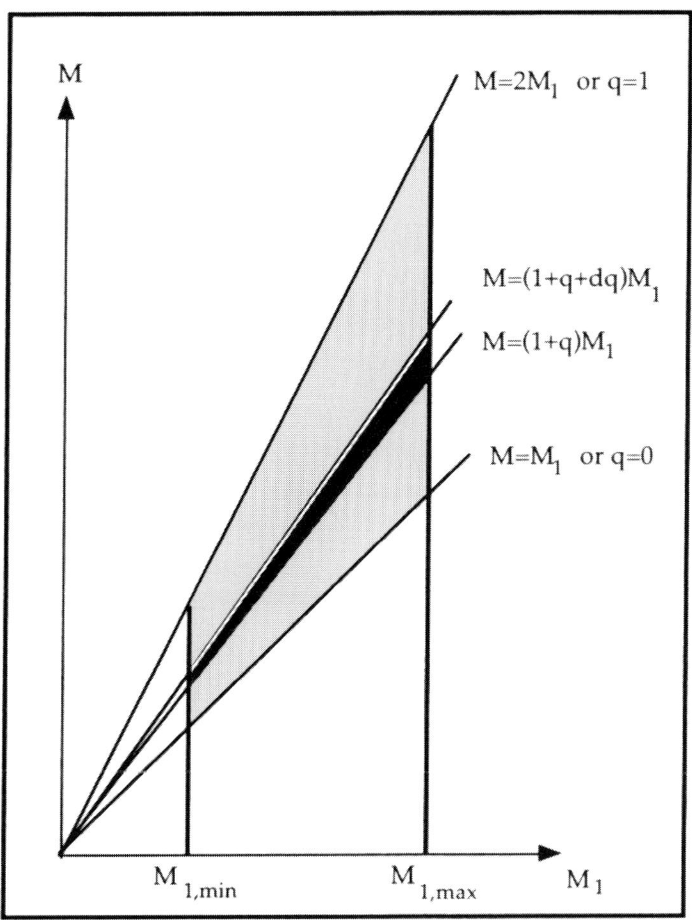

Figure 2 A binary with a primary mass $\in [M_{1,min} - M_{1,max}]$ is located in a domain in the $(M_1 - M)$-plane. M $\in [M_1$ - 2 $\times$ $M_1]$ is the mass of the cloud which became a binary under the action of selfgravitation.

$M_1 \in [7 - 16.7]\ M_{\odot}$ for early B type primaries (Vanbeveren et al. 1998). $M_1 \in [2.5 - 7]\ M_{\odot}$ for late B type primaries (Schaller et al. 1992).

Each binary is located in the $[M_1 - M]$-plane, in which straight lines defined by $M = M_1(q+1)$ are loci of constant q. Each point in the domain is weighted with its probability of occurrence. The fraction of binaries with primary mass $\in [M_{1,min} - M_{1,max}]$ and q $\in [q - q + dq]$ is the weighted heavily hatched surface divided by the weighted smoothly hatched surface (Figure 2).

The model assumes that the mass of the binary star forming clouds follows a normalised IMF:

$$\zeta(M) = A_W M^{-\alpha_W} \quad with \quad A_W = \frac{1-\alpha_W}{M_{max}^{1-\alpha_W} - M_{min}^{1-\alpha_W}} \tag{2.3}$$

The probability that a primary with mass M_1 is produced from a cloud with mass M is given by:

$$\xi(M_1, M) = f(M) M_1^{-\alpha_P} \quad if \quad M_1 \in [\frac{M}{2}, M]$$
$$\xi(M_1, M) = 0 \quad if \quad M_1 \notin [\frac{M}{2}, M] \tag{2.4}$$

Since $\int_{\frac{M}{2}}^{M} f(M) M_1^{-\alpha_P} \, dM_1 = 1$, we find that

$$f(M) = \frac{\alpha_P - 1}{[(0.5)^{1-\alpha_P} - 1]} \; M^{\alpha_P - 1} \tag{2.5}$$

The probability of occurrence of a binary; i.e. the weight of each point in the hatched domain of figure 2 is thus given as: $\xi(M_1, M).\zeta(M)$.

With the aid of expressions (2.3) to (2.5) one finds that the IMF for the primaries follows the same power law as the IMF for the cloud fragments:

$$\xi(M_1) = \int_{M_1}^{2M_1} \xi(M_1, M) \; \zeta(M) \; dM \propto M_1^{-\alpha_W} \tag{2.6}$$

The theoretical mass-ratio distribution is found by determining first the probability that a binary shows a mass-ratio smaller than $q^\star$:$P(q < q^\star)$. By integrating over the suitable domain [i.e. below the straight line M = $(1 + q^\star)\, M_1$] one finds:

$$P(q^\star) \propto \int_{M_{1,min}}^{M_{1,max}} \int_{M=M_1}^{M=(1+q^\star)M_1} M_1^{-\alpha_P} dM_1 M^{\alpha_P - \alpha_W - 1} dM$$
$$\propto [(1+q^\star)^{\alpha_P - \alpha_W} - 1] \tag{2.7}$$

The theoretical mass-ratio distribution is obtained by taking the derivative of expression (2.7):

$$\Psi(q) \propto (1+q)^{\alpha_P - \alpha_W - 1} \tag{2.8}$$

Expression (2.8) is in accordance with the distribution as derived by Kuiper (1935) [$\Psi(q) \propto (1+q)^{-2}$], satisfies a sound physical model and needs no artificial constant q_o in order to avoid a divergence for q $\rightarrow$ 0, as was done by Hogeveen (1991): $\Psi(q) \propto (1+q)^{-\alpha}$ $(q > q_o)$; $\Psi(q) = C$ $(q < q_o)$.

It was shown by Vanbeveren (1982) that the slopes of the IMF of single stars and of primaries of binaries (-α_W in this model) need not to be equal. The binary frequency can give information about these different slopes. Since we will make an independent statement of the binary frequency in section 4 and that we are working here in a population of SBs only, we assume α_W to be not much different from the exponent in the IMF for single stars:

Salpeter (1955) proposed $\alpha = 2.35$. Rana (1987) argued $\alpha = 1.8$ for M > 1.6 $M_\odot$. Elmegreen (1997) finds $\alpha \in [2.5-3]$ for field stars and $\in$ $[2-2.5]$ for stars in clusters. Since α_W will not differ substantially from α, we allow α_W to vary in the interval $[1.8-2.7]$.

Since the physics of binary star formation in a star forming cloud is not well understood, the exponent α_P is unknown. We have taken $\alpha_P \in [-5;+5]$ where a large positive value favours the formation of the lowest masses of the primary ($M_1 \approx \frac{M}{2}$, i.e. $q \approx 1$); a large negative value favours the formation of the highest masses of the primary ($M_1 \approx M$, i.e. $q \approx 0$); whereas $\alpha_P = 0$ gives equal chances to high and low mass primaries to be formed.

The theoretical q-distribution is transformed into an observed one after application of the selection effects. The exponents α_W and α_P are chosen such as to reproduce the observed q-distribution best.

Simulating the population of binaries. In order to evaluate the impact of the selection effects we generate a synthetic population of 50.000 binaries using a distribution for a number of independent variables:

- The distribution of the cloud mass M follows expression (2.3) in which α_W varies in the interval $[1.8-2.7]$
- The distribution of the primary mass M_1 follows expression (2.4) in which α_P varies in the interval $[-5;+5]$
- The distribution of the period P follows, according to Popova et al. (1982): $\Pi(P) = \frac{A}{P}$ with $\int_{P_{min}}^{P_{max}} \Pi(P)dP = 1$
- The distribution of the eccentricity e follows, according to Hogeveen (1991): $\epsilon(e) = Ae^{-\alpha_e}$ with $\int_{e_{min}}^{e_{max}} \epsilon(e)de = 1$

- The inclination angle i follows the random distribution: $I(i) = Asini$ with $\int_0^{\pi} I(i)di = 1$
- The stars are taken to be distributed homogeneously in a sphere: $D(r) = Ar^2$ with $\int_{r_{min}}^{r_{max}} D(r)dr = 1$

From these quantities a number of dependent variables can be calculated easily:

- The secondary mass $M_2 = M - M_1$ and the mass-ratio q $= \frac{M_2}{M_1}$
- The absolute luminosities ($L/L_\odot$) of the components are determined by their mass. This is done using Schaller et al. (1992) for stars with M > 0,8 $M_\odot$. The middle of the main sequence was taken for this luminosity calibration. Allen (1973) was used for the lowest mass stars to which secondaries with a B type primary can belong. The absolute bolometric magnitude is determined from the absolute luminosity, using the standard relation $M_{bol} = 4.75 - 2.5\, log \frac{L}{L_\odot}$
- The absolute visual magnitude M_V was calculated from M_{bol}, using the bolometric corrections from Vanbeveren et al. (1998) for the high mass stars (M > 7 $M_\odot$) and Allen (1973) for the lower mass stars. The apparent visual magnitude is calculated from the standard relation $m_{V1,2} = M_{V1,2} - 5 + 5\, log\, r(pc)$.
- The semi major axis of a system is given by the third Kepler law $a(R_\odot) = 4.20677[(M_1 + M_2)P^2]^{\frac{1}{3}}$:
- The semi axes of the components are found from the relations: $a_1 + a_2 = a$ and $\frac{a_1}{a_2} = \frac{M_1}{M_2}$
- From this, one finds the radial velocity amplitude as $K_{1,2}\ [\frac{km}{s}] = \frac{50.6131\, a_{1,2}\, [R_\odot]\, sin\, i}{(1-e^2)^{\frac{1}{2}}\, P[days]}$:

4. MODELING THE OBSERVED POPULATION OF SBS

It is our aim to transform the theoretical population of binaries from section 3 into an observed population of SBs. First, we subtract the visual binaries from our theoretical population. For every binary in our sample the distance and the semi major axis of the system are known. From this, one derives the angular separation of the components. A small number of visual binaries with angular separations > 0.05" have been removed from our population, which in consequence contains SBs only.

The probability for a SB to be recognized as SB2. Since the observations divide the homogeneous group op SBs into two categories, we have to separate SB2s from SB1s in our simulation. A SB will be recognized as a SB2 when the difference in apparent visual magnitudes (Δm_V) is "not to large".

Batten et al. (1989) report 331 SB2s for which Δm_V has been estimated. Table 3 shows this distribution over Δm_V-bins, 0.5 mag wide. The 31 SB2s with an early B type primary and 33 with a late B type primary are given separately.

Table 3 Most SB2s have $\Delta m_V \leq 1$; some have been detected above this limit. The fact that the number of SB2s is not monotonuously decreasing with Δm_V is an artifact from the low numbers available in this statistics.

Δm_V	*[0-0.5]*	*[0.5-1]*	*[1-1.5]*	*[1.5-2]*	*[2-2.5]*
Early B	12	6	8	2	3
Late B	14	3	5	3	3
All	147	66	48	28	22
Δm_V	*[2.5-3]*	*[3-3.5]*	*[3.5-4]*	it [4-4.5]	*[>4.5]*
Early B	0	0	0	0	0
Late B	0	2	1	2	0
All	10	6	1	3	0

Table 3 can be used to determine $P_{SB2}(\Delta m_V)$: the probability as a function of Δm_V for a SB to be detected as a SB2. $P_{SB2}(\Delta m_V)$ will be determined for binaries with an early B and late B primary separately. The procedure goes,for instance for binaries with an early B type primary, as follows:

- Take $P_{SB2}(\Delta m_V = 0) = 1$
- From the simulation of section 3, one can draw a (q-Δm_V)- graph for every combination of (α_W,α_P) and for systems with early resp. late B primaries. Figure 3 shows this relation for 10.000 binaries with an early B primary, using α_W= 1.8 and α_P=2.8. The components of all systems with q > $q_{max} \approx 0.83$ differ by less than 0.5 in visual magnitude, together with a known fraction having q $\in$

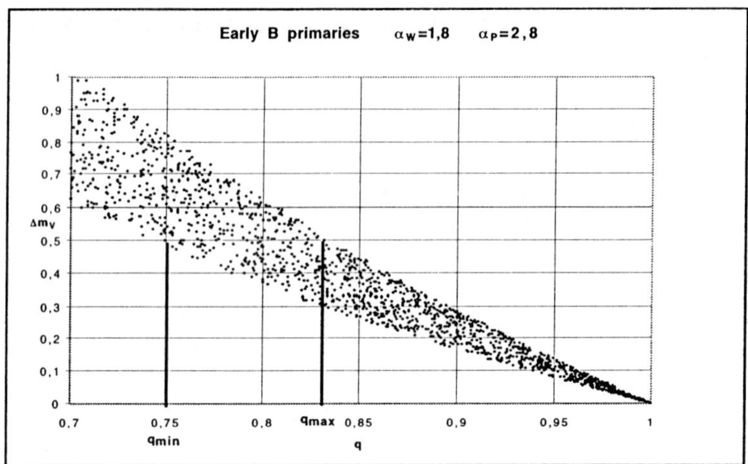

Figure 3 A typical [q-Δ m_V]-relation for binaries with an early B type primary. In this simulation only 10.000 binaries were generated in order not to over-populate the figure with dots.

$[q_{min} \approx 0.75\ q_{max}]$. The values of q_{min} and q_{max} are only slightly dependent from the model; i.e. (α_W,α_P)-combination.

- The observed population of every q-bin can be determined with the method of section 2. We find for the binaries with an early B type primary with q > q_{max}, 14 (out of 39) SB2s and 3.2275 (out of 74) SB1s. With q > q_{min}, we find 20 SB2s and 5.3535 SB1s.

- The simulation shown in Figure 3 contains N_1 binaries with Δm_V = [0 – 0.5], N_2 binaries with q=[q_{max}-1], N_3 with q=[q_{min}-q_{max}] out of which only a part N_4 belongs to Δm_V = [0 – 0.5]: $N_1 = N_2 + N_4$. Since this simulation has to reproduce the observed numbers, one finds in Δm_V = [0 – 0.5], e.g. 16 SB2s and 4 SB1s; i.e. $P_{SB2}(\Delta m_V{=}0.25) = 0.8$.

- This procedure can be repeated, yielding a $P_{SB2}(\Delta m_V) \neq 0$, for every Δm_V for which table 3 shows a non-vanishing presence of SB2s.

After removal of the visual binaries, this procedure divides the simulated binary population between SB2 and SB1.

Selection criteria for SB1s. All the SB2s of our simulated population are now observable. Some of the SB1s of our population will however not be in the catalogue.

A real SB1 will not be detected as such if its radial velocity amplitude K_1 is below some critical value K_0. For binaries with a late B primary we have taken K_0= 5 km/s. Spectral lines of early B stars are broader, so that we have taken 15 km/s for K_0 in this case.

5. COMPARING MODELS WITH OBSERVATIONS

After removal of visual binaries and SB1s that will be missed by the observations, our simulation using 50.000 binaries yields a simulated observed mass-ratio distribution for the population of SBs as well as for SB1s and SB2s separately.

The simulations that reproduce the observations of table 2 best, were found using the Kolmogorov-Smirnov one sample test (see e.g. Siegel and Castelan, 1989). The results are mainly dependent upon the choice of the parameters α_W and α_P, respectively in the intervals [1.8-2.7] and [-5;+5]. The results do not change much with the other parameters, for which we have made the realistic choice shown in table 4.

Table 4 Parameters which were used in our simulation of 50.000 binaries

Parameter	*Unit*	*Early B primary*	*Late B primary*
$M_{1,min}$	$M_\odot$	7	2.5
$M_{1,max}$	$M_\odot$	16.7	7
K_0	km/s	15	5
M_{min}	$M_\odot$	$M_{1,min}$	$M_{1,min}$
M_{max}	$M_\odot$	$2 \times M_{1,max}$	$2 \times M_{1,max}$
P_{min}	days	1	1
P_{max}	days	3000	3000
e_{min}		0.02	0.02
e_{max}		0.7	0.7
r_{min}	pc	1	1
r_{max}	pc	1000	1000

The most likely models (combination of α_W and α_P-values) were those fitting best with the observed mass-ratio distribution of SBs, SB2s and SB1s simultaneously and reproducing as good as possible the observed fraction of SB2s, which are from table 2 found to be $\frac{N(SB2)}{N(SB1)} \approx 0.5$ for binaries with a B-type primary. This observed fraction is most difficult to be reproduced by the models. Table 5 shows the best models which reproduce the observations almost equally well.

Table 5 Models which explain the observed distribution of SBs, SB1s and SB2s best for binaries with a B type primary. The three first columns are for binaries with an early B primary; the last three columns refer to binaries with a late B primary

α_W	α_P	$\frac{N(SB2)}{N(SB1)}$	α_W	α_P	$\frac{N(SB2)}{N(SB1)}$
1.8	-0.5	0.382	1.8	1.5	0.303
2	-0.5	0.360	2	1.5	0.282
2.35	0	0.369	2.35	2	0.292
2.5	0	0.354	2.5	2	0.281
2.7	0.5	0.379	2.7	2	0.268

Combining relation (2.8) with the models of table 5, we find for that the mass-ratio of interacting binaries with a B type primary obey the following distributions:

$$\Psi(q) \propto (1+q)^{-3.37} early\ B\ primaries \tag{2.9}$$

$$\Psi(q) \propto (1+q)^{-1.47} late\ B\ primaries \tag{2.10}$$

Figure 4 and Figure 5 show the agreement between the model and the observed mass-ratio distributions, for binaries with an early B and late B primary, respectively.

Finally, it has been established that a simulation with 50.000 binaries shows very little statistic fluctuation and that the theoretical distributions Ψ(q) differ very little from that distribution Ψ'(q) as generated by the random numbers.

Figure 6 compares the mass-ratio distributions from our models [expressions (2.9) and (2.10)] with those as obtained by Hogeveen (1991) and Kuiper (1935). The model of Hogeveen is proposed for all stars with a B primary and the model is Kuiper applies to all binaries.

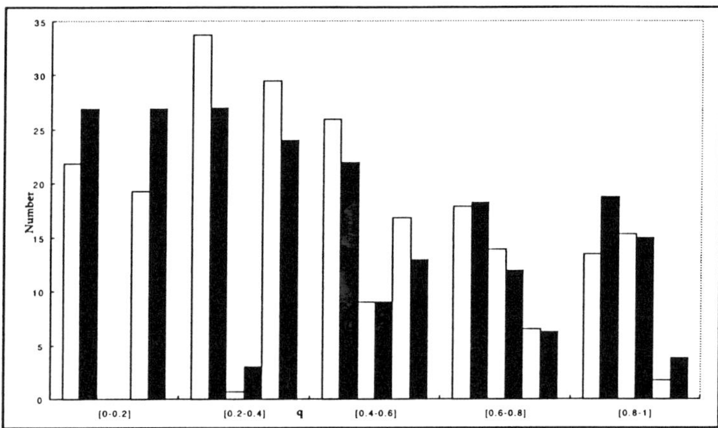

Figure 4 The distribution of 113 SBs, 39 SB2s and 74 SB1s with an early B type primary, from left to right in each of the five mass ratio bins. In this case the model with α_W = 1.8 and α_P = -0.5 is shown. The model is "white" and the observations "black".

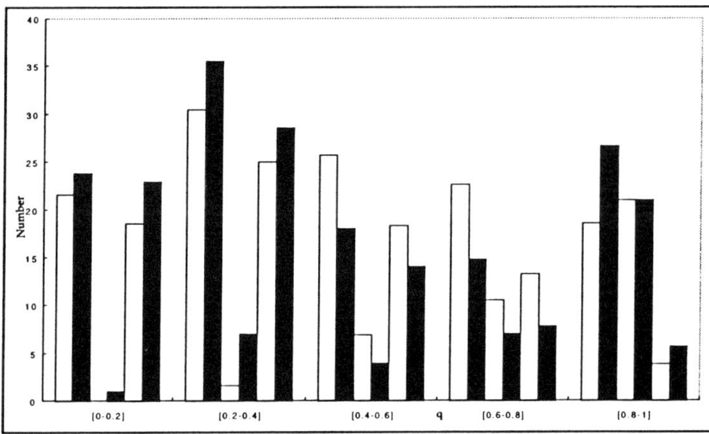

Figure 5 The distribution of 119 SBs, 40 SB2s and 79 SB1s with an late B type primary, from left to right in each of the five mass ratio bins. In this case the model with α_W = 2.5 and α_P = 2 is shown. The model is "white" and the observations "black"..

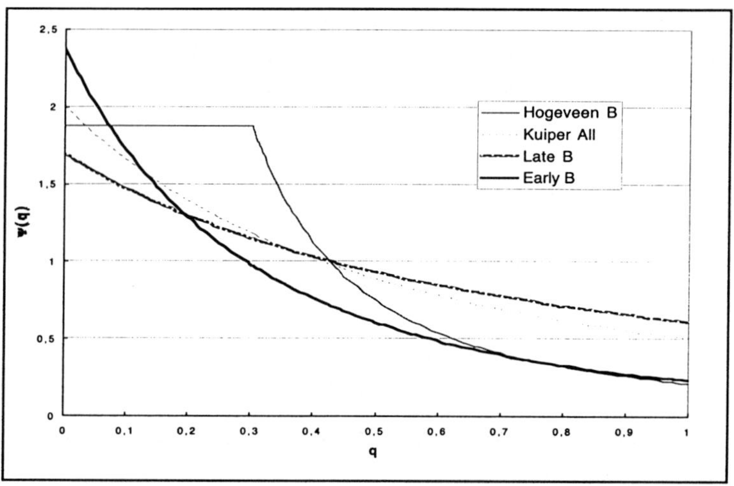

Figure 6 The mass-ratio distributions for non-evolved binaries with an early B and late B primary are compared to the model of Hogeveen, which applies to binaries with a B type primary and the model of Kuiper which was applicable to all binaries.

6. ESTIMATING THE FRACTION OF BINARIES

The fits between the observed mass-ratios and the model, shown in figures 4 and 5, are satisfactory. Table 5 shows however that the number of observed SB1s ($\frac{N(SB1)}{N(SB2)} \approx 2$) is lower than the number which is produced by the models ($\frac{N(SB1)}{N(SB2)} \approx 3$). This would mean that only 2 observable SB1s out of 3 have been observed yet.

The best models from our Monte Carlo simulation, together with our selection criteria to establish if a system is observed as a spectroscopic binary, show the number of SB1s that will never be observed as such. The low inclination angle and the low mass companion cause the radial velocity amplitude to be to low for such binaries. If one removes the systems with $M_2 \leq 0{,}08\ M_\odot$ from our simulation, one finds approximately 70 % real binaries more for every system observed with an early B primary and 20 % more for every binary that has been observed with a late B primary.

The real number of SB1s could thus very well be more than twice as much as the observed number.

References

Abt,H.& Levy,S.: 1985, Astrophys. J. Suppl. 59, 229

Allen,C.: 1973, "Astrophysical Quantities", The Athlone Press
Batten,A., Fletcher,J.& McCarthy,D.: 1989, Pupl. Dom. Astrophys. Obs. XVII: "The 8th catalogue of the orbital elements of spectroscopic binary systems"
Elmegreen,B.: 1997,Astrophys. J. 486, 944
Hogeveen,S.: 1991, "The Mass-Ratio Distribution of Binary Stars", Ph. D. dissertation, University of Amsterdam
Kuiper,G.: 1935, Pupl. Astr. Soc. Pacific, 47, 15-41
Popova,E., Tutukov,A.& Yungelson,L.: 1982, Astrophys. Space Sci. 88, 55
Rana,N.:1987, Astron. Astrophys. 184, 104 Salpeter,E.: 1955, Astrophys.J. 121, 161
Schaller,G., Scharer,D., Meynet,G.& Maeder,A.: 1992, Astron. Astrophys. Suppl. 96, 268
Siegel,S.& Castellan.: 1989, "Nonparametric statistics", McGraw-Hill
Vanbeveren,D.: 1982, Astr. Astrophys. 115, 65-68
Vanbeveren,D., Van Rensbergen, W.& De Loore,C.: 1998, "The Brightest Binaries", Kluwer Acad. Publ.

MASSIVE BINARIES

Speckle Interferometry of O, B, and WR stars

Brian D. Mason
U.S. Naval Observatory, 3450 Massachusetts Ave., NW, Washington, DC 20392, U.S.A.
bdm@draco.usno.navy.mil

Douglas R. Gies
Center for High Angular Resolution Astronomy, Georgia State University, Atlanta, GA 30303, U.S.A.
gies@chara.gsu.edu

William I. Hartkopf
U.S. Naval Observatory
wih@draco.usno.navy.mil

Keywords: binaries : general — binaries : spectroscopic — binaries: visual

Abstract Due to the typically large distances and low space densities of massive stars, combined (astrometric and spectroscopic) studies of hot stars are lacking. This is unfortunate, since combining orbital solutions from these techniques provides an important means of measuring mass and distance. However, in the last decade the situation has improved considerably as significant efforts led to the systematic investigation of the brightest (and closest) O stars. While the period overlap between binaries detected by these techniques is not likely to be fully explored until optical interferometry has matured, the results obtained thus far have been quite tantalizing. We also made speckle interferometric surveys of the Wolf-Rayet stars and luminous B giants and supergiants at the time of the O star survey, and we summarize here our results on the incidence of astrometric binaries among these related groups.

D. Vanbeveren (ed.), The Influence of Binaries on Stellar Population Studies, 37–52.

1. O STARS

The starting point for models of massive binary star evolution is a set of reliable distributions for binary period and mass ratio plus an estimate of the overall fraction of massive stars found in binary or multiple systems. There are considerable uncertainties surrounding all these parameters in large measure because of the observational selection effects surrounding different kinds of observations and sample completeness. Over the last few years we have undertaken several surveys of the massive stars using the technique of speckle interferometry to find new visual binaries down to the resolution limit of 4-m telescopes (approximately 30 milliarcsecond). This work has led to the discovery of many new binaries including some close enough that the prospect exists for combined spectroscopic — astrometric orbital solutions (which yield both masses and system distance). Our most complete survey was made of Galactic O-type stars mainly brighter than $V = 8$, and we begin here with an updated discussion of the O-stars which extends the main survey results of Mason et al. (1998; hereafter Paper I).

1.1. ASTROMETRY UPDATE

An initial step in this analysis of O stars was to update the astrometry for visual systems in Paper I with their current status in the Washington Double Star Catalog[1] (hereafter, WDS). While updating all system information is certainly advisable, of greatest concern were those systems with only one astrometric observation, as unconfirmed binaries may be spurious for one reason or another (wrong object, false detection, etc.). These single detections are summarized in Table 1.

Table 1 Resolved doubles with $N = 1$ in Mason et al. (1998).

Number of Observations	Δm_V (mag)	Number of Systems
$N > 1$	—	25
$N = 1$	$\Delta m \geq 5$	17
$N = 1$	$3 \leq \Delta m < 5$	12
$N = 1$	$\Delta m < 3$	4
$N = 1$	*unknown*	9

[1] http://ad.usno.navy.mil/ad/wds/wds.html, also, see Worley & Douglass 1997

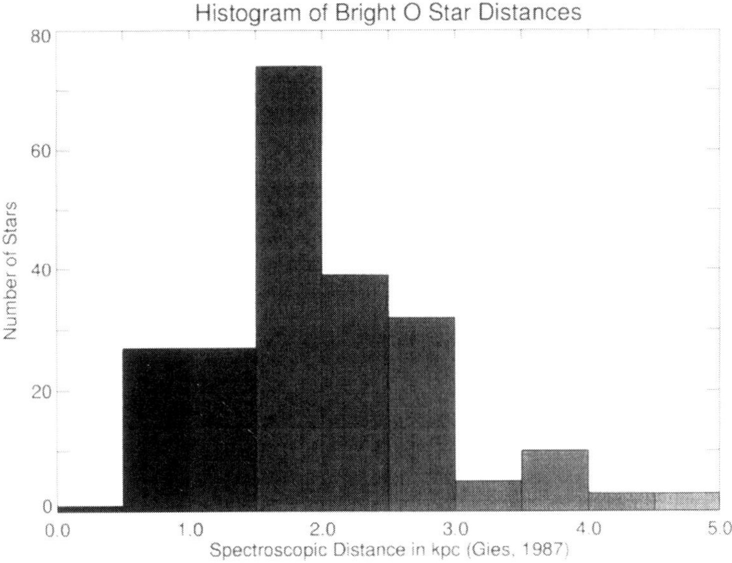

Figure 1 Histogram of spectroscopic O star distances.

Table 1 (specifically, columns 10–13) from the online O Star Speckle Survey[2] has been updated to reflect the current astrometric information on O stars. This supersedes all astrometric information given in Paper I.

1.2. DISTANCES TO O STARS

Column 14 of Table 1 of Paper I presented the spectroscopic distance to O stars from Gies (1987) and Humphreys & McElroy (1984). These distances are presented in histogram form in Figure 1. The distribution peaks just below 2 kpc.

Figure 2 is a plot of spectroscopic parallax (triangles) and Hipparcos parallax and error (line) for the 34 systems (stacked in the y-dimension) where $\frac{\pi}{\sigma\pi} > 2$ (145 other systems have the ratio ≤ 2). The Hipparcos parallaxes appear to be systematically larger than the spectroscopic parallaxes. It should be noted that Høg et al. (2000a, 2000b) find Hipparcos proper motion errors underestimated by about 30% for double stars. A similar problem may exist with the parallax. Be that as it may, the source of the systematic error is uncertain. What is not uncertain is that, while distances for O stars in clusters where main-sequence fitting

[2]http://ad.usno.navy.mil/dsl/Ostars/ostars.html

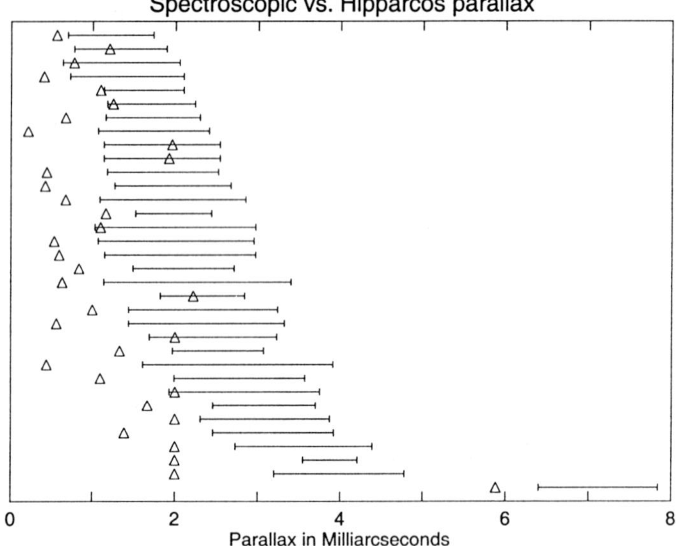

Figure 2 Spectroscopic distances and Hipparcos distance

is possible may not be too bad, distances to O stars are only known in the most gross sense.

1.3. NEW ASTROMETRIC WORK

Several O stars have been the subject of significant work since Paper I. These are detailed below.

15 Mon. Since the original 1988 resolution first reported by Gies et al. (1993), the binary 15 Mon has shown significant evolution in orbital elements due to the paucity of original data. Figure 3 is a plot of the relative orbit. In this figure, the scale is in arcseconds and the shaded circle centered on the origin is the 30 milliarcsecond (mas) resolution limit of a 4-m telescope in V.

Available data come from speckle interferometry (filled circles) and HST-FGS (**H**), and are connected to the current orbit calculation by O−C lines. Earlier orbit calculations by Gies et al. (1993, 1997) are shown as dashed curves. Speckle points are from 1988 (CFHT) and 1993 (KPNO & CTIO). HST-FGS measures are from 1996, 1997, and 1999. An HST-FGS observation is scheduled for April 2001 as well as in cycles 9, 10, and 11 and speckle observations are scheduled for 2001. Also shown are the current (2000.6; i.e., at the epoch of the Brussels

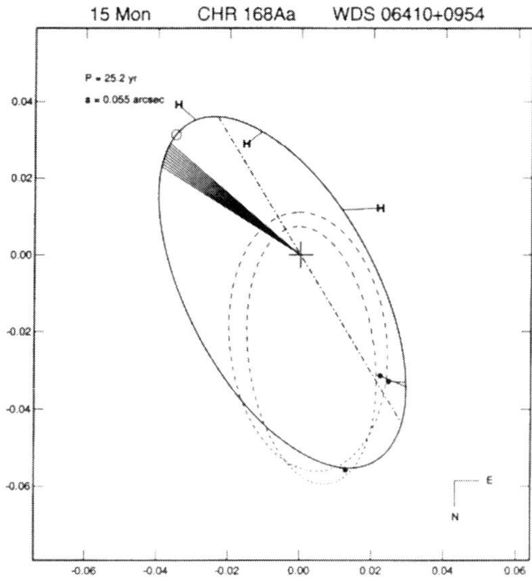

Figure 3 Relative orbit of 15 Mon

meeting) position, shown as an open circle. Radial lines from the origin to the orbit indicate predicted orbital motion during 2001.

HD 193322. Figure 4 shows of the relative orbit of CHARA 96 (HD 193322) from Hartkopf et al. (1993). At the time of the 1993 orbit it was unresolved from a 4-m telescope, but is now near the maximum predicted separation for this 31.3 year period. The positions for 2001 (when a speckle observation is scheduled) are shown here as well as the position at the time of the Brussels meeting. Symbols are as Figure 3.

In a spectroscopic investigation of this system by McKibben et al. (1998) the A component was revealed to be an unresolved spectroscopic binary with a period of 311 days.

Theta Orionis. The Orion Trapezium system has many close, and mostly unconfirmed IR companions from Prosser et al. (1994), Simon, Close, & Beck (1999), and Weigelt et al. (1999). The new companion noted by Weigelt et al. is to θ Orionis C and separations from 33 and 37 mas were measured in 1997 and 1998. The Δm of this companion was determined to be 1.46 in H-band and 1.24 in K. The new companion is believed to be a very young, intermediate or low mass ($\mathcal{M} < 6\mathcal{M}_{\odot}$) star. While it is unclear what the Δm is at visual wavelengths, the expected magnitude difference in visual band is <4.5; this possibly accounts for

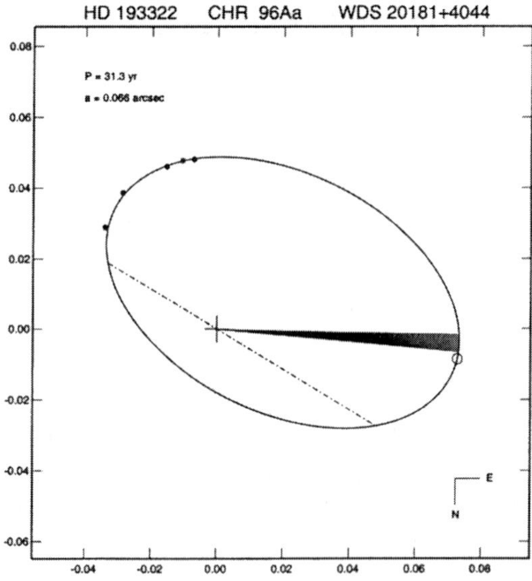

Figure 4 Relative orbit of HD 193322

the lack of detection in the speckle survey of Paper I. It is unclear how many of these new companions are physical, or just nearby components of the cluster. Repeated observation and the establishment of Keplerian or non-Keplerian motion will be required to ascertain this.

Zeta Orionis. A striking example, and a useful demonstration (if one were actually needed) of the utility of optical interferometry was the detection of the new 4th magnitude companion to ζ Orionis A by Hummel et al. (2000). Hanbury Brown, Davis, & Allen (1974) first reported the likely non-singular nature of ζ Orionis based on correlations significantly below 1 obtained at the Intensity Interferometer at Narrabri. However, it was unresolved. The duplicity of ζ Orionis remained doubtful in spectroscopic analyses by Bohannan & Garmany (1978) and Levato et al. (1988), although the latter reported a small (< 35 km/s) radial velocity variation.

Hummel et al. utilized the Navy Prototype Optical Interferometer (NPOI) in 1998 and 1999 to resolve the new companion a total of 12 times at separations ranging from 42 to 47 mas and a Δm of 2.0 (in wavelengths ranging from 520 to 850 nm). While an orbital analysis of the ζ Orionis A system has not been made, it is possible that the non-detections cited in Paper I may place limits on the relative astrometry of the system at those epochs.

1.4. BINARY FREQUENCY

While the discoveries of ζ Orionis Aa,Ab and θ Orionis Ca,Cb (as well as other companions of this complex system) are quite intriguing, they occur in systems already designated as "visual multiple systems," in Table 3 of Paper I, so do not effect the statistics.

Paper I did, however, miss two visual binaries, one optical and one physical:

- **10440−5932** = CPD−58 2620 : Was noted as having a constant radial velocity in Penny et al. (1993) The known visual component not listed in Paper I is SEE 123.
- **22469+5805** = HD 215835 = DH Cep : The companion missed here is clearly not the 2.1-d spectroscopic companion of Penny, Gies, & Bagnuolo (1997), but a more distant companion designated HJ 1810 in the WDS.

Hipparcos "problem" O Stars. The Hipparcos satellite made measurements of over 9,734 known double stars, 3,406 new double stars, and 11,687 unresolved, but possible double stars. These 11,687 objects, designated "problem" stars, are objects whose duplicity status ranges from near certain (unresolved, but with orbit solutions) to dubious (anomalous results, some simply carrying the catch-all note "suspected non-single"). These objects are designated in the Hipparcos (ESA 1997) Catalogue with a G, O, V, or X in column H59 or an S in column H61. Mason et al. (1999, 2001) have discussed the applicability of speckle interferometry to investigate these systems, the status of these "problem" stars, compared with their astrometric status is given in Table 2.

In Table 2, column 1 categorizes the O stars into one of three main groups, cluster/association members, field stars, or runaways. The status of astrometric companions, whether known or single is given in column 2, while the number of "problem" stars is given in column 3. The final column lists by HD number the specific "problem" stars. The known systems typically fall into one of three categories. While most either have a larger Δm than Hipparcos measured and some were closer than Hipparcos measured, others were relatively wide and had a more moderate Δm. These would mostly have fallen into the category of a Hipparcos "double-entry" system, however, the known double was not identified in the Hipparcos Input Catalogue as double, were treated as single stars. Stray light may have caused the problem and led to the conclusion given in the Hipparcos Catalogue.

Of note are the those components designated single in the astrometric survey. Of the three cluster objects, their spectroscopic status is notable.

Table 2 Hipparcos O type "Problem" Stars

O group status	Astrometric Survey Status	Number of Systems	HD number
Cluster/ Association	known	9	36456, 47043, 46150, 71304, 75821 96254, 148937, 152270, 206183
	single	3	100099, 101131, 226868
Field	known	0	
	single	2	48149, 169515
Runaway	known	0	
	single	1	198846

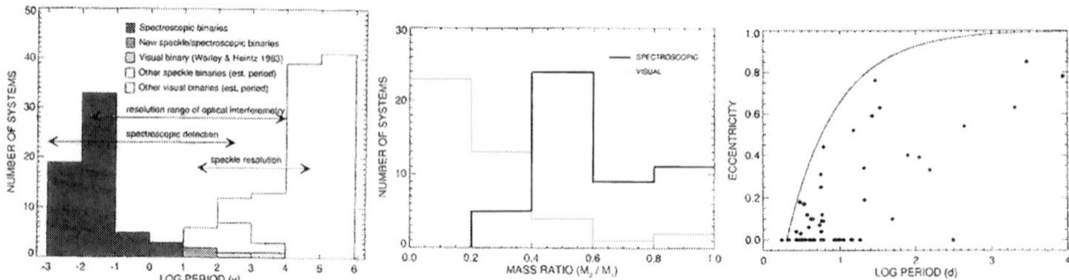

Figure 5 Statistical figures from Paper I.

- the three cluster objects were classified SB1? (suspect single-lined binary, radial velocity excusion is excess of 35 km s^{-1}), SB2? (double-lined, but no orbit), SB1OE (single-lined binary with orbit and eclipsing or ellipsoidal variable),

- the two field objects were classified USB (unknown spectroscopic status) and SB2OE (double-lined binary with orbit and eclipsing or ellipsoidal variable), and

- the runaway object was classified SB2OE.

While the statistics of Table 3 do not change, but they do add credence to those questionable spectroscopic binaries. A modified version of Table 3 (from Paper I) is provided.

1.5. OTHER STATISTICS

Other statistical information from Paper I have not been updated, and the Figures from that paper are here presented in a three-part Figure 5.

Table 3 Binary Frequency

Group (Number)	Cluster/Association (144)	Field (44)	Runaway (19)
A. Visual Multiplicity			
Visual Binary	31	7	0
Visual Multiple System	38	2	0
Total	42%	20%	0%
Optical	5	4	1
Single	90	31	18
Total	58%	80%	100%
B. Spectroscopic Properties			
SB2O	26	4	1
SB1O	20	2	0
SBE	4	2	0
SB2?	13	4	1
SB1?	27	8	3
Less SB?	34%	20%	5%
Total	61%	50%	26%
Constant	57	20	14
Total	39%	50%	74%
Unknown	17	4	0
C. Fraction with Any Companion			
Less SB?	59%	35%	5%
Total	75%	58%	26%

The left panel provides the number distribution of orbital periods for all binaries in the speckle survey of the O stars. The bimodal distribution is probably a selection effect. While the gap is being filled in to a certain extent by operational optical interferometers (e.g., NPOI; see Section 1.3.4 above), at present these instruments are limited to the brighter targets (and the O stars are fainter in the red and near-IR colors used in most interferometers). Both speckle/spectroscopic binaries (15 Mon and HD 193322, plotted in the center of the figure) appear in the new "5th Catalog of Orbits of Visual Binary Stars" (Hartkopf, Mason, & Worley, 2000)[3].

The center panel is an illustration of the number distribution of mass ratio for the visual and spectroscopic binaries in the survey. Visual binaries include those with probable physical orbits, and a single bright primary (i.e., strictly double systems), and the primary of luminosity class V-III. The mass ratio is based on the magnitude difference using tables from Howarth & Prinja (1989), and is thus model dependent. Spectroscopic binaries include both SB2O and SB1O systems. The mass-ratios for the latter are based on the statistical methods of Mazeh & Goldberg (1992), and the SB1O sample excludes those with very evolved companions and the triple HD 152623 (where the interpretation of the spectroscopy suffers from line blending of the features of the spectroscopic primary with the "stationary" lines of the third star).

The right panel provides the distribution of orbital eccentricity as a function of $\log P$ (days). The solid line traces the boundary for a periastron separation greater than 23 $R_\odot$ for a binary with a total mass of 40 $\mathcal{M}_\odot$. All spectroscopic binaries are plotted. From the figure, it is noted that

1 there is an absence of high-e/short-P systems, confirming the expectation that binaries with periods $<$ 4 days tend to circularize due to tidal interaction, and

2 there is a lack of low-e/long-P systems. This may be due to circumbinary disks driving eccentricity to higher values (an effect seen in low-mass binaries), though the number of systems is small.

2. WR STARS

As in Section 1, the starting point for the discussion of Wolf-Rayet stars is a paper in the literature, specifically, the survey of Hartkopf et al.

[3] web version available now at http://ad.usno.navy.mil/ad/wds/hmw5.html, cd version in mid-2001

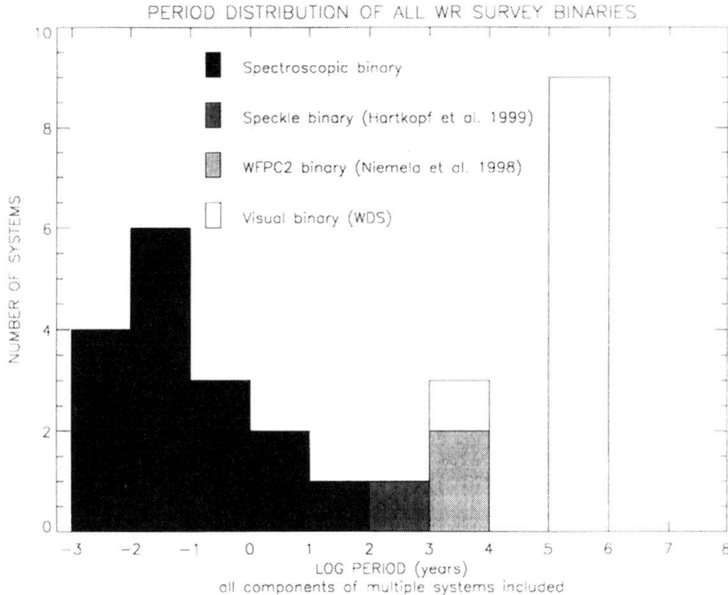

Figure 6 Number distribution of orbital periods for all binaries in the WR survey.

(1999; here after Paper II). Not provided in Paper II is a period distribution for known WR binaries. This is shown in Figure 6.

The bimodal distribution is, again, a selection effect. All visual binary periods are estimates based on literature distances (possibly wildly inaccurate) and a mass sum of 30 $\mathcal{M}_\odot$(probably wrong, but possibly good enough for statistical purposes). The number of WR systems is quite small, and what are needed most are astrometric measures of short-P systems, as the known visual binaries have very long periods.

Figure 7 is a plot of the relative positions of CHR 247 (θ Muscae). Symbols are as Figure 3. The speckle measure is from Paper II, and the HST-FGS measure is from D. Wallace (private comm.). Also, Grant Hill (private comm.) in an analysis of the 18-d short period system finds "stationary" spectral lines of an O-supergiant (narrow C III λ5696) which presumably originates in the speckle companion. While there is significant uncertainty in the motion of the system, speckle observations will at least be attempted in 2001.

3. B GIANTS AND SUPERGIANTS

Unlike the O and WR sections, no results from the luminous B star survey have yet been published. Table 4 presents the systems which were observed. These observations were made at the time of the O and

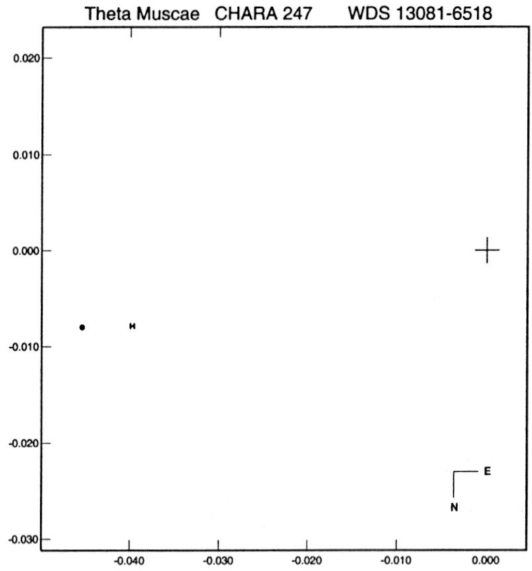

Figure 7 Measures of θ Muscae = CHR 247.

Table 4 B giants and supergiants surveyed for duplicity

Bright Star Catalogue number							
1131	2595	2699	3456	3825	4611	5027	6356
1713	2596	2789	3494	3940	4644	5036	
1903	2618	2815	3571	4135	4653	5358	
2004	2627	2827	3654	4147	4806	6260	
2187	2653	3090	3703	4250	4817	6261	
2294	2657	3203	3708	4338	4887	6262	

WR surveys, however, no companions, new or known, were measured in the separation range 30 mas $< \rho < 2''$. By their very nature, companions of B giants would be difficult to find. Due to their luminosity, only stellar companions of approximately the same MK class would be noticed in V-band. Coupled with the very short lifetime of B giants, the chances of observing a binary is quite small.

Despite this, significant work has been done in the past year on known B giants that were not in the sample above. A significant recent addition is the three dimensional orbit of the triple system 64 Ori by Scarfe, Barlow & Fekel (2000). Unfortunately, the 13.0-yr pair is only single-

Table 5 BIII orbits

Discovery Designation		HD	P (yr)	a ($''$)	Reference
MCA	24	41040	12.98	0.0471	Scarfe et al. 2000
BTZ	1	45542	13.00	0.1826	Mason 1997
FIN	322	49643	171.4	0.261	Seymour et al. 2000
HU	1594	87652	174.	0.371	Seymour et al. 2000
HU	200	196662	420.	0.48	Heintz 1998

lined, so no orbital parallax is possible. As Table 5 shows, all B giant orbits contained in the 5th Catalog are recent additions. The situation is improving, but more work is needed.

4. FUTURE WORK

4.1. THE USNO CMOS SYSTEM

There remain a plethora of systems which have not been detected because of a large magnitude difference, Δm. USNO astronomers have begun a program to search for faint companions using both a CCD and a CMOS type detector. Tests with the Lick astrograph and a conventional CCD have demonstrated the capability to resolve systems as close as $2''$ and Δm values of up to 5. The CMOS is a logarithmic, continuous readout detector system which allows a Δm of 7 to be reached. However, the continuous readout demands extremely accurate tracking. After testing is complete, a regular program to survey the bright stars for faint companions can be initiated. See Winter (2000) for more information.

4.2. ADAPTIVE OPTICS OF MASSIVE STARS

Low mass companions to massive stars are notoriously difficult to detect due to the large magnitude difference. Adaptive optics observations of these stars can aid in a similar manner to the CMOS system in that large magnitude differences can routinely be detected (see Turner et al. 2001). A program specifically designed to look for these companions is planned for early 2001 using the AMOS system on Haleakala (ten Brummelaar 2000).

4.3. OPTICAL INTERFEROMETRY

Still in its infancy, optical interferometry has the potential to directly resolve many of these systems (as demonstrated in Section 1.3.4 above); these instruments are, at present, magnitude limited, however.

4.4. FAME

One of the most significant areas where improvement can be made in massive star research is the determination of their distances. As shown in Figure 1, bright O star distances peak at about 2 kpc. The FAME satellite (scheduled for launch in 2004, see Johnston 2000) will measure positions, proper motions, and parallaxes to 40 million stars, as well as determining their companion properties from astrometric signatures. At a distance of 2 kpc, the error is expected to be less than 10%. The determination of O star distances for mass determination of single-lined spectrocopic and visual binaries and the calibration of the absolute magnitude scale for hot stars is but one outcome of this very cost effective mission (costing roughly, $4 U.S. per star).

Acknowledgments

We thank the U.S. Naval Observatory for their continued support of the Double Star Program. Support for this work was provided by NASA through grant numbers 7484 and 8392 from the Space Telescope Science Institute, which is operated by AURA, Inc., under NASA contract NAS5-26555.

References

Bohannan, B., & Garmany, C.D. 1978, ApJ, 223, 908

ten Brummelaar, T.A. 2000, private communication

ESA 1997, The Hipparcos and Tycho Catalogues (ESA SP-1200) (Noordwijk: ESA)

Gies, D.R. 1987, ApJS, 64, 545

Gies, D.R., Mason, B.D., Hartkopf, W.I., McAlister, H.A., Frazin, R.A., Hahula, M.E. Penny, L.R., Thaller, M.E., Fullerton, A.W., & Shara, M.M. 1993, AJ, 106, 2072

Gies, D.R., Mason, B.D., Bagnuolo, Jr., W.G., Hahula, M.E., Hartkopf, W.I., McAlister, H.A., Thaller, M.L., McKibben, W.P., & Penny, L.R. 1997, ApJL, 475, 49

Hanbury Brown, R., Davis, J., & Allen, L.R. 1974, MNRAS, 167, 121

Hartkopf, W.I., Gies, D.R., Mason, B.D., Bagnuolo, W.G., & McAlister, H.A. 1993, BAAS, 182, 4511

Hartkopf, W.I., Mason, B.D., Gies, D.R., ten Brummelaar, T., McAlister, H.A., Moffat, A.F.J., Shara, M.M., & Wallace, D.J. 1999, AJ, 118, 509

Hartkopf, W.I., Mason, B.D. & Worley 2000, Fifth Catalog of Orbits of Visual Binary Stars (http://ad.usno.navy.mil/ad/wds/hmw5.html)

Heintz, W.D. 1998, ApJS, 117, 587

Howarth, I.D., & Prinja, R.K. 1989, ApJS, 69, 527

Høg, E., Fabricius, C., Makarov, V.V., Bastian, U., Schwekendiek, P., Wicenec, A., Urban, S., Corbin, T., & Wycoff, G. 2000a, A&A, 357, 367

Høg, E., Fabricius, C., Makarov, V.V., Urban, S., Corbin, T., Wycoff, G., Bastian, U., Schwekendiek, P., & Wicenec, A. 2000b, A&A, 355, L19

Hummel, C.A., White, N.M., Elias, N.M., II, Hajian, A.R., & Nordgren, T.E. 2000, ApJ, 540, L91

Humphreys, R.M., & McElroy, D.B. 1984, ApJ, 284, 565

Johnston, K.J. 2000, "The Future of Space Astrometry," in Towards Models and Constants for Sub-Microarcsecond Astrometry, Proceedings of IAU Colloquium 180, U.S. Naval Observatory, K.J. Johnston, D.D. McCarthy, B.J. Luzum, and G.H. Kaplan, Eds., p. 392

Levato, H., Morell, N., Garcia, B., & Malaroda, S. 1988, ApJS, 68, 319

Mason, B.D. 1997, AJ, 114, 808

Mason, B.D., Gies, D.R., Hartkopf, W.I., Bagnuolo, W.G., Jr., ten Brummelaar, T., & McAlister, H.A. 1998, AJ, 115, 821

Mason, B.D., Hartkopf, W.I., Holdenried, E.R., & Rafferty, T.J. 2001, (*in preparation*)

Mason, B.D., Martin, C., Hartkopf, W.I., Barry, D.J., Germain, M.E., Douglass, G.G., Worley, C.E., Wycoff, G.L., ten Brummelaar, T., & Franz, O.G. 1999, AJ, 117, 1890

Mazeh, T., & Goldberg, D. 1992, ApJ, 394, 592

McKibben, W.P., Bagnuolo, W.G., Jr., Gies, D.R., Hahula, M.E., Hartkopf, W.I., Roberts, L.C., Jr., Bolton, C.T., Fullerton, A.W., Mason, B.D., Penny, L.R., & Thaller, M.L. 1998, PASP, 110, 900

Penny, L.R., Gies, D.R., & Bagnuolo, W.G., Jr. 1997, ApJ, 483, 439

Penny, L.R., Gies, D.R., Hartkopf, W.I., Mason, B.D., & Turner, N.H. 1993, PASP, 105, 588

Prosser, C.F., Stauffer, J.R., Hartmann, L., Soderblom, D.R., Jones, B.F., Werner, M.W., & McCaughrean, M.J. 1994, ApJ, 421, 517

Scarfe, C.D., Barlow, D.J., & Fekel, F.C. 2000, AJ, 119, 2415

Seymour, D.M., Mason, B.D., Hartkopf, W.I., & Wycoff, G.L. 2001, *in progress*

Simon, M., Close, L.M., & Beck, T.L. 1999, AJ, 117, 1375

Turner, N.H., ten Brummelaar, T.A., McAlister, H.A., Mason, B.D., Hartkopf, W.I., & Roberts, L.C., Jr. 2001, AJ, (*submitted*)

Tuthill, P.G., Monnier, J.D., Danchi, W.C., Wishnow, E.H., & Haniff, C.A. 2000, PASP, 112, 555

Winter, L. 2000, "SIM Grid Star Observations: Astrometry with a New High Dynamic Range Imaging Device," in Towards Models and Constants for Sub-Microarcsecond Astrometry, Proceedings of IAU Colloquium 180, U.S. Naval Observatory, K.J. Johnston, D.D. McCarthy, B.J. Luzum, and G.H. Kaplan, Eds., p. 380

Weigelt, G., Balega, Y., Preibisch, T., Schertl, D., Schöller, M., Zinnecker, H. 1999, A&A, 347, L15

Worley, C.E., & Douglass, G.G. 1997, A&AS, 125, 523

II

OBSERVATIONS OF EVOLVED BINARIES

CATACLYSMIC VARIABLES, SUPER-SOFT SOURCES AND DOUBLE-DEGENERATES

Tom R. Marsh
Department of Physics and Astronomy, Southampton University, UK
trm@soumac1.phys.soton.ac.uk

Keywords: Cataclysmic variables, super-soft X-ray sources, double-degenerates, type Ia supernovae

Abstract I review observations of cataclysmic variable stars, super-soft X-ray sources and double-degenerate binary stars with particular focus upon the numbers of these systems, all of which have been proposed as Type Ia supernova progenitors. All these systems are of interest as examples of binary star evolution. Cataclysmic variables and (at least some) super-soft sources have very closely related evolutionary paths. Double-degenerates appear to be remarkably common end-points for binary star evolution and are probably the most significant source of low frequency ($< 0.1\,\mathrm{Hz}$) gravitational waves in the Galaxy; if on merging they collapse to neutron stars rather than explode as Type Ia supernovae, they may even be detectable with ground-based gravitational wave detectors.

1. INTRODUCTION

There is general agreement that Type Ia supernovae arise as the result of exploding white dwarfs. The most favoured model is the Chandrasekhar mass model in which a CO white dwarf exceeds the Chandrasekhar limit with the subsequent collapse igniting fusion in its core, leading to total disruption of the star. Key properties explained by this model are the absence of hydrogen and helium in the spectra of Type Ia supernovae (Leibundgut, 2000) and their occurrence in elliptical galaxies (old stellar populations). There has been less agreement on how to get a white dwarf to the point of collapse as it is harder than one might imagine to add material to a white dwarf. In single degenerate models one white dwarf accretes hydrogen-rich material from a companion until collapse occurs, while in the double-degenerate model, two white dwarfs merge and the combined object collapses and explodes. Modellers of the explosions,

D. Vanbeveren (ed.), The Influence of Binaries on Stellar Population Studies, 55–71.

led by the Tokyo group, have tended to favour the single degenerate model because the double degenerate case may well lead to collapse to a neutron star rather than an explosion. The double-degenerate model has survived however because the single degenerate model poses difficulties in adding mass without inducing nova explosions or burying the white dwarf altogether (Nomoto, 1982) (section 2). However, in recent years this problem seems to have been solved and it is probably fair to say that most people favour the single degenerate model at present (Livio, 2000; Branch, 2000).

Another approach to take is to consider what plausible candidates we know of, and whether they meet the requirements necessary to produce Type Ia supernovae. I will follow this line in this review looking at three potential progenitor classes: the cataclysmic variable stars, super-soft X-ray sources and double-degenerate binary stars. Much of the interest in the first two classes centres upon their use as examples of accretion. This I will only touch upon, as the focus of this meeting is more upon the global significance of differing types of binary stars to stellar population studies. Apart from the supernova aspect, the main significance of these binaries is probably as tests of binary evolution theory, which is covered in other contributions, so I will mainly focus upon how many of these systems we think there are.

I start with a brief discussion of the difficulty of adding material to white dwarfs.

2. ACCRETION ONTO WHITE DWARFS

It is now well established that classical novae are explosions of hydrogen accumulated upon the surface of white dwarfs through accretion. This occurs in cataclysmic variable stars. If hydrogen is added to a white dwarf, it builds up until when something like 10^{-5} to $10^{-4}\,\mathcal{M}_\odot$ of matter has accumulated, the base of the accreted layer starts to undergo fusion which, under degenerate conditions, runs away in a process known as a shell flash. The most violent of these eject the accreted material, as occurs during classical nova explosions. In fact, the material ejected in nova explosions is observed to be enhanced in CNO elements to an extent that suggests that some of the underlying white dwarf's material is ejected as well. Thus we have the curious situation that in the long term accretion can actually lead to a *reduction* of the white dwarf's mass.

A critical feature here is the rate at which mass is added. If added at a high rate, it is possible for gentler and even steady nuclear fusion to occur. Too high a rate buries the white dwarf entirely. All this was

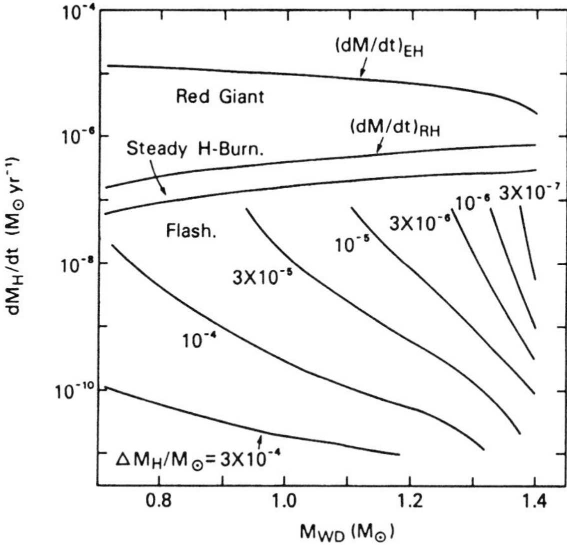

Figure 1 This plot from Nomoto (1982) shows the amount of material in solar masses needed to produce a nova explosion as a function of white dwarf mass and accretion rate. Steady nuclear burning is possible in the band from $10^{-7}\,\mathcal{M}_\odot\,\mathrm{yr}^{-1}$ to $10^{-6}\,\mathcal{M}_\odot\,\mathrm{yr}^{-1}$. Above this an envelope forms, burying the white dwarf.

worked out many years ago and is illustrated in Fig 1 taken from Nomoto (1982).

Even the "steady burning" regime above hides the fact that a layer of helium builds up which may itself experience a violent shell flash (Cassisi, Iben & Tornambe, 1998; Kato & Hachisu, 1999). Thus the obstacles to the accumulation of material are many, and it starts to look more difficult to produce Type Ia supernovae by collapse of white dwarfs than one might first have guessed. Nevertheless, as mentioned earlier, this is becoming more and more the favourite model. Therefore I now look at some of the observed systems in which this process takes place. I start with the most familiar of these, the cataclysmic variable stars.

3. CATACLYSMIC VARIABLE STARS

3.1. GENERAL BACKGROUND

Cataclysmic variable stars (CVs) are long-lived systems in which a low mass star transfers mass to a white dwarf companion (Fig. 2). They are long lived ($> 10^9$ yr) because the mass donor star is less massive than the white dwarf which ensures stability of mass transfer on dynamical and thermal timescales. With low mass, Roche-lobe filling donor stars,

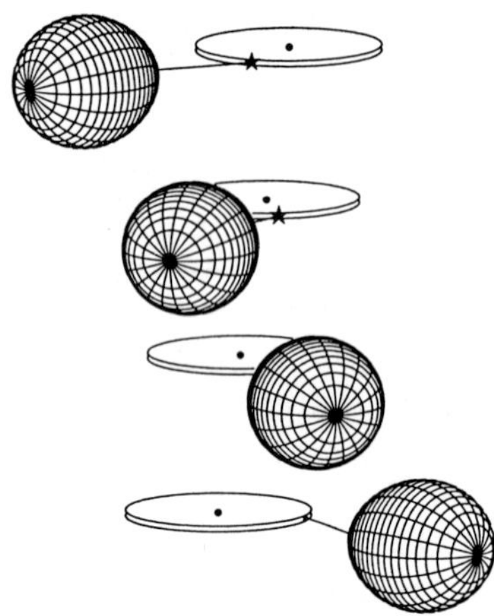

Figure 2 A schematic picture of a cataclysmic variable star.

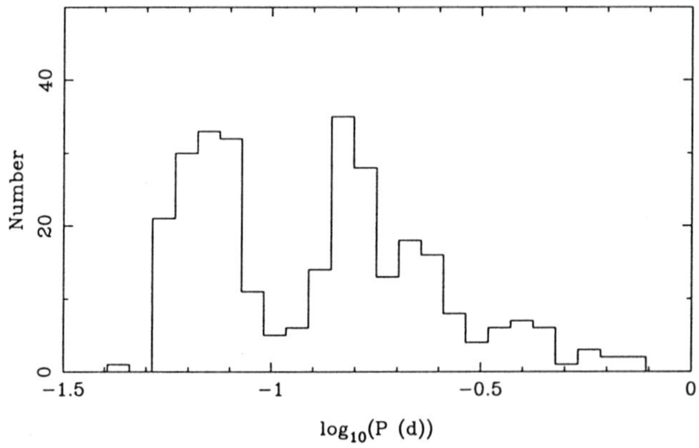

Figure 3 The orbital period distribution of cataclysmic variable stars (Ritter & Kolb, 1998).

the periods of CVs are short, ranging from around 80 minutes to 2 days, with the bulk of systems having periods below 10 hours (Fig. 3). CVs chief claim to fame is as examples of accretion because the low luminosities of the component stars often allows accretion-generated light to dominate. They show a plethora of beautiful effects from limit-cycle instabilities and spiral shocks of the discs to magnetically-controlled accretion in some systems. For general information on CVs and many other references, the reader is referred to Warner (1995) and Ritter & Kolb (1998).

CVs are probably minor contributors to the overall luminosity and reprocessing of matter in stellar populations. If we assume that there are

$\sim 10^7$ CVs in the Galaxy accreting at a mean rate of $10^{-9}\,\mathcal{M}_\odot\,\mathrm{yr}^{-1}$, then the total accretion luminosity of CVs in the Galaxy is of order $2\times10^7\,\mathrm{L}_\odot$. This is small, although since CVs emitt much of their luminosity at short wavelengths, it is not quite as negligible as it might seem.

Mass transfer in CVs is driven by angular momentum loss through gravitational waves and magnetic stellar wind braking. It ranges from around $10^{-8}\,\mathcal{M}_\odot\,\mathrm{yr}^{-1}$ downwards which puts it in the regime of the violent shell flashes discussed in the previous section: CVs are in fact the progenitor systems of the classical novae. There are thought to be of order 35 classical novae per year in the Galaxy, each ejecting something like $2\times10^{-4}\,\mathcal{M}_\odot$ of material into the interstellar medium, giving a total rate of $0.007\,\mathcal{M}_\odot\,\mathrm{yr}^{-1}$. The total is again small compared to the total expected from other stars, but novae may contribute significantly to some isotopes such as ^{7}Li, ^{15}N, ^{17}O, ^{26}Al and ^{22}Na and are probably necessary to explain some abundance ratios of meteoritic material as they are able to produce dust of unusual composition (Gehrz et al., 1998; Amari et al., 2001).

3.2. CVS AS TYPE IA PROGENITORS

The one area where CVs could have a very significant impact on entire stellar populations is if they produce Type Ia supernovae. If again we consider $\sim 10^7$ CVs accreting at a mean rate of $10^{-9}\,\mathcal{M}_\odot\,\mathrm{yr}^{-1}$, then the total accretion rate is $10^{-2}\,\mathcal{M}_\odot\,\mathrm{yr}^{-1}$, and thus, if a typical white dwarf needs to accrete about $1\,\mathcal{M}_\odot$ to reach the Chandrasekhar limit, CVs potentially can produce 1 supernova per century, more than enough to match the probable Type Ia rate of 2 or 3 per millennium. However, this calculation is hopelessly optimistic because of nova explosions. Instead white dwarfs only gain mass in the small fraction of systems with massive white dwarfs and high accretion rates ($M_{\mathrm{WD}} > 1.25\,\mathcal{M}_\odot$, $\dot{M} > 10^{-7}\,\mathcal{M}_\odot\,\mathrm{yr}^{-1}$) (Prialnik & Kovetz, 1995). These two conditions also lead to short recurrence times and the sub-class of CV known as the recurrent novae, which include the stars U Sco and T Pyx, are widely thought to be just such systems (Hachisu et al., 2000). Such systems seem to satisfy all the requirements for the Chandrasekhar collapse model. However, the restricted region of parameter space allowed for them seems to make it unlikely that they are the only, or even the main, progenitor class of Type Ia SNe (della Valle & Livio, 1996).

3.3. HOW MANY CVS?

The number of CVs in the Galaxy remains very much an open question. There has been a long standing discrepancy between observational

Table 1 Distances of CVs with parallaxes compared to earlier estimates

Star	Parallax pc	Parallax reference	Warner (1987)	Patterson (1984)
U Gem	96 ± 5	Harrison et al. (2000)	81	78
SS Cyg	166 ± 13	Harrison et al. (2000)	76	95
SS Aur	200 ± 26	Harrison et al. (2000)	200	190
RW Tri	340 ± 35	McArthur et al. (1999)	224	250
IX Vel	96 ± 10	Duerbeck (1999)	51	—

and theoretical estimates, the latter being typically about a factor of 10 higher than the former. In his comprehensive 1984 paper Patterson (1984) deduced a space density in the solar neighbourhood of about $6 \times 10^{-6}\,\mathrm{CVs\,pc^{-3}}$. Taking an effective volume of $2 \times 10^{11}\,\mathrm{pc}^3$ for the Galaxy then suggests that there are about 10^6 CVs. In a more recent paper, Patterson revised his estimate upwards somewhat to around $10^{-5}\,\mathrm{CVs\,pc^{-3}}$, but this is still the same to within the uncertainties. Patterson's estimates require distances which are hard to come by for CVs. However, there are now some reliable parallaxes for CVs which can be compared to Patterson's estimates and those from Warner's work on absolute magnitudes (Warner, 1987). I do so in Table 1. The conclusions from this comparison are mixed – the estimates for U Gem and SS Aur were fairly good, while those for IX Vel and SS Cyg were too low. The discrepancy in the case of SS Cyg, the best known CV of them all, is perhaps the most alarming. If the trend of increasing distances is continued, the observational estimates of the space density will lower still further, although it is too early to know this for sure.

Theoretical estimates of the space density based upon binary evolution range from 2 to $20 \times 10^{-5}\,\mathrm{CVs\,pc^{-3}}$, consistently higher than observational estimates (Politano, 1996; de Kool, 1992; Han, Podsiadlowski & Eggleton, 1995). Given the many uncertain assumptions that go into such estimates, one can perhaps dismiss the difference. However, there is another clue that we are not seeing all the CVs theoretically predicted. The angular momentum loss that drives mass transfer in CVs means that they mostly evolve towards short periods until at around $P = 80\,\mathrm{min}$ the mass donor starts to become degenerate and the period increases again. The mass transfer rate drops rapidly after this point. Theoretically about 70% of CVs are thought to have passed the period minimum (Kolb, 1993) whereas it is not clear that *any* observed systems have (Patterson, 1998). It therefore seems possible that most CVs either do not exist in the state predicted theoretically or are not easily detected in this state.

4. SUPER-SOFT X-RAY SOURCES

Evolutionary models of CVs produce many dud systems. Some systems are too widely separated ever to interact; in others interaction only occurs when the mass donor has evolved up the giant branch – we know these as symbiotic stars. Finally, in other systems mass transfer can be dynamically or thermally unstable. The last of these possibilities took on a new significance with the identification of a new class of X-ray sources in the early 1990s (Truemper et al., 1991; Greiner, Hasinger & Kahabka, 1991). Dubbed the "Super Soft Sources" (SSSs), these objects have *much* softer X-ray spectra than typical X-ray binaries, with the peak emission at energies of order 40 eV. Some of the sources had been known to be soft sources for several years, but it took the extra sensitivity of ROSAT to soft X-rays to see that there was something radically different about them.

The SSSs are not a very homogeneous set of objects and include novae, symbiotic stars, recurrent novae as well as the well-known LMC binaries Cal 83 and Cal 87. Nevertheless it seems likely that in many cases we are seeing nuclear burning on the surface of a white dwarf, and this was the basic model developed soon after the identification of the class (van den Heuvel et al., 1992). The key constraints are luminosities of the order 10,000 $L_\odot$, emitted with an effective temperature of order 400,000 K which together imply emission from an object comparable in radius to a white dwarf accreting at a rate of $\sim 2 \times 10^{-6}\,\mathcal{M}_\odot\,\mathrm{yr}^{-1}$. However the latter is such a high rate that one expects the X-rays to be blocked. This led to the key new feature: the accreted material undergoes steady fusion. Since the energy released per unit mass by fusion is about 30 times that released by accretion onto a white dwarf, this reduces the necessary accretion rate to a more manageable $10^{-7}\,\mathcal{M}_\odot\,\mathrm{yr}^{-1}$.

4.1. EVOLUTIONARY STATE OF THE SUPER-SOFT SOURCES

The accretion rate needed to power the SSSs is one or two orders of magnitude higher than those seen in CVs. This is possible if the mass transfer is thermally unstable, which led to the suggestion of Roche lobe overflow from main-sequence or sub-giant donors somewhat more massive than the white dwarf (van den Heuvel et al., 1992), say 1.5 to 3 $\mathcal{M}_\odot$. Moreover, the accretion rate is just that needed for steady nuclear burning and mass gain by the white dwarf.

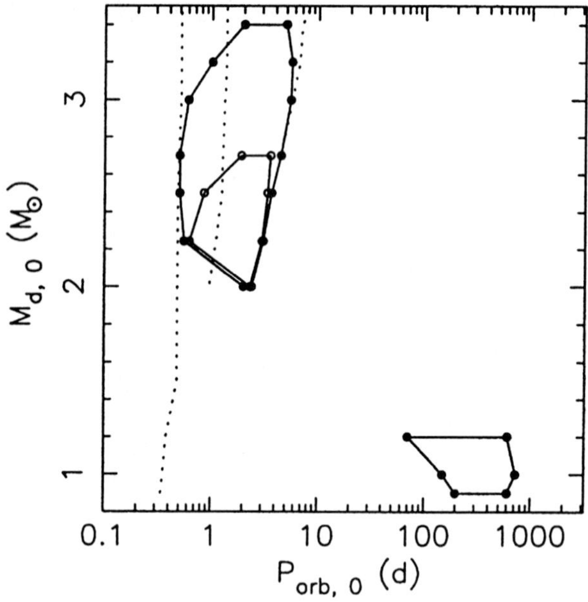

Figure 4 The combinations of orbital period and donor mass that can lead to steady fusion on a white dwarf (Li & van den Heuvel, 1997).

As Type Ia progenitors, van den Heuvel et al's model fell short because 1.5 to 3 $\mathcal{M}_\odot$ stars do not live long enough to produce Type Ia events in old stellar populations. A possible solution to this problem is that at high accretion rates the white dwarf develops a wind which stabilises mass transfer over a wider range of mass ratios than realised before (Hachisu, Kato & Nomoto, 1996). This model was applied by Li & van den Heuvel (1997) to find a new region of parameter space that could lead to SSSs and Type Ia SNe (see Fig. 4). The wind model opens up the possibility of Type Ia progenitors in which red giants transfer mass to white dwarfs in binary stars with orbital periods of order a year. Crucially, the mass of the red giant can be of order a solar mass making long-lived progenitors possible. Observational counterparts of these are perhaps the symbiotic stars and some of the long period recurrent novae such as T CrB and RS Oph. Some symbiotic stars such as AG Dra and RR Tel are classified as super-soft sources which supports this model; so too is the recurrent nova U Sco (when in outburst) which I discussed earlier.

4.2. THE NUMBERS OF SUPER-SOFT SOURCES

Super-soft sources are detected by their X-ray spectra but as they are so soft, their fluxes are extremely sensitive to interstellar absorption. This in turn makes it exceedingly difficult to estimate their numbers. This is illustrated by the fact that we know more of these sources in M31 than we do in our own galaxy. Nevertheless there have been attempts to estimate total numbers from the objects observed for the Magellanic Clouds, M31 and the Galaxy (Kahabka, 1999; di Stefano & Rappaport, 1994). Estimated numbers for M31 range from around 1,000 to 15,000 (extrapolated from 26 observed sources!), with approximately half this number in the Milky Way. The Magellanic clouds are nearer and suffer relatively little extinction, and so the extrapolation from observed to total numbers is less drastic, from 6 to 30 for the LMC for example (di Stefano & Rappaport, 1994).

Clearly these numbers are uncertain, but they are in the main consistent with Type Ia rates. In the future more sensitive X-ray observations of M31 may reduce the uncertainty, although absorption will always be a problem with these sources and absorption intrinsic to the sources may make any model of galactic absorption unreliable.

5. DOUBLE-DEGENERATE BINARY STARS

As various authors have pointed out recently, with the discovery of sources in which white dwarfs are apparently undergoing steady fusion on their surfaces, the need for a double-degenerate route to Type Ia SNe appears to have receded (Livio, 2000; Branch, 2000). Moreover, it has been long suggested that the merging of two CO white dwarfs will lead to fusion to an O-Ne-Mg composition followed by collapse to a neutron star with little, if any, ejection of matter. Nevertheless, while the case may be swinging away from double-degenerates as Type Ia progenitors, it would be premature to dismiss them entirely. Moreover, as I hope to show, they are of considerable interest as a surprisingly common end product of binary star evolution and significant gravitational wave sources.

By "double-degenerates" (DDs) I mean two detached white dwarfs in orbits close enough that interaction must once have occurred. I exclude semi-detached double-degenerate binary stars (the "AM CVn" systems) although these may well be largely descended from their detached counterparts (Nelemans et al., 2001). When close enough, gravitational radiation losses are sufficient to drive the two white dwarfs together until the least massive of them fills its Roche lobe. In some cases the initial contact will lead to runaway mass transfer and the disruption of the least massive white dwarf into a massive disc around its companion. As I have indicated, what happens after this is uncertain, and therefore my focus here will be more upon what leads up to two white dwarfs in a close orbit and what we know about such systems observationally.

5.1. FINDING DOUBLE DEGENERATES

Searches for DDs were first undertaken in the late 1980s and early 1990s with a view to finding the elusive Type Ia progenitors (Robinson & Shafter, 1987; Foss, Wade & Green, 1991; Bragaglia et al., 1990). These surveys turned up one system 0957-666 (Bragaglia et al., 1990) which however had too long an orbital period to be representative of Type Ia progenitors ($P = 1.15\,\mathrm{d}$). Another DD found earlier, L870-2 (Saffer, Liebert & Olszewski, 1988), also had too long an orbital period ($P = 1.56\,\mathrm{d}$). What is meant by "too long" a period? In the context of white dwarf mergers, only systems with short enough periods to merge within the age of the Galaxy can represent the class of Type Ia progenitors. The merger time τ is given by

$$\tau(\mathrm{yr}) = 1.00 \times 10^7 \frac{(M_1 + M_2)^{1/3}}{M_1 M_2} P^{8/3}$$

where the orbital period is in units of hours and the masses are in solar masses. For $M_1 = M_2 = 0.7\,\mathcal{M}_\odot$, the merger time $\tau < 10^{10}\,\mathrm{yr}$ for $P < 9.8\,\mathrm{h}$. Neither of the first two DDs discovered satisfy this restriction, although it was later found that the first measurement of the orbital period 0957-666 was in error, and that its correct value, $P = 88\,\mathrm{min}$, implies that merging will occur in only 200 million years (Moran, Marsh & Bragaglia, 1997).

The early failures to find suitable systems led to a view that there were not enough DDs to match the necessary supernova rates. However it has been pointed out that observations are not sensitive to very short-period DDs which merge soon after their formation (Iben, Tutukov & Yungelson, 1997). Moreover, the supernova rates adopted by investigators in the early searches were rather high, exaggerating the discrepancy between the lack of short period DDs and the required merger rate (Robinson & Shafter, 1987). In fact perhaps only 1 in 400 white dwarfs need be a Type Ia progenitor (Iben, Tutukov & Yungelson, 1997) and given that the total sample size of the early searches was of order 100, the failure to find any candidates was hardly surprising.

I started looking for DDs with a different motivation entirely. My aim was to solve the mystery of white dwarfs which appeared to have too low a mass to have resulted from the evolution of single stars (Bergeron, Saffer & Liebert, 1992). Standard evolution to a white dwarf requires

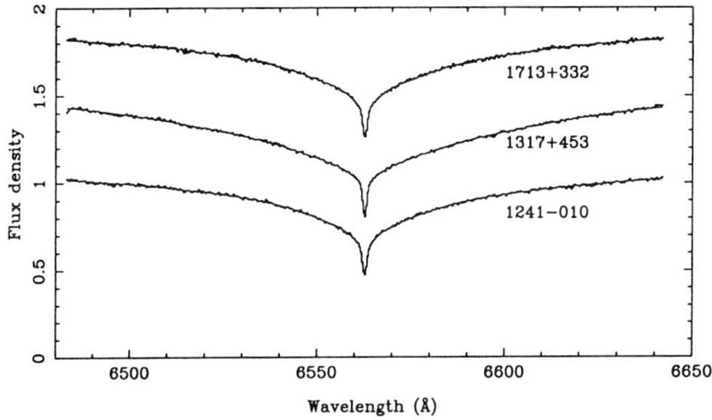

Figure 5 Mean Hα profiles and multi-gaussian fits of three double-degenerates (Marsh, Dhillon & Duck, 1995).

a star to reach the AGB at which stage it loses its envelope revealing the core. This must take place within the lifetime of the Galaxy, which implies a minimum mass of order $0.5\,\mathcal{M}_{\odot}$ for white dwarfs formed from single stars. However, with the development of good atmosphere models for white dwarfs, spectral fitting established the existence of a substantial fraction ($\sim 10\%$) of white dwarfs with masses below $0.45\,\mathcal{M}_{\odot}$ (Bergeron, Saffer & Liebert, 1992). The obvious explanation is evolution within a binary star since this can prematurely terminate a star's advancement up the red giant branch leading to a low mass white dwarf composed (usually) of helium. A search for radial velocity variations in low mass white dwarfs revealed 5 new DDs out of 7 targets (Marsh, Dhillon & Duck, 1995) and several other discoveries have been made since then (Marsh, 1995; Holberg et al., 1995; Maxted et al., 2000).

The success of this work compared to the earlier searches can be put down to two reasons: first, especially low mass white dwarfs were the priority targets, and did indeed turn out to be largely a result of binary evolution; second, the radial velocity variations focussed mainly upon Hα which in white dwarfs has a narrow core not seen in the other Balmer lines. This makes it possible to detect much smaller shifts than could the earlier surveys. This is illustrated by the profiles of three DDs shown in Fig. 5 (Marsh, Dhillon & Duck, 1995).

A consequence of the focus upon low mass white dwarfs is that none of the DDs found so far has a total mass exceeding the Chandrasekhar limit; some fall very far short, e.g. WD0957-666 which has a *total* mass of $\sim 0.7\,\mathcal{M}_{\odot}$. Thus the lack of short period, massive DDs is no surprise – we have yet to search a large enough sample of white dwarfs to find one.

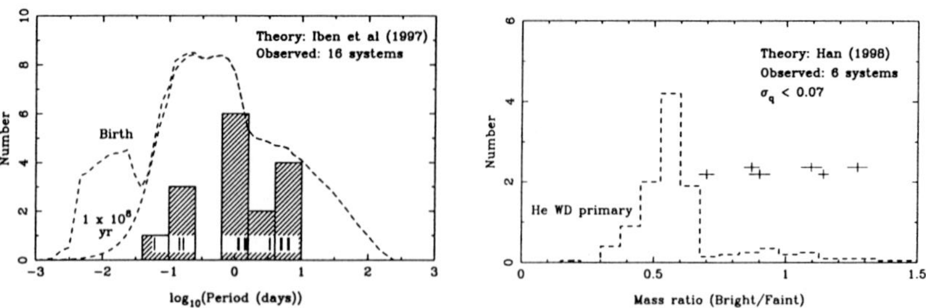

Figure 6 The observed periods and mass ratios of double degenerate binaries are compared to the predictions of Iben, Tutukov & Yungelson (1997) and Han (1998).

5.2. OBSERVED PROPERTIES OF DOUBLE DEGENERATES

Double white dwarfs provide a rather remarkable population for testing binary evolution theory. White dwarfs are unique in that their temperatures and gravities (from spectroscopy alone) can be translated into masses and ages. It is also sometimes possible to see both components and measure precise mass ratios. Along with the orbital period, and with relatively slight (and calculable) selection effects, rather stringent tests of theory become possible. This is to be compared with the case of cataclysmic variables where although many more systems are known, strong selection effects heavily distort the period distribution (Kolb, 1993).

We now know the orbital periods of some 16 DDs and the mass ratios of 6 of these. Fig. 6 shows the observed periods and mass ratios compared to theory. This figure is a good example of the power of the parameter constraints available for these systems: while the periods contain only a hint of possible problems with an apparent gap between periods of a few hours and periods longer than a day, the mass ratios hardly match up at all. In particular there is a pronounced tendency for the observed mass ratios to be more equal than expected. Such differences are not trivial because the mass ratio for example can determine whether on the next Roche lobe overflow the mass transfer is stable or not, and thus whether the system survives as a mass transferring double degenerate.

Theory predicts unequal mass ratios for the simple reason that the second white dwarf to form is usually expected to be formed in a tighter binary than its companion. Thus its progenitor giant should have been smaller, resulting in a smaller mass as a result. Thus when the mass ratio is expressed as the ratio of the younger divided by the older white dwarf, we expect mass ratios less than unity. The observations suggest that the systems do not in fact shrink much. This simple but stark disagreement has led to the suggestion that these systems do not undergo the standard common envelope shrinkage (Nelemans et al., 2000).

5.3. THE NUMBERS OF DOUBLE DEGENERATES

Although the known number of double degenerates is small, they are probably one of the most numerous populations of close binary star. We are helped here by the extensive efforts that have gone into tying down the white dwarf space density (Liebert, Dahn & Monet, 1988; Oswalt et al., 1996; Knox, Hawkins & Hambly, 1999). Including both isolated white dwarfs and those in common proper motion pairs, the total space density is of order $10^{-2}\,\mathrm{pc}^{-3}$. On the other hand, observations and theory suggest that about 10% of white dwarfs are detached double-degenerates, although a factor of 2 or 3 uncertainty in this number cannot be ruled out (Bergeron, Saffer & Liebert, 1992; Bragaglia, Renzini & Bergeron, 1995; Iben, Tutukov & Yungelson, 1997; Saffer, Livio & Yungelson, 1998; Maxted & Marsh, 1999). Thus a best guess at the DD space density is $10^{-3}\,\mathrm{pc}^{-3}$. This can be compared to the highest estimate for CVs of $2\times10^{-4}\,\mathrm{pc}^{-3}$ and the observationally-based estimate of $10^{-5}\,\mathrm{pc}^{-3}$ that I quoted earlier. Clearly, DDs are a huge population of potentially very close binary stars. The small number known can be put down to the difficulty in finding them and because the majority are old and faint.

The large number of DDs has two interesting consequences. First, whether or not they explode, there must be a fairly high rate of white dwarf binaries driven into contact. Han (1998) and Nelemans et al. (2001) estimate about 1 such event every 30 to 50 years in the Galaxy (regardless of mass and normalised to a formation rate of 1 primordial binary per year). The rate of mergers of pairs with a combined mass in excess of the Chandrasekhar limit comes out at about 1 every 300 to 500 years, in reasonable agreement with estimated Type Ia rates. DD mergers are very likely to produce the semi-detached double degenerates, and possibly other unusual objects such as R CrB stars and isolated millisecond pulsars.

The second consequence of the large numbers and mergers of DDs is that many of them must be in very short period orbits now, and thus be strong gravitational wave sources. For instance for a merger rate of $0.02\,\mathrm{yr}^{-1}$, and typical masses $M_1 = M_2 = 0.3\,\mathcal{M}_\odot$, there will be 300,000 DDs with orbital period $P < 30\,\mathrm{min}$. Thus they are likely to be the "main-sequence stars" of gravitational wave astronomy as the most numerous and long-lived sources in the Galaxy Hils, Bender & Webbink

(1990). DDs merge at orbital periods of a few tens to a hundred seconds or so and therefore in they do not reach high enough frequencies to be detected by ground-based detectors during their in-spiral phase, however they may be a dominant source of background for space-based detectors such as LISA. There are so many in fact, that at gravitational wave frequencies below $\sim 2 \times 10^{-3}$ Hz (corresponding to $P \sim 1000$ sec) they will form an unresolved background noise, which may affect the ability of space-based detectors to detect a cosmological background for example (Hils, Bender & Webbink, 1990). Conversely however, we may be able to learn a great deal about DDs from gravitational waves, and will for instance, be able to measure the numbers of short period systems and predict merger rates. Finally it is worth noting that if super-Chandrasekhar mergers lead to collapse to neutron stars rather than Type Ia SNe, the neutron stars can be expected to rotate very rapidly and may be detectable by the ground-based gravitational wave detectors currently under construction (Lai & Shapiro, 1995).

5.4. SDB/WHITE DWARF BINARIES

There are a growing number of close binaries containing sdB stars (Moran et al., 1999; Maxted et al., 2000). It is likely that many of these have white dwarf companions (Orosz & Wade, 1999). These are probably direct ancestors of DDs, although they may only contribute a relatively modest fraction to the total since the sdB phase is a rare one. In evolutionary terms, sdB stars are EHB (extreme horizontal branch) and post-EHB stars and are of interest since they may contribute significantly to the total UV output of old stellar populations (Brown, Ferguson & Davidsen, 1995).

In the context of this review, these objects are well worthy of mention as one of them, KPD1930+2752, is the first short period (2.28 h) merger candidate which is also probably more massive than the Chandrasekhar limit (Maxted, Marsh & North, 2000). The evidence for this comes from the large radial velocity amplitude of the sdB star (Fig. 7). The sdB component in this system falls right in the same part of the $\log g$ – $T_{\rm eff}$ as other sdB stars. Evolutionary models convincingly explain the locations of these stars if they have masses close to $0.5\,\mathcal{M}_\odot$ (Dorman, Rood & O'Connell, 1993). Assuming that this is the case leads to a minimum total mass for the system of 1.47 h. This star is important in at least establishing that it is possible to produce systems of the sort needed for the double degenerate merger model, no matter what other difficulties may present themselves.

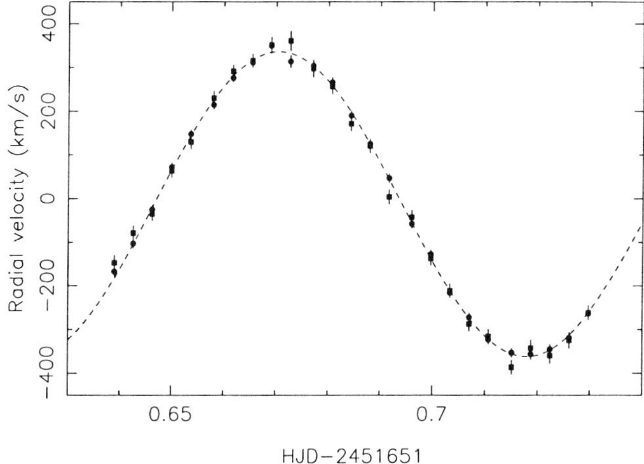

Figure 7 The radial velocity curve of the sdB component of KPD1930+2752 due to a white dwarf companion (Maxted, Marsh & North, 2000).

6. SUMMARY

The low and intermediate mass close binary systems discussed in this review mostly have a modest effect upon stellar populations with the significant qualifier that they are almost certainly the hosts of Type Ia supernovae. The latter have a significant impact upon galactic chemical evolution. The exact nature of the progenitor is uncertain, but developments over the past decade have swung opinion away from merging double white dwarfs towards single degenerate models in which white dwarfs steadily accrete and fuse hydrogen-rich material at rates of order $10^{-7}\,\mathcal{M}_{\odot}\,\mathrm{yr}^{-1}$. These systems probably correspond to the Super-Soft X-ray sources and related objects such as recurrent novae and symbiotic stars. Recurrent novae are a sub-class of cataclysmic variable stars, of which there are of order 2×10^{6} to 4×10^{7} in the Galaxy. Most of these do not lead to Type Ia supernovae because classical nova explosions erode the mass of the white dwarf components in these systems; these explosions may affect the abundances of some rare isotopes but have little overall upon chemical processing in the Galaxy.

There is now good evidence for a very significant population of double-degenerate binary stars, numbering about 200 million in the Galaxy. Only 16 of these have measured orbital periods, and just 6 have mass ratios, and yet already their observed properties have required significant alteration of theoretical evolution models: they are a good test population for binary evolution. Double degenerates are likely to be a significant source of gravitational waves for space-based detectors such as LISA, and may produce detectable events in ground-based detectors if their mergers lead to accretion-induced collapse and rapidly rotating neutron stars as opposed to Type Ia supernovae.

References

Amari, S. et al., 2001, *ApJ* , in press, astro.

Bergeron, P., Saffer, R.A., Liebert, J., 1992, *ApJ* , 394, 228.

Bragaglia, A., Greggio, L., Renzini, A., D'Odorico, S., 1990, *ApJ* (Letters) , 365, L13.

Bragaglia, A., Renzini, A., Bergeron, P., 1995, *ApJ* , 443, 735.

Branch, D., 2000, 11th Annual Astrophysics Conference in Maryland, Young Supernova Remnants, eds Holt, S.S., Hwang, U.

Brown, T. M., Ferguson, H. C., Davidsen, A. F., 1995, *ApJ* (Letters) , 454, L15.

Cassisi, S., Iben, I. J., Tornambe, A., 1998, *ApJ* , 496, 376.

de Kool, M., 1992, *A&A* , 261, 188.

della Valle, M., Livio, M., 1996, *ApJ* , 473, 240.

di Stefano, R., Rappaport, S., 1994, *ApJ* , 437, 733.

Dorman, B., Rood, R.T., O'Connell, R.W., 1993, *ApJ* , 419, 596.

Duerbeck, H. W., 1999, Informational Bulletin on Variable Stars, 4731, 1.

Foss, D., Wade, R.A., Green, R.F., 1991, *ApJ* , 374, 281.

Gehrz, R. D., Truran, J. W., Williams, R. E., Starrfield, S., 1998, *PASP* , 110, 3.

Greiner, J., Hasinger, G., Kahabka, P., 1991, *A&A* , 246, L17.

Hachisu, I., Kato, M., Nomoto, K., 1996, *ApJ* (Letters) , 470, L97.

Hachisu, I., Kato, M., Kato, T., Matsumoto, K., Nomoto, K., 2000, *ApJ* (Letters) , 534, L189.

Han, Z., Podsiadlowski, P., Eggleton, P. P., 1995, *MNRAS* , 272, 800.

Han, Z., 1998, *MNRAS* , 296, 1019.

Harrison, T. E., McNamara, B. J., Szkody, P., Gilliland, R. L., 2000, *AJ* , 120, 2649.

Hils, D., Bender, P.L., Webbink, R.F., 1990, *ApJ* , 360, 75.

Holberg, J. B., Saffer, R. A., Tweedy, R. W., Barstow, M. A., 1995, *ApJ* (Letters) , 452, L133.

Iben, I., Jr., Tutukov, A.V., Yungelson, L.R., 1997, *ApJ* , 475, 291.

Kahabka, P., 1999, *A&A* , 344, 459.

Kato, M., Hachisu, I., 1999, *ApJ* (Letters) , 513, L41.

Knox, R.A., Hawkins, M.R.S., Hambly, N.C., 1999, *MNRAS* , 306, 736.

Kolb, U., 1993, *A&A* , 271, 149.

Lai, D., Shapiro, S. L., 1995, *ApJ* , 442, 259.

Leibundgut, B., 2000, *The A&A Review* , 10, 179.
Li, X. -., van den Heuvel, E. P. J., 1997, *A&A* , 322, L9.
Liebert, J., Dahn, C.C., Monet, D.G., 1988, *ApJ* , 332, 891.
Livio, M., 2000, The greatest explosions since the Big Bang, Supernovae and Gamma-ray bursts, eds Livio, M., Panagia, N., Sahu, K..
Marsh, T.R., Dhillon, V.S., Duck, S.R., 1995, *MNRAS* , 275, 828.
Marsh, T.R., 1995, *MNRAS* , 275, L1.
Maxted, P.F.L., Marsh, T.R., 1999, *MNRAS* , 307, 122.
Maxted, P.F.L., Moran, C.K.J., Marsh, T.R., Gatti, A.A., 2000, *MNRAS* , 311, 877.
Maxted, P. F. L., Marsh, T. R., Moran, C. K. J., Han, Z., 2000, *MNRAS* , 314, 334.
Maxted, P. F. L., Marsh, T. R., North, R. C., 2000, *MNRAS* , 317, L41.
McArthur, B. E. et al., 1999, *ApJ* (Letters) , 520, L59.
Moran, C., Marsh, T.R., Bragaglia, A., 1997, *MNRAS* , 288, 538.
Moran, C., Maxted, P., Marsh, T.R., Saffer, R.A., Livio, M., 1999, *MNRAS* , 304, 535.
Nelemans, G., Verbunt, F., Yungelson, L. R., Portegies Zwart, S. F., 2000, *A&A* , 360, 1011.
Nelemans, G., Yungelson, L. R., Portegies Zwart, S. F., Verbunt, F., 2001, *A&A* , in press, astro.
Nomoto, K., 1982, *ApJ* , 253, 798.
Orosz, J.A., Wade, R.A., 1999, *MNRAS* , 310, 773.
Oswalt, T.D., Smith, J.A., Wood, M.A., Hintzen, P., 1996, *Nature* , 382, 692.
Patterson, J., 1984, *ApJS* , 54, 443.
Patterson, J., 1998, *PASP* , 110, 1132.
Politano, M., 1996, *ApJ* , 465, 338.
Prialnik, D., Kovetz, A., 1995, *ApJ* , 445, 789.
Ritter, H., Kolb, U., 1998, *A&AS* , 129, 83.
Robinson, E.L., Shafter, A.W., 1987, *ApJ* , 322, 296.
Saffer, R.A., Liebert, J., Olszewski, E.W., 1988, *ApJ* , 334, 947.
Saffer, R.A., Livio, M., Yungelson, L.R., 1998, *ApJ* , 502, 394.
Truemper, J., Hasinger, G., Aschenbach, B., Braeuninger, H., Briel, U.G., 1991, *Nature* , 349, 579.
van den Heuvel, E.P.J., Bhattacharya, D., Nomoto, K., Rappaport, S.A., 1992, *A&A* , 262, 97.
Warner, B., 1987, *MNRAS* , 227, 23.
Warner, B., 1995, Cataclysmic variable stars, Cambridge Astrophysics Series, Cambridge, New York: Cambridge University Press.

OBSERVING GLOBULAR CLUSTER BINARIES WITH GRAVITATIONAL RADIATION

Matthew Benacquista
Montana State University-Billings
Billings, Montana, 59101 USA
benacquista@msubillings.edu

Keywords: gravitational radiation, globular clusters, binaries

Abstract Ultracompact binaries are expected to be the best candidates for continuous gravitational radiation in the frequency range of space-based gravity wave detectors such as LISA. The dynamics of globular cluster evolution are expected to enhance the population of such binaries in globular clusters. We present the nature of the globular cluster population of binaries which will be detectable by LISA as well as the fraction of the galactic globular cluster system which can be explored by LISA.

1. INTRODUCTION

With the recent progress in ground based gravitational wave detectors such as LIGO and VIRGO and the expected development of a space based detector (LISA), gravitational wave astronomy may soon be a viable observational field. Ground based detectors will explore the frequency band containing burst sources such as inspiraling neutron star or black hole binaries. Continuous sources such as ultra-compact binaries containing evolved objects will be seen in LISA's frequency band. It is important to explore the properties of these potential sources so that proper data analysis techniques can be developed in anticipation of the actual mission.

At the low end of LISA's frequency band, below about 1 mHz, the large disk population of close white dwarf binaries is expected to produce a confusion-limited background as each frequency bin will contain many sources Hils and Bender, 2000. Thus, only nearby sources will stand out above this background. If a source lies outside of the plane of the galactic disk, it may be possible to extract its signal from the background due

D. Vanbeveren (ed.), The Influence of Binaries on Stellar Population Studies, 73–78.

to the disk. Thus, globular clusters may prove to be the ideal hosts for sources of gravitational radiation.

2. THE LISA MISSION

The LISA mission is a space-based laser interferometer with armlengths of 5×10^6km arranged in the form of an equilateral triangle situated in heliocentric orbit at 1 AU and 20° behind the earth. The plane of the equilateral triangle makes an angle of 60° with the plane of the ecliptic. By judicious choice of the orbits, the plane of the interferometer precesses once per orbit. The LISA detector acts as two semi-independent interferometers which are sensitive to both the plus and cross polarizations. The nominal life of the mission is three years. Danzmann *et al.*, 1998

3. SIGNAL ANALYSIS

A gravitational wave is detected by LISA as a strain amplitude from the changing armlengths. The gravitational wave produced by a close binary system can be approximated by a continuous, monochromatic wave with a frequency, f, that is twice the orbital frequency, $f_{\rm orb}$ of the binary. This wave produces a signal in one interferometer given by:

$$h(t) = \frac{\sqrt{3}}{2}(A_+F^+ \cos(2\pi ft) + A_\times F^\times \sin(2\pi ft)), \qquad (5.1)$$

where the form factors, F^+ and $F^\times$, are functions of the direction angles to the source with respect to the plane of the detector. The polarization amplitudes are:

$$A_+ = \frac{G^{5/3}}{rc^4}\frac{M_1M_2}{(M_1+M_2)^{1/3}}f^{2/3}(1+\cos^2 i) \qquad (5.2)$$

$$A_\times = -2\frac{G^{5/3}}{rc^4}\frac{M_1M_2}{(M_1+M_2)^{1/3}}f^{2/3}\cos i. \qquad (5.3)$$

We define the "chirp mass" to be:

$$\mathcal{M}^{5/3}\frac{M_1M_2}{(M_1+M_2)^{1/3}} \qquad (5.4)$$

so that the only mass dependence of $h(t)$ is by $\mathcal{M}$. The chirp mass for values of component masses is shown in Fig 1. The motion of LISA about the sun imposes a periodic Doppler shift on the overall frequency of the signal. The important noise sources in the frequency range of interest are both instrumental and galactic in origin. The instrumental

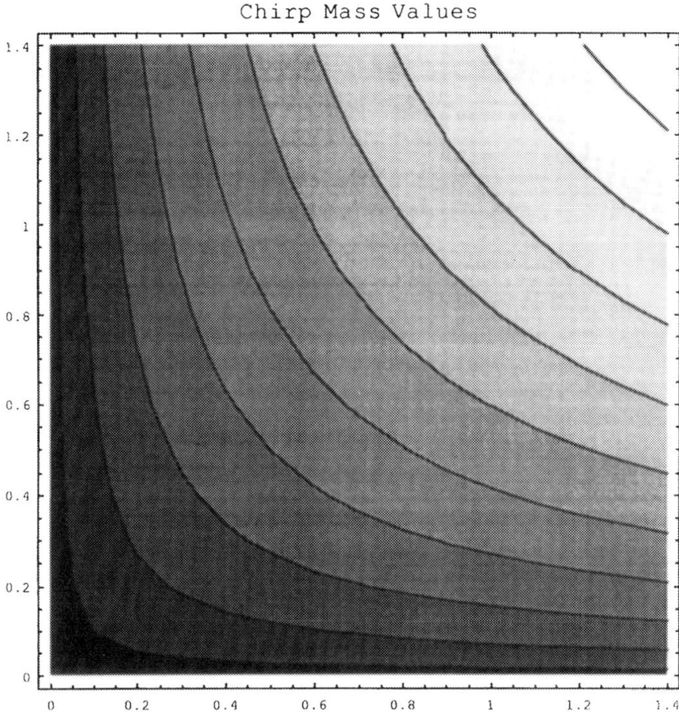

Figure 1 Chirp Mass Values. The contours are of constant chirp mass values for differing component masses which range from 0 to 1.4 $M_{\odot}$

noise is expected to arise from acceleration noise in the positions of the spacecraft for frequencies below about 3 mHz and shot noise in the lasers above that value Larson *et al.*, 2000. Coincidentally, the population of close white dwarf binaries in the galactic disk is expected to produce a confusion-limited background noise which is also significant below about 3 mHz. At frequencies below about 1 mHz, the space density of galactic binaries is high enough so that the background noise will be isotropic. Above 1 mHz, the background noise should reflect the shape of the disk. These noise levels for LISA and the confusion limited background are shown in Fig. 2. The likelihood of detecting globular cluster binaries relies upon the signal rising above these noise sources. We characterize the signal to noise ratio as Cutler, 1998:

$$\rho^2 = 4\int_0^{\infty} \frac{\tilde{h}_I^* \tilde{h}_I + \tilde{h}_{II}^* \tilde{h}_{II}}{S_n(f)} df \tag{5.5}$$

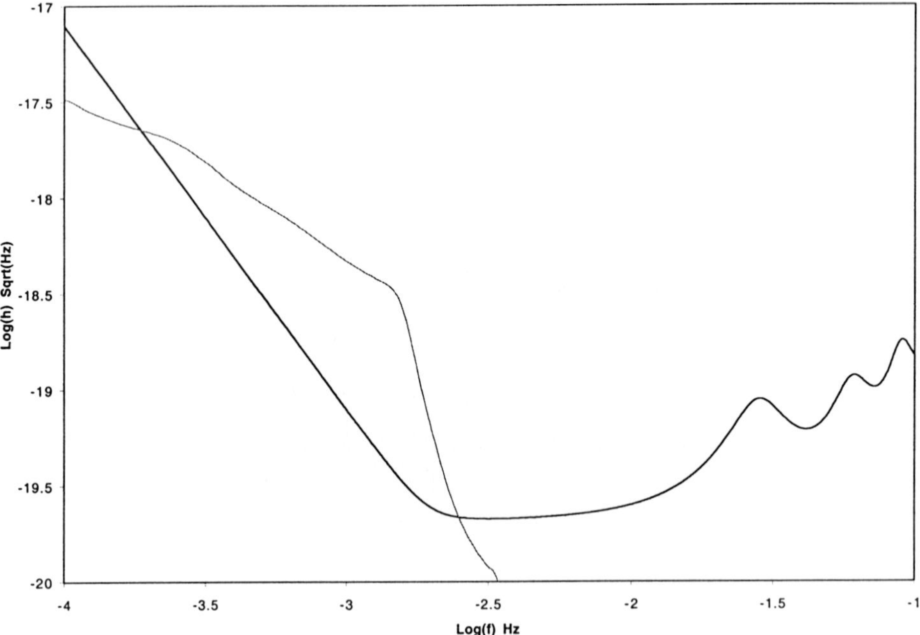

Figure 2 Instrumental noise and background noise. The instrument noise of LISA is shown along with the expected confusion-limited background due to the disk population of close white dwarf binaries.

where I and II indicate the two semi-independent interferometers, $S_n(f)$ is the spectral density of the instrumental and background noise, and the Fourier transform of the signal is:

$$\tilde{h}(f) = \int_{-\infty}^{\infty} e^{2\pi i f t} h(t) dt. \tag{5.6}$$

4. BINARY POPULATION REQUIREMENTS

In order for a signal to be identified with a binary source in a particular globular cluster, the angular resolution must be comparable to the size of the globular cluster. The angular resolution of LISA depends upon the Doppler shift in frequency being measured as a phase variation during a one-year orbit about the sun. The minimum observation time, T, is generally taken to be one year, which implies a frequency bin-width of $\Delta f = 1/T$. The Doppler shift scales as $f(v/c) = f(2\pi R/cT)$. Consequently, the wavelength must be less than $2R = 2\text{AU}$ so that the minimum gravitational wave frequency is $\sim 1000s$. This corresponds to a maximum orbital period of $P_{\text{orb}} < 2000s$. The angular resolution of LISA also depends upon the ability to reliably extract the signal from the noise. This, in turn, requires that the system have a clean signal which can be well modeled. A detached system with no significant mass transfer, can be well modeled by a point-mass quadrupole whose period evolution is determined solely by general relativity. Thus, we also require that the detectable population consist only of detached systems. Naturally, this requirement can be lifted as better models for the gravitational radiation from interacting systems are developed. Out of this population, a system is considered detectable if $\rho > 2$. Since the angular resolution also scales as $1/\rho$, we also look at systems with $\rho > 5$ for a more conservative estimate of the detectable population.

5. RESULTS AND CONCLUSIONS

Using globular cluster position and distance from Harris, 1996, we determined the signal to noise ratio for NS-WD binaries with WD masses in the range of 0.1 to 0.4 $M_{\odot}$ and frequencies ranging from 1 mHz up to the onset of Roche lobe overflow. Because of the degeneracy in the chirp mass, these calculations are also valid for double white dwarf binaries with masses as low as $0.3M_{\odot}$ each. Note that the signal strength for a binary with $i = 0$ is about twice that for one with $i = \pi/2$. Globular clusters were then classified by the strength of the detected signal for binaries with favorable or unfavorable angles of inclination. There are 7 clusters with some binaries having $i = \pi/2$ and $\rho \geq 5$ and 17 for which only binaries with $i = 0$ have $\rho \geq 5$. The binaries associated with these clusters would have sufficient angular resolution to clearly identify the binary with the cluster. There are 72 clusters with $i = \pi/2$ and $\rho \geq 2$ and 36 with only $i = 0$ and $\rho \geq 2$. These clusters could house binaries whose angular resolution would be a few square degrees, and the identification of the binary with its host cluster may be a little more suspect. There are only 15 clusters which are too distant to detect any binaries with $\rho \geq 2$. Thus, almost 90% of the galactic globular cluster system can be observed with LISA for a wide range of compact binary systems if $P_{\rm orb} < 2000s$. Such systems would be very difficult to detect in globular clusters by any means other than gravitational radiation. If a population of these binaries is predicted by dynamical evolution scenarios, LISA will prove to be a valuable tool for observing binaries in globular clusters.

Acknowledgments

This work has been supported by NASA EPSCoR grant NCCW-0058 and Montana EPSCoR grant NCC5-240

References

Cutler, C., (1998) Phys. Rev. D 57, 7089.

Danzmann, K., *et al.* (The LISA Study Team), (1998) *LISA Pre-Phase A Report*, Garching, Germany: Max-Planck-Institut für Quantenoptik.

Harris, W.E., (1996) AJ 112, 1487.

Hils, D. and Bender, P.L., (2000) ApJ 537, 334.

Larson, S.L., Hiscock, W.A., and Hellings, R.W., (2000) Phys. Rev. D 62, 062001.

THE ALGOL-TYPE BINARIES

Geraldine J. Peters
Space Sciences Center, University of Southern California
Los Angeles, CA 90089-1341; USA
gjpeters@mucen.usc.edu

Keywords: Stars:Binaries:Close,Algols - Stars: Circumstellar Matter - Stars: Mass Loss - Stars:emission-line,Be - Accretion Disks

Abstract The nature of the circumstellar material in interacting binaries of the Algol-type, including the accretion disk, gas stream, high temperature plasma, and domains of outflow is reviewed. Mass transfer is compared in six similar systems with early B primaries and short-moderate periods that approximately represent a series of snapshots of such binaries during their slow evolutionary phase. Evidence for moderate systemic mass loss during the current and earlier stages is presented.

1. INTRODUCTION

A contemporary definition of an Algol-type binary was given by Alan H. Batten (1989a) in his review talk at IAU Colloquium No. 107 (*Algols*):

> "A binary in which the less massive stellar component fills its Roche lobe and the other, which does not, is not degenerate".

Of course this class of interacting binaries was named after the system Algol (β Per) itself, whose light variability was discovered by John Goodricke (1783). The primary star in these semi-detached binary systems is usually a B or A-type main sequence object, while the cooler secondary is an A–K III-IV star. The Algol phase is the slow stage of mass transfer ($\sim 10^{-5} - 10^{-7} M_{\odot}$ yr^{-1}) after the mass reversal in the component stars has taken place. Algols as interacting binaries have been featured in five IAU Symposia/Colloquia, IAUS 51 (Batten 1973), IAUS 73 (Eggleton, Mitton, & Wehlan 1976), IAUC 88 (Plavec, Popper, & Ulrich 1980), IAUC 107 (Batten 1989b), & IAUS 151 (Kondo, Sisteró, & Polidan 1992) and several other topical conferences, e.g. NATO ASI (Eggleton & Pringle 1985), the 105th ASP Meeting (Shafter 1994), and various Pacific Rim Conferences (Leung 1997).

D. Vanbeveren (ed.), The Influence of Binaries on Stellar Population Studies, 79–94.

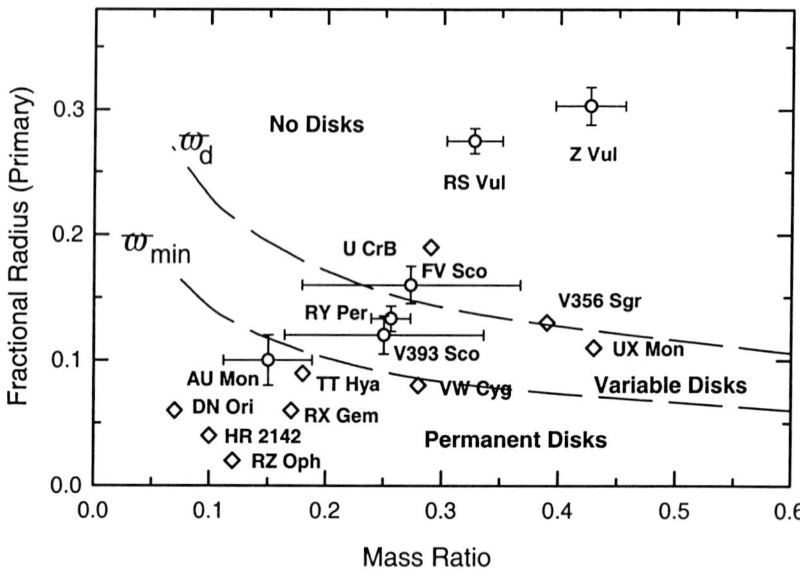

Figure 1 The locations of selected Algol systems in the r–q diagram. The systems that are plotted with *open circles* and error bars are specifically discussed in this paper. For comparison some other well-known systems are plotted with *open diamonds.*

Current knowledge on the circumstellar (CS) material and mass transfer in Algol binaries will be reviewed in this paper. At least four types of CS material have been identified in Algol binaries: 1) the accretion disk, 2) gas stream (or domain of infall), 3) high temperature plasma, and 4) various domains of mass outflow. Each of these components will be individually discussed then empirical models for the CS material in six short-moderate period systems that contain early B primaries will be intercompared.

The nature of the CS material in Algol binaries can be understood with the aid of the so-called r–q diagram in which the fractional radius of the mass gainer (R_p/a) is plotted versus the mass ratio, $q = M_{loser}/M_{gainer}$, and compared with the theoretical computations of gas stream hydrodynamics of Lubow and Shu (1975). Such a display is presented in Figure 1. The upper *dashed* curve, ϖ_d, delineates the fractional radius of the dense accretion disk for systems of different mass ratios. The lower *dashed* curve, ϖ_{min}, shows the minimum distance that a stream particle will achieve relative to the center of the gainer. If the system falls above the upper curve, the gas stream will strike the photosphere of the gainer and any Hα emission seen from the CS material will be transient. If the system falls below the lower curve, then one will see a prominent accretion disk that emits strongly in Hα. The amount of material in the accretion disk can vary, especially for systems near the lower curve. The most variable accretion disks are found in systems that fall between the two curves, since the inner part of a wide gas stream will strike or graze the photosphere of the primary while the outer edge will feed an accretion disk. For the systems plotted above the $\varpi_{\rm min}$ line, the angle between the impacting gas stream and the photosphere of the B star varies from $\sim 50^o$ to the normal to tangential.

2. THE ACCRETION DISK

In the optical spectral region, an accretion disk in an Algol binary is revealed by the presence of a double-peaked Hα emission feature that primarily emanates from the lower density regions of the disk. Spectroscopic surveys done about two decades ago revealed that permanent Hα emission is commonplace in Algol systems with periods larger than 6 days whereas shorter-period Algols display at best transient emission (Peters 1980). Of course this cutoff period is only an approximation as the important criterion is the location of the system in the r–q diagram. Some systems with periods less than 6 days can accommodate a disk (e.g. SW Cyg, TZ Eri), while others of longer period (e.g V356 Sgr) are devoid of an Hα-emitting disk. A comprehensive study of the Hα emission in 18 Algols has recently been published by Richards & Albright (1999).

The distribution of Hα-emitting material in Algol binaries with short–moderate periods can be effectively seen from the use of Doppler tomograms (Richards, Albright, & Bowles 1995, RAB95, Albright & Richards 1996) that are produced using a back-projection code (Marsh et al. 1990, Robinson et al. 1993). The technique for generating the tomograms is described in RAB95. A good comparison of representative systems has been prepared by Richards (Figure 2). In this display a polar view of a scaled-model of each binary is contrasted with its tomogram. The periods range from 2.87–6.95 days and the impact angle that the gas stream makes with gainer's photosphere varies from steep–tangential. Regions of greatest emissivity of Hα appear as the darker shades of gray in the tomograms. It is evident that in the systems with the shortest periods the emitting region lies between the stars. Two epochs are shown for the short-period Algol U CrB: just after a flare-up in the mass transfer rate (upper tomogram) and during quiescence. Note the presence of a transient disk-like structure after the active event. In quiescence the gas stream dominates. In the longer-period systems, the tomograms reveal a disk-like structure with material that displays Keplerian motion and extends from the photosphere of the gainer to its Roche surface.

The extended nature of the disk is also suggested from Hα occultation investigations (Peters 1989). The phase-dependent behavior of the Hα emission profile in most Algols that undergo a total eclipse indicates that the low-density portion of the accretion disk extends out to at least 95% of the gainer's Roche lobe and that it is highly flattened with a vertical height of <20% of its radial size. It is typical for occultation of the disk to begin around phase 0.75–0.80, as is evident in TT Hya (Figure 3). Hydrodynamic simulations of mass flow in Algols predict an asymmetry in the disk material with an extension on the *leading* hemisphere (Richards & Ratliff 1998) of the binary. This is clearly seen in the Hα occultation data for TT Hya as well as many other systems with low density disks, since the secondary star begins blocking out the emission from the disk material as early as $\phi \sim 0.75$ but disk eclipse appears to end around $\phi \sim 0.15$. Estimates for the density in the disk vary with the method used to determine it. Uncertainties in each method are at least two orders of magnitude, as one must make many assumptions concerning dimensions, geometry, ionization, LTE/NLTE, stratification, velocity fields, etc. The volume emission measure, N_e^2V, from Hα in disk systems of moderate period typically implies a mean electron density of 10^{10} cm^{-3}. A lower disk density ($10^8 - 10^9$ cm^{-3}) was inferred for the 11^d Algol AU Mon from *ORFEUS-SPAS 2* observations of Fe III absorption lines (Peters & Polidan 1998).

The most prominent disks are found in systems with large semi-major axes that are currently undergoing a high rate ($\sim 10^{-5}$–$10^{-6} M_\odot$ yr^{-1}) of mass transfer. Such systems were first called the "W Serpentis" systems, or "Serpentids" by Mirek Plavec (1980). In the UV these Algols display a conspicuous continuum and line emission from high–low ionization species, e.g. N V, C IV, Si IV, C II, and Fe II,III, (Plavec, Weiland, & Koch 1982, Plavec 1989).

The disks in the W Ser binaries are optically thick with densities comparable to those in the upper photospheres of B stars. Using *HST*/GHRS observations of the Si IV emission features at phases 0.0 and 0.5, Weiland et al. (1995) have estimated the physical parameters in the disk of the prototype system W Ser itself. From Fe line absorption in the Si IV emission profiles, they infer a disk size of $\sim 3 \times 10^{11}$ cm, $T_{ex} \sim 7,000$ K, and $N_e \sim 2 \times 10^{11}$ cm^{-3}. Plavec et al. (1982) concluded that the density in the disk in SX Cas is ~10 times higher. Weiland et al. found evidence for a density enhancement on the *trailing* side of the W Ser system where the gas stream impacts the disk. V/R asymmetries in the Hα emission

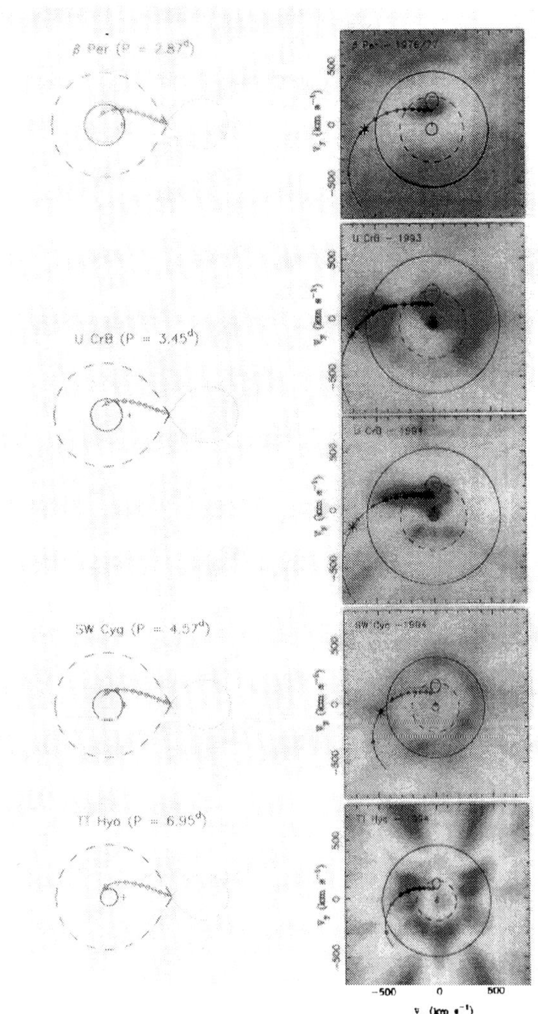

Figure 2 Hα Doppler tomograms of representative Algol systems of short–moderate period show the geometry of the CS material versus period. A polar view of each binary is shown on the left with the Roche surface of the primary indicated with a *dashed* curve. In the gray-scale tomograms on the right the region of most intense Hα emission is indicated by the darkest gray tone. In both spacial and velocity maps the trajectory of the gas stream is indicated. Keplerian motion near the photosphere of the mass gainer and at its Roche surface limit are indicated by the outer *solid* and inner *dashed* circles in the tomograms. Illustration courtesy of M. Richards from *Mercury*, 29, 34, 2000.

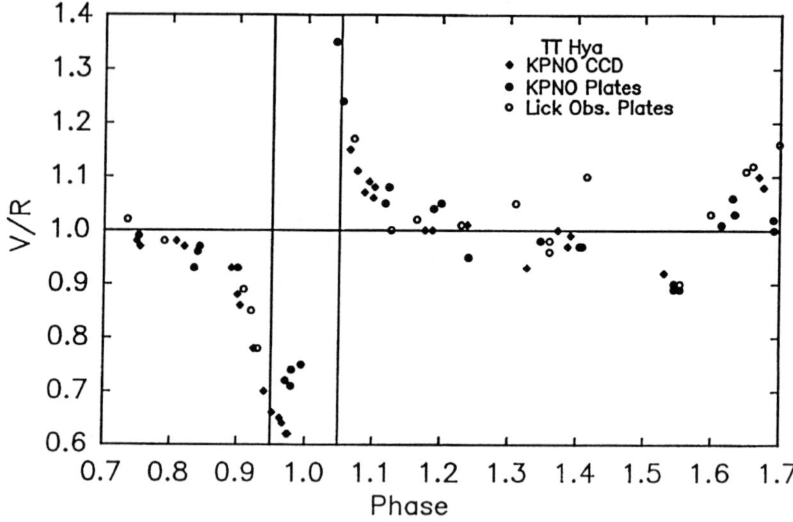

Figure 3 The ratio of the intensity of the violet-shifted lobe (V) of the double-peaked Hα emission line in TT Hya (tomogram shown in Fig. 2) to its red-shifted (R) component versus phase. V and R are measured relative to the continuum (I_V/I_{cont}, I_R/I_{cont}. The vertical bars show the limits of the stellar eclipse. Disk eclipse is evident from $0.75 < \phi < 0.15$. These observations suggests that $R_{disk} \simeq 0.95\ R_{RocheLobe}$ but is asymmetric with more extension on the *leading* hemisphere of the gainer. Illustration from Peters (1989).

feature in RZ Oph (P $= 262^d$) at the conjunction points suggest that the latter condition also exists in this system (Peters 1989).

3. THE GAS STREAM

The gas stream, or the domain of material infall, is detected from red-shifted absorption lines usually observed between phases 0.75–0.95. The absorption can appear in the form of either discrete or "inverse" wind features. A typical example of the latter is shown in Figure 4. The trajectory of the gas stream is basically a free-fall path modified by the Coriolis force and appears generally the same whether simple three-body calculations are made or hydrodynamical simulations. The stream becomes more focused as it approaches the gainer (Lubow & Shu 1975, LS75) From LS75, the gas stream particles are expected to have a velocity $\sim 400 - 450$ km s^{-1} when they reach or impact the primary, but normally we observe the flow in projection. Some examples of stream trajectories are shown in Figure 2. The vertical height of the stream is uncertain. LS75 assumed it was $\sim \epsilon$, the ratio of the isothermal sound

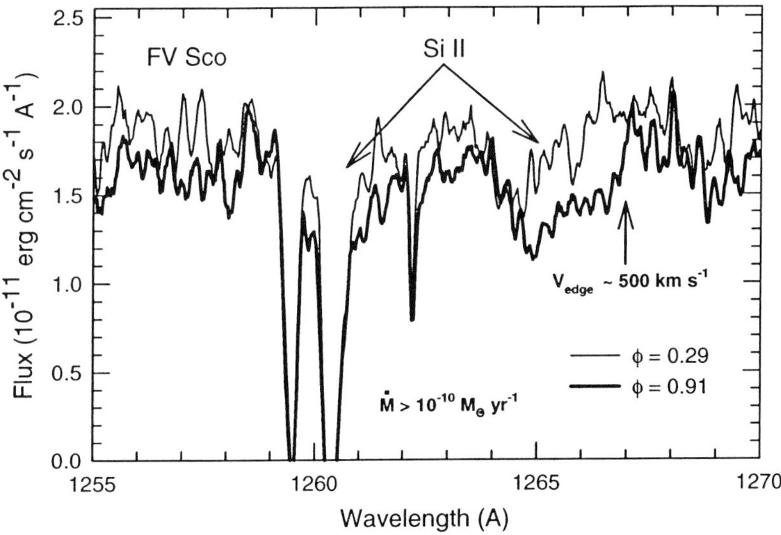

Figure 4 The gas stream absorption in FV Sco is evident from this comparison of the profiles of the Si II resonance doublet at $\lambda 1260, 65$ Å observed near the quadrature point on the *leading* hemisphere of the system and those at a phase in which the line-of-sight to the primary passes through the infalling material. The sharp features at 1260Å are interstellar lines. $\lambda 1265$ does not contain an IS component, as it originates from a level that is 287 cm^{-1} above ground.

speed to the product of the systemic angular velocity and the binary separation. However, in a later 3D study, Lubow & Shu (1976) concluded that the scale height is several time the hydrostatic value and the stream can actually flow above/below the disk and produce a turbulent wake. In computing the mass transfer rate from the observed red-shifted absorption components, one usually makes the assumption that the gas stream is geometrically thick and in projection covers the surface of the gainer. This and other assumptions concerning path length, LTE, line optical depth, etc. contribute to a large uncertainty in this quantity. For FV Sco, we find a mass infall rate $> 10^{-10}$ $M_\odot$ yr^{-1} from the Si II resonance lines. A smaller mass transfer rate ($> 10^{-12}$ $M_\odot$ yr^{-1}) was inferred for TT Hya from *ORFEUS-SPAS 2* observations (Peters & Polidan 1998) of the gas stream components in N I 1085 Å and NII 1135 Å. These observations implied a stream temperature of $\sim 17,000$ k that is about 1.5 T_{phot}. Interestingly, a mismatch between the tomogram from this star and their hydrodynamical simulation led Richards and Ratliff (1998) to suggest that the gas stream might be hotter than the 8,000 K that they assumed.

4. THE HIGH TEMPERATURE ACCRETION REGION

The unexpected discovery of prominent absorption and emission lines from N V, C IV, and Si IV in the *IUE* spectra of non-peculiar Algol binaries in the early 1980s revealed the presence of a high-temperature component to the CS material in these objects (Peters & Polidan 1984, PP84; Plavec 1983). From a study of the absorption features, PP84 concluded that these lines are formed by resonance scattering in a plasma of high temperature ($T_{ion} \sim 10^5$K), intermediate density ($N_e \sim 10^9 cm^{-3}$), and moderate-high carbon depletion. PP84 called this plasma the High Temperature Accretion Region, or HTAR. The HTAR, which appears to be most prominent on the *trailing* ($0.5 \leq \phi \leq 0.95$) hemisphere, is most likely turbulent and heated by shocks caused from the impacting gas stream. Indeed observations show that the HTAR prevails in the systems in the upper two-thirds of the r–q diagram that cannot form stable disks. A comparison of systems containing early B primaries reveals that the azimuthal extent of the HTAR depends on the period of the binary (cf. §6). It is still uncertain if the CS material in which the superionized emission lines are formed is the same, adjacent to, overlaps, or is distinct from the HTAR. This will be discussed further below.

5. DOMAINS OF OUTFLOW

Two noteworthy domains of mass outflow have been identified in Algol binaries: 1) mass loss at/near phase 0.5 and 2) bipolar flows of material perpendicular to the orbital plane. A typical example of mass loss near phase 0.5 is presented in Figure 5. This type of mass loss, sometimes called an enhanced "wind", can be in the direction of the Lagrangian point L_2 or around a "splash" region (discussed below). It is usually observed in the Si IV resonance lines, but sometimes the presence of N V absorption indicates a region of higher ion temperature.

Recently there have been several spectropolarimetric observations of Algols from space-borne telescopes that suggest the existence of bipolar flows. Using *ASTRO-2*/WUPPE and ground-based data, Hoffman, Nordsieck, & Fox (1998) concluded that the plasma that produces the Balmer and UV emission lines and NUV continuum in β Lyr is oriented 90° to the source of the visual polarization, which is most likely the disk. Spectropolarimetry of V356 Sgr and TT Hya from the *HST*/FOS has revealed conspicuous polarization (~4% & 20%, respectively) in a non-stellar continuum observed during totality that also suggest prominent bipolar flows in these systems (Lynch et al. 1996, Polidan, et al., in preparation). Perhaps the UV emission lines from superionized species

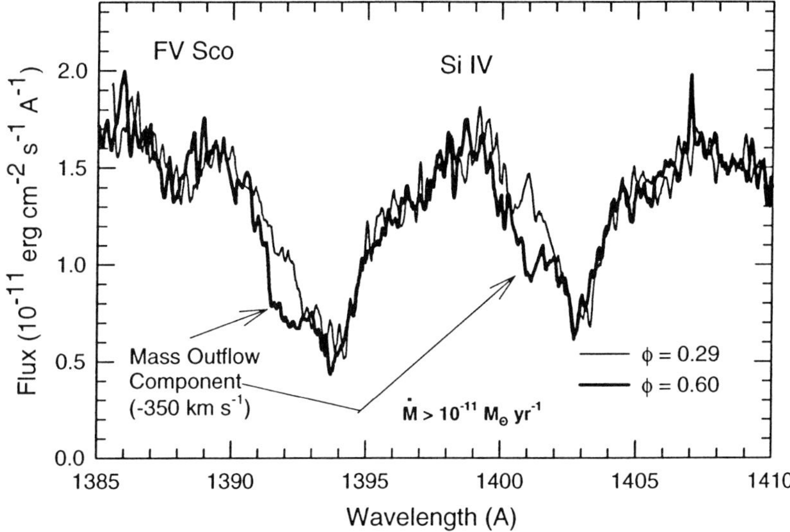

Figure 5 Superposition of the Si IV profiles in FV Sco observed at phases 0.60 and 0.29 reveal prominent mass loss in the direction of an apparent "splash" caused when the gas stream impacts the gainer's photosphere nearly tangentially (cf. §6). The edge velocity implies that the mass is probably lost from the system as the flow is most likely seen in projection. The computed rate is an order of magnitude lower than the inferred mass transfer rate from Si II (c.f. Figure 4), but uncertainties in the degree of ionization, amount of clumpiness, photospheric coverage, etc. produce an error of at least this amount.

originate in bipolar flow regions but the absorption lines associated with the HTAR are only a result of gas stream impact in the equatorial plane.

6. PUTTING IT ALL TOGETHER

From the standpoint of stellar evolution, the behavior of the mass transfer in the Algols of short-intermediate period that cannot form stable disks are probably the most interesting as these are the ones that lose mass during the slow phase of mass transfer. The degree of systemic mass loss depends on where/if the gas stream impacts the gainer's photosphere. In this section we compare the plasmas in six systems containing early B gainers in which the stream impact angle ranges from steep ($\sim 50°$ (Z Vul) to tangential (V393 Sco). Except for one object, RS Vul, we are looking at systems with nearly identical component stars (B3-4V + A-FIII) but with different periods, so we approximately have a series of snapshots of one kind of binary as its separation increases during the latter stages of its evolution. All of the systems discussed are probably

Table 1 Algols with Early B Primaries.

System	HD No.	Sp. Types	Period (days)	Phase Coverage
Z Vul	181987	B3V + A2III	2.46	0.16,0.31,0.53,0.56,0.69,0.88
RS Vul	180939	B5V + G0III:	4.48	0.18,0.40,0.42,0.62,0.81
FV Sco	155550	B4V + AIII:	5.73	0.11,0.29,0.42,0.60,0.78,0.91
RY Per	17034	B3V + F6III:	6.86	0.15,0.16,0.31,0.46,0.60, 0.75,0.86,0.93
V393 Sco	161741	B3V + AIII:	7.71	0.23,0.45,0.58,0.73,0.86,0.99
AU Mon	50846	B3Ve + F8III	11.11	47 SWP images,good phase coverage

undergoing Case B mass transfer, although Z Vul has sometimes been considered a Case A candidate (Plavec 1970). The observations used for the study are *IUE* HIRES images processed with the NEWSIPS software from the MAST archives at STScI. Information on each system and the phase coverage of the *IUE* data used for the study are given in Table 1. Empirical models for the systems considered are presented in Figure 6.

Clear evidence for mass infall was observed in all but the shortest-period system in which we expect the gas stream to be more difficult to detect as it should be more focused. The dominant ions were Si II and similar species. The binaries with the shortest periods, Z Vul and RS Vul are direct impact systems. In RS Vul, where the gas stream impacts at a more shallow angle, the ionization temperature in, and azimuthal extent of the HTAR are larger. There is no evidence for systemic mass loss in either of these systems. It is not uncommon to observe enhanced absorption from either low or high-ionization species in the phase interval 0.10–0.20 in systems of short or moderate period. This is probably a result of an interaction between the disk and gas stream. Evidence of a concentration of material between the components of an Algol binary was first reported in β Per (Richards, Bolton, & Mochnacki, 1989, Richards 1993).

We find the most interesting mass flow in FV Sco and RY Per in which the gas stream impacts the gainer's photosphere almost tangentially, but still appears to make contact with the star. In both cases there is evidence for substantial mass outflow near phase 0.6 from additional violet-shifted absorption in the Si IV or N V resonance lines (cf. Figure 5). I call this the *splash zone*, as it appears that the gas stream is reflected off of the star's photosphere and the shock from this encounter also

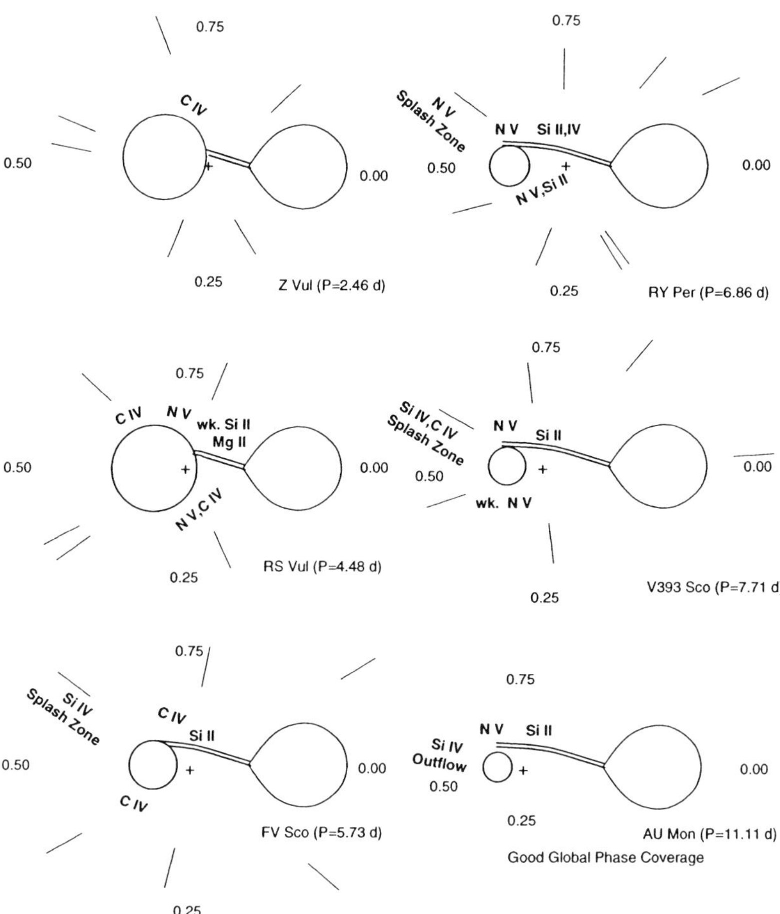

Figure 6 Polar views of the B-type Algols considered in this paper. Each system is scaled to the mass loser shown on the right, that is assumed to fill its critical Roche surface. The center of mass is indicated by a *cross* and the radial lines emanating from this point show the phases at which the *IUE* observations were made. The approximate locations of the ions formed in the gas stream, HTAR, and outflow/*splash* regions are shown in bold print.

heats the CS plasma at this azimuth. In V393 Sco, a slightly wider system that contains a gainer with nearly the same spectral type as RY Per, we also observe a splash zone, but the overall level of ionization in the CS material is much higher. N V absorption is seen at all non-eclipse phases observed. The heating might be caused from the apparent tangential impact of the gas stream, or perhaps a magnetic field that has been inferred to exist between the component stars from observed radio emission (Stewart et al. 1989) is important in this case. Weak, variable Hα emission is seen in this binary. The wider system AU Mon shows a persistent Hα-emitting disk, mass infall, an azimuthally-confined HTAR, and modest mass outflow near phase 0.5, all of which vary cyclically with the star's long-term (411^{d}, Lorenzi 1985) light cycle (Peters 1994).

The UV observations reveal that for systems with early B primaries there is noticeable mass loss in the phase interval 0.50–0.65 if the period is between $5.5 \leq \phi \leq 12.0$ days. For FV Sco uncertainties in the mass flow computations present a possible scenario that most, if not nearly all of the mass that is transferred, albeit small at this time, is funneled directly to the ISM. There is no evidence for a stable disk in this binary as Hα emission is not observed. In due time as the mass of the A-type secondary is reduced the period will lengthen (if most of the $\sim 0.5M_{\odot}$ that is still left to be lost by the secondary reaches the gainer) and the system should resemble AU Mon, which currently has an FIII secondary with a small mass of 0.75 $M_{\odot}$ (Popper 1989).

Despite the small systemic mass loss seen in snapshots of these early B Algols, the most important issue is whether significant mass is lost in outbursts or continuously during the rapid period of mass transfer. The mass loss at phase 0.5 in AU Mon is enhanced when the star is bright (Peters 1994). Even direct impact systems like U Cep display mass loss during an outburst (McCluskey, Kondo, & Olson 1988). To assess whether the B stars discussed here have lost significant mass earlier in their evolution, I have considered the calculations of De Greve (1986), DEG86. Using masses for Algols in Giuricin & Mardirossian (1981), De Greve showed that the majority of the systems appear to have undergone non-conservative mass transfer. In Figure 7 the current locations of the secondaries in the B-type Algols considered here are shown in the $q--M_l$ diagram of DEG86. The masses are from Popper (1980), Z Vul, Popper (1989), RS Vul & AU Mon, Olson & Plavec (1997), RY Per, and Peters, FV Sco & V393 Sco (the mass calibration in Harmanec 1988 was used). From the calculations in DEG86 whose results are reproduced in Figure 7, it appears that moderate non-conservative mass transfer has taken place in our B-type Algols. The history for AU Mon is somewhat ambiguous due to the small mass of its secondary. Most of the mass loss

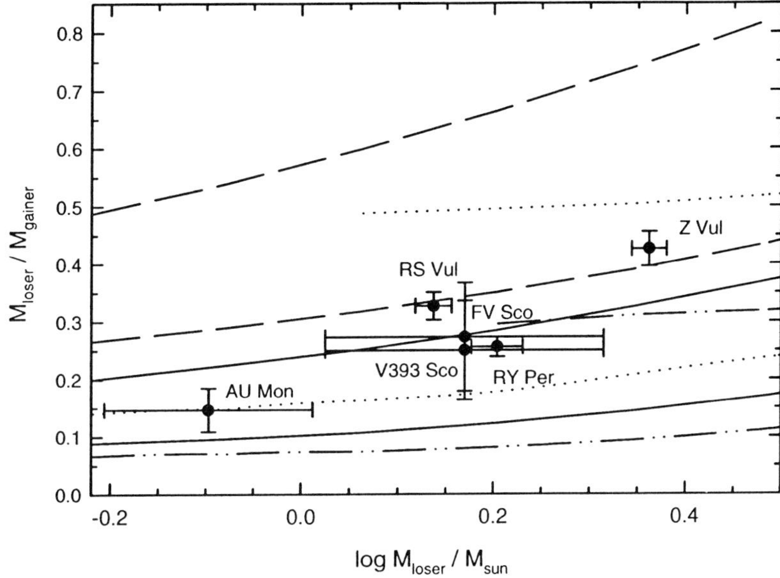

Figure 7 The locations of the B-type Algols discussed in this paper in the $q - M_l$ diagram of De Greve (1986, Figs. 5.2 & 5.3). The various lines are from stellar evolution calculations for the case of conservative mass transfer (*solid & dashed* lines) and no accretion (*dot-dashed & dotted* lines. The former are for an initial q of 0.5; the latter are for 1.0. The upper lines indicate the location of the mass loser at the time of minimum luminosity; the lower bound corresponds to the end of mass transfer.

in these systems probably occurred during rapid mass transfer but we cannot rule out frequent periodic or random outburst events.

7. CONCLUDING REMARKS

In this paper I have briefly reviewed some of our current knowledge on the nature of the CS material in the Algol binaries and intercompared it in six systems that contain early B gainers. Systemic mass loss appears to be important in the binary's evolution. To assess the overall importance of Algol evolution in a population of B stars, one must determine the number of non-eclipsing systems. These are difficult to identify, but a few ($< 5-10\%$, Van Bever & Vanbeveren 1997) might be found among the Be stars (e.g. CX Dra, Richards et al. 2000). Currently there are less than ten confirmed "pole-on" Algol counterparts.

Many details concerning mass transfer in Algol binaries are yet uncertain. There is still debate on the prevalence of Case A mass transfer. The carbon deficiencies that have been found in and around Algol primaries (Peters & Polidan 1984, Dobias & Plavec 1985, Tomkin et al. 1993) lend support for Case B evolution but only a few systems have been investigated. Some Algols, apparently in their latter stages of mass transfer, show evidence of a detached secondary (e.g. DN Ori, Etzel & Olson 1995, AU Mon, Peters 1994). Forthcoming should be the results from current studies on the effect of magnetic activity on mass transfer. Perhaps dynamo action exists (Sarna et al. 1998) and magnetic fields on the photosphere of the loser (Etzel & Olson 1993, Richards & Albright 1993) regulate mass transfer and even lift material to the vicinity of the Roche surface in the above-mentioned systems . The structure of the gas stream and outflowing material should be further investigated from observations. The generics of mass transfer are now understood, but the specifics are yet to come.

Acknowledgments

It is a pleasure to thank Mercedes Richards for fruitful discussions and for providing Figure 2. My research on the B-type Algols was partially supported by NASA grants NSG-5422, NAG5-8978, & NAG5-9124.

References

Albright, G. E., & Richards, M. T. 1996, ApJ, 459, L99

Batten, A. H. 1973, IAU Symp. 51, Extended Atmospheres and Circumstellar Matter in Spectroscopic Binary Systems, ed. A. H. Batten (Dordrecht: Reidel)

———. 1989a, Sp.Sci.Rev., 50, 1.

———. 1989b, IAU Colloq. 107, Algols, ed. A. H. Batten (Dordrecht: Kluwer)

De Greve, J. P. 1986, Space.Sci.Rev., 43, 139 (DEG86)

Dobias, J. J., & Plavec, M. J. 1985, 97, 138

Eggleton, P., Mitton, S., & Whelan, J. 1976, IAU Symp. 73, Structure and Evolution of Close Binary Systems, ed. P. Eggleton, S. Mitton, & J. Whelan (Dordrecht: Reidel)

Eggleton, P. P., & Pringle, J. E. 1985, Interacting Binaries, ed. P. P. Eggleton, & J. E. Pringle (Dordrecht: D. Reidel)

Etzel, P. B., & Olson, E. C. 1993, AJ, 106, 342

———. 1995, AJ, 110, 1809

Goodricke, J. 1783, Phil.Trans.Roy.Soc, 73, 474

Giuricin G., & Mardirossian, F. 1981, ApJS, 46, 1

Harmanec, P. 1988, BAC, 39, 329

Hoffman, J. L., Nordsieck, K. H., & Fox, G. K. 1998, AJ, 115, 1576

Kondo, Y., Sisteró, & Polidan, R. S. 1992, IAU Symp. 151, Evolutionary Processes in Interacting Binary Stars, ed. Y. Kondo, R. F. Sisteró, & R. S. Polidan (Dordrecht: Kluwer)

Leung, K. C. 1997, The Third Pacific Rim Conference on Recent Developments on Binary Star Research, ASP CS-130, ed. K. C. Leung (San Francisco: ASP)

Lorenzi, L. 1985, IBVS, No. 2704

Lubow, S. H., & Shu, F. H. 1975, ApJ, 198, 383 (LS75)

———. 1976, ApJ, 207, L53

Lynch, D. E., Polidan, R. S., Keyes, C. D., Nordsieck, K. H., & Peters, G. J. 1996, BAAS, 28, 1374

Marsh, T. R., Horne, K., Schlegel, E. M., Honeycutt, R. K., & Kaitchuck, R. H. 1990, ApJ, 364, 637

McCluskey, G. E., Jr., Kondo, Y., & Olson, E. C. 1988, ApJ, 332, 1019

Olson, E. C., & Plavec, M. J. 1997, AJ, 113, 425

Peters, G. J. 1980, in IAU Symp. 88, Close Binary Stars: Observations and Interpretation, ed. M. J. Plavec, Popper, D. M., & Ulrich, R. K. (Dordrecht: Reidel), 287

———. 1989, Space.Sci.Rev, 50, 9

———. 1994, in Interacting Binary Stars, ASP CS-56, ed. A. W. Shafter (San Francisco: ASP),384

Peters, G. J., & Polidan, R. S. 1984, ApJ, 283,745

———. 1998, ApJ, 500, L17

Plavec, M. 1970, PASP, 82, 957

———. 1980, in IAU Symp. 88, Close Binary Stars: Observations and Interpretation, ed. M. J. Plavec, Popper, D. M., & Ulrich, R. K. (Dordrecht: Reidel), 251

———. 1983, ApJ, 275, 251

———. 1989, Space.Sci.Rev, 50, 95

Plavec, M. J., Popper, D. M., & Ulrich, R. K. 1980, IAU Symp. 88, Close Binary Stars: Observations and Interpretation, ed. M. J. Plavec, Popper, D. M., & Ulrich, R. K. (Dordrecht: Reidel)

Plavec, M. J., Weiland, J. L., & Koch, R. H. 1982, ApJ, 256, 206

Popper, D. M. 1980, Ann.Rev.Astr.Ap., 18, 115

———. 1989, ApJS, 71, 595

Richards, M. T. 1993, ApJS, 86, 255

———. 2000, Mercury, 29, 34

Richards, M. T., & Alright, G. E. 1993, ApJS, 88, 199

———. 1999, APJS, 123, 537

Richards, M. T., Albright, G. E., & Bowles, L. M. 1995, ApJ, 438, L103 (RAB95)

Richards, M. T., Bolton, C. T, & Mochnacki, S. W., 1989, Space.Sci.Rev, 50, 358

Richards, M. T., Koubský, Šimon, V., Peters, G. J., Hirata, R., Škoda, & Masuda, S. 2000, ApJ, 531, 1003

Richards, M. T., & Ratliff, M. A. 1998, ApJ, 493, 326

Robinson, E. L., March, T. R., & Smak, J. I. 1993, in Accretion Disks in Compact Stellar systems, ed. J. C. Wheeler (Singapore: World Scientific), 75

Sarna, M. J., Yerli, S. K., & Muslimov, A. G. 1998, MNRAS, 297, 760

Shafter, A. W. 1994, Interacting Binary Stars, ASP CS-56, ed. A. W. Shafter (San Francisco: ASP)

Stewart, R. T., Slee, O. B., White, G. L, Budding, E., Coates, D. W., Thompson, K., & Bunton, J. D. 1989, ApJ, 342, 463

Tomkin, J., Lambert, D. L., & Lemke, M. 1993, MNRAS, 265, 581

Van Bever, J. & Vanbeveren, D. 1997, A&A, 322, 116

Weiland, J. L., Shore, S. N., Beaver, E. A., Lyons, R. W., & Rosenblatt, E. I. 1995, ApJ, 447, 401

Be STARS: SINGLE AND BINARY COMPONENTS

Douglas R. Gies
Center for High Angular Resolution Astronomy, Department of Physics and Astronomy, Georgia State University, Atlanta, GA 30303, U.S.A.
gies@chara.gsu.edu

Keywords: Stars: emission-line, Be — binaries: spectroscopic — stars: evolution

Abstract There is growing evidence that a significant number of the rapidly rotating Be stars were spun up through mass transfer in massive close binary stars. We observe Be stars in several kinds of binaries predicted by theory: (1) Be + a cool, Roche-filling companion (so-called "hot Algols"), (2) Be + He star (the stripped down core of the mass donor), and (3) Be + neutron star (Be X-ray binaries). There are no known examples of Be + white dwarf binaries (although EUVE observations of the system λ Sco may have revealed the first rapidly rotating B star + white dwarf binary). Here I review the observational data on Be stars in binaries for comparison with model predictions. Because the detection of hot, faint (He star and white dwarf) companions is so difficult, we still cannot estimate accurately the fraction of such systems among the Be star population. Thus, the actual percentage of Be stars formed by binary mass transfer processes remains unknown.

1. INTRODUCTION TO Be STARS

One of the main consequences of mass transfer in close binary stars is an increase in the rotational velocity of the mass gainer. We find among the massive stars a population of rapid rotators known as Be stars which are distinguished from normal B-type stars by the presence of strong and variable hydrogen Balmer emission lines. Because the detection of these stars is relatively straight forward (through spectroscopy or narrow-band imaging), there has been an irresistible urge among many astronomers to identify the Be stars as the manifestation of close binary processes (despite the non-detection of a companion in most cases). My goal in this paper is to review the evidence linking the Be stars to binary

D. Vanbeveren (ed.), The Influence of Binaries on Stellar Population Studies, 95–110.

evolutionary processes and to motivate observers and theoreticians alike to reconsider what can be learned from the Be stars.

A good working definition of a Be star comes from Jaschek et al. (1981): a non-supergiant B-type star whose spectrum has, or had at some time, one or more Balmer lines in emission. The origin of the emission lines was a matter of heated debate for many decades (Slettebak 1988), but recent work has demonstrated clearly that the emission lines form in a circumstellar disk. The most dramatic evidence comes from the direct resolution of the emitting disks by radio interferometry (Dougherty & Taylor 1992) and by Hα optical interferometry (Quirrenbach et al. 1994, 1997; Stee 2000). These observations establish that the emission originates in a flattened cloud centered on the Be star. Even before these observations, however, there was strong circumstantial evidence for disks from the extent of linear polarization (Quirrenbach et al. 1997; Bjorkman 2000a), the infrared excess (Waters 1986; Waters et al. 2000), and the double-peaked appearance of the emission lines (Hummel 2000). Detailed studies of variability in certain Be stars (Telting et al. 1998; Telting 2000) indicates that there is a net outflow of gas in the disk, so that the disk gas originates in the Be star itself.

Be stars comprise from $\approx$ 18% of Galactic field B-type stars (Abt 1987) up to $\approx$ 39% of B-type stars in the SMC (Maeder, Grebel, & Mermilliod 1999). Be stars in clusters first appear in clusters $\approx 10^7$ years old and become rare in clusters older than 10^8 years (Mermilliod 1982; Fabregat & Torrejón 2000). Their positions in the Hertzsprung-Russell diagram span from the zero-age main sequence to the terminal age main sequence (Mermilliod 1982).

The key problem that has eluded solution for nearly a century is the origin of their circumstellar disks. Most Be stars have broad spectral lines, clearly indicating that they are rapid rotators, and Struve (1931) first proposed that all Be stars suffer mass loss from their equatorial zones due to the high rotational velocities. However, subsequent work by Slettebak (1979) demonstrated that while Be stars are indeed rapid rotators their equatorial velocities are generally below the critical rate where centripetal forces balance gravitational forces. Furthermore, the spectra of Be stars are notoriously time variable so that constant fast rotation alone is insufficient to explain their variable circumstellar disks. There are many well documented cases (e.g., γ Cas; Doazan et al. 1983) in which the spectrum transforms from a Be type to a Be "shell" spectrum (in which narrow absorption cores appear which are presumably formed far out in the disk), then to a normal B-type spectrum (lacking emission), and then back to a new Be outburst phase on timescales of years to decades. Furthermore, almost all Be stars display rapid photometric (Balona 2000) and spectroscopic variability (Baade 2000; Smith 2000) on timescales of hours to days.

It appears that rapid rotation is a necessary but not sufficient condition to create a Be star (since we do find rapidly rotators among B-type stars, usually indicated by a letter "n" appended to their spectral classification, which are not known Be stars). Thus, we are forced to seek some additional cause or causes which may act in concert with rapid rotation to produce the circumstellar disks. There is still no consensus about the defining processes but several promising explanations are being explored (Bjorkman 2000b):

- Bjorkman & Cassinelli (1993) have proposed a wind compressed disk model in which stellar wind flows from northern and southern hemispheres follow trajectories which cross at the equator and form an enhanced density region. Although this mechanism must occur at some level, Owocki, Cranmer, & Gayley (1996) found that models that include non-radial forces result in equatorial densities reduced to levels far below those observed (see Bjorkman 2000b).

- Many Be stars display absorption line variations that can be explained by photospheric nonradial pulsations (Baade 2000), and these pulsations may help to promote mass loss into an equatorial disk. The most striking example of this process is seen in the Be star μ Cen. Rivinius et al. (1998) have demonstrated that Be outbursts in this star occur at times of constructive interference between the 6 pulsational modes observed.

- Smith (2000) has found evidence of transient heating events in Be star atmospheres which can be interpreted as mass ejections presumably related to magnetic loops (or more generally mass loss by magneto-centrifugal acceleration; see Balona 2000).

- During the months prior to this meeting, the B-star δ Sco brightened and developed emission lines at the time of a close periastron passage of a binary companion (Fabregat et al. 2000; Hartkopf, Mason, & McAlister 1996). Thus, episodic tidal effects may also be a means of creating a Be disk around a rapid rotator.

Whatever the mechanisms of disk formation that actually occur, the central question remains how these B stars obtained their rapid spin. There are two probable explanations: (1) spin up at the end of core hydrogen burning, or (2) spin up by mass transfer in a close binary.

Models of massive star evolution that include rotation (Meynet & Maeder 2000; Heger & Langer 2000) show that there is a core contraction at the end of the main sequence phase that results in an increase in the rotational velocity of the outer layers. However, Mermilliod (1982) and others have shown that Be stars occupy the entire band of the main sequence, and are not confined to the terminal age portion.

The idea that Be stars are spun up by mass transfer in binaries has come in and out of fashion over the course of the last few decades. Kříž & Harmanec (1975) originally made the bold suggestion that all Be stars were spun up by binary mass transfer, and that their disks are formed by the mass transfer process (as accretion disks). However, if Be stars are active mass gainers, then their Roche-filling companions should be cool and large, and the companions would be detectable in the near-IR where they would contribute a larger fraction of the flux. A number of careful searches for cool companions were made and these were generally (but not totally) unsuccessful (Baade 1992; Waters, Coté, & Pols 1991; Floquet et al. 1989). The lack of Roche-filling companions led Baade (1992) to conclude that "... the explanation of the 'Be phenomenon' should not be sought from binarity," or, more specifically, Be disks are generally not formed by accretion from a companion.

Nevertheless, there are now many observational clues that some considerable fraction of the Be stars are rapid rotators because they are *post-mass transfer binaries.* The first line of evidence came from the discovery that most of the massive X-ray binaries consist of Be stars with neutron star companions. Rappaport & van den Heuvel (1982) proposed an evolutionary scenario for these Be X-ray binaries (BeXRBs) in which the mass donor transfers both mass and angular momentum to the gainer, and the donor, if massive enough, eventually explodes as a supernova leaving a neutron star orbiting a rapidly rotating Be star. Pols et al. (1991) showed that other kinds of Be companions should exist (stripped-down helium star cores and white dwarfs) that are intrinsically faint and difficult to detect. It is only in the last few years that we have finally detected these predicted Be binaries with faint companions.

In the following sections, I review the predictions from binary models about the numbers and kinds of Be binaries that may exist, and then I summarize the exciting progress that has occurred in the observation of these binaries. Because the detection of some types of companions is so difficult, it is still premature to determine what fraction of the rapid rotators (and Be stars) results from binary mass transfer.

2. BINARY EVOLUTION MODELS

Models of the evolution of intermediate mass close binaries are now well established (and less beleaguered by issues related to systemic mass loss which are important in the most massive stars). One evolutionary path is illustrated in Figure 1 (from a model by van den Heuvel & Rappaport 1987) which shows many of the common stages of development. The model begins with a close binary system of two main sequence B-type stars. The initially more massive star is the first to expand to fill its Roche lobe, and Roche lobe overflow begins (usually assumed to occur during H shell burning, or Case B). The gas accreted by the mass gainer causes that star to spin up. In fact, as little as $0.1 M_{\odot}$ of accreted gas is sufficient to attain break-up velocity (Packet 1981). This spin up process is observed in lower mass Algol binaries where the gainers are often found to spin faster than the orbital synchronous rate (Šimon 1999; Etzel & Olson 1993). The system separation decreases until the mass ratio is inverted, and then the system widens (increasing the orbital period) until mass transfer is cut off. The mass transfer rate begins very high and then is followed by an extended period of reduced mass transfer (van der Linden 1987), in which the donor finally loses most of its envelope. The end result is the small, stripped down core of the mass donor (often referred to as helium star) orbiting a rapidly rotating gainer (usually assumed to be a Be star).

There are several possible outcomes of the binary interaction:

- In small mass ratio systems ($q = M_2/M_1 < 0.2$), the mass accretion rate is too large for the gainer to accept, and the gainer swells into a contact configuration. Such systems then experience a common envelope phase in which the gainer spirals into and merges with the donor. The rejeuvenated star might appear as a single rapid rotator (Be star).

- If the remnant He star mass exceeds $2.2 M_{\odot}$ (after a possible second mass exchange phase), then the remnant will explode as a peculiar supernova (lacking hydrogen lines in its spectrum). If the total mass lost in the explosion is relatively small, then the neutron star (NS) remnant will remain bound to the gainer (producing a BeXRB). On the other hand, if there is significant mass loss or if the remnant obtains a substantial asymmetric "kick" velocity during the SN, then the binary will be disrupted leaving a single rapid rotator (Be star) and separate NS.

- Finally, if the He star mass is $< 2.2 M_{\odot}$, then the system will continue to widen as the He star loses mass by a second phase of mass transfer (Case BB) or by stellar wind. The system will eventually appear as a rapid rotator (Be star) with an orbiting white dwarf (WD).

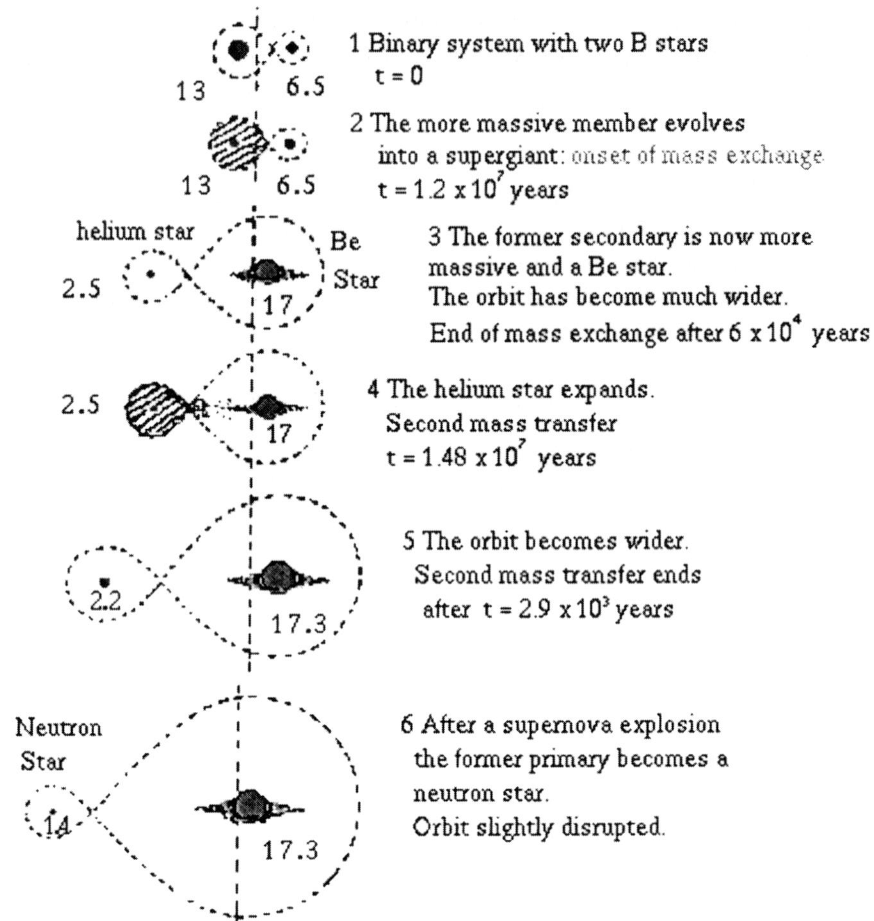

Figure 1 An evolutionary scenario beginning with a pair of 13 and $6.5 M_{\odot}$ stars in a 2.6 d orbit and ending with a BeXRB with a 30.6 d orbit (from van den Heuvel & Rappaport 1987; courtesy of I. Negueruela, BeppoSAX/ESA; see http://www.sdc.asi.it/ignacio/bex.html).

Numerical models of this process (Pols et al. 1991; Portegies Zwart 1995; Lipunov, Postnov, & Prokhorov 1996; Van Bever & Vanbeveren 1997) require input parameters describing the binary frequency, distributions of mass, mass ratio, and period, the fraction of mass accreted versus lost to system, the angular momentum lost, the kick velocity during the supernova, and the relative lifetimes in the various stages. Large uncertainties remain about the correct assumptions for all these parameters, and consequently, the final results about the numbers of binary products can be quite diverse. A common result is that there should be significant numbers of Be + WD and Be + He systems relative to the more readily observed Be + NS systems (cf. Fig. 1 of Portegies Zwart 1995). The predicted number of mass gainers as a fraction of the observed population of Be stars varies from 5 – 20% (Van Bever & Vanbeveren 1997) to ≈ 50% (Pols et al. 1991). Taken at face value, these results suggest that a significant fraction of the Be stars must have obtained their rapid spin by means other than mass transfer in a binary. However, given the uncertainties surrounding these calculations, we should not rule out binary mass transfer as the main cause of spin up.

3. OBSERVATIONAL RESULTS

Abt & Cardona (1984) explored the binary frequency of Be and B stars over a wide range in orbital period. They found that both groups generally have the same binary frequency for orbital periods greater than 1 year, a result confirmed in a speckle interferometric survey of Be stars by Mason et al. (1997). However, Abt & Cardona found a marked deficit of short period binaries (especially with $P < 0.1$ yr) among the Be stars compared to the B stars. They suggest that the Be stars are born as rapid rotators and maintain their fast spin because of a lack of tidal braking from close companions. However, recent models of the evolution of rotating stars (Meynet & Maeder 2000; Heger & Langer 2000) indicate that most single stars spin down during their main sequence lifetime.

The perceived absence of short-period systems is partially due to the practical difficulties of measuring radial velocities in Be stars: their photospheric lines are broad and shallow, usually show distortions related to rapid variability, and are often affected by shell absorption, disk emission, and/or gas streams (cf. Harmanec 1987). Furthermore, if theories of close binary evolution are correct, then the companions will have small masses, and the orbital semi-amplitudes of the Be star primaries will be small (only a few km s^{-1}). These difficulties explain why there are relatively few orbital solutions from optical spectra for BeXRBs (Negueruela 1998; Wang & Gies 1998). Nevertheless, hard and determined observational work (particularly by Harmanec, Tarasov, Koubsky, Peters, and others) has led to the discovery of a significant number of Be binaries with periods shorter than 10 yr (the limit for interaction) that were unknown to Abt & Cardona (1984). The following subsections summarize our current knowledge of binaries among the Be stars.

3.1. KNOWN BINARIES AMONG CLASSICAL Be STARS

In a recent literature search (Gies 2000), I found 27 classical Be stars in binaries with orbits (and 13 more suspected or needing confirmation). Two new single-lined spectroscopic binary systems have been discovered since then: π Aqr (P = 84 d; Bjorkman, Miroshnichenko, & Krugov 2000) and γ Cas (P = 204 d; Harmanec et al. 2001). The fact that these two are among the brightest and best studied Be stars in the sky testifies to the difficulties in detecting such low velocity amplitude binaries. The total numbers of binaries are summarized in Table 1 according to the following groups: binaries with bright companions, binaries with faint or undetected companions, suspected binaries, and BeXRBs (Coe 2000). Notes and references for the individual targets can be found in Gies (2000) and Coe (2000).

Table 1 Galactic Be Stars in Binaries ($P < 10$ yr)

Category	Number	Example
Bright Companions (Algols)	16	CX Dra (B2.5 Ve + F5 III)
Faint/Undetected Companions	13	ϕ Per (B0.5 III-Ve + sdO)
Possible Binaries	13	FY CMa (B0 IVep)
Be X-ray Binaries	> 24	HDE 245770 (O9.5 III-Ve)

The kinds of binaries observed generally support the binary formation scenario for Be stars:

- There are 16 systems with cool, Roche-filling companions which represent the stage of active mass transfer (Tarasov 2000). Many of these systems have accretion disks that result from mass transfer as originally proposed by Křiž & Harmanec (1975).
- There are 13 systems with faint or undetected companions which plausibly include the post-mass transfer stages of Be + He and Be + WD stars (with long periods, small eccentricities, and few if any eclipsing systems).
- Significant by their absence is the complete lack of Be + B or Be + Be binaries in the list. If the Be phase was a common stage in the evolution of single stars, then we would expect to find many normal B-type companions.

- Note that the list is not drawn from a complete or uniform sample; the BeXRBs, for example, are faint while the group with faint companions represents some of the brightest stars in the sky. This suggests that many more Be + faint companion systems await discovery.

3.2. Be + HELIUM STAR BINARIES: ϕ PERSEI AND 59 CYGNI

The first positive detection of a He star companion to a Be star was recently made by Gies et al. (1998) through the identification of its spectral lines in high signal-to-noise UV spectra made with the Goddard High Resolution Spectrograph and *Hubble Space Telescope*. These high excitation absorption lines were found to move in anti-phase to the known radial velocity curve of the Be star ($P = 126.6731$ d; Božić et al. 1995), and Gies et al. used measured radial velocities of both components to arrive at a double-lined spectroscopic orbital solution. They estimated that the system inclination was large, $i > 63°$, based on limits on the axial ratio of the resolved circumstellar disk (Quirrenbach et al. 1994) and the statistics of the occurrence of shell type spectra (Hanuschik 1996). Gies et al. then calculated masses of $8.9 < M_p/M_\odot < 9.6$ and $1.09 < M_s/M_\odot < 1.18$ for the Be star and He star, respectively. They used the new radial velocity curves to extract the individual component spectra from the ensemble of observed spectra using a Doppler tomographic reconstruction algorithm (Bagnuolo et al. 1994). The reconstructed spectra were matched using spectrum synthesis methods to arrive at the spectral classifications, projected rotational velocities, temperatures, gravities, and radii listed in Table 2.

Table 2 Parameters for the Components of ϕ Per (Gies et al. 1998)

Parameter	Primary	Secondary
Spectral Classification	B0.5 III-Ve	sdO
$V \sin i$ (km s^{-1})	450	0
$T_{\rm eff}$ (°K)	29300	53000
$\log g$ (c.g.s.)	3.7	4.2
R ($R_\odot$)	7	1.3

The hot subdwarf that orbits ϕ Per is clearly the remains of the originally more massive star that transferred both mass and angular momen-

Figure 2 A painting of the ϕ Per system by Bill Pounds.

tum to the Be star (one of the fastest rotators in the sky). Thus, we now have in ϕ Per a firm example of a Be star formed by close binary evolution. The stars are well separated and there is no evidence of active mass transfer at present, so the Be star's disk results from outflow from the Be star (a painting of the system by Bill Pounds appears in Fig. 2). Vanbeveren et al. (1998) present a conservative evolutionary scenario for the system that starts with a $6M_{\odot} + 5M_{\odot}$ pair in a 13.5 d orbit. This picture may need revision since the current luminosity of the subdwarf is more in line with the predictions for the remnant of a 8 $M_{\odot}$ progenitor (see Fig. 12.2 in Vanbeveren et al. 1998). In fact, the secondary of ϕ Per is the brightest hot subdwarf in the sky but was lost in the glare of the Be star prior to the *HST* observations. Interestingly, the Be star is also overluminous for its mass (see Fig. 2 of Schönberner & Harmanec 1995) which may indicate that the Be star has undergone large scale mixing as the result of the mass transfer (see a different interpretation in Harmanec 2000).

Rivinius & Štefl (2000) have recently presented a radial velocity curve for the Be binary, 59 Cyg (P = 28.17 d, and a small mass function indicating a low mass companion). Rivinius & Štefl find several optical emission lines which move in anti-phase motion as they do in ϕ Per, and these suggest the presence of a hot companion. I have verified their orbit with velocity measurements from UV high dispersion spectra in the *IUE* archive. The mass ratio is unknown at present, but a preliminary tomographic reconstruction with a mass ratio $M_1/M_2 = 3$ and a flux ratio $F_2/(F_1 + F_2) = 0.1$ shows evidence of a secondary UV spectrum indicative of a very hot companion. Thus, 59 Cyg is probably the second known example of a Be + He system.

3.3. Be + WHITE DWARF BINARIES

Unfortunately, there are no known representatives of this class (Meurs et al. 1992) despite the predictions that many should exist (Waters et al. 1989; Raguzova 2000). This is perhaps not surprising, since WD companions are probably very difficult to detect by conventional means. However, there have been several reports in the last few years of WD companions to normal B-type stars discovered through *EUVE* observations of the extreme UV and soft X-ray flux from the WD companion (Burleigh & Barstow 1999, 2000; Vennes 2000; Berghöfer, Vennes, & Dupuis 2000). The only known B + WD system with a separation small enough to be interacting is λ Sco, and Berghöfer et al. (2000) suggest that this consists of a $1.3M_\odot$ WD plus a B1.5 IV primary in a $P = 5.96$ d orbit. It interesting to note that although the primary is not a known Be star, it is a fast rotator (4× synchronous; De Mey et al. 1997). Thus, if confirmed, λ Sco may represent the first recognized example of a B + WD system formed through binary mass transfer.

3.4. Be + NEUTRON STAR BINARIES

The BeXRBs form the largest subgroup of massive X-ray binaries (Coe 2000). The binary periods lie in the range 16 d – 400 d, and the orbits are wide and eccentric (and possibly inclined to the disk around the Be star). The BeXRBs display large variations in X-ray flux, and frequently X-ray outbursts are observed that are associated with the passage of the NS through the disk of the Be star. The primaries mainly have spectral types in the range B0 – B2 and luminosity classes in the range III – V. Thus, the BeXRBs have primaries at the high mass end of the distribution of Be stars as a whole. Supernovae that occur in binaries with lower mass primaries may lead to disruption of the binary in most cases. The evidence collected to date indicates that the known BeXRBs have companions that are neutron stars (and not black holes or white dwarfs; Negueruela 1998).

3.5. RUNAWAY Be STARS

Rinehart (2000) compared *Hipparcos* proper motions of the Be and B-type stars, and showed that they have very similar distributions. This work was extended to a study of proper motions and radial velocities in a sample of 344 Be stars by Berger & Gies (2001) who found that although most Be stars have slow random speeds, there is a fraction ($3-7\%$) which have peculiar space motions > 40 km s^{-1}.

The binary scenario for the formation of Be stars predicts that the surviving post-SN systems will achieve a "runaway" velocity given by (Nelemans, Tauris, & van den Heuvel 1999)

$$v_{\rm sys} = 213 \left(\frac{\Delta M}{M_\odot}\right) \left(\frac{m}{M_\odot}\right) \left(\frac{P_{\rm re-circ}}{\rm day}\right)^{-\frac{1}{3}} \left(\frac{M_{\rm rem}+m}{M_\odot}\right)^{-\frac{5}{3}} \text{km s}^{-1} \quad (7.1)$$

where ΔM is the mass lost during the supernova, m is the mass of the surviving companion, $M_{\rm rem}$ is the mass of the neutron star remnant, and $P_{\rm re-circ}$ is the orbital period after subsequent re-circularization of the orbit. Based on the typical parameters for the BeXRBs and their orbits (Coe 2000), we expect to find space velocities that are proportional to the mass lost in the SN, $v_{\rm sys} \approx 8\ (\Delta M/M_\odot)$ km s^{-1}.

If Be stars are formed in close binaries, then the small fraction of Be runaway stars and the relatively slow systemic velocities of most BeXRBs (Chevalier & Ilovaisky 1998) indicate that those in which a SN occurred suffered significant mass loss prior to the SN event ($\Delta M/M_\odot$ of order unity). However, Be binaries that avoided a SN (those with He or WD companions) also avoided receiving any runaway velocity boost. Thus, the low velocity distribution characteristic of most Be stars is consistent with the binary formation scenario if (1) a substantial fraction avoided a SN, and (2) the progenitors lost most of their mass prior to the SN in systems where a SN did occur.

4. CONCLUSIONS

There are now many examples known of Be stars in the kinds of binary systems predicted by models including (1) Be + a cool, Roche filling companion (hot Algols), (2) Be + He star (ϕ Per, 59 Cyg), and (3) Be + neutron star (BeXRBs). Thus, we are now confidant that some of the rapidly rotating massive stars (the source of Be stars) can be formed through binary processes.

What is still unknown is whether or not all Be stars are spun up this way. Current models of binary evolution (Van Bever & Vanbeveren 1997) suggest that only a modest fraction of the observed Be population are formed by the binary route (so that the majority form from stars spun up at the end of core hydrogen burning or from stars born with large spins). The observational side of this question is still fraught with uncertainty. The recent discovery of the He star companion to ϕ Per (Gies et al. 1998), one of the brightest and best studied Be stars in the sky, shows how difficult it can be to find faint, low mass companions (WD, He star). Furthermore, we know that there are some binary paths that can lead to truly single Be stars (systems disrupted in a SN explosion and Be stars formed by mergers).

Despite these difficulties, we can expect progress in the observational picture in the near future. There is evidence that Be stars first appear in clusters some 10^7 yr after birth (Fabregat & Torrejón 1999), and we would like to know if there is also a general increase in the distribution of rapid rotators at that time. Several groups are currently involved in measuring projected rotational velocities for large numbers of B stars in clusters, and such data will provide important clues on the time evolution of rotation in massive stars. The systems with suspected low mass companions are especially worthy of intensive spectroscopic observation, and some of the nearby systems may be resolved directly with the new generation of optical interferometers. Since the suspected companions are likely hot, spectroscopic studies in the FUV and EUV will be especially important. If a large number of hot companions exist, they would be significant contributors to the EUV flux budget of galaxies, and understanding these systems will be important in our interpretation of the optical and IR spectra of high redshift galaxies.

Acknowledgments

I am grateful for the advice of many colleagues including W. Bagnuolo, D. Berger, P. Harmanec, W. Hartkopf, A. Kaye, B. Mason, L. Penny, G. Peters, Th. Rivinius, M. Smith, and M. Thaller. This work was supported by NASA grant NAG5-2979. Institutional support has been provided from the GSU College of Arts and Sciences and from the Research Program Enhancement fund of the Board of Regents of the University System of Georgia, administered through the GSU Office of the Vice President for Research and Sponsored Programs.

References

Abt, H. A. 1987, in Physics of Be stars, ed. A. Slettebak & T. P. Snow (Cambridge: Cambridge Univ.), 470

Abt, H. A., & Cardona, O. 1984, ApJ, 285, 190

Baade, D. 1992, in Evolutionary Processes in Interacting Binary Stars, ed. Y. Kondo, R. Sistero, & R. S. Polidan (Dordrecht: Kluwer Academic), 147

Baade, D. 2000, in The Be Phenomenon in Early-type Stars, IAU Colloquium 175, A.S.P. Conf. Series 214, ed. M. A. Smith, H. F. Henrichs, & J. Fabregat (San Francisco: A.S.P.), 178

Bagnuolo, W. G., Jr., Gies, D. R., Hahula, M. E., Wiemker, R., & Wiggs, M. S. 1994, ApJ, 423, 446

Balona, L. 2000, in The Be Phenomenon in Early-type Stars, IAU Colloquium 175, A.S.P. Conf. Series 214, ed. M. A. Smith, H. F. Henrichs, & J. Fabregat (San Francisco: A.S.P.), 1

Berger, D. H., & Gies, D. R. 2001, ApJ, in press

Berghöfer, T. W., Vennes, S., & Dupuis, J. 2000, ApJ, 538, 854

Bjorkman, J. E. 2000b, in The Be Phenomenon in Early-type Stars, IAU Colloquium 175, A.S.P. Conf. Series 214, ed. M. A. Smith, H. F. Henrichs, & J. Fabregat (San Francisco: A.S.P.), 435

Bjorkman, J. E., & Cassinelli, J. P. 1993, ApJ, 409, 429

Bjorkman, K. S. 2000a, in The Be Phenomenon in Early-type Stars, IAU Colloquium 175, A.S.P. Conf. Series 214, ed. M. A. Smith, H. F. Henrichs, & J. Fabregat (San Francisco: A.S.P.), 384

Bjorkman, K. S., Miroshnichenko, A. S., & Krugov, V. D. 2000, AAS Meeting 196, #05.03

Božić, H., Harmanec, P., Horn, J., Koubský, P., Scholz, G., McDavid, D., Hubert, A. M., & Hubert, H. 1995, A&A, 304, 235

Burleigh, M. R., & Barstow, M. A. 1999, A&A, 341, 795

Burleigh, M. R., & Barstow, M. A. 2000, A&A, 359, 977

Chevalier, C. & Ilovaisky, S. A. 1998, A&A, 330, 201

Coe, M. J. 2000, in The Be Phenomenon in Early-type Stars, IAU Colloquium 175, A.S.P. Conf. Series 214, ed. M. A. Smith, H. F. Henrichs, & J. Fabregat (San Francisco: A.S.P.), 656

De Mey, K., Aerts, C., Waelkens, C., Cranmer, S. R., Schrijvers, C., Telting, J. H., Daems, K., & Meeus, G. 1997, A&A, 324, 1096

Doazan, V., Franco, M., Sedmak, G., Stalio, R., & Rusconi, L. 1983, A&A, 128, 171

Dougherty, S. M., & Taylor, A. R. 1992, Nature, 359, 808

Etzel, P. B., & Olson, E. C. 1993, AJ, 106, 1200

Fabregat, J., Reig, P., & Otero, S. 2000, IAU Circ., 7461, 1

Fabregat, J., & Torrejón, J. M. 2000, A&A, 357, 451

Floquet, M., Hubert, A. M., Chauville, J., Chatzichristou, H., & Maillard, J.-P. 1989, A&A, 214, 295

Gies, D. R. 2000, in The Be Phenomenon in Early-type Stars, IAU Colloquium 175, A.S.P. Conf. Series 214, ed. M. A. Smith, H. F. Henrichs, & J. Fabregat (San Francisco: A.S.P.), 668

Gies, D. R., Bagnuolo, W. G., Jr., Ferrara, E. C., Kaye, A. B., Thaller, M. L., Penny, L. R., & Peters, G. J. 1998, ApJ, 493, 440

Hanuschik, R. W. 1996, A&A, 308, 170
Harmanec, P. 1987, in Physics of Be Stars, ed. A. Slettebak & T. P. Snow (Cambridge: Cambridge Univ.), 339
Harmanec, P. 2000, in The Be Phenomenon in Early-type Stars, IAU Colloquium 175, A.S.P. Conf. Series 214, ed. M. A. Smith, H. F. Henrichs, & J. Fabregat (San Francisco: A.S.P.), 13
Harmanec, P., et al. 2000, A&A, in press
Hartkopf, W. I., Mason, B. D., & McAlister, H. A. 1996, AJ, 111, 370
Heger, A., & Langer, N. 2000, ApJ, 544, 1016
Hummel, W. 2000, in The Be Phenomenon in Early-type Stars, IAU Colloquium 175, A.S.P. Conf. Series 214, ed. M. A. Smith, H. F. Henrichs, & J. Fabregat (San Francisco: A.S.P.), 396
Jaschek, M., Slettebak, A., & Jaschek, C. 1981, Be Star Newsletter, 4, 9
Kříž, S., & Harmanec, P. 1975, Bull. Astr. Inst. Cz., 26, 65
Lipunov, V. M., Postnov, K. A., & Prokhorov, M. E. 1996, A&A, 310, 489
Maeder, A., Grebel, E. K., & Mermilliod, J.-C. 1999, A&A, 346, 459
Mason, B. D., ten Brummelaar, T., Gies, D. R., Hartkopf, W. I., & Thaller, M. L. 1997, AJ, 114, 2112
Mermilliod, J.-C. 1982, A&A, 109, 48
Meurs, E. J. A., et al. 1992, A&A, 265, L41
Meynet, G., & Maeder, A. 2000, A&A, 361, 101
Negueruela, I. 1998, A&A, 338, 505
Nelemans, G., Tauris, T. M., & van den Heuvel, E. P. J. 1999, A&A, 352, L87
Owocki, S. P., Cranmer, S. R., & Gayley, K. G. 1996, ApJ, 472, L115
Packet, W. 1981, A&A, 102, 17
Pols, O. R., Coté, J., Waters, L. B. F. M., & Heise, J. 1991, A&A, 241, 419
Portegies Zwart, S. F. 1995, A&A, 296, 691
Quirrenbach, A., Buscher, D. F., Mozurkewich, D., Hummel, C. A., & Armstrong, J. T. 1994, A&A, 283, L13
Quirrenbach, A., et al. 1997, ApJ, 479, 477
Raguzova, N. V. 2000, in The Be Phenomenon in Early-type Stars, IAU Colloquium 175, A.S.P. Conf. Series 214, ed. M. A. Smith, H. F. Henrichs, & J. Fabregat (San Francisco: A.S.P.), 693
Rappaport, S. A., & van den Heuvel, E. P. J. 1982, in Be Stars, ed. M. Jaschek & H. G. Groth (Dordrecht: Reidel), 327
Rinehart, S. A. 2000, MNRAS, 312, 429
Rivinius, Th., Baade, D., Štefl, S., Stahl, O., Wolf, B., & Kaufer, A. 1998, Be Star Newsletter, 33, 15

Rivinius, Th., & Štefl, S. 2000, in The Be Phenomenon in Early-type Stars, IAU Colloquium 175, A.S.P. Conf. Series 214, ed. M. A. Smith, H. F. Henrichs, & J. Fabregat (San Francisco: A.S.P.), 581

Schönberner, D., & Harmanec, P. 1995, A&A, 294, 509

Šimon, V. 1999, A&AS, 134, 1

Slettebak, A. 1979, Space Sci. Rev., 23, 541

Slettebak, A. 1988, PASP, 100, 770

Smith, M. A. 2000, in The Be Phenomenon in Early-type Stars, IAU Colloquium 175, A.S.P. Conf. Series 214, ed. M. A. Smith, H. F. Henrichs, & J. Fabregat (San Francisco: A.S.P.), 292

Stee, P. 2000, in The Be Phenomenon in Early-type Stars, IAU Colloquium 175, A.S.P. Conf. Series 214, ed. M. A. Smith, H. F. Henrichs, & J. Fabregat (San Francisco: A.S.P.), 129

Struve, O. 1931, ApJ, 73, 94

Tarasov, A. E. 2000, in The Be Phenomenon in Early-type Stars, IAU Colloquium 175, A.S.P. Conf. Series 214, ed. M. A. Smith, H. F. Henrichs, & J. Fabregat (San Francisco: A.S.P.), 644

Telting, J. H. 2000, in The Be Phenomenon in Early-type Stars, IAU Colloquium 175, A.S.P. Conf. Series 214, ed. M. A. Smith, H. F. Henrichs, & J. Fabregat (San Francisco: A.S.P.), 422

Telting, J. H., Waters, L. B. F. M., Roche, P., Boogert, A. C. A., Clark, J. S., de Martino, D., & Persi, P. 1998, MNRAS, 296, 785

Van Bever, J., & Vanbeveren, D. 1997, A&A, 322, 116

Vanbeveren, D., Van Rensbergen, W., & De Loore, C. 1998, The Brightest Binaries (Dordrecht: Kluwer Academic)

van den Heuvel, E. P. J., & Rappaport, S. A. 1987, in Physics of Be Stars, ed. A. Slettebak & T. P. Snow (Cambridge: Cambridge Univ.), 291

van der Linden, T. J. 1987, A&A, 178, 170

Vennes, S. 2000, A&A, 354, 995

Wang, Z., & Gies, D. R. 1998, PASP, 110, 1310

Waters, L. B. F. M. 1986, A&A, 162, 121

Waters, L. B. F. M., Coté, J., & Pols, O. R. 1991, A&A, 250, 437

Waters, L. B. F. M., Pols, O. R., Hogeveen, S. J., Coté, J., & van den Heuvel, E. P. J. 1989, A&A, 220, L1

Waters, L. B. F. M., et al. 2000, in The Be Phenomenon in Early-type Stars, IAU Colloquium 175, A.S.P. Conf. Series 214, ed. M. A. Smith, H. F. Henrichs, & J. Fabregat (San Francisco: A.S.P.), 145

CHEMICALLY-PECULIAR RED GIANTS: UNCOVERING THE BINARY INTRUDERS

Sophie Van Eck
European Southern Observatory
Karl-Schwarzschild-Str. 2, D-85748 Garching, Germany
svaneck@astro.ulb.ac.be

A. Jorissen
Institut d'Astronomie et d'Astrophysique, Université Libre de Bruxelles
Campus Plaine, C.P. 226, Boulevard du Triomphe, B-1050 Brussels, Belgium
ajorisse@astro.ulb.ac.be

Keywords: Barium stars, S stars, binaries (spectroscopic), AGB stars

Abstract It is now widely recognised that peculiar red giants such as S stars are overgrown with "extrinsic" stars. Those binary masqueraders share the same abundance peculiarities as the prototypes of their spectral family (as judged from low-resolution spectra), but have a totally different evolutionary history (chemical peculiarities originating from mass transfer across a binary system instead of internal nucleosynthesis). Some characteristic properties of these extrinsic stars are investigated and the importance of un-masking them in stellar population studies is stressed.

1. INTRODUCTION

Because thermally-pulsing asymptotic giant branch stars (TPAGB stars) are efficient producers of carbon and heavy elements, all stars observed to exhibit carbon and heavy element overabundances (i.e., peculiar red giant stars, PRG) were traditionally assigned to the TPAGB phase. Indeed, during this phase, the s-process of nucleosynthesis efficiently produces heavy elements, while "third dredge-up" episodes bring them to the stellar surface. However, an increasing number of arguments recently suggested that the PRG family is overgrown with intruders, i.e., stars that look like TPAGB stars but that are definitely not TPAGB

D. Vanbeveren (ed.), The Influence of Binaries on Stellar Population Studies, 111–116.

stars. Unlike TPAGB stars, those intruder stars did not produce heavy elements by themselves, but rather accreted them in the past from a companion TPAGB star (now a dim white dwarf). The barium stars are examples of such masqueraders. The situation for S stars is more complex, since some of them really belong to the TPAGB phase of evolution, whereas others are binary intruders. Such binary intruders are generically called "extrinsic" stars, as opposed to the "intrinsic", genuine TPAGB stars.

2. ORBITAL ELEMENTS OF EXTRINSIC STARS

A very useful tool for studying binary evolution is the (e, log$P$) diagram (where e and P stand for the eccentricity and the orbital period of the system, respectively). The sample of G and K giants from open clusters (from Mermilliod, 1996) presented in Fig. 1 is used as a reference to which the PRG binaries can be compared. Indeed, the cluster giants presumably belong to pre-mass-transfer binary systems.

The (e, log$P$) diagram of barium stars is drastically different from that of cluster giants; in particular:

- the maximum period for a circular orbit is $\sim$ 4400 d for barium stars, as compared to $\sim$ 350 d for cluster giants;

- at a given orbital period, the maximum eccentricity found among barium systems is much smaller than for cluster giants.

The (e, log$P$) diagrams of S and CH stars, despite the smaller sample size, are very similar to that of barium stars. The differences between the (e,log$P$) diagrams of PRG and cluster binaries reflect the fact that the orbits of the PRG systems have been shaped by the mass-transfer episode responsible for their chemical peculiarities.

Binarity is clearly the rule among extrinsic stars, since the fraction of binaries among barium and extrinsic[1] S stars amounts to 69/77 (90%) and 24/28 (86%), respectively, after 10 years of radial-velocity monitoring (Jorissen et al., 1998; Udry et al., 1998). Given the efficiency for binary detection, these numbers are compatible with a 100% binary frequency.

[1] as defined by spectroscopic criteria, see Sect. 4

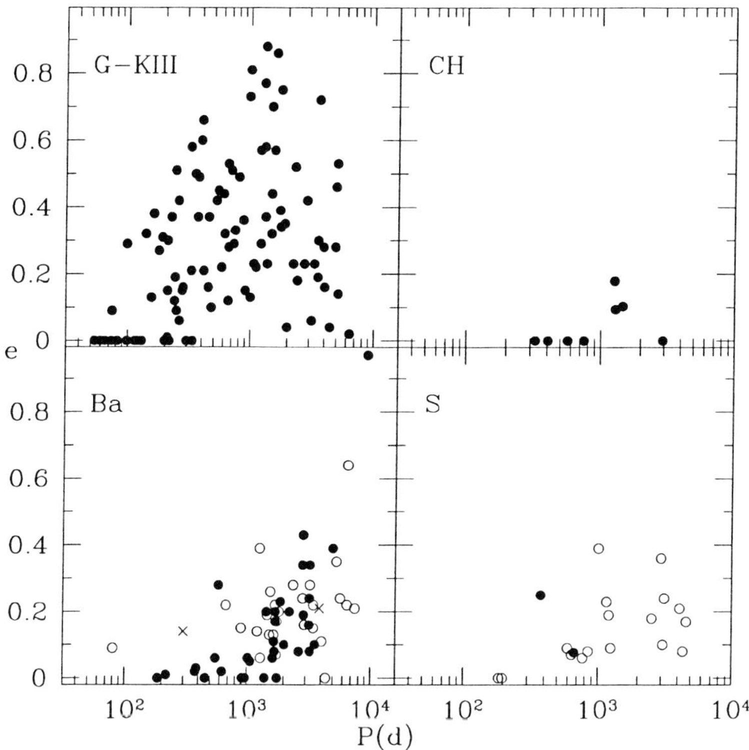

Figure 1 The (e, log$P$) diagram for different samples of red giant stars (after Jorissen et al., 1998). Lower left panel: barium stars (open circles: mild Ba stars; filled circles: strong Ba stars; crosses: triple hierarchical system BD+18°118). Upper left panel: binaries containing G or K giants in open clusters (Mermilliod, 1996). Upper right panel: CH stars (McClure and Woodsworth, 1990). Lower right panel: S stars

3. LUMINOSITIES OF S STARS

The Hertzsprung-Russell diagram of S stars constructed from HIPPARCOS parallaxes (Van Eck et al., 1998) further confirms the intrinsic/extrinsic dichotomy among S stars, since:

- extrinsic S stars are hotter and intrinsically fainter than intrinsic S stars;
- extrinsic S stars are low-mass stars populating either the first giant branch or the early AGB, and in any case are found below the luminosity threshold marking the onset of thermal pulses on the

AGB. It fully confirms that the chemical peculiarities of extrinsic stars cannot originate from internal nucleosynthesis and dredge-up processes occurring along the AGB. On the contrary, intrinsic S stars are found just above the TPAGB luminosity threshold.

Comparison with SMC samples of carbon (C) stars (Westerlund et al., 1991; Westerlund et al., 1995) reveals that galactic S stars and SMC C stars span the same luminosity range. Since the lower-luminosity S stars are *extrinsic S stars*, it is suggested that the lower-luminosity SMC C stars are likewise *extrinsic C stars*, i.e., they belong to binary systems and have accreted their carbon from a (now defunct) companion.

More generally, it is of key importance to know whether a given stellar sample is polluted by extrinsic stars: in that case, the derived properties could be strongly biased. For example, the third dredge-up luminosity threshold, commonly derived as the transition luminosity between M and S stars, may be totally in error if extrinsic S stars have not been removed from the studied sample. In the same vein, the ratio of S stars to M or C stars must be corrected for the presence of extrinsic S stars, when trying to deduce from this ratio the respective durations of the early AGB and the TPAGB. The magnitude of the error induced on these parameters depends of course on the proportion of extrinsic stars among the considered stellar class.

4. THE PROPORTION OF EXTRINSIC STARS AMONG S STARS

The properties of S stars have been investigated thanks to a large observing program devoted to the well-defined Henize sample (205 S stars south of $\delta = -25°$ and brighter than $R = 10.5$, covering all galactic latitudes; Henize, 1960; Van Eck and Jorissen, 1999, 2000a, 2000b).

The more stringent spectroscopic difference between extrinsic and intrinsic stars resides in their technetium (Tc) content. Tc is an especially interesting chemical element because all its isotopes are radioactive and the half-life of ^{99}Tc, the isotope produced by the s-process, is 2.1×10^5y, which is comparable to the TPAGB duration for low-mass stars like S stars. Therefore Tc is expected to be present in intrinsic S stars (the genuine TPAGB S stars), but not in extrinsic S stars (because in the latter stars Tc has had time to decay since the epoch of the mass transfer). The two example spectra of Fig. 2 clearly illustrate this dichotomy.

Several other observational parameters are efficient to some extent in distinguishing intrinsic S stars from their extrinsic masqueraders (UBV, $JHKL$ and IRAS photometry, radial-velocity standard deviation, combination of band-strength indices derived from low-resolution spectra).

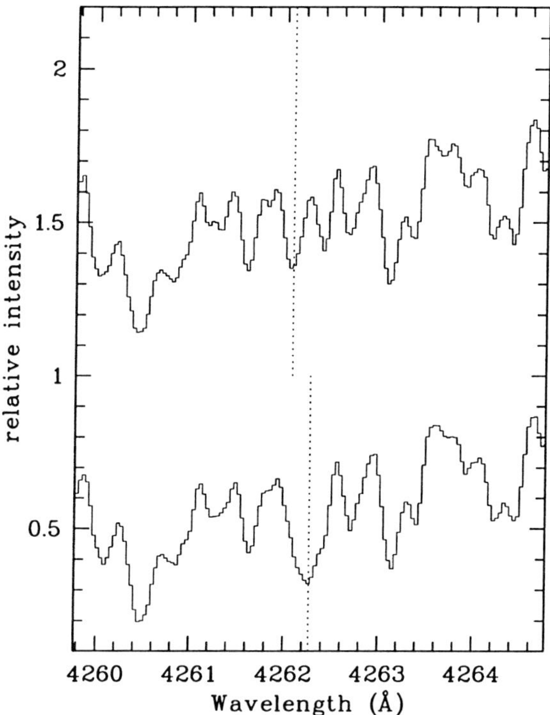

Figure 2 Spectra around the λ4262.27Å line of Tc I. The top spectrum is Hen 3, a Tc-poor S star; the bottom spectrum is Hen 39, a Tc-rich S star. The centroid wavelength of the blend located around 4262.2Å (represented by a dotted line) is clearly different in Tc-poor and Tc-rich stars

In order to guarantee a classification as objective as possible and handling at the same time a large number of parameters, multivariate classification has been applied to the Henize sample. This classification leads to a frequency of extrinsic stars among S stars around 40% in a volume-limited sample, but values as low as 30% and as high as 70% cannot be excluded given the large remaining uncertainties.

Finally, intrinsic S stars appear much more concentrated towards the galactic plane than extrinsic S stars, with exponential scale heights $z_{int} = 200 \pm 100$ pc and $z_{ext} = 600 \pm 100$ pc.

5. CONCLUSIONS

The presence of a substantial number of extrinsic stars among red giant stars has been firmly demonstrated during the past 20 years. Some families of extrinsic stars probably remain to be uncovered (especially among C stars). High-resolution spectroscopy is the most straightforward way to unmask the intruder stars, by looking for technetium-poor stars. The 8-meter class telescopes (such as the VLT + UVES at ESO) now allow to obtain such data for stars in external systems, and thus to better constrain the properties of extrinsic versus intrinsic stars.

References

Henize, K. (1960).
AJ, 65:491.

Jorissen, A., Van Eck, S., Mayor, M., and Udry, S. (1998). *A&A*, 332:877.

McClure, R. D. and Woodsworth, A. W. (1990). *ApJ*, 352:709–723.

Mermilliod, J.-C. (1996). In Milone, G. and Mermilliod, J.-C., editors, *ASP Conf. Ser. 90: The Origins, Evolution, and Destinies of Binary Stars in Clusters*, page 95.

Udry, S., Jorissen, A., Mayor, M., and Van Eck, S. (1998). *A&AS*, 131:25–41.

Van Eck, S. and Jorissen, A. (1999). *A&A*, 345:127–136.

Van Eck, S. and Jorissen, A. (2000a). *A&AS*, 145:51–65.

Van Eck, S. and Jorissen, A. (2000b). *A&A*, 360:196–212.

Van Eck, S., Jorissen, A., Udry, S., Mayor, M., and Pernier, B. (1998). *A&A*, 329:971–985.

Westerlund, B. E., Azzopardi, M., Breysacher, J., and Rebeirot, E. (1995). *A&A*, 303:107.

Westerlund, B. E., Azzopardi, M., Rebeirot, E., and Breysacher, J. (1991). *A&AS*, 91:425 (with erratum in A&AS 92, 683).

ECCENTRICITIES OF THE BARIUM STARS

John Lattanzio
Department of Mathematics & Statistics
Monash University, Clayton, Vic, 3800, Australia
john.lattanzio@maths.monash.edu.au

Amanda Karakas
Department of Mathematics & Statistics
Monash University, Clayton, Vic, 3800, Australia
amanda.karakas@maths.monash.edu.au

Christopher Tout
Institute of Astronomy
The Observatories, Madingley Road, Cambridge CB3 0HA
cat@ast.cam.ac.uk

Keywords: nucleosynthesis, population synthesis, AGB, barium stars

Abstract We investigate the eccentricities of Ba stars using a rapid binary evolution algorithm, including full tidal evolution, mass loss and accretion, nucleosynthesis and dredge-up on the AGB. We show that the eccentricities and orbital periods are consistent with wind accretion and weak viscous tidal dissipation, but we overproduce high-period Ba stars.

1. INTRODUCTION AND OBSERVATIONS

Barium stars were first identified by Bidelman & Keenan (1951) as a class of chemically peculiar Pop I red giants comprising about 1% of the total population of G and K giants. The discovery that all observed Ba stars belong to binary systems (McClure, Fletcher & Nemec 1980, McClure 1984, Jorissen & Mayor 1988, McClure & Woodsworth 1990) with eccentricities much lower than normal giants is evidence that the Ba overabundances might result from accretion from a former asymptotic giant branch (AGB) companion (Webbink 1986, Boffin & Jorissen 1988).

D. Vanbeveren (ed.), The Influence of Binaries on Stellar Population Studies, 117–124.

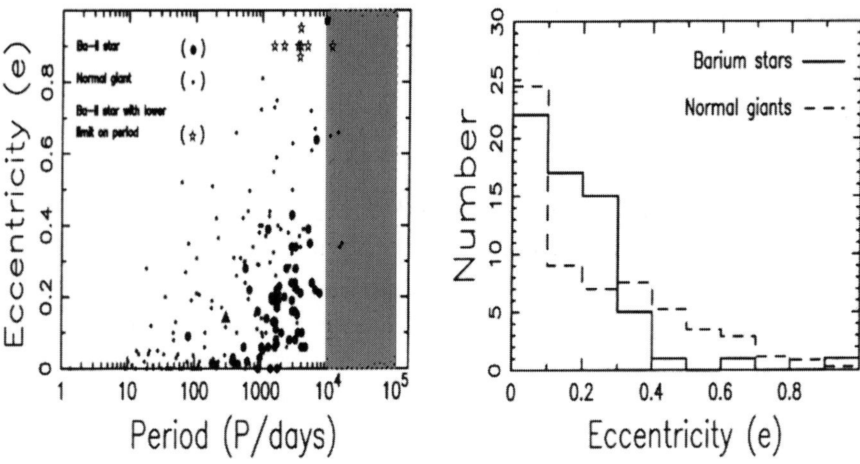

Figure 1 The observed (e, P) diagram for a sample of 62 Ba stars and 8 with lower-limits for P and unknown e (open stars) arbitrarily plotted at $e = 0.9$. The 213 spectroscopically normal giants (small dots) are from Boffin et al. (1993). Note that the e histogram has the number of normal giants scaled to that of Ba stars.

Two distinct mass-transfer processes are possible. The first is Roche-lobe overflow, resulting in rapid circularization of the orbit. The non-zero eccentricities of the Ba stars rule this out. The second mechanism is the transfer of mass by a stellar wind. We test this with a binary evolution algorithm developed by Tout et al. (1997), which includes tidal evolution, wind accretion and nucleosynthesis. We carry out Monte-Carlo simulations, much as described by Han et al.(1995), but we improve on their estimate of the Ba enhancement and add a tidal evolution model.

A decade of observations revealed that all Ba stars are in binary systems with P typically between 100 d and 10 yr. Jorissen et al. (1998) recently reported Ba stars with P outside this range. Figure 1 shows the observed (e, P) diagram and e distributions of Ba stars from Jorissen et al. (1998) and a similar set of normal giants from Boffin, Cerf & Paulus (1993). The average eccentricities are 0.17 for the Ba stars and 0.23 for the normal giants. A Kolmogorov-Smirnov test reveals that the probability of the Ba star eccentricities being chosen from the same distribution as the normal giants is $\simeq 10^{-3}$.

2. MODEL PARAMETERS

Formulae for the computation of stellar evolution may be found in Tout et al. (1997). Here we briefly describe the additions made for this work– see Karakas, Tout & Lattanzio (2000) for details.

Tout & Eggleton (1988) added a term to Reimers mass-loss to allow for tidal effects of a companion star:

$$\dot{M}_{\mathrm{TE}}/M_{\odot}\,\mathrm{yr}^{-1} = \dot{M}_{\mathrm{R}} \left\{ 1 + B \times \min\left[\left(\frac{R}{R_L} \right)^6, \frac{1}{2^6} \right] \right\}, \tag{9.1}$$

where R_L is the Roche lobe around the mass-losing star and B is a parameter. A fraction of the mass lost in winds from the primary can be accreted by the secondary. We use

$$\dot{M}_2 = -\min\left\{ 1, \frac{1}{\sqrt{1-e^2}} \left(\frac{GM_2}{V_{\mathrm{wind}}^2} \right)^2 \frac{\alpha_{\mathrm{acc}}}{2a^2\,(1+v^2)^{3/2}} \right\} \dot{M}_1, \tag{9.2}$$

where $v = V_{\mathrm{orb}}/V_{\mathrm{wind}}$, V_{wind} is the wind velocity from the primary which we set to be its escape velocity. The parameter α_{acc} is an accretion efficiency which we vary between 0.05 to 1.5.

In our syntheses we are only interested in the production of Ba, so we do not include nucleosynthesis prior to the TP-AGB phase. The core mass at the first thermal pulse is taken from Lattanzio (1989) as $0.562\,M_{\odot}$. We take the inter-pulse period to be a constant $t_{\mathrm{ip}} = 10^4\,\mathrm{yr}$; we varied this number and its effect was minimal. We assume that third dredge-up begins when M_{c} exceeds $M_{\mathrm{c}}^{\mathrm{min}}$, which we take as:

$$M_{\mathrm{c}}^{\mathrm{min}} = \begin{cases} 0.61\,M_{\odot}, & M_0 \leq 3.0\,M_{\odot} \\ 0.795\,M_{\odot}, & M_0 > 3.0\,M_{\odot}. \end{cases} \tag{9.3}$$

where M_0 is the initial mass of the star. The value for $M_0 \leq 3.0\,M_{\odot}$ is consistent with low-mass models of Lattanzio (1989) and Straniero et al. (1997) while that for $M_0 > 3.0\,M_{\odot}$ is from Frost (1997). Between pulses M_{c} will grow by ΔM_{c} from H burning. If there is dredge-up then a mass $\Delta M_{\mathrm{dredge}} = \lambda \Delta M_{\mathrm{c}}$ will be mixed to the envelope, where λ is the dredge-up parameter. For $M_0 < 3.0\,M_{\odot}$ the calculations of Straniero et al. (1997) indicate a linear relation between λ and M_0 while for $M_0 \geq 3.0\,M_{\odot}$ Frost (1997) favours a constant λ:

$$\lambda = \begin{cases} 0.2 + 0.0866 M_0, & M_0 < 3.0\,M_{\odot}, \\ 0.9, & M_0 \geq 3.0\,M_{\odot}. \end{cases} \tag{9.4}$$

These values of λ are not continuous, and reflect the lack of reliable dredge-up calculations (Frost & Lattanzio 1996). Now if $M_{\mathrm{c}} > M_{\mathrm{c}}^{\mathrm{min}}$, the core mass M_{c} increases over an interpulse phase by ΔM_{c} then decresses by $\Delta M_{\mathrm{dredge}}$ during dredge-up while the envelope mass M_{env} will increase by $\Delta M_{\mathrm{dredge}}$ but decrease by ΔM_{c} as well as due to mass-loss. Barium is produced in the intershell zone and we take $[\mathrm{Ba_{sh}/Fe}] \approx$

$\log(\mathrm{Ba_{sh}}/\mathrm{Ba_\odot}) = 2.6$ (Gallino, private communication). The mass of Ba added to the envelope during dredge-up is then $\Delta \mathrm{M_{Ba}} = \mathrm{Ba_{sh}} \Delta M_{\mathrm{dredge}}$. The abundance of Ba in the primary's envelope is thus easily calculated. The total mass transferred from the primary to the secondary is M_{acc}. The pre-Ba star evolves to become a red giant, and the accreted Ba is diluted into the envelope of the secondary. Subsequently the surface Ba abundance remains constant (see Karakas et al 2000 for details).

Tidal forces drive a binary system toward its lowest energy state for a given angular momentum. This is one in which the orbit is circular, the spins of the stars are synchronized with the orbit and the spin axes are parallel to the orbital axis (Hut 1980). Although the orbits of Ba stars are eccentric they are less so than ordinary giant binaries (Figure 1). We conclude that tidal circularization has begun but not proceeded to circularity. If our model for the formation of Ba stars is correct then we must be able to include the effects of tides and recover the distribution of Ba-star eccentricities alongside their other properties. The circularization timescale is given by Verbunt & Phinney (1995) or Rasio et al (1996) as

$$\frac{1}{\tau_{\mathrm{circ}}} = \frac{f'}{\tau_{\mathrm{conv}}} \frac{M_{\mathrm{env}}}{M} q(1+q) \left(\frac{R}{a}\right)^8, \tag{9.5}$$

where $q = M_1/M_2$ is the mass ratio, $M = M_1 + M_2$ is the total mass of the system, and τ_{conv} is the timescale on which the largest convective cells turn over. Rasio et al. split f' into

$$f' = f \min\left[\left(\frac{P_{\mathrm{tid}}}{2\tau_{\mathrm{conv}}}\right)^2, 1\right], \tag{9.6}$$

where P_{tid} is the tidal pumping timescale. The factor f we vary in the absence of a complete theory of tidal damping. Both theory (Zahn 1977) and observations (Verbunt and Phinney 1995) yield $f \simeq 1$.

For each Monte-Carlo simulation we model 10^6 binaries with initial separations, masses and eccentricities determined from the following distributions. The initial separation is flat in $\log a$ between 0.1 and 10^4 a.u.; the initial distribution of e is linear between 0 and 0.99. The masses of the two stars in each binary are generated independently from the formula of Eggleton, Fitchett & Tout (1989), which leads to a mass function similar to that of Miller & Scalo (1979). The mass range chosen for both stars is $0.08 \leq M_1, M_2 \leq 8\,M_\odot$. We evolve each pair of stars from the ZAMS to the age of the galaxy, $T_{\mathrm{gal}} = 1.2 \times 10^{10}$ y.

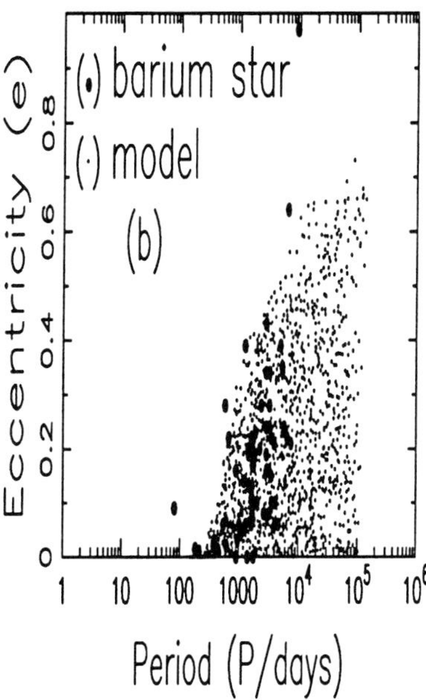

Figure 2 Simulated (e, P) diagram for the Ba stars produced using $f = 0.075$, $B = 7.5\,10^2$ and $\alpha_{acc} = 0.075$ (Model 10).

3. RESULTS AND DISCUSSION

We evolved over 600 models, each containing 10^6 binaries, varying B, f and α_{acc}. We tested our models with the Kolmogorov-Smirnov (KS) test to calculate a probability of association between the theoretical and observed P and e distributions (for Ba stars with P up to 10^4 d to match the observed range of Ba star periods). Most of our models overproduced the fraction of Ba stars among the red giants compared with observations (1%). Details may be found in Karakas et al (2000), but here we discuss only our best fit, Model 10, with $f = 0.075$, $B = 750$ and $\alpha_{acc} = 0.075$. This has a KS probability of 44% for the e distribution and 13% for P, and 1.2% Ba stars. Figure 2 shows the (e, P) diagram for Model 10 which is seen as a good fit to the data. Most models required f much smaller than expected. Figure 3 shows the e and P histograms for Model 10 and the observations. Note that we have scaled to the number of observed Ba stars. The e histogram follows the observed distribution quite closely but the P histogram is significantly different for periods greater than 10^4 d.

The P distributions were highly dependent on α_{acc} and B and mostly independent of f. Reducing α_{acc} from 0.10 to 0.05 greatly reduced the number of high-period Ba systems while increasing B from zero to 10^4 increased the number of low-period, zero-eccentricity systems. Thus

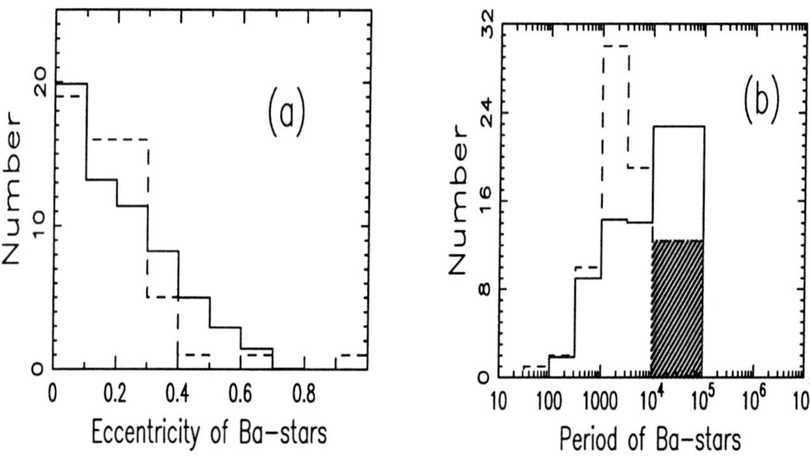

Figure 3 Histograms showing the distribution of (a) e and (b) P for Model 10 (unbroken line) compared with observations (dashed line). Note that the numbers in (a) have been scaled to the number of observed Ba stars given in Jorissen et al. (1998), and the numbers in (b) have been scaled to the number of Ba stars with known periods or upper limit period estimates in Jorissen et al. (1998).

we require B greater than zero but less than 10^4, and a low value of α_{acc}, to match the observed periods. The eccentricity was found to be highly sensitive to the tidal strength f and moderately dependent on B and α_{acc}. There is no unique set of the three parameters, B, f and α_{acc} that gives a significantly better fit than other combinations. From a large number of models we can rule out strong and very weak tidal strengths, so that $0.01 < f < 1.0$. We require a tidal wind enhancement of $10^2 < B \leq 2 \times 10^3$. For accretion efficiency, we find values close to that expected by Theuns et al. (1996), where $0.01 < \alpha_{acc} < 0.2$. Models with $\alpha_{acc} \geq 0.2$ produce too many high-period systems.

From observations we expect no more than 20% of Ba stars to have periods between 10^4 and 10^5 days and 0% of Ba stars to have periods greater than 10^5 days. Only models with with $\alpha_{acc} \leq 0.01$ adequately satisfied this criterion but the low KS probabilities found for these models ruled them out. Another motivation for choosing Model 10 as the best model was that the number of high-period Ba stars (over 10^5 d) was effectively zero ($\approx 1\%$) owing to its low value of α_{acc}; models with α_{acc} larger than 0.2 produced too many Ba stars with periods greater 10^5 d, up to 17.4% where α_{acc} is equal to 1.

All models, even those that were good fits to the observations, produce too many Ba stars in the range $10^4 < P < 10^5$ d. We could not solve this problem by varying parameters. This may indicate a problem with the wind accretion model, or that some aspect of our modelling or parameterizations is inadequate.

The model Ba stars span a large mass range, from $0.9\,M_\odot$ to $6\,M_\odot$ with an average $\simeq 1.6, M_\odot$. The companion stars also span a fairly large range from $0.6\,M_\odot$ to $1.3\,M_\odot$ with an average $\simeq 0.8\,M_\odot$.

4. CONCLUSIONS

The eccentricity distribution of Ba stars is consistent with the stellar wind accretion and tidal evolution models. We require a tidally-enhanced wind and weaker tides than expected (by a factor of ten) to match the eccentricity distribution. The period distribution is sensitive to changes in the efficiency of wind accretion for long period systems and changes in the wind tidal enhancement parameter for the short period systems. Models that fit well the observed e and P distributions below 10,000 d are prone to produce too many high-period Ba stars, a problem we could not reconcile with any combination of our parameters.

References

Bidelman W. P., Keenan P. C., 1951, ApJ 114, 473
Boffin H. M. J., Cerf N., Paulus G., 1993, A&A 271, 125
Boffin H. M. J., Jorissen A., 1988, A&A 205, 155
Eggleton P. P., Fitchett M. J, Tout C. A., 1989, ApJ 354, 387
Frost C. A., 1997, PhD thesis, Monash University
Frost C. A., Lattanzio J. C., 1996, ApJ 473, 383
Han Z., Eggleton P. P., Podsiadlowski P., Tout C. A, 1995, MNRAS 277,1442
Hut P., 1980, A&A 92, 167
Jorissen A., Mayor M., 1988, A&A 198, 187
Jorissen A., Van Eck S., Mayor M., Udry S., 1998, A&A 332, 877
Karakas A. I., Lattanzio J. C., Tout C. A, 2000, MNRAS 316, 689
Lattanzio J. C., 1989, ApJ 347, 989
McClure R. D., 1984, PASP 96, 117
McClure R. D., Woodsworth A. W., 1990, ApJ 352, 709
McClure R. D., Fletcher J. M., Nemec J. M., 1980, ApJL 238, L35
Miller G. E., Scalo J. M., 1979, ApJSupp 41, 513
Rasio F. A., Tout C. A., Lubow S. H., Livio M., 1996, ApJ 470, 1187
Straniero O., Chieffi A., Limongi M., Busso M., Gallino R., Arlandini C., 1997, ApJ 478, 332
Theuns T., Boffin H. M., Jorissen A., 1996, MNRAS 280, 1264
Tout C. A., Aarseth S. J., Pols O. R., Eggleton P. P, 1997, MNRAS 291, 732

Tout C. A., Eggleton P. P, 1988, MNRAS 231, 823

Verbunt F., Phinney E. S., 1995, A&A 293, 709

Webbink R. F., 1986, in *Critical Observations verses Physical models for Close Binary Systems*, eds. K.-C. Leung, & D.S. Zhai (Gordon & Breach: New York), 403

Zahn J. -P., 1977, A&A 57, 383

HIGH-MASS X-RAY BINARIES AND OB-RUNAWAY STARS

Lex Kaper

Astronomical Institute, University of Amsterdam,
Kruislaan 403, 1098 SJ Amsterdam, The Netherlands
lexk@astro.uva.nl

Keywords: Massive stars, OB runaways, X-ray binaries, stellar winds, ISM

Abstract High-mass X-ray binaries (HMXBs) represent an important phase in the evolution of massive binary systems. HMXBs provide unique diagnostics to test massive-star evolution, to probe the physics of radiation-driven winds, to study the process of mass accretion, and to measure fundamental parameters of compact objects. As a consequence of the supernova explosion that produced the neutron star (or black hole) in these systems, HMXBs have high space velocities and thus are runaways. Alternatively, OB-runaway stars can be ejected from a cluster through dynamical interactions. Observations obtained with the *Hipparcos* satellite indicate that both scenarios are at work. Only for a minority of the OB runaways (and HMXBs) a wind bow shock has been detected. This might be explained by the varying local conditions of the interstellar medium.

1. THE EVOLUTION OF MASSIVE BINARIES

In a high-mass X-ray binary (HMXB) a massive, OB-type star is orbited by a compact, accreting X-ray source: a neutron star or a black hole. Mass is accreted either from the stellar wind of the massive companion (resulting in a modest X-ray luminosity), or flows with a higher rate to the compact star when the massive star fills its Roche lobe. In the latter case a much higher X-ray luminosity is achieved. Two classes of HMXBs can be distinguished based on the nature of the massive star: (i) the OB-supergiant systems, and (ii) the Be/X-ray binaries (for a recent catalogue of HMXBs, see Liu et al. 2000). The first class consists of the most massive systems, some of them containing a black hole (e.g. Cyg X-1). The second class comprises the majority (~80 %) of the

D. Vanbeveren (ed.), The Influence of Binaries on Stellar Population Studies, 125–140.

HMXBs; most of them are detected as X-ray transients. The transient character is explained by the periodic increase in X-ray luminosity when the neutron star, in its eccentric orbit, passes through periastron and accretes from the dense equatorial Be-star disk. In the remainder of this paper, we will concentrate on the OB-supergiant systems.

HMXBs are the descendants of massive binaries in which the initially most massive star (the primary) has become a neutron star or a black hole (Van den Heuvel & Heise 1972; for extensive reviews on binary evolution, see e.g. Van den Heuvel 1993, Vanbeveren et al. 1998). That the system remains bound after the supernova is due to a phase of mass transfer that occurs in the system when the primary grows larger than its Roche lobe. As a consequence, the secondary becomes the most massive of the two, so that with the supernova explosion of the primary less than half of the total system mass is lost and the system remains bound (Boersma 1961). It is assumed here that the supernova explosion is symmetric; in case of an asymmetry, the additional kick exerted on the compact remnant can cause the disruption of the system. Anyway, due to the loss of material (and momentum), the system gets a substantial "runaway" velocity (Blaauw 1961): on the order of 50 km s^{-1} for the most massive systems (and about 15 km s^{-1} for the less massive Be/X-ray binaries, cf. Van den Heuvel et al. 2001).

As long as the secondary is a massive main sequence star, accretion of its relatively tenuous stellar wind onto the compact companion does not result in an observable X-ray flux. Only when the secondary becomes a supergiant and starts filling its Roche lobe – this scenario refers to Case B binary evolution; in Case A the secondary fills its Roche lobe already on the main sequence – the accretion flow is dense enough to power a strong X-ray source; now the system has become a HMXB. This phase lasts for a relatively short period of time ($\sim$10,000 year): as soon as the Roche-lobe overflow commences, the orbit will shrink leading to an even higher mass transfer rate and a further tightening of the orbit. At some point, the X-ray source will be completely swamped with material optically thick in X-rays and/or penetrate the mantle of the secondary, followed by a rapid spiral-in. When also the secondary explodes as a supernova, a neutron-star binary (like the Hulse-Taylor binary pulsar) or two single neutron stars remain. The detailed evolution of massive binaries is still a matter of debate (e.g. Wellstein & Langer 1999, and several contributions to these proceedings).

In the following we will discuss the observational consequences of the evolutionary scenario outlined above for the properties of high-mass X-ray binaries. Firstly, we will compare the stellar parameters of the OB supergiants in HMXBs to those of single massive stars. These stars

Table 1 High-mass X-ray binaries with OB supergiant companion in the Milky Way and the Magellanic Clouds (ordered according to right ascension). The name corresponds to the X-ray source, the spectral type to the OB supergiant. For the systems hosting an X-ray pulsar the masses of both binary components can be measured (given an estimate of the inclination of the system). The last two systems contain a black-hole candidate. The radius of the OB supergiant can be determined in case X-ray eclipses are observed. The system parameters were taken from Van Kerkwijk et al. (1995), except for 2S0114+650 (Reig et al. 1996), Vela X-1 (Barziv et al. 2001), GX301-2 (Kaper & Najarro 2001), 4U1700-37 (Heap & Corcoran 1992), 4U1907+09 (Van Kerkwijk et al. 1989), LMC X-1 (Hutchings et al. 1987), Cyg X-1 (Herrero et al. 1995). The rapid X-ray pulsars are found in Roche-lobe overflow systems. The space velocity (or only the radial component) has been measured for some HMXBs and has been corrected for the solar peculiar motion and differential galactic rotation.

Name	Sp. Type	M_{OB} ($M_\odot$)	R ($R_\odot$)	M_X ($M_\odot$)	P_{pulse} (s)	v_{space} (km s^{-1})
2S0114+650	B1 Ia	~16		~1.7	860[a]	32
SMC X-1	B0 Ib	17.2	15	1.6	0.71	
LMC X-4	O7 III-IV	15.8	8	1.5	13.5	
Vela X-1	B0.5 Ib	23.9	30	1.9	283	35
Cen X-3	O6.5 II-III	18.4	11	1.1	4.84	42[b]
GX301-2	B1.5 Ia$^+$	>40		>1.3	696	~5[b]
4U1538-52	B0 Iab	16.4	15	1.1	529	85[b]
4U1700-37	O6.5 Iaf+	~52[c]	22	~1.8[c]		76
4U1907+09	early B I	>9		>0.7	438	
LMC X-1	O7-9 III			4-10		
Cyg X-1	O9.7 Iab	17.8		10		~20[d]

[a] A spin period of 2.7h is proposed by Corbet et al. (1999); [b] Only the radial-velocity component has been measured; [c] Rubin et al. (1996) give 30 $M_\odot$ and 2.6 $M_\odot$ for the mass of the O supergiant and X-ray source, respectively; [d] The space velocity of Cyg X-1 is with respect to the OB-star population in its local environment.

gained a significant fraction of their present mass from the progenitor of the X-ray source; this matter likely is chemically enriched. Not only matter, but also angular momentum is transfered. Therefore, both the internal and the atmospheric properties of these OB stars might differ from those of single massive stars. Secondly, the binarity of these systems provides the opportunity to measure fundamental parameters (such as the mass) of neutron stars and black holes.

Thirdly, the impact of the X-ray source on the structure of the OB-supergiant's wind is discussed. The X-ray source creates a Strömgren zone of high ionization in which the radiative acceleration of the wind

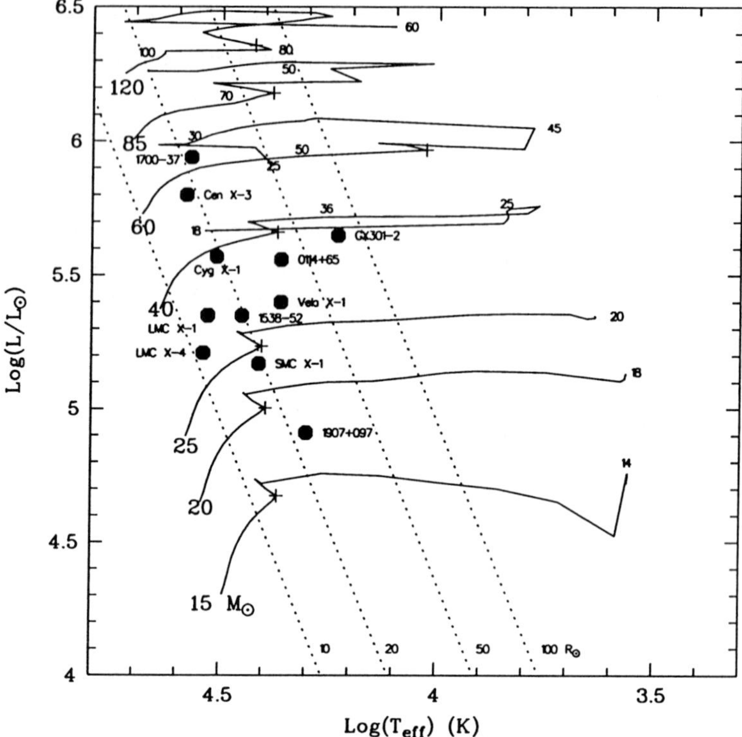

Figure 1 The location of the OB supergiants in HMXBs (filled circles) in the Hertzsprung-Russell diagram. The evolutionary tracks of Lejeune & Schaerer (2000) for stars of given mass are drawn in (up to the phase of core-helium burning). The numbers along the tracks (smaller font) show the decrease in mass of the star with time due to mass loss. The plus signs indicate the end of core-hydrogen burning. The diagonal dotted lines are lines of constant radius.

is quenched. This leads to the formation of a photo-ionization wake trailing the X-ray source in its orbit. In the most extreme case, only in the X-ray shadow a radiatively driven wind emerges: a so-called shadow wind.

The last section deals with the runaway nature of HMXBs. Although on theoretical grounds high space velocities are expected, the observational proof that HMXBs are runaway objects has just recently been obtained. *Hipparcos* results indicate that both the binary supernova scenario and the cluster ejection mechanism produce OB-runaway stars.

2. THE OB SUPERGIANTS IN HMXBS

Table 1 gives an overview of the OB-supergiant systems in the Milky Way and the Magellanic Clouds. The orbital periods range from 1.4 (LMC X-4) to 41.5 days (GX301-2). A relatively long pulse period (minutes) indicates that the X-ray source is a neutron star in a wind-fed system. The Roche-lobe overflow systems (e.g. Cen X-3) host rapid X-ray pulsars, most likely surrounded by an accretion disk. In these systems not only more mass, but also more angular momentum is transfered to the X-ray source, explaining the high (Eddington) X-ray luminosity and rapid spin period.

Figure 1 shows the distribution of the sample of OB supergiants in HMXBs (Table 1) in the Hertzsprung-Russell diagram. The spectral type of the OB supergiant is used to determine the effective temperature of the star and to estimate the interstellar reddening. With the latter, an estimate of the distance, and the bolometric correction (which also depends on the spectral type), the luminosity is derived. For several systems the distance is known with reasonable accuracy, especially the Magellanic Cloud systems and systems for which a parent OB association has been identified (§ 5). For comparison, the evolutionary tracks of (single) massive stars are drawn in (Lejeune & Schaerer 2000), which take into account the stellar-wind mass loss. The mass on the zero-age main sequence is indicated (and the reduced mass due to mass loss is displayed at some locations along the tracks).

The mass of the OB supergiant in a HMXB can be accurately measured when the system hosts an X-ray pulsar. With the delay in pulse-arrival time and the radial-velocity curve of the OB supergiant the orbits of both stars are determined, and thereby their masses if the orbital inclination is known. The latter is well constrained if the system is eclipsing. In those systems, also the radius of the OB supergiant can be derived with high precision from the duration of the X-ray eclipse. The masses and radii are listed in Table 1. The mass ratio sets the size of the Roche lobe (e.g. Eggleton 1983); it turns out that the measured radii of the OB supergiants are in very good agreement with the estimated size of the Roche lobe.

Earlier studies (e.g. Conti 1978, Rappaport & Joss 1983) suggested that the OB supergiants in HMXBs are too luminous for their mass. This is clearly demonstrated in Figure 1: e.g. the O6.5 giant companion of Cen X-3 has a mass of 19 $M_\odot$, while its luminosity corresponds to that of a star of more than 50 $M_\odot$. Besides this, the radius corresponding to the luminosity and effective temperature (R_{eff}) is larger than its measured (Roche-lobe) radius. Thus, apart from being undermassive, the OB

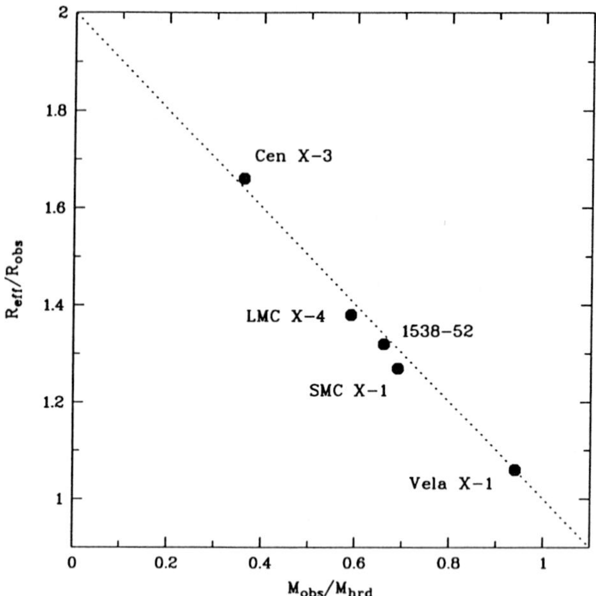

Figure 2 The ratio of the observed mass of OB supergiants (in HMXBs hosting an eclipsing X-ray pulsar) and the mass estimated from their position in the HRD is compared to the ratio of the effective temperature radius and the observed radius. The difference between observed and expected mass and radius seems to be bigger for systems that are presently in a phase of Roche-lobe overflow.

supergiants in HMXBs also seem to be *undersized* for their luminosity and temperature. Figure 2 shows that the OB supergiants (in systems with an eclipsing X-ray pulsar) with the largest deviation in radius, are also the most "undermassive" ones. The wind-fed system 4U1700-37 is eclipsing, but the X-ray source is not pulsating; the measured radius of its O6.5 Iaf companion is in good agreement with R_{eff}, but the supergiant's mass cannot be accurately determined. For a similar reason, the non-eclipsing HMXBs hosting an X-ray pulsar are not included in Fig. 2, because the radius is not well known.

However, the effective temperature of a star is determined by its luminosity and size. Apparently, the spectral type – effective temperature calibration of single stars cannot be applied to the OB supergiants in HMXBs (presuming that the distances to these objects are not systematically too large). For the five objects in Fig. 2, the deviation in radius and mass is worst for the stars with earliest spectral type, but this might be a coincidence (e.g. 4U1700-37 has an O6.5 companion, but has R_{eff}/R_{obs} close to one). The discrepancy between R_{eff} and R_{obs} can be removed by systematically increasing T_{eff} by 10-25 %, while keeping the luminosity constant (the bolometric correction does not change very much). But this does not explain why the stars are undermassive.

Probably the trend observed in Fig. 2 is related to the phenomenon of Roche-lobe overflow. The OB star tries to become a supergiant, but at some point reaches its critical Roche lobe and starts to transfer mass to its companion. While the luminosity of the star is determined by the core (which does not notice much of what is happening to the outer mantle), the star likes to be bigger than allowed by its Roche lobe and is peeled off. This would explain why systems like Vela X-1 (and 4U1700-37, GX301-2) which are wind-fed X-ray sources not (yet) experiencing Roche-lobe overflow, are not showing large discrepancies in mass and radius.

Regarding the chemical enrichment of the massive stars in HMXBs, a thorough analysis is still lacking. The ultraviolet spectrum of HD77581, the companion of Vela X-1, shows very strong N V and Al III resonance lines, typical for nitrogen enhanced B stars, the BN stars (Kaper et al. 1993). Atmospheric models of HDE226868, the O supergiant companion of Cyg X-1, by Herrero et al. (1995) indicate a helium abundance higher than expected for its evolutionary state (a helium discrepancy, assuming single star evolution). However, according to Herrero et al. (1992) also rapid rotators and very luminous stars, of which several are single, show helium enhancement. Blaauw (1993) and Hoogerwerf et al. (2000) used these findings to suggest that the helium enhancement and rapid rotation are observational characteristics of OB runaways; unfortunately, we are still dealing with small-number statistics.

3. NEUTRON STARS, BLACK HOLES, AND GAMMA-RAY BURSTS

The accurate measurement of neutron-star masses is essential for our understanding of the equation of state (EOS) of matter at supra-nuclear densities, and of the mechanism of core collapse of massive stars. The EOS can, so far, only be studied on the basis of theoretical models. These remain very uncertain and subject to hot dispute. Brown & Bethe (1994) have strongly argued that the EOS must be “soft”, i.e., that matter would be relatively compressible, due to kaon condensation (kaons are bosons which do not contribute to the Fermi pressure). Heavy ion collision experiments (in Hamburg, DESY) confirm their predictions, but at densities still quite a bit lower than those appropriate for neutron stars. If the theory were correct, one of the astrophysical implications would be that neutron stars cannot have a mass larger than 1.55 $M_\odot$; for larger masses, the object would collapse into a black hole.

The masses of X-ray pulsars (Table 1) can also be used to test the predictions of supernova models. For instance, Timmes et al. (1996) present model calculations from which they find that Type II supernovae (massive, single stars) give a bimodal compact-object mass distribution, with peaks at 1.28 and 1.73 $M_\odot$, while Type Ib supernovae (such as produced by stars in binaries, which are stripped of their envelopes) will produce neutron stars within a small range around 1.32 $M_\odot$. The massive neutron star in Vela X-1 with a mass of 1.9 ± 0.2 $M_\odot$ would be consistent with the second peak in this distribution (Barziv et al. 2001).

Presently, all accurate mass determinations besides that of Vela X-1 have been for neutron stars that were almost certainly formed from Type Ib supernovae and that have accreted little since. The most accurate have been derived for the binary radio pulsars. All of these are consistent with a small range near $1.35\,M_\odot$ (Thorsett & Chakrabarty 1999). That this mass is so close to the Chandrasekhar mass is thought to be due to the maximum mass of the degenerate iron core, just before supernova. Apparently, the supernova explosion in this case prevents additional material from accretion onto the proto-neutron star, thus also preventing the formation of a black hole. However, the (binary) radio pulsars are likely to originate from binaries of moderate mass which might not be able to produce massive neutron stars, nor black holes. This would be consistent with the observed absence of black holes in Be/X-ray binaries. If massive neutron stars exist, they are more likely to have formed in systems of high initial mass. These systems are capable of producing black holes, thus a fall-back mechanism must exist (because these stars will also have a degenerate iron core before collapse). The alternative is that black holes are not produced by normal supernovae but in gamma-ray bursts (hypernovae), such as SN1998bw/GRB980425 (Galama et al. 1998, Iwamoto et al. 1998).

Our hypothesis is that when a fall-back mechanism exists, the masses of neutron stars in the most massive HMXBs will be evenly distributed over a relatively large mass interval, up to a maximum mass set by the neutron star equation of state. Alternatively, black holes are produced in gamma-ray bursts, and massive neutron stars like in Vela X-1 are explained by the second of the two peaks in the mass distribution such as predicted by Timmes et al. (1996).

4. INTERACTION BETWEEN X-RAY SOURCE AND STELLAR WIND

The X-ray source ionizes a significant fraction of the ambient stellar wind (Fig. 3). The presence of a Strömgren zone in the system becomes

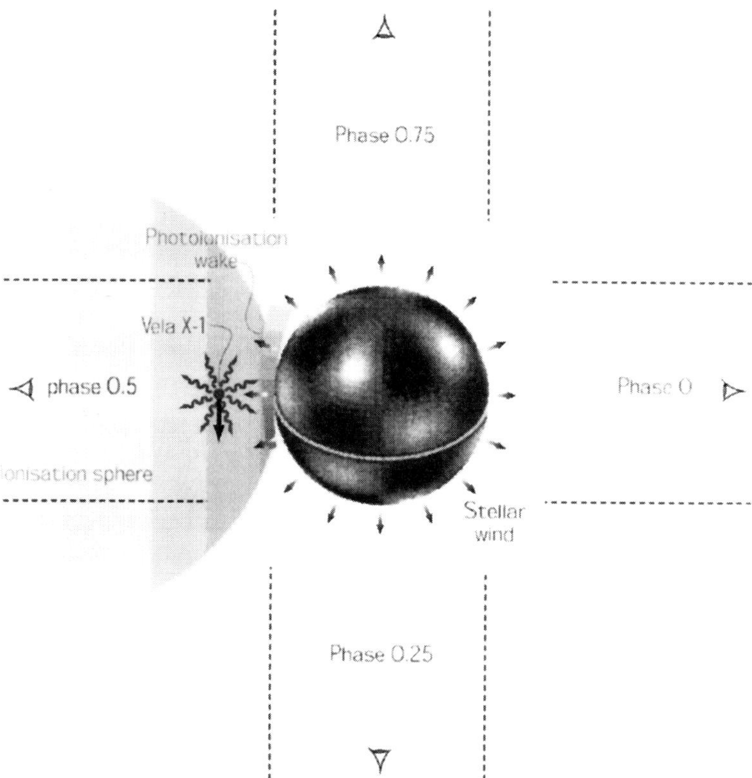

Figure 3 A sketch of the HMXB Vela X-1. The X-ray source creates a Strömgren zone in the stellar wind. A photo-ionization wake is formed at the trailing border of the Strömgren zone where the stagnant flow meets the fast, accelerating wind.

apparent from the orbital modulation of strong ultraviolet resonance lines (e.g. from N V, Si IV, and C IV, cf. Kaper et al. 1993) formed in the stellar wind. When the X-ray source is in the line of sight ($\phi = 0.5$), the blue-shifted absorption part of the P-Cygni profile is strongly reduced in strength, because ions like C IV are removed by the X-rays. The orbital modulation of UV resonance lines is observed in most HMXBs with OB supergiant companion for which ultraviolet spectra are available. However, in HD153919/4U1700-37, the system for which Hatchett & McCray (1978) originally predicted this effect, the P-Cygni profiles do not vary with orbital phase. When the non-monotonic velocity structure of the stellar wind is taken into account, the absence of the HM-effect in 4U1700-37 can be explained (Van Loon et al. 2001). For a review on stellar winds in HMXBs, see Kaper (1998).

A consequence of the high degree of ionization inside the Strömgren zone is that in this region the radiative acceleration is quenched. The highly ionized plasma becomes effectively transparent for ultraviolet photons emitted by the OB companion that accelerate the stellar wind. Due to the revolution of the system, the stagnant flow inside the ionization zone meets the rapidly accelerating stellar wind at the trailing border. Here, a strong shock is formed (Fig. 3), the so-called photo-ionization wake (Blondin et al. 1990). The relatively large extent of the photo-ionization wake results in an observable blue-shifted absorption feature in strong optical lines at late orbital phases (Kaper et al. 1994).

The Roche-lobe overflow systems (SMC X-1, LMC X-4, Cen X-3) contain a very strong X-ray source. The X-ray luminosity approaches the bolometric luminosity of the OB companion, so that the X-ray flux dominates the ionization equilibrium in a large fraction of the stellar wind. Actually, only in the X-ray shadow behind the OB companion a normal stellar wind can develop, the shadow wind (Blondin 1994). Ultraviolet spectra of LMC X-4 (Boroson et al. 1999) clearly show the presence of a shadow wind in this system. High-resolution UV spectra of LMC X-4 obtained with HST/STIS reveal the signature of a photo-ionization wake, there where the fast shadow wind rams into the highly ionized Strömgren zone (Kaper et al. 2001).

The spin behaviour of the X-ray pulsar provides information on the accretion flow in the system. In Roche-lobe overflow systems the X-ray pulsar is spun up due to the relatively high accretion rate and the corresponding transfer of angular momentum. The spin-up and spin-down episodes observed in some wind-fed systems (e.g. Vela X-1) indicate that sometimes a small accretion disk is formed, the direction of rotation depending on details (e.g. instability) of the accretion flow (Blondin et al. 1990). Variations in the accretion flow are also apparent from the strong X-ray flaring behaviour observed in HMXBs.

5. THE ORIGIN OF OB-RUNAWAY STARS

About 20% of the O stars, and a smaller fraction of the B stars, have a space velocity much higher than observed on average for the OB-star population in the Milky Way (about 10 km s^{-1}, Stone 1979). Some OB stars have a space velocity exceeding 100 km s^{-1}. Blaauw (1961) called these stars runaway stars, because at least for some of them the reconstructed path through space suggests an origin in a nearby OB association. How did these massive stars obtain such a high space velocity? Blaauw proposed that the supernova explosion of a massive companion in a binary results in a high velocity of the remaining massive star; the "modern" version of this Blaauw scenario is described in § 1, which predicts that all HMXBs are runaway systems. It took several decades

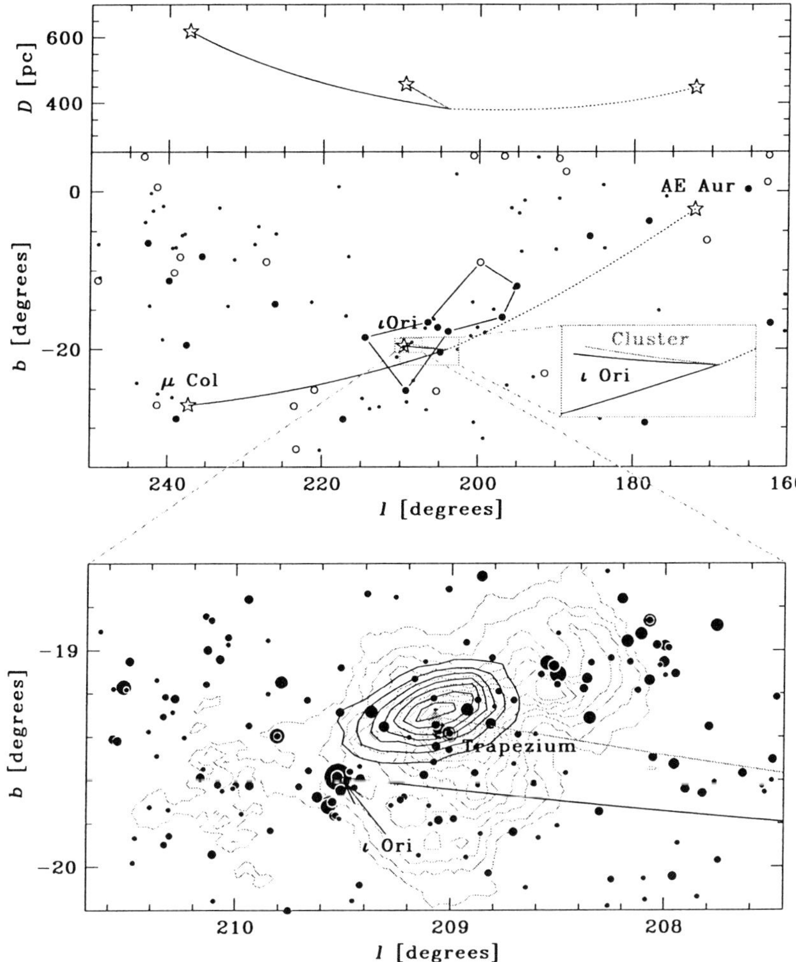

Figure 4 Top & middle: The orbits, calculated back in time, of the runaways AE Aur (dotted line) and μ Col (solid line), and the binary ι Ori based on Hipparcos observations. The top panel shows the distance versus galactic longitude, the middle panel displays the orbits projected on the sky in galactic coordinates. The stars met $\sim$ 2.5 Myr ago. Filled circles denote O and B stars, open circles represent other spectral types. Bottom: A blow up of the central region shows the predicted position of the parent cluster (black contours) together with all stars in the Tycho Catalogue (ESA 1997) in the field down to V = 12.4 mag. The brightest star is ι Ori; also the Trapezium is indicated. The grey contours (IRAS 100 μm flux) outline the Orion Nebula. Figure taken from Hoogerwerf et al. (2000).

to obtain the observational evidence proving that this scenario actually works (Van Oijen 1989, Van Rensbergen et al. 1996, Kaper et al. 1997); now, *Hipparcos* measurements have definitely confirmed the runaway nature of HMXBs (Chevalier & Ilovaisky 1998, Kaper et al. 1999); Table 1 lists the derived space velocities.

An alternative scenario for the formation of OB runaways is dynamical ejection from a compact cluster (Poveda et al. 1967, Portegies Zwart 2000). Due to the dynamical interaction between binaries (and single stars), now and then a massive star is ejected from an OB association. The probability for ejection is higher in dense starclusters; given the expansion of OB associations, one expects that dynamical ejection is most effective in young OB associations. An impressive example demonstrating the cluster ejection scenario is provided by the OB runaways AE Aur and μ Col. Blaauw & Morgan (1954) noticed that these two stars move in almost opposite directions away from the Ori OB1 association and have the same kinematical age (i.e. the travel time from the association to its present location). Using *Hipparcos* observations, Hoogerwerf et al. (2000) managed to reconstruct the kinematical history of these two runaways in great detail and could show that both stars and the massive binary system ι Ori were at the same place in Ori OB1 $\sim$ 2.5 million years ago (Fig. 4). The dynamical interaction between two binary systems apparently led to the disruption of one of them and the subsequent ejection of the two members (in opposite directions due to momentum conservation).

Based on the *Hipparcos* proper motion of HD153919 (4U1700-37), Ankay et al. (2001) demonstrate that this HMXB very likely originates from the OB association Sco OB1 at a distance of 2 kpc (Fig. 5). The kinematical age of the system is 2±0.5 Myr, which sets the time of the supernova producing the compact star in the system. The present age of Sco OB1 is 6±2 Myr (the "nucleus" of the association, NGC 6231, is probably younger). These observations can be used to derive a constraint on the initial mass of the progenitor of 4U1700-37. The turn-off mass of a cluster of 4 Myr is about 25 $M_\odot$. If the system was born as a member of NGC 6231, the progenitor of 4U1700-37 likely was an even more massive star ($\sim$ 40 $M_\odot$).

Hoogerwerf et al. (2000) studied the kinematical history of a sample of nearby runaways (17) and their (candidate) parent OB associations using *Hipparcos* data. A comparison of the kinematical age of the runaway and the age of the cluster is used to discriminate between the two formation scenarios. In case of dynamical ejection, the kinematical age of the runaway should be about equal to the age of the association. In the binary supernova scenario, the evolution of the binary system implies that the kinematical age of the runaway is significantly shorter than the age of the cluster. Furthermore, the OB runaway will be a blue straggler in the HRD of the cluster, because the OB runaway is rejuve-

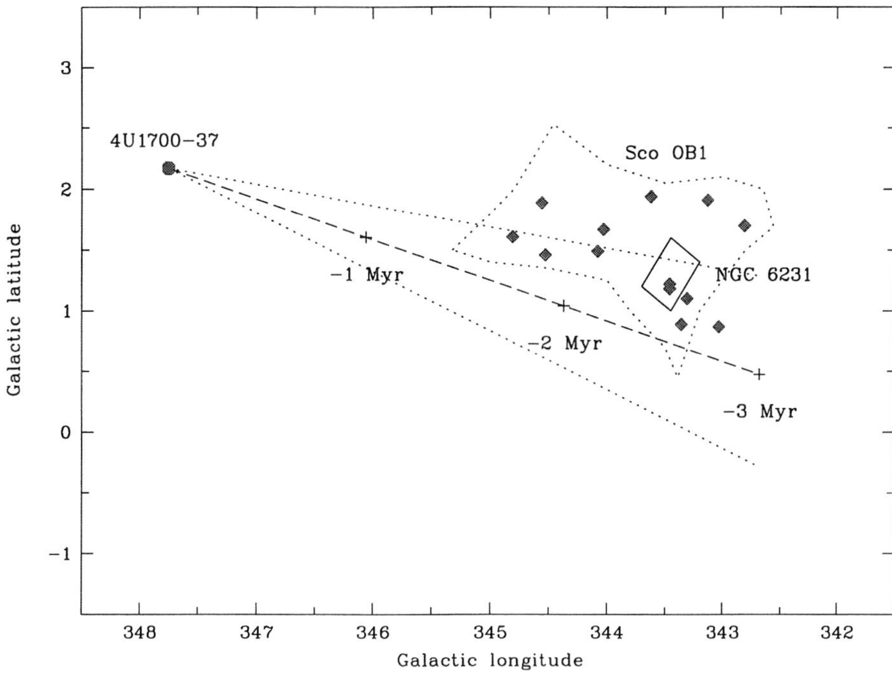

Figure 5 The reconstructed path of the runaway HMXB 4U1700-37 intersects with the location of Sco OB1; the error cone is indicated by the dotted straight lines. The Hipparcos confirmed members are shown as filled diamonds. The proper motion of 4U1700-37 is with respect to the average proper motion of Sco OB1. The corresponding kinematical age of 4U1700-37 is 2 ± 0.5 million year. The current angular separation between 4U1700-37 and NGC 6231 (at 2 kpc) corresponds to a distance of about 150 pc.

nated (by gaining mass) during the first phase of mass transfer in the system. Hoogerwerf et al. conclude that the two mechanisms produce about equal amounts of runaways. This conclusion is in agreement with theoretical predictions (e.g. De Donder et al. 1997, Portegies Zwart 2000).

OB runaways can be used as probes of the interstellar medium; their stellar wind interacts with the ambient medium reflecting the motion of the star. Where the ram pressure of wind and interstellar medium balance, a wind bow shock is formed if the runaway is moving with a supersonic velocity. Van Buren et al. (1995) detected wind bow shocks around several OB runaways, with a detection rate of about 30 %. Kaper et al. (1997) discovered a wind bow shock around the HMXB Vela X-1. Huthoff & Kaper (2001) searched for the presence of wind bow shocks around other HMXBs, but could not add a significant detection. The non-detection of wind bow shocks for the majority of runaways is likely related to the varying physical conditions of the ISM. For example, inside a hot superbubble the sound speed is about 100 km s^{-1}, so that the runaway would not move supersonically and a bow shock is not formed. These superbubbles are created by the combined action of stellar winds and supernovae of the massive stars in OB associations. Huthoff & Kaper demonstrate that the OB runaways with a detected wind bow shock concentrate in regions in between superbubbles.

Acknowledgments

It is a pleasure to thank Askin Ankay, Ed van den Heuvel, Gijs Nelemans, Freek Huthoff, Ronnie Hoogerwerf, Jos de Bruijne, and Norbert Langer for stimulating discussions. LK is supported by a fellowship of the Royal Academy of Sciences in the Netherlands.

References

Ankay, A., Kaper, L., De Bruijne, J.H.J., et al. 2001, submitted to A&A

Barziv, O., Kaper, L., Van Kerkwijk, M.H., et al. 2001, submitted to A&A

Blaauw, A. 1961, Bull. Astr. Inst. Neth. 15, 265

Blaauw, A. 1993, in ASP Conf. Series, Volume 35, p. 207

Blaauw, A., Morgan, W.W. 1954, ApJ 119, 625

Blondin, J.M. 1994, ApJ 435, 756

Blondin, J.M., Kallman, T.R., Fryxell, B.A., Taam, R.E. 1990, ApJ 356, 591

Boersma, J. 1961, Bull. Astr. Inst. Neth. 15, 291

Boroson, B., Kallman, T., McCray, R., Vrtilek, S.D., Raymond, J. 1999, ApJ 519, 191

Brown, G.E, Bethe, H.A. 1994, ApJ 423, 659

Chevalier, C., Ilovaisky, S.A. 1998, A&A 330, 201

Conti, P.S. 1978, A&A 63, 225

Corbet, R.H.D., Finley, J.P., Peele, A.G. 1999, ApJ 511, 876

De Donder, E., Vanbeveren, D., Van Bever, J. 1997, A&A 318, 812

Eggleton, P. 1983, ApJ 268, 368

Galama, T.J., Vreeswijk, P.M., Van Paradijs, J., et al. 1998, Nat 395, 670

Heap, S.R., Corcoran, M.F. 1992, ApJ 387, 340

Herrero, A., Kudritzki, R.P., Vilchez, J.M. et al. 1992, A&A 261, 209
Herrero, A., Kudritzki, R.P. Gabler, R., et al. 1995, A&A 297, 556
Hoogerwerf, R., de Bruijne, J.H.J., de Zeeuw, P.T. 2000, ApJ 544, L133
Hutchings, J.B., Crampton, D., Cowley, A.P., et al. 1987, AJ 94, 340
Huthoff, F., Kaper, L. 2001, in Proc. ESO workshop on Black Holes in Binaries and Galactic Nuclei, Eds. Kaper, Van den Heuvel, Woudt, ESO Conf. Proc., Springer, in press
Iwamoto, K., Mazzali, P.A., Nomoto, K., et al. 1998, Nat 395, 672
Kaper, L. 1998, in Proc. "Boulder-Munich II", ASP Conf. Ser. 131, Ed. Howarth, p. 427
Kaper, L., Comerón, F., Barziv, O. 1999, in Proc. "Wolf-Rayet Phenomena in Massive Stars and Starburst Galaxies", IAU Symp. 193, Eds. Van der Hucht, Koenigsberger, Eenens, p. 316
Kaper, L., Hammerschlag-Hensberge, G., Van Loon, J.Th. 1993, A&A 279, 485
Kaper, L., Hammerschlag-Hensberge, G., Zuiderwijk, E.J. 1994, A&A 289, 846
Kaper, L., Van Loon, J.Th., Augusteijn, T., et al. 1997, ApJ 475, L37
Kaper, L., Hammerschlag-Hensberge, G., De Koter, A., et al. 2001, to be submitted to A&A
Kaper, L., Najarro, F. 2001, to be submitted to A&A
Lejeune, Th., Schaerer, D. 2000, A&A in press (astro-ph/0011497)
Liu, Q.Z., Van Paradijs, J., Van den Heuvel, E.P.J. 2000, A&AS 147, 25
Portegies Zwart, S.F. 2000, ApJ 544, 437
Poveda, A., Ruiz, J., Allen, C. 1967, Bol. Obs. Tonantzintla y Tacubaya 4, 860
Rappaport, S.A., Joss, P.C. 1983, in *Accretion-driven stellar X-ray sources*, Eds. Lewin, Van den Heuvel, Cambridge Univ. Press, p. 1
Reig, P., Chakrabarty, D., Coe, M.J., et al. 1996, A&A 311, 879
Rubin, B.C., Finger, M.H., Harmon, B.A., et al. 1996, ApJ 459, 259
Stone, R.C. 1979, ApJ 232, 520
Thorsett, S.E., Chakrabarty, D. 1999, ApJ 512, 288
Timmes, F.X., Woosley, S.E., Weaver, T.A. 1996, ApJ 457, 834
Vanbeveren, D., De Loore, C., Van Rensbergen, W. 1998, A&A Rev. 9, 63
Van Buren, D., Noriega-Crespo, A., Dgani, R. 1995, AJ 110, 2914
Van den Heuvel, E.P.J. 1993, in *Saas-Fee Advanced Course on Interacting Binaries* (Springer-Verlag), p. 263
Van den Heuvel, E.P.J., Heise, J. 1972, Nat. Phys. Sci. 239, 67
Van den Heuvel, E.P.J., Portegies Zwart, S.F., Bhattacharya, D., Kaper, L. 2001, A&A in press (astro-ph 0005245)

Van Kerkwijk, M.H., Van Oijen, J.G.J., Van den Heuvel, E.P.J. 1989, A&A 209, 173

Van Kerkwijk, M.H., Van Paradijs, J., Zuiderwijk, E.J. 1995, A&A 303, 497

Van Oijen, J.G.J. 1989, A&A 217, 115

Van Rensbergen, W., Vanbeveren, D., De Loore, C. 1996, A&A 305, 825

Wellstein, S., Langer, N. 1999, A&A 350, 148

THE OBSERVED GALACTIC WOLF-RAYET STAR DISTRIBUTION IN RELATION WITH METALLICITY

Karel A. van der Hucht
Space Research Organization Netherlands,
Sorbonnelaan 2, NL-3584 CA Utrecht, the Netherlands
K.A.van.der.Hucht@sron.nl

Keywords: Galaxy: structure;
Stars: distances; Stars: evolution; Stars: statistics; Stars: Wolf-Rayet

Abstract The recent *VIIth Catalogue of Galactic Wolf-Rayet Stars* lists 227 Population I WR stars, comprising 127 WN, 87 WC, 10 WN/WC and 3 WO stars. Additional discoveries bring the census to 234 WR stars, including 22 WNL and 11 WCL stars within 50 pc of the Galactic Center.

A re-determination of the optical photometric distances and the galactic distribution of WR stars shows in the solar neighbourhood a projected surface density of 2.7 WR stars per kpc^2, a $N_{\mathrm{WC}}/N_{\mathrm{WN}}$ number ratio of 1.3, and a WR binary frequency of 40 %.

The galactocentric distance (R_{WR}) distribution per subtype shows $\overline{R_{\mathrm{WN}}}$ and $\overline{R_{\mathrm{WN}}}$ decreasing with later WN and WC subtypes. The observed trend is more indicative of WNE → WCE and WNL → WCL subtype evolution, rather than of WNL → WNE and WCL → WCE subtype evolution.

Compared with other Local Group galaxies and in relation with metallicity, the Milky Way has within 3 kpc from the Sun a number ratio $N_{\mathrm{WC}}/N_{\mathrm{WN}} = 1.3$, which is a factor of ~ 2 above the expected trend. This could mean that some 30 galactic WN stars in the $d < 3$ kpc volume are still hiding.

1. INTRODUCTION

WR stars are characterized by strong He, N, C and O emission lines, originating in their hot stellar winds with terminal velocities in the range $v_\infty \simeq 400$–$5000\,\mathrm{km\,s^{-1}}$, which drive mass loss rates of the order of $\dot{M} = 10^{-5}\,\mathrm{M_\odot yr^{-1}}$. In these WR winds also their free-free μm-to-cm continuum emission originates, which allows mass loss rate determina-

D. Vanbeveren (ed.), The Influence of Binaries on Stellar Population Studies, 141–156.

tions, given estimates of wind abundances, ionization, clumping rates, terminal wind velocities and distances of the WR stars (Wright & Barlow 1975; Crowther 1999). Recent studies of coeval open clusters and OB associations suggest that the minimum initial stellar mass for reaching the WR phase is 18 $M_\odot$ in the Milky Way (Massey *et al.* 2001), 30 $M_\odot$ in the LMC and 70 $M_\odot$ in the SMC (Massey *et al.* 2000). Observed galactic WR masses in binary systems fall in the range 2.3-55 $M_\odot$ (van der Hucht 2001).

It is important to discover and monitor WR stars, while each of them is a unique physics laboratory, a tracer of star formation in spiral arms, and a representative of the one but final phase in the evolution of massive stars, probably followed by a Type Ib/Ic supernova explosion (*e.g.*, Langer & El Eid 1986; Woosley *et al.* 1993; Maeder & Conti 1994; Langer & Woosley 1996; Chevalier & Li 2000; Kaper & Cherepashchuk 2001; Chu 2001), either single in origin or binary (Langer & Heger 1999; Wellstein & Langer 1999). One cannot hope to understand, *e.g.*, the formation of black holes without understanding the late stages of massive star evolution. And where statistically the next galactic supernova is overdue for already more than a century, it is of obvious relevance to gather detailed knowledge about each of its potential progenitors.

Comprehensive reviews of the WR phenomenon have been presented by, *e.g.*, van der Hucht (1992), Maeder & Conti (1994), and in Proc. IAU Symp. No. 143 (van der Hucht & Hidayat 1991), in Proc. IAU Symp. No. 163 (van der Hucht & Williams 1995), in Proc. 33rd Liège Intern. Astroph. Coll. (Vreux *et al.* 1996), and in Proc. IAU Symp. No. 193 (van der Hucht, Koenigsberger & Eenens 1999).

The *VIIth Catalogue of Galactic Wolf-Rayet Stars* (van der Hucht 2001, replacing the *VIth Catalogue* of van der Hucht *et al.* 1981), brings the number of known galactic Population I WR stars to 227, derives photometric distances and shows their galactic distribution. An additional seven new WR stars have been discovered at near-IR wavelengths by Genzel *et al.* (2000) and Paumard *et al.* (2001). The galactic WR census is very much incomplete due to the ubiquitous interstellar extinction in the plane of the Milky Way. Especially at IR wavelengths numerous WR stars are waiting to be discovered.

2. DISTANCES OF GALACTIC WOLF-RAYET STARS

To derive heliocentric photometric distances of WR stars, van der Hucht (2001) re-investigated the absolute visual magnitudes M_v and intrinsic colours $(b-v)_o$ of WR stars in open clusters and OB associations, in

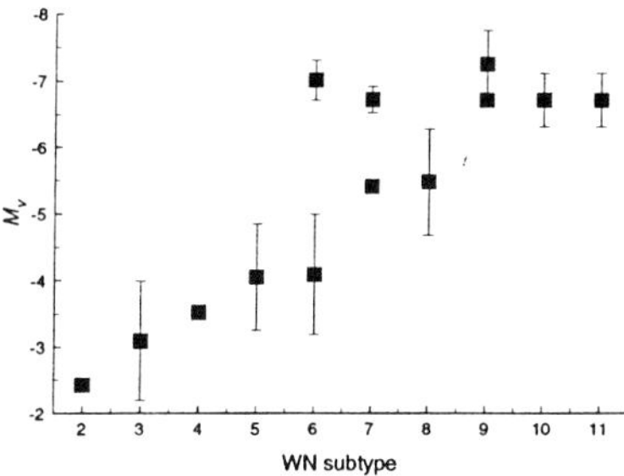

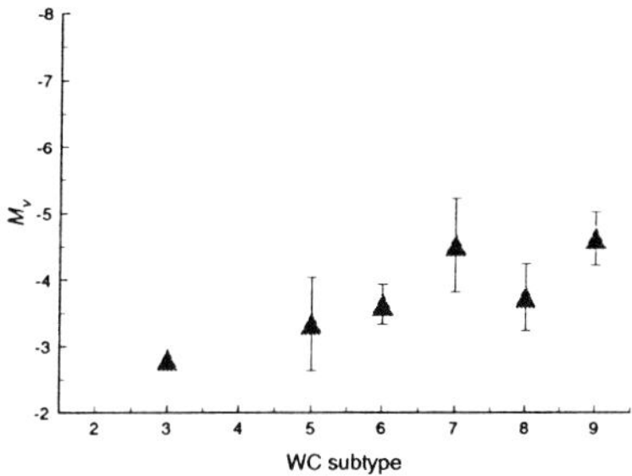

Figure 1 Average absolute visual magnitude $\overline{M_v(\mathrm{WR})}$ versus WR subtype. *Left:* the upper values at WN6, WN7 and WN9 refer to, respectively, WN6ha WN7ha and WN9ha stars. *Right:* the point at WC3 refers to WO stars only. Error bars denote the standard deviations of the averages. From van der Hucht (2001).

the filter system of Smith (1968a), correcting for contributions from binary companions where warranted. The resulting average M_v-values are quoted in Fig. 1 as a function of WR sub-type.

Subsequently, photometric distances of fields WR stars followed directly from their apparent *bv* magnitudes and subtype-dependent intrinsic parameters M_v and $(b\text{–}v)_\circ$, again corrected for companion contributions in case of binarity. The errors in the resulting photometric distances of WR field stars correspond at least to the range of M_v-values for the particular subtype (see Fig. 1). It was estimated that the average error in the WR field star distances is $\sim 50\,\%$.

The distance of the Galactic Center was adopted at $d_{\mathrm{GC}} = 8.0 \pm 0.5$ kpc (Reid 1993; McNamara *et al.* 2000). For the 26 WR objects in the *VIIth Catalogue* within 50 pc of the Galactic Center (15 WNL and 11 WCL stars) that distance was adopted, as well as $A_V = 29 \pm 5$ mag (Figer *et al.* 1999a), *i.e.*, $A_v = 32$ mag.

3. WOLF-RAYET STAR LONGITUDE DISTRIBUTION

3.1. GALACTIC STRUCTURE

The WR galactic distribution (l^{II},d) is plotted in Fig. 2, which reflects a galactic spiral structure (compare with the galactic distribution of H II regions by Georgelin & Georgelin 1976, fig. 11, up-dated by Russeil 1998, fig. 8-4, and Russeil 2001, fig. 4).

As stated earlier by Conti & Vacca (1990), the overall similarity of Fig. 2 and previous diagrams for WR stars by Smith (1968b, fig. 1), van

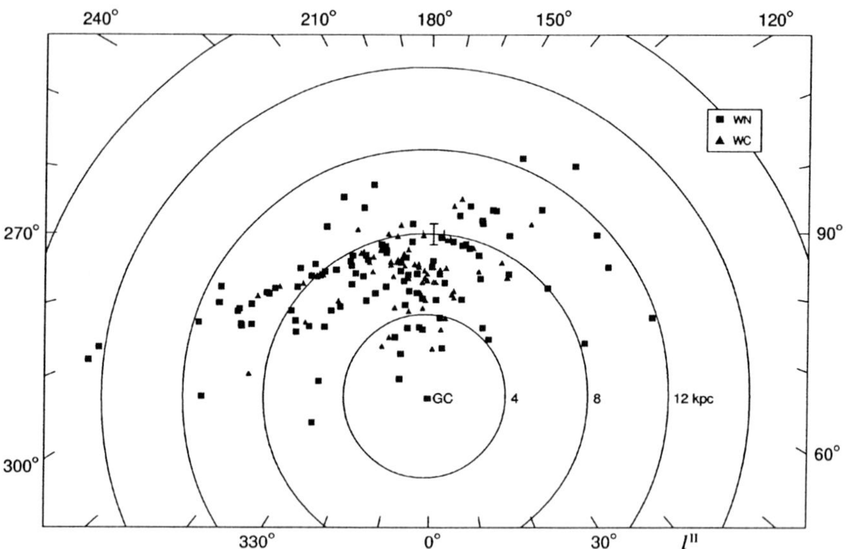

Figure 2 The galactic Wolf-Rayet star distribution (l^{II},d) projected on the galactic plane for the stars in the *VIIth Catalogue*. The Sun is indicated by +. The distance from the Sun to the Galactic Center is adopted at 8.0 kpc (Reid 1993). The point at the Galactic Center harbours 33 WR stars. From van der Hucht (2001).

der Hucht *et al.* (1988, fig. 3) and Conti & Vacca (1990, fig. 1) demonstrates that the photometric distance determination procedure is fairly 'robust', on average. However, distances for individual WR stars can differ substantially. We confirm earlier conclusions by Roberts (1962), Smith (1968b), Conti *et al.* (1983), van der Hucht *et al.* (1988) and Conti & Vacca (1990), that the galactocentric distance distribution of WR stars is subtype-dependent, notably with the WCL stars confined towards the Galactic Center, and that the galactic anti-center quadrant (Orion spur) is devoid of WR stars.

In the volume with $d < 3$ kpc the *VIIth Catalogue* lists 77 WR stars (16 more than the census of van der Hucht *et al.* 1988) with a number ratio $N_{\rm WN}$:$N_{\rm WC}$:$N_{\rm WO} = 34$:42:1. Since the WN and WC phases are consecutive in the evolution of massive stars, this implies that in the solar neighbourhood the average WC-phase life-time is ~ 1.3 times longer than the average WN-phase life-time.

When divided in shells around the Galactic Center, we count in the $d < 3$ kpc volume a number ratio $N_{\rm WC}/N_{\rm WN} = 0.9$ for $7 < R_{\rm WR} < 9$ kpc, while $N_{\rm WC}/N_{\rm WN} = 1.8$ for $5 < R_{\rm WR} < 7$ kpc. Projected on the galactic plane the WR surface density in that volume is $\sigma_{\rm WR} = 2.7\,{\rm kpc}^{-2}$.

Table 1 Numbers, number ratios, and surface densities σ_{WR} of the 64 WR stars within $d < 2.5$ kpc as a function of galactocentric distance R (adopting $R_{\odot} = 8.0$ kpc).

R (kpc)	N_{WR}	N_{WN}	N_{WC}	N_{WO}	N_{WC}/N_{WN}	N_{WN}^{binary}	N_{WC}^{binary}	N_{WR}^{binary}	area (kpc^2)	σ_{WR} (kpc^{-2})
5.5 - 7	33	10	23		2.3	1	9	30 %	4.44	7.4
7 - 9	28	14	13	1	0.9	5	9	50 %	9.76	2.9
9 - 10.5	3	1	2		2.0		1	33 %	5.44	0.6
total	64	25	38	1	1.5	6 (24 %)	19 (50 %)	39 %	19.63	3.3

Table 2 Numbers, number ratios, and surface densities σ_{WR} of the 77 WR stars within $d < 3$ kpc as a function of galactocentric distance R (adopting $R_{\odot} = 8.0$ kpc).

R (kpc)	N_{WR}	N_{WN}	N_{WC}	N_{WO}	N_{WC}/N_{WN}	N_{WN}^{binary}	N_{WC}^{binary}	N_{WR}^{binary}	area (kpc^2)	σ_{WR} (kpc^{-2})
5 - 7	39	14	25		1.8	3	10	33 %	9.17	4.3
7 - 9	31	16	14	1	0.9	5	10	48 %	11.83	2.6
9 - 11	7	4	3		0.8	2	1	43 %	7.27	1.0
total	77	34	42	1	1.3	10 (29 %)	21 (50 %)	40 %	28.27	2.7

The observed WR binary frequency in the $d < 3$ kpc volume (31 WR binaries) is 40 %. The WC binary frequency in that volume (50 %) is almost twice the WN binary frequency (29 %).

The overall subtype distribution and binary frequency breakdown is listed in Table 1 for the volume with $d < 2.5$ kpc, and in Table 2 for the volume with $d < 3$ kpc.

3.2. WOLF-RAYET SUBTYPE DISTRIBUTION AND EVOLUTION

The *VIIth Catalogue* allows to investigate the WR galactocentric distances $R_{\rm WR}({\rm kpc}) = \sqrt{d^2 + 64 + 16d\cos b^{\rm II}\cos l^{\rm II}}$. Fig. 3 shows the WR galactocentric distribution per subtype. In the column at the right side of Fig. 3, the WN stars show a remarkable trend for $\overline{R_{\rm WN}}$ values per subtype to decrease with later WN subtype. The WC stars show a similar trend for $\overline{R_{\rm WC}}$ values per WC subtype to decrease with later WC subtype up to WC6. For WC6-4 stars $\overline{R_{\rm WC}}$ is about constant, at the value of $\overline{R_{\rm WN5}}$. All WN9-11 stars and WC9 stars are located within the solar circle ($R_\odot = 8$ kpc). We do not observe WC stars beyond $R = 10$ kpc, but WN3-8 stars are present there.

Fig. 3 poses no strict constraints on WR subtype evolution, but shows that evolution from the WN2-3 phase to the WC9 phase is unlikely. Fig. 3 does not encourage WNL → WNE or WCL → WCE subtype evolution (favoured by Moffat 1995). The observed similar $R_{\rm WN}$ and $R_{\rm WC}$ trends do allow WNE → WCE and WNL → WCL subtype evolution.

4. WOLF-RAYET STAR LATITUDE DISTRIBUTION

The *VIIth Catalogue* allows to investigate the WR star separation from the galactic plane $z({\rm pc}) = d({\rm pc})\sin b^{\rm II}$. In Fig. 4 the ($l^{\rm II}$,$z$) distribution shows that out of the galactic plane a fair number of WR stars with large z-distances is visible. Those are worthy of further attention as possible run-away stars. The WR distribution in Fig. 4 is clearly asymmetric with respect to the galactic plane, in the sense that more stars are found at negative latitudes and with a concentration in the fourth quadrant. This aspect of the WR distribution was discussed earlier by van der Hucht *et al.* (1988) and Conti & Vacca (1990), and that of O-B2 stars by Reed (1996, 2000). The asymmetry may be caused by tidal interaction during the prehistoric passage of LMC and SMC (Fujimoto & Sofue 1977).

In Fig. 5 the ($d\cos b^{\rm II}$,z) distribution shows that beyond 6 kpc incompleteness (due to IS extinction in the galactic plane) becomes increasingly dominant. Fig. 5 also shows that WN stars at larger distances are easier observed than WC stars. WC stars are confined to the strip $z = +250,–300$ pc around the galactic plane. WN stars with $d < 6$ kpc are also confined to the strip $z = +250,–300$ pc around the galactic plane, but beyond 6 kpc three WN stars are found at $z > 250$ pc and eight WN stars are found at $z < –300$ pc.

Considering only the 64 WR stars within $d < 2.5$ kpc, we note in Fig. 5 that they are predominantly restricted to the strip $z = +100,–200$ pc with

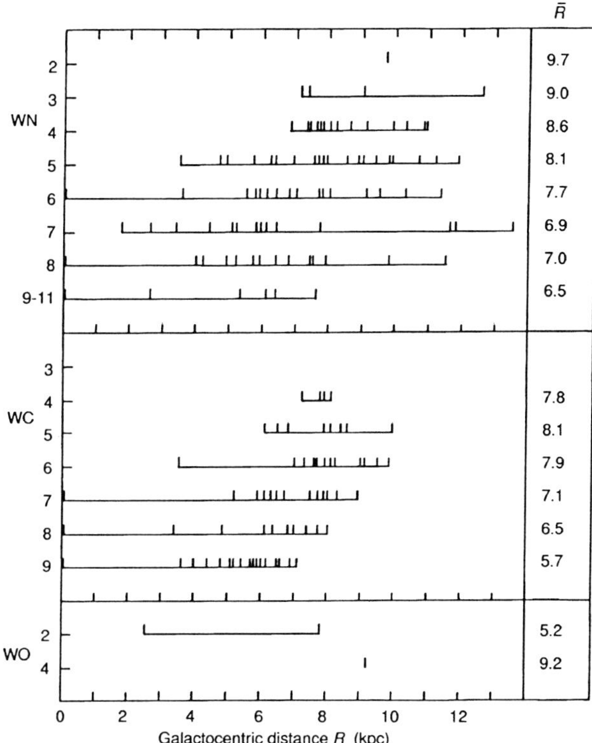

Figure 3 The galactocentric Wolf-Rayet subtype distribution. *Right:* the average $R_{\rm WR}$ per subtype, excluding the 33 WR stars at the Galactic Center. $R_\odot = 8.0$ kpc. From van der Hucht (2001).

a scale height of $h_{\rm WR} \simeq 50$ pc, of the same order of magnitude as that found for O-B5 stars (Reed 2000; Maíz-Apellániz 2001). Four stars have $z < -150$ pc, *i.e.*, WR 6 (WN4), WR 40 (WN8h), WR 57 (WC8) and WR 103 (WC9d+?), and could be runaway candidates.

For the 64 WR stars within $d < 2.5$ kpc we find that $\overline{z} = -17$ pc, which should be representative for the Sun's location *above* the galactic plane. The Sun's location above the galactic plane has been determined at $z_\odot = 20.5 \pm 3.5$ pc by Humphreys & Larsen (1995) from star counts in twelve Palomar Sky Survey fields, at $z_\odot = 10$–12 pc by Reed (1997) for samples of 3600-12000 O-B2 stars, and at $z_\odot = 24.2 \pm 0.4$ pc by Maíz-Apellániz (2001) for 3382 O-B5 stars with parallaxes obtained from the *Hipparcos* catalog.

For the 64 WR stars we find that $\overline{|z|} = 49$ pc (for the 39 presumably single WR stars: 50 pc), in reasonable correspondence with $\overline{|z|} = 45$ pc found for O-type stars by Garmany *et al.* (1982).

5. AN ESTIMATE OF THE TOTAL NUMBER OF WOLF-RAYET STARS IN THE MILKY WAY

Since we can observe roughly only one quadrant of the Milky Way as projected on the galactic plane, it is clear that the minimum total number of galactic WR stars is just four times the number of WR stars in the *VIIth Catalogue*, *i.e.*, $N_{\rm min}^{\rm WR} \simeq 1{,}000$.

Attempts have been made to estimate the total number of WR stars in the Milky Way. *E.g.*, Prantzos & Cassé (1986), by scaling an earlier observed WR density distribution (Hidayat *et al.* 1982) to that of molecular hydrogen (Sanders *et al.* 1984) and adopting that the number ratio of WR stars to O-type stars is metallicity-dependent as $N_{\rm WR}/N_{\rm O} \propto Z^{1.7}$ (Maeder 1984), estimated that $N_{\rm total}^{\rm WR} \simeq 8{,}000$. The present total number of O-B2 stars in the Milky Way has been estimated at $\sim 60{,}000$ (Reed 2000).

Van der Hucht (2001) estimated the present total number of WR stars in the Milky Way, on the basis of the observed galactic WR surface density distribution $\sigma_{\rm WR}(R)$ only. Assuming that the WR inventory in the shell $R = 7$–12 kpc is complete, the data in Fig. 6 imply that $\log \sigma_{\rm WR}(R) = -0.26\,R + 2.58$ and that $N_{\rm total}^{\rm WR} \simeq 6{,}500$, with $\sim 3{,}600$ WR

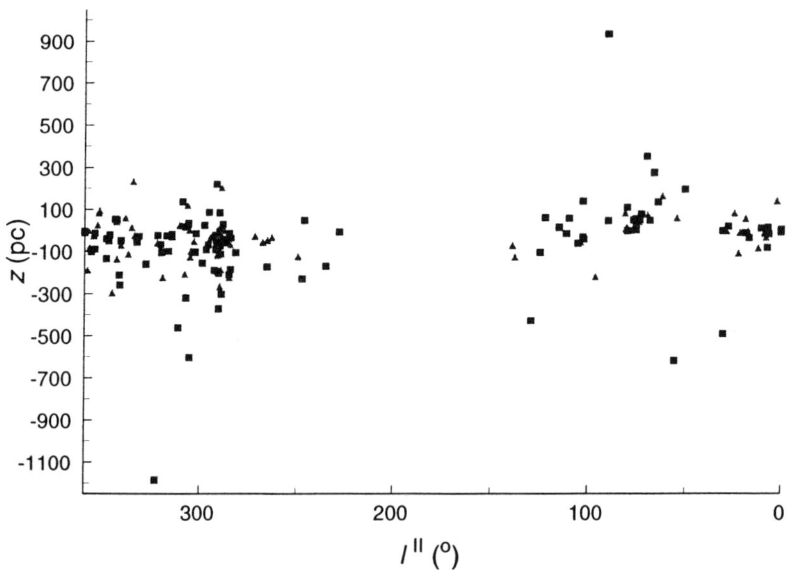

Figure 4 The ($l^{\rm II}$,z) distribution of galactic WR stars. Symbols as in Fig. 2. From van der Hucht (2001).

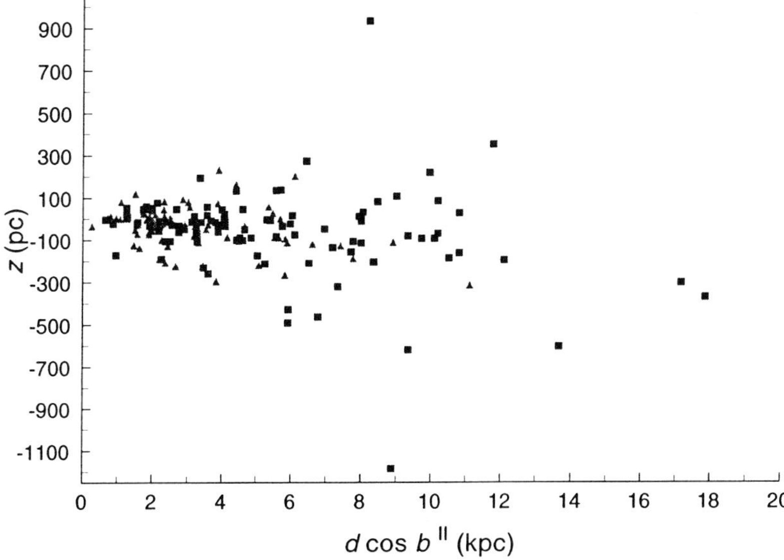

Figure 5 The $(d\cos b^{\rm II},z)$ distribution of galactic WR stars. Symbols as in Fig. 2. From van der Hucht (2001).

stars within 3 kpc from the Galactic Center (where only 37 WR stars have been discovered so far).

Adopting as the average WR star life-time $t_{\rm WR} \simeq 4\times10^5$ yr (Maeder & Meynet 1994), we can expect for the average time-interval $\tau_{\rm WR}$ between two consecutive galactic WR deaths (possibly as supernovae) that $\tau = t_{\rm WR}/N^{\rm WR}_{\rm total} \simeq 60$ yr, or, in the observable quadrant of the Milky Way $\tau \simeq 250$ yr.

6. WOLF-RAYET STARS IN THE GALACTIC CENTER

The Galactic Center (GC) region proves to be an area rich in WR stars. The *VIIth Catalogue* lists within 50 pc from the GC 15 WNL and 11 WCL stars, at near-IR wavelengths discovered by Blum *et al.* (1995); by Krabbe *et al.* (1995) in the Galactic Center Cluster; by Figer *et al.* (1995, 1996, 1999a, 1999b) in the Quintuplet Cluster (AFGL 2004); and by Cotera *et al.* (1999) in the Arches Cluster (near the radio emission region G 0.12+0.02), although some of them are questioned by Paumard *et al.* (2001). Earlier classifications of so-called He I stars have been revised by Crowther (private communication) in the *VIIth Catalogue* into WN9-11 stars. In the meantime, the number of WR stars discovered in the GC region has increased with seven additional WNL stars (Genzel

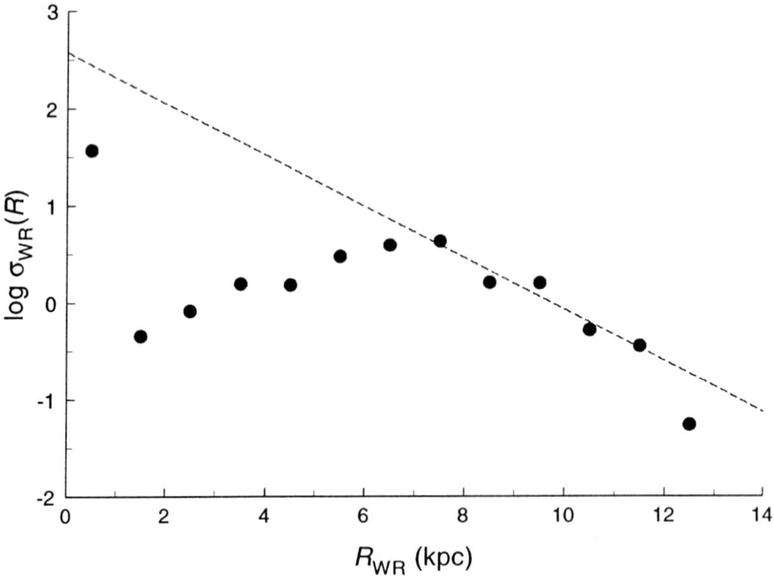

Figure 6 The galactic WR surface density distribution $\sigma_{\mathrm{WR}}(R)$ versus galactocentric distance R_{WR} for the stars in the *VIIth Catalogue*. The dashed line, fitting the five data points with $R_{\mathrm{WR}} = 7$-12 kpc, corresponds to $\log \sigma_{\mathrm{WR}}(R) = -0.26\, R_{\mathrm{WR}} + 2.58$.

et al. 2000: 13S SE, 16CC and MPE−8.3,–5.7; Paumard *et al.* 2001: ID 180, IRS 7E2, HeI N3 and HeI N2). Moreover, in the GC region five additional WCL stars may be present, of which the heated circumstellar dust radiates strongly in the near-IR and masks their WC IR emission line spectra, thus preventing proper spectral classification. Preliminary classifications coin them *cocoon* stars (Moneti *et al.* 1992, 1994, 2001) or *DWCL?* stars (Figer *et al.* 1996, 1999a).

The surface density of the GC WR stars in the $R < 50$ pc volume is $\sigma_{\mathrm{WR}} \simeq 4200\,\mathrm{kpc}^{-2}$; their number ratio $N_{\mathrm{WC}}/N_{\mathrm{WN}} = 0.50$ (or 0.73 if the five DWCL? stars are included).

7. WOLF-RAYET STARS IN LOCAL GROUP GALAXIES

WR stars have been discovered in numerous galaxies, in the Local Group and beyond. Recently, Breysacher *et al.* (1999) presented the *IVth Catalogue of Population I Wolf-Rayet Stars in the Large Magellanic Cloud.* The WR census in other Local Group galaxies has been reviewed

Table 3 Wolf-Rayet stars discovered in Local Group galaxies.

Local Group galaxies	d^a (Mpc)	type[a]	N_{WR}	N_{WC}/N_{WN}	references
Milky Way		Sbc[b]	234	0.6	Hu01,GP00,PM01
LMC	0.052	Ir III-IV	135	0.2	BAT99,MW00
SMC	0.063	Ir IV-V	11	0.1	AB79,MV91,MD01
NGC 6822	0.50	Ir IV-V	4	0.0	AM85,MJ98
IC 10	0.66	Ir IV:	28[c]	2.1	MA95,MJ98,RS01
IC 1613	0.72	Ir V	1		AM85,MJ98
M 31	0.76	Sb I-II	49[d]	0.7	MS87,MJ98,BT99,GW99
M 33	0.79	Sc II-III	141[e]	0.4	AM85,DM93,MJ98,GDP00

Notes: a: van den Bergh 1999; *b*: Russeil, priv. comm.; *c*: and 6 WN candidates; *d*: and 6 WR candidates; *e*: and 20 WN candidates.
References: AB97: Azzopardi & Breysacher 1979; AM85: Armandroff & Massey 1985; BAT99: Breysacher *et al.* 1999; BT99: Bransford *et al.* 1999; DM93: Drissen *et al.* 1993; GDP00: González Delgado & Pérez 2000; GP00: Genzel *et al.* 2000; GW99: Galarza *et al.* 1999; Hu01: van der Hucht 2001; MA95: Massey & Armandroff 1995; MD01: Massey & Duffy 2001; MJ98: Massey & Johnson 1998; MS87: Moffat & Shara 1987; MV91: Morgan *et al.* 1991; MW00: Massey *et al.* 2000; PM01: Paumard *et al.* 2001; RS01: Royer *et al.* 2001.

by Massey & Johnson (1998). Those and other recent up-dates are quoted in Table 3.

Evolutionary studies (*e.g.*, Maeder *et al.* 1980, Maeder 1991, Maeder & Meynet 1994) predict that the metallicity Z, through its effect on mass loss rates $\dot{M}$ of O-type stars and supergiants is responsible for the large differences of the WR populations in galaxies with active star formation. The observed number ratios N_{WR}/N_O, N_{WC}/N_{WR} and N_{WC}/N_{WN} in Local Group galaxies of various metallicities put severe constraints on the dependence of $\dot{M}$ on Z.

Since the galactic O-star census has not been up-dated since Garmany *et al.* (1982), we cannot determine the N_{WR}/N_O ratio in the solar neighbourhoud in a sensible way, unfortunately. However, we can compare N_{WC}/N_{WN} ratios for different Local Group galaxies, as done in Table 3, adapted from Massey & Johnson (1998, their table 5). They found that N_{WC}/N_{WN} in Local Group galaxies correlates with oxygen abundance. We up-date their numbers here in Table 4, confirming the trend and deviations from it found by Massey & Johnson (1998, their fig. 8): while the SMC, LMC, NGC 6822, M 33 and M 31 show a linear relation $N_{WC}/N_{WN} \propto \log(O/H)+12$, the Milky Way and IC 10 are clearly

Table 4 Wolf-Rayet numbers, number ratios N_{WC}/N_{WN} and surface densities σ_{WR} in Local Group galaxies, compared with oxygen abundances and metallicities (adapted from Massey & Johnson 1998, with up-dates).

Local Group galaxies		N_{WR}	N_{WC}/N_{WN}	σ_{WR} (kpc^{-2})	12+log(O/H)	Z^a
Milky Way	$d<3$ kpc	77	1.26	2.7	$8.85^b \pm 0.1$	
- $R=9$-11 kpc	$d<3$ kpc	7	0.75	1.0	$8.75^b \pm 0.2$	0.013
- $R=7$-9 kpc	$d<3$ kpc	31	0.94	2.6	$8.85^b \pm 0.2$	0.020
- $R=5$-7 kpc	$d<3$ kpc	39	1.79	4.3	$8.88^b \pm 0.2$	0.029
- Galactic Center	$R<50$ pc	33	0.50	4200	$8.90^b \pm 0.2$	
LMC		135	0.23	2.6	8.37	0.006
SMC		11	0.10	1.2	8.13	0.002
NGC 6822		4	0.00	0.6	8.25	0.005
IC 10		28	2.11	9.5	8.25	
IC 1613		1	(1 WC)	0.7	7.85	0.002
M 31		49	0.67	0.7	9.00	0.035
M 33		141	0.44	4.1		0.013

References: *a*: Maeder & Meynet 1994; *b*: Smartt *et al.* 2001.

discrepant. The up-dated numbers in Tables 3 and 4 demonstrate that those discrepancies appear to be even larger:
- in the Milky Way within 3 kpc from the Sun, the N_{WC}/N_{WN} ratio is a factor of 2.5 above the trend if 12+log(O/H) = 8.70 as adopted by Massey & Johnson (1998), who concluded that WN stars are under-represented in that volume. Or, if we adopt 12+log(O/H) = 8.85±0.2 (Smartt *et al.* 2001), then the N_{WC}/N_{WN} ratio is a factor of 1.9 above the trend. This would imply that within $d<3$ kpc some 30 galactic WN stars are still hiding;
- in IC 10, the N_{WC}/N_{WN} ratio discrepancy is a factor 20 above the trend, as confirmed by Royer *et al.* (2001), which is probably related to its current status of starburst galaxy.

8. SUMMARY AND CONCLUSIONS

The *VIIth Catalogue of Galactic Wolf-Rayet Stars* lists 227 Population I WR stars, comprising 127 WN stars, 87 WC stars, 10 WN/WC stars and 3 WO stars. Recent new discoveries of seven GC WR stars bring the total number to 234 known galactic WR stars, including 22 WNL and 11 WCL stars within 50 pc of the Galactic Center.

A re-determination of optical photometric distances and the galactic distribution of the WR stars in the *VIIth Catalogue* shows in the volume $d < 3\,\mathrm{kpc}$ a projected surface density $\sigma_{\mathrm{WR}} = 2.7\,\mathrm{kpc}^{-2}$, a number ratio $N_{\mathrm{WC}}/N_{\mathrm{WN}} = 1.3$, and a WR binary frequency of 40 %.

The galactocentric distance (R_{WR}) distribution per subtype shows $\overline{R_{\mathrm{WR}}}$ decreasing with later WR subtype for both the WN and WC sequences. The observed trend is more indicative of WNE $\rightarrow$ WCE and WNL $\rightarrow$ WCL subtype evolution, rather than of WNL $\rightarrow$ WNE and WCL $\rightarrow$ WCE subtype evolution.

Compared with other Local Group galaxies and in relation with metallicity, the Milky Way has within 3 kpc from the Sun a number ratio $N_{\mathrm{WC}}/N_{\mathrm{WN}} = 1.3$, a factor of ~ 2 above the expected trend. This could mean that some 30 galactic WN stars in the $d < 3\,\mathrm{kpc}$ volume are still hiding.

References

Azzopardi, M., Breysacher, J. 1979, *A&A* 75, 120
Armandroff, T.E., Massey, P. 1985, *ApJ* 291, 685
van den Bergh, S. 1999, *The A&A Review* 9, 273
Blum, R.D., Sellgren, K., DePoy, D.L. 1995 *ApJ* (Letters) 440, L17
Bransford, M.A., Thilker, D.A., Walterbos, R.A.M, King, N.L. 1999, *AJ* 118, 1635
Breysacher, J., Azzopardi, M., Testor, G. 1999, *A&AS* 137, 117
Chevalier, R.A., Li, Z.-Y. 2000, *ApJ* 536, 195
Chu, Y.-H. 2001, in: A.F.J. Moffat & N. St-Louis (eds.), Interacting Winds from Massive Stars, Proc. Int. Workshop, Les Îles -de-la-Madeleine (Québec, Canada) 10-14 July 2000, *ASP-CS* submitted
Conti, P.S., Garmany, C.D., de Loore, C., Vanbeveren, D. 1983, *ApJ* 274, 302
Conti, P.S., Vacca, W.D. 1990, *AJ* 100, 431
Cotera, A.S., Simpson, J.P., Erickson, E.F., Colgan, S.W.J., Burton, M.G., Allen, D.A. 1999, *ApJ* 510, 747
Crowther, P.A. 1999, in: K.A. van der Hucht, G. Koenigsberger & P.R.J. Eenens (eds.), Wolf-Rayet Phenomena in Massive Stars and Starburst Galaxies, *Proc. IAU Symp.* No. 193 (San Francisco: ASP), p. 116
Drissen, L., Moffat, A.F.J., Shara, M.M. 1993, *AJ* 105, 1400
Figer, D.F., McLean, I.S., Morris, M. 1995, *ApJ* (Letters) 447, L29
Figer, D.F., Morris, M., McLean, I.S. 1996, in: R. Gredel (ed.), The Galactic Center, *ASP-CS* 102, 263
Figer, D.F., McLean, I.S., Morris, M. 1999, *ApJ* 514, 202

Figer, D.F., Morris, M., Geballe, T.R., Rich, R.M., Serabyn, E., McLean, I.S., Puetter, R.C., Yahil, A. 1999b, *ApJ* 525, 759

Fujimoto, M., Sofue, Y. 1977, *A&A* 61, 199

Galarza, V.C., Walterbos, R.A.M., Braun, R. 1999, *AJ* 118, 2775

Garmany, C.D., Conti, P.S., Chiosi, C. 1982, *ApJ* 263, 777

Genzel, R., Pichon, C., Eckart, A., Gerhard, O.E., Ott, T. 2000, *MNRAS* 317, 348

Georgelin, Y.M., Georgelin, Y.P. 1976, *A&A* 49, 57

González Delgado, R.M., Pérez, E. 2000, *MNRAS* 317, 64

Hidayat, B., Supelli, K., van der Hucht, K.A. 1982, in: C. de Loore & A.J. Willis (eds.), Wolf-Rayet Stars: Observations, Physics, Evolution, *Proc. IAU Symp.* No. 99 (Dordrecht: Reidel), p. 27

van der Hucht, K.A., Conti, P.S., Lundström, I., Stenholm, B. 1981, *SSR* 28, 227 `[VIth Catalogue]`

van der Hucht, K.A., Hidayat, B., Admiranto, A.G., Supelli, K.R., Doom, C. 1988, *A&A* 199, 217

van der Hucht, K.A., Hidayat, B. (eds.) 1991, Wolf-Rayet Stars and Interrelations with Other Massive Stars in Galaxies, *Proc. IAU Symp.* No. 143 (Dordrecht: Kluwer)

van der Hucht, K.A. 1992, *The A&A Review* 4, 123

van der Hucht, K.A. 2001, *New Astronomy Reviews* 45, 135 `[VIIth Catalogue]`

van der Hucht, K.A., Williams, P.M. (eds.) 1995, Wolf-Rayet Stars: Binaries, Colliding Winds, Evolution, *Proc. IAU Symp.* No. 163 (Dordrecht: Kluwer)

van der Hucht, K.A., Koenigsberger, G., Eenens, P.J.R. (eds.) 1999, Wolf-Rayet Phenomena in Massive Stars and Starbursts Galaxies, *Proc. IAU Symp.* No. 193 (San Francisco: ASP)

Humphreys, R.M., Larsen, J.A. 1995, *AJ* 110, 2183

Kaper, L., Cherepashchuk, A. 2001, in: L. Kaper, E.P.J. van den Heuvel & P.A. Woudt (eds.), Black Holes in Binaries and Galactic Nuclei, *Proc. ESO Astrophys. Symp.* (Berlin: Springer), p 277

Krabbe, A., Genzel, R., Eckart, A., Najarro, F., Lutz, D., Cameron, M., Kroker, H., Tacconi-Garman, L.E., Thatte, N., Weitzel, L., Drapatz, S., Geballe, T., Sternberg, A., Kudritzki, R. 1995, *ApJ* (Letters) 447, L95

Langer, N., El Eid, M.F. 1986, *A&A* 167, 265

Langer, N., Woosley, S.E. 1996, in: C. Leitherer, U. Fritze-von Alvensleben, J. Huchra (eds.), From Stars to Galaxies. The Impact of Stellar Physics on Galaxy Evolution, *ASP-CS* 98, 220

Langer, N., Heger, A. 1999, in: K.A. van der Hucht, G. Koenigsberger & P.R.J. Eenens (eds.), Wolf-Rayet Phenomena in Massive Stars and

Starburst Galaxies, *Proc. IAU Symp.* No. 193 (San Francisco: ASP), p. 187
Maeder, A., Lequeux, J., Azzopardi, M. 1980, *A&A* (Letters) 90, L17
Maeder, A. 1984, *Adv. Space Res.* 4, 55
Maeder, A. 1991, *A&A* 242, 93
Maeder, A., Conti, P.S. 1994, *ARAA* 32, 227
Maeder, A. & Meynet, G. 1994, *A&A* 287, 803
Maíz-Apellániz, J. 2001, *AJ* in press (May 2001)
Massey, P., Armandroff, T.E. 1995, *AJ* 109, 2470
Massey, P., Johnson, O. 1998, *ApJ* 505, 793
Massey, P., Waterhouse, E., DeGioia-Eastwood, K. 2000, *AJ* 119, 2214
Massey, P., DeGioia-Eastwood, K., Waterhouse, E. 2001, *AJ* 121, 1050
Massey, P., Duffy, A.S. 2001, *ApJ* in press
McNamara, D.H., Madsen, J.B., Barnes, J., Ericksen, B.F. 2000, *PASP* 112, 202
Moffat, A.F.J., Shara, M.M. 1987, *ApJ* 320, 266
Moffat, A.F.J. 1995, in: K.A. van der Hucht & P.M. Williams (eds.), Wolf-Rayet Stars: Binaries, Colliding Winds, Evolution, *Proc. IAU Symp.* No. 163 (Dordrecht: Kluwer), p. 213
Moneti, A., Glass, I.S., Moorwood, A.F.M. 1992, *MNRAS* 258, 705
Moneti, A., Glass, I.S., Moorwood, A.F.M. 1994, *MNRAS* 268, 194
Moneti, A., Stolovy, S., Blommaert, J.A.D.L., Figer, D.F., Najarro, F. 2001, *A&A* 366, 106
Morgan, D.H., Vassiliadis, E., Dopita, M.A. 1991, *MNRAS* 251, 51P
Paumard, T., Maillard, J.P., Morris, M., Rigaut, F. 2001, *A&A* in press (astro-ph/0011215)
Prantzos, N., Cassé, M. 1986, *ApJ* 307, 324
Ramírez, S.V., Sellgren, K., Carr, J.S., Balachandran, S.C., Blum, R., Terndrup, D.M., Steed, A. 2000, *ApJ* 537, 205
Reed, B.C. 1996, *AJ* 111, 804
Reed, B.C. 1997, *PASP* 109, 1145
Reed, B.C. 2000, *AJ* 120, 314
Reid, P. 1993, *ARAA* 31, 345
Roberts, M.S. 1962, *AJ* 67, 79
Royer, P., Smartt, S.J., Manfroid, J., Vreux, J.-M., 2001, *A&A* 366, L1
Russeil, D. 1998, Étude multispectrale des régions d'hydrogène ionisé dans notre Galaxie, thesis Univ. de Provence, Aix-Marseille I.
Russeil, D. 2001, *A&A* submitted
Sanders, D.B., Solomon, P.M., Scoville, N.Z. 1984, *ApJ* 276, 182
Smartt, S.J., Venn, K.A., Dufton, P.L., Lennon, D.J., Rolleston, W.R.J., Keenan, F.P. 2001, *A&A* 367, 86
Smith, L.F. 1968a, *MNRAS* 138, 109

Smith, L.F. 1968b, *MNRAS* 141, 317
Vreux, J.-M., Detal, A., Fraipont-Caro, D., Gosset, E., Rauw, G. (eds.) 1996, Wolf-Rayet Stars in the Framework of Stellar Evolution, *Proc. 33rd Liège Int. Astroph. Coll.* (Liège: Univ. of Liège)
Wellstein, S., Langer, N. 1999, *A&A* 350, 148
Woosley, S.E., Langer, N., Weaver, T.A. 1993, *ApJ* 411, 823
Wright, A.E., Barlow, M.J. 1975, *MNRAS* 170, 41

SYSTEMATIC SEARCH FOR WOLF-RAYET BINARIES IN THE MAGELLANIC CLOUDS: PRELIMINARY RESULTS

Cédric Foellmi and Anthony F.J. Moffat

Université de Montréal, C.P. 6128 succ. Centre-ville, Montréal (Qc) H3C 3J7, Canada and Observatoire du Mont Mégantic.

foellmi@astro.umontreal.ca, moffat@astro.umontreal.ca

Abstract The Magellanic Clouds represent the best laboratories to study the metallicity effect on the late evolution of massive stars. The low ambient metallicity is expected to lower the mass-loss rate by stellar winds. Therefore, it increases the minimum mass to form Wolf-Rayet (WR) stars. Binarity is then expected to be responsible of the formation of a large fraction of WR stars in the Magellanic Clouds. We present here the rationale behind a systematic search for radial velocity variations in a subsample of WR stars both in the Small Magellanic Cloud (SMC) and the Large Magellanic Cloud (LMC). Some very preliminary results of this study are also presented and discussed.

1. INTRODUCTION

The Magellanic Clouds are most appropriate to study how metallicity (Z) affects the evolution of massive stars. In a low ambient metallicity environment compared to our Galaxy, stellar winds (and then mass-loss) are expected to be much lower. The consequence is a large increase of the minimum initial mass required to form a Wolf-Rayet (WR) star, assuming that only stellar winds are responsible for such a formation. However, the question arises: Can a massive companion in a close binary system enhance the mass-loss and allow the primary to become a WR star? If correct, the binary frequency among the WR star population in the Magellanic Clouds should be higher than that of the Galaxy. Furthermore, the evaluation of the binary frequency in the Magellanic Clouds (MC) could have a strong impact on our understanding of young starbursts.

D. Vanbeveren (ed.), The Influence of Binaries on Stellar Population Studies, 157–163.

2. SCIENTIFIC BACKGROUND

More precisely, this project encompasses two main aspects, each on a different scale. By looking carefully at the WR binary frequency in the SMC and the LMC, this is the first serious empirical attempt to understand how massive binaries should modify our understanding of late stages of stellar evolution, as well as of massive populations. The SMC ($Z \sim Z_{\odot}/10$) is the best case, although it contains only 9 WR stars (Azzopardi & Breysacher 1980; Morgan et al. 1991). Note that between the conference and the publication of these proceedings two additional WR stars have been found in the SMC (Massey & Duffy 2000) and therefore are presently not included in this study. The LMC ($Z \sim Z_{\odot}/3$) provides a better statistical environment, with 135 WR stars (Breysacher et al. 1999; Massey et al. 2000).

1. Small scale: stellar evolution.

One might expect the short-period binary channel to be an important way to form WR stars in the MC via the Roche-lobe overflow (RLOF) mechanism (100% in the SMC and 52% in the LMC; Maeder & Meynet 1994). Although the RLOF has never been observed directly, the evolution of massive stars in binaries with initial periods less than ~1000d is supposed to be much different from that for single stars (see e.g. Vanbeveren et al. 1998). Very long period (> 1000d) massive binaries are not expected to experience RLOF (Vanbeveren et al. 1998). Massive binaries of initial period 200d < P < 1000d are expected to end up as WR + OB stars with P < 200d (Wellstein & Langer 1999).

2. Large scale: stellar populations.

If the binary frequency among the WR population is non-negligible, short-period binaries would play an important role in the global characteristics of a massive population (because they will increase the number of WR stars, Schaerer & Vacca 1998). Young, extragalactic starbursts are expected to show WR characteristics (de Koter et al. 1997) during a much longer time if the binary frequency is large: from c. 3-6 Myr with only single stars, up to 12 Myr with a 30% binary frequency (Schaerer & Vacca 1998). Furthermore, the estimation of the ages of starbursts using WR signatures, might be completely wrong if we do not take into account the role of binaries (van Bever & Vanbeveren 1998).

The age determination and the duration of the phase in which WR stars dominate are strongly correlated with the binary content. A high binary frequency in the Clouds, if revealed, should also influence what we expect about starbursts at high redshifts, most of which are supposed to have a low metallicity. A non-zero fraction of binaries should also be incorporated in any synthesis code describing massive populations.

3. PROJECT

This project is the first to gather repeated spectroscopic data spread over several contiguous weeks (for 3 different epochs spread over 2 years) on a complete sub-population of WR stars in the MC in order to detect short-period binaries (i.e., $P < 200d$, those that should show evidence of an evolutionary interaction between the two stars) as well as medium- to moderately long-period binaries (up to 2 years).

We focus our study only on the WNE stars for the following reasons. In the LMC the binary frequency among the WC stars is already known: 13% (Bartzakos et al. 2000). Most of the WN stars are of subtype WNE (WN2-5, 67 among 110 WN stars). The binary frequency in the WNL (WN6-9) sub-sample has already been partially investigated by Moffat (1989). In addition, WNE stars are expected to show a larger binary frequency than WNL or WC stars (Maeder & Meynet 1994). Because the WNE stars in the 30 Dor region are all of H-rich subtype WN5h and are probably more closely related to WNL stars, we focus our observations only on the 62 remaining mostly H-poor WNE stars. We observed so far a complete half of the sample in the LMC (33 WNE stars), down to the magnitude 15.0. In the SMC only one star is not a WN (AB8), and is a binary (WO4+O4, Moffat et al. 1985). Although there are only 8 WN stars in the SMC and a few of them are already known to belong to a binary (or a multiple system), we observed them all.

We have so far gathered about 550 spectra from CASLEO (Argentina) and SAAO (South Africa), spread over two epochs (1998Nov. and 1999Dec. - 2000Jan.). Each run consisted of 2-3 contiguous weeks in order to cover a time range as long as possible. We recently obtained observing time at the beginning of 2001 at CASLEO and SAAO to carry out a third and last epoch of observation. The spectra are of moderately low resolution ($\sim$ 5 Å), and cover the wavelength range between $\sim$3500 Å and $\sim$7000 Å. We used the strongest emission line (HeIIλ4686) to measure the radial velocity (RV) for each spectrum although other emission lines (if strong enough, like HeIIλ5412) will be used later to confirm the RV behavior. If possible, we will also use absorbtion lines from the companion. The average precision is $\sigma_{obs} \approx$ 30km/s. Spread over three epochs each separated by a year, our observations will also allow a detection of long timescale RV variations.

4. PRELIMINARY RESULTS

Very preliminary results for the SMC are summarized in Table 1, where we indicate the star's name, other ID, the spectral type, the V magnitude, the binary status, a period estimation "by eye" for new binaries and the dispersion around the mean radial velocity from our observations (σ). Table 2 gives statistical information based on the fifth catalogue of population I WR stars in the LMC (134 WR stars, Breysacher et al. 1999), with the addition of the new LMC WR star (of subtype WNE, Massey et al. 2000). The table has been supplemented by the number of WNE stars observed within this project, and the number of stars classified by us binaries or suspected binaries ("Bin" and "Bin?", respectively). New values of these statistics, taken from our study, are indicated in bold characters. All other values are taken directly from the catalogue without care taken to recent spectral reclassification. These tables contain very preliminary results, and do not constitute a definitive determination. The (new) binary status has been assigned first from the examination of RV variations. An empirical threshold for detection of binaries by RV variations based on the accuracy of our observations was fixed to $\sigma \approx$ 40-50 km/s. Any star with RV variations above this limit (and so far taking into account data from only one epoch) is certainly a short-period binary. Below this limit we cannot decide. As a preliminary additional indication of binarity we have also examinated the line profile variations. Indeed, our low spectral resolution allows us only to detect large profile variations which correspond to wind-wind collision effects in a short period binary system (with a period P less than ~200 days). The line profile variations are seen on the top of the line, and do not significantly affect our determination of the radial velocities.

Table 1 SMC WR stars.

Star	Other ID	Spectral Type	V (mag)	P (d)	σ (km/s)
AB1	AV2a	WN3 + O4	15.48	> 24	136
AB2	AV39a	O3If*/WN6-A	14.43	> 28	52
AB3	AV60a	WN3 + O4	14.55	9?	166
AB4	AV81, Sk41	WN6-A	13.43	16?	43
AB5	HD5980	WN4 + O7I	11.88	19.266	66
AB6	Sk108, AV332	WN3 + O6.5I	12.36	6.861	197
AB7	AV336a	WN3 + O7	13.16	19.64	144
AB8	Sk188	WO3 + O4V	12.7	16.63	156
MG9	Morgan et al. (1991)	WN2.5 + O5	15.7	-	77

The two new WNE stars of Massey & Duffy (both WN3+abs) have not been included here. Periods for AB5, AB6, AB7 and AB8 are taken from the literature (even if we recover these values from our data).

Suspected (as opposed to certain) binaries constitute stars where a large but apparently non-coherent RV behavior is visible, with no clear evidence for line profile variations. These stars will deserve our particular attention during the next observing runs. Note that medium- and long-timescale spectral behavior has not been checked yet.

The frequencies given in Table 2 are computed taking into account that only half of the sample of WNE stars has been observed. Two values of the binary frequency in the WNE subsample and, as a consequence, in the whole LMC population, are given. The first value gives the frequency if no binaries are present in the second (and not observed) half of the sample, and the second value gives the frequency if we suppose that the same number of binaries is present in both halves of the sample. We also computed in the same way the binary frequency taking the sum of the binaries and suspected binaries.

5. DISCUSSION

From Table 1 we see that the WR binary frequency in the SMC (obviously excluding for now the two new WR stars, which we have not observed) is $\approx 89 \pm 30\%$, comparable with the frequency expected from the theory (124±34%, see Bartzakos et al. 2000, for details). The binary nature of MG9 remains questionable, within the accuracy of our data. A question mark also remains about the binary status of AB4. Within small number statistics it seems that a binary scenario is necessary for the formation of WR stars in the SMC. Later we will investigate the ages based on cluster membership, to test for the extended age scenario of Schaerer & Vacca (1998).

Table 2 LMC WR stars.

	WNL	WNE(all)	WNE(Obs)	WCL	WCE	total
# Stars	43	67	33	2	23	135
# Bin	6	1	**7**	1	3	**17**
Frequency	14%	1.5%	**11-21%**	-	13%	**12-25%**
# Bin?	1	15	**13**	1	5	**20**
# Bin + # Bin?	7	16	**20**	2	8	**37**
Frequency	16%	24%	**30-60%**	-	35%	**27-54%**

Statistics based on Breysacher et al. (1999). New values are in bold (see text). Stars with "slash" classification have been classified using the "WR part" of their classification, i.e. O3If*/WN6-A means WN6, and then WNL.
Values for WC binaries are corrected from Bartzakos et al. (2000).

From Table 2 it seems that the binary frequency among the population I WR stars in the LMC is much lower than expected from the theory. The only way to attain the 52±14% value expected (see Bartzakos et al. 2000, for details) would be to confirm that all our suspected binaries are

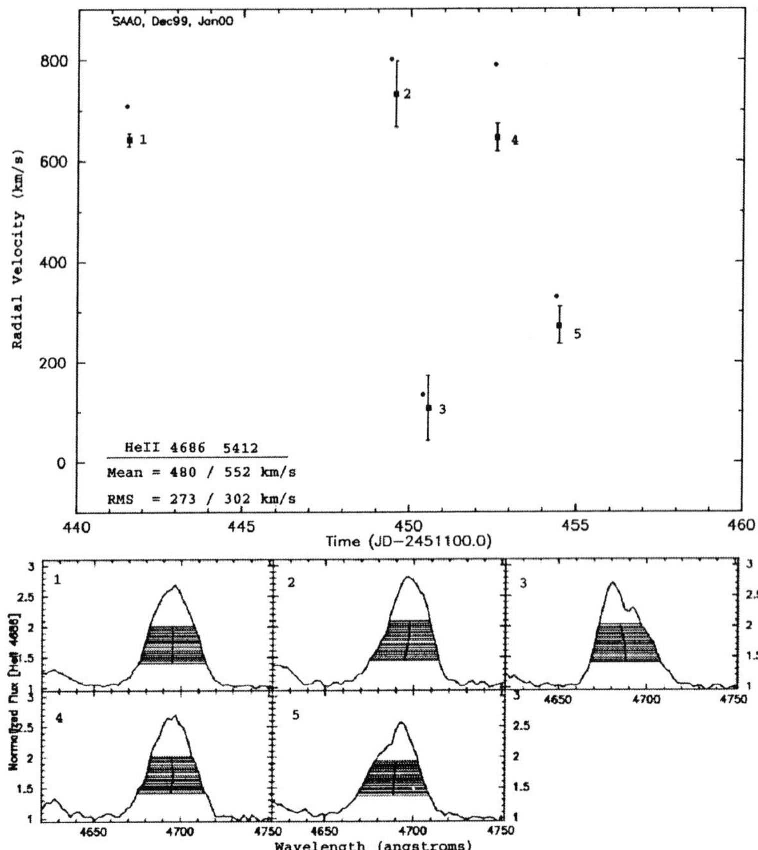

Figure 1 Example: LMC star BAT99-129, of spectral type WN4. Upper figure: radial velocity vs time. Squares indicate RV measured from HeIIλ4686. Circles from HeIIλ5412. Lower figure: line profile of HeIIλ4686 for each observation. The continuum has been normalized to unity. The horizontal lines yield different bisectors, indicated by a wavey vertical line, whose mean position gives the adopted RV. Despite the apparently non-coherent RV behavior, this star has been classified "Bin" because of its exceptionaly large RV variations and clear line profile variations.

in fact true binaries, and then to have the same number of binaries in the half-sample we did not yet observe. The probable binary frequency in the LMC is around 30% (taking into account that the WNL sample has been only partially observed), and thus would be comparable to the frequency in our Galaxy (38%, van der Hucht 2000).

Acknowledgments

C.F. gratefully aknowledges Virpi S. Niemela and Michael M. Shara for their strong support to obtain long observing runs both in CASLEO (Argentina) and SAAO (South Africa). A.F.J.M. acknowledges financial support from NSERC (Canada) and FCAR (Qubec). C.F. was a visiting astronomer, Complejo Astrónomico El Leoncito operated under agreement between the Consejo Nacional de Investigaciones Científicas y Técnicas de la República Argentina and the National Universities of La Plata, Córdoba and San Juan. The autors acknowledge use of the CCD and data acquisition system supported under U.S. National Science Foundation grant AST-90-15827 to R.M. Rich. This paper uses observations made from South African Astronomical Observatory (SAAO).

References

Azzopardi, M. & Breysacher, J., 1980, A&A Suppl. Ser., **39**, 19

Bartzakos, P., Moffat, A. F. J. & Niemela, V. S., 2000, MNRAS, in press

Breysacher, J., Azzopardi, M. & Testor, G., 1999, A&A Suppl. Ser., **137**, 117

de Koter, A., Heap, S. R. & Hubeny, I., 1997, ApJ, **477**, 792

Maeder, A. & Meynet, G., 1994, A&A, **287**, 803

Massey, P. & Duffy, A. S., 2000, ApJ, submitted

Massey, P., Waterhouse, E. & DeGioia-Eastwood, K., 2000, AJ, **119**, 2214

Moffat, Anthony F. J., 1989, ApJ, **347**, 373

Morgan, D. H., Vassiliadis, E. & Dopita, M. A., 1991, MNRAS, **251**, 51P

Schaerer, D. & Vacca, W. D., 1998, ApJ, **497**, 618

van Bever, J. & Vanbeveren, D., 1998, A&A, **334**, 21

Vanbeveren, D., van Rensbergen, W. & de Loore, C., 1998, "The Brightest Binaries", Kluwer Academic

van der Hucht, K. A., 2000, New Astron., in press

Wellstein, S. & Langer, N., 1999, A&A, **350**, 148

AN X-RAY EXAMINATION OF RUNAWAY STARS

E.J.A. Meurs, G. Fennell, L. Norci
Dunsink Observatory, Castleknock, Dublin 15, Ireland
ejam@dunsink.dias.ie

Abstract We have analysed ROSAT X-ray data available for 21 OB runaway stars, in order to search for evidence of accreting compact companions. The 11 detected runaways have 0.1–2.4 keV luminosities of order 10^{32} erg s^{-1}. This is too low for high-mass X-ray binaries, but fully consistent with high-energy emission from the early-type stars. Also the Hardness Ratios indicate stellar sources rather than X-ray binaries. Thus, like radio and optical searches (and a previous X-ray study with Einstein data) not revealing a substantial fraction of binary runaway systems, also the present results suggest ejection from a dense OB cluster rather than Supernova kick velocities as the main mechanism for producing the runaway phenomenon.

1. INTRODUCTION

Runaway stars are recognizable among the OB stars on the basis of large peculiar velocities or also substantial Galactic heights; both these characteristics refer to abnormally large space velocities. For many runaway stars their movement can be traced back to the OB association where they must have been born. Explanations for their large velocities focus on either kick velocities received at the time of a Supernova explosion in a binary system (Zwicky 1957; Blaauw 1961) or slingshot velocities due to close stellar encounters taking place during an early, dense stage in the evolution of OB associations (Poveda et al. 1967).

Supernova explosions occurring in binaries are in many cases not expected to disrupt the binary and they would live further with one component as a Neutron Star. For the cluster ejection mechanism mostly single stars would leave the associations. As a discriminant between these two possible avenues for producing runaway stars, knowledge about the presence of collapsed companions to such objects is obviously of interest.

D. Vanbeveren (ed.), The Influence of Binaries on Stellar Population Studies, 165–171.

2. SEARCHES FOR COLLAPSED COMPANIONS TO RUNAWAY STARS

Three search methods for finding collapsed companions have been employed: optical spectroscopy, X-ray observations and radio measurements. Detailed optical spectroscopy by Gies & Bolton (1986) did not reveal any single-lined binaries among 20 reliable runaway cases. They consider therefore the cluster ejection mechanism as the more likely explanation. An X-ray search by Kumar et al. (1983) resulted in 5 detections out of 9 runaway stars observed, but the X-ray luminosities are too low for X-ray binaries while appropriate for early-type stars. If the level of X-ray emission is due to a collapsed companion, then these Neutron Stars must be orbiting at large distances from the OB stars. Philp et al. (1996) and Sayer et al. (1996) have been looking for pulsars with radio searches that were particularly sensitive to pulsars with long periods in eccentric orbits. In neither of these two radio searches any pulsar emission was detected associated with the runaway targets.

3. NEW X-RAY SEARCH

X-ray observations thus constitute a useful diagnostic for assessing the presence of compact companions to OB runaway stars. The work of Kumar et al. (1983) employed however Einstein data, while since then ROSAT data have become available. The newer ROSAT data cover more runaway targets, provide better fluxes and offer improved positional and spectral information. We have therefore used the ROSAT Public Archive for a new X-ray study of runaway stars.

The objects in the lists of Philp et al. (1996), Kumar et al. (1983), Sayer et al. (1996) and Gies & Bolton (1986) have been checked against available ROSAT fields. In total 74 runaway candidates are included in these lists, ROSAT data are available for 21 of these. Most of the exposures were taken with the Position Sensitive Proportional Counter (PSPC), with imaging and spectral capabilities. Three of the exposures had been obtained with the High Resolution Imager (HRI), which provides the highest spatial resolution but features almost no spectral resolution. In cases with a couple or a few frames, the one with the longest exposure time was examined. All data were reduced and analysed according to standard procedures, using the EXSAS software package (Zimmermann et al. 1998). Detections are pursued down to a Maximum Likelihood level of 6, which corresponds to about 3 gaussian σ.

Table 1 Detected OB runaway stars

HD nr	*Other name*	*Sp. type*	*Exp. time (ksec)*	*Offset (')*	Δ_{XO} *(")*	*Note*
16429		O9.5III	16.615	3.4	8	
24912	ξ Per	O7.5III	1.604	0.1	2	
30614	α Cam	O9.5Ia	3.879	0.2	1	
34078	AE Aur	O9.5V	2.473	0.2	7	
38666	μ Col	O9.5V	3.908	0.3	2	
45995		B2.5V	5.566	18.3	2	
66811	ζ Pup	O4	55.566	0.3	1	
149757	ζ Oph	O9.5V	9.622	0.2	3	HRI
188001	9 Sge	O7.5Ia	7.492	0.1	3	
203064	68 Cyg	O7.5III	10.087	23.6	7	
210839	λ Cep	O6Iab	2.431	0.3	8	
220057		B3IV+B3V	15.630	5.7	21	?

4. ROSAT RESULTS

A list of detected sources is given in Table 1, non-detections are in Table 2. The offset in arcmin in these tables refers to the distance of the runaway target from the fieldcentre. The numbers of X-ray observations (21) and detections (11 or 12, see below) roughly double those resulting from the earlier Einstein study. Most of the detected runaway stars are O stars, for the non-detections the B stars are predominant. The differences between X-ray and optical positions (Δ_{XO}, in arcsec) are mostly of the order of a couple of arcsec; the four cases with 7" or 8" may easily arise from systematic boresight errors. Only the 21" for HD220057 is probably getting too big a difference to be considered consistent in view of the optical and X-ray positional accuracies; the Note column in Table 1 indicates therefore a question mark for this star and this source is left out from the further discussion of results.

Fluxes have been calculated for the detected stars under the assumption of a relatively soft spectrum, a thermal bremsstrahlung spectrum with kT = 1 keV. In order to convert the fluxes to luminosities, the distances to the runaway stars were determined from their Hipparchos parallaxes for distances smaller than 1 kpc and from their absolute magnitudes and reddening otherwise (all have spectral types assigned). The fluxes and luminosities have been determined for the ROSAT range (0.1–2.4 keV). Figure 1 presents the luminosity distribution of the detected

Table 2 Non-detections for OB runaway stars

Name	*Sp. type*	*Exp. time (ksec)*	*Offset (')*	*Note*
HD3950	B1III	5.194	17.3	HRI
HD39680	O6	0.961	11.5	HRI
HD192281	O5	9.310	46.8	
HD235807	B0.5IV	4.960	40.6	
AGK+60 1562	B9III	15.630	27.8	
BD+59 186	B5II	9.670	21.1	
BSD 1117	B5V	9.670	26.3	
BD+60 169	B1V	16.965	54.6	
BD+45 3260	O9V	3.471	37.5	

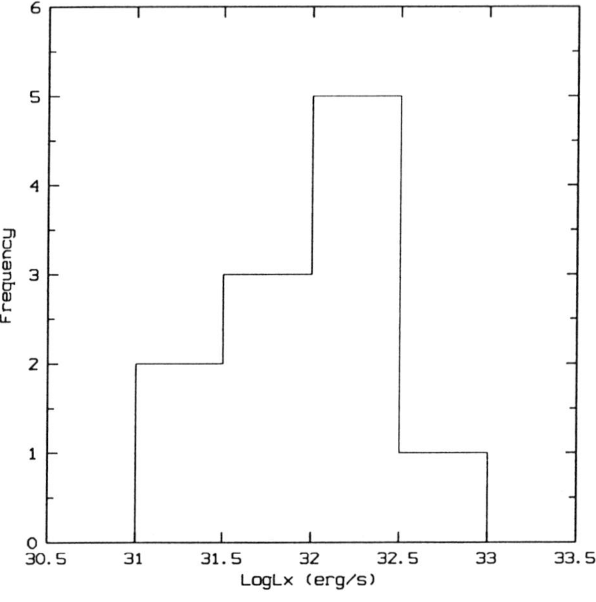

Figure 1 Histogram of $\log(L_X)$ for detected OB runaway stars.

runaway stars. The derived values lie all close together, around 10^{32} erg s^{-1} and are typical of OB stars (see e.g. Figure 4 in Berghöfer et al. 1997). Known X-ray binaries occupy a flux range that starts well above the values determined here.

For most of the detections, the observed number of counts is not sufficient for detailed spectral modelling. The spectral distribution can in such cases be investigated with hardness ratios. These are defined as

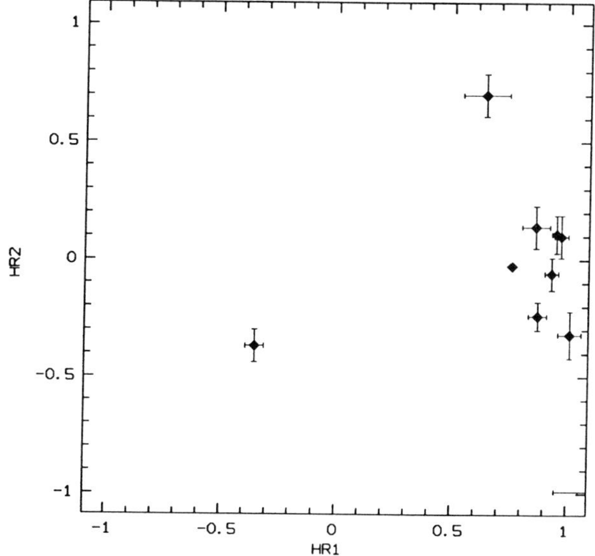

Figure 2 Distribution of HR1 versus HR2 values.

HR_1 = (B-A)/(B+A) and HR_2 = (D-C)/(D+C) where A(11-41), B(52-201), C(52-90) and D(91-201) are counts summed over the pulse height channels quoted in the parentheses, roughly corresponding to energies of 0.11, 0.42, 0.52, 0.91 and 2.02 keV. Hardness ratios for the detections are plotted in the HR1–HR2 diagram in Figure 2 (not including the one HRI detection). Most of them cluster in an area around HR1 $\approx$ 0.9, HR2 $\approx$ 0.0. A comparison with the Hardness Ratios distribution of some broad classes of object (Motch et al. 1998) shows that the runaway data points may indeed be consistent with stars, rather than with X-ray binaries. Although the bulk of the normal stars exhibits lower HR1 values (around 0.1), examination of the OB star data of Berghöfer et al. indicates that the HR1 values of these early type stars actually are concentrated towards HR1 = 1.0, as is also seen in Figure 2. The object at HR1 $\approx$ -0.4, HR2 $\approx$ -0.4 has a very soft HR1 value as could be caused by a White Dwarf companion. The other outlier in Figure 2, at HR1 $\approx$ 0.6, HR2 $\approx$ 0.7 is the only case where a mixture of normal OB star emission and Neutron Star companion could be considered.

In just a few instances the number of counts does allow a spectral model fit. Their spectra were modelled with 1- and 2-temperature fits in order to check that the adopted 1 keV thermal spectra for the flux calculations are justified. Soft spectra are indeed found, with kT values typically ranging from 0.2–1.1 keV.

5. DISCUSSION: COLLAPSED COMPANIONS OR NOT

The ROSAT luminosities and Hardness Ratios thus suggest that the runaway OB stars themselves are the sources of the observed X-ray emission. There is no unambiguous evidence for compact companions.

Any Neutron Star companions would have to orbit at uncommonly large distances from their OB star (Kumar et al. 1983). Proper X-ray binaries exhibit much closer orbits; since it has been argued that all these possess runaway characteristics (Van Oijen 1989), there would in that case have to be a noticeable dichotomy in the binary period distribution of runaway stars.

Yet, the possibility of Neutron Star companions at substantial distances from the OB components illustrates that it is not straightforward to use the searches for compact companions as clear discriminants between the two proposed explanations for the high velocities. In fact, while the searches supply little concrete evidence for Neutron Star companions, the case of the X-ray binaries demonstrates that Supernova kicks do occur and that such binaries can stay together. Other phenomena that may hamper the interpretation of the various search results are the possibility of asymmetric Supernova explosions (often causing disruption of the binary) and mechanisms that prevent the emission of X-ray or radio radiation.

One may expect that detailed examination of individual runaway cases and their origins could be of greater importance for establishing which mechanism is the most relevant one. Since such studies show that actually both mechanisms are at work, cluster ejection and supernova kicks (examples in respectively Gies & Bolton 1986 and Hoogerwerf et al. 2000), it makes more sense to adopt scenarios in which both mechanisms have a natural place: first cluster ejection dominates, during the early dense stage of OB associations, and later on the Supernovae take over. If only one mechanism were relevant for producing OB runaway stars, then an interesting qualitative expectation would be that for the cluster ejection the OB associations should be surrounded by runaway cases at always comparatively large distances, in accordance roughly with the age of the association and the speeds of movement (if the runaway velocities do not vary too much for the ejections), while the Supernova induced runaways could be found at any distance from an association, until even very recently.

References

Berghöfer, T.W., et al. (1997), AA, 322, 167

Blaauw, A. 1961, Bull. Astron. Inst. Netherlands, 15, 265
Gies, D.R., Bolton, C.T. 1986, ApJS 61, 419
Hoogerwerf, R., et al. 2000, ApJL, in press
Kumar, C.K., et al. 1983, ApJ 272, 219
Motch, C., et al. 1998, AAS 132, 341
Van Oijen, J.G.J. 1989, AA 217, 115
Philp, C.J., et al. 1996, AJ 111, 1220
Poveda, A., et al. 1967, Bol. Obs. Tonantzintla y Tacubaya, 4, 86
Sayer, R.W., et al. 1996, ApJ 461, 357
Zimmermann, U., et al. 1998, EXSAS User's Guide (Garching: MPE)
Zwicky, F. 1957, Morphological Astronomy (Berlin: Springer)

HST FGS1R OBSERVATIONS OF 18 RADIO EMITTING WOLF-RAYET STARS

A Search for the Origins of Non-Thermal Radio Emission

Debra J. Wallace
Center for High Angular Resolution Astronomy, Department of Physics and Astronomy, Georgia State University, Atlanta, GA, 30303, U.S.A.
wallace@chara.gsu.edu

Edmund Nelan and Claus Leitherer
Space Telescope Science Institute, Baltimore, MD, 21212, U.S.A.
nelan@stsci.edu; leitherer@stsci.edu

Douglas R. Gies
Center for High Angular Resolution Astronomy, Department of Physics and Astronomy, Georgia State University, Atlanta, GA, 30303, U.S.A.
gies@chara.gsu.edu

Anthony F. J. Moffat
Département de Physique, Université de Montréal, C. P. 6128, Succ Centre-Ville, Montréal, Québec, H3C 3J7, Canada, and Observatoire du mont Mégantic
moffat@ASTRO.UMontreal.CA

Michael M. Shara
Department of Astrophysics, American Museum of Natural History, New York, NY, 10024, U.S.A.
mshara@amnh.org

D. Vanbeveren (ed.), The Influence of Binaries on Stellar Population Studies, 173–180.

Keywords: Wolf-Rayet Stars–Binaries–Non-thermal Radio Emission

Abstract Non-thermal radio emission in Wolf-Rayet (WR) stars is hypothesized to be synchrotron emission from shocks in the wind. For single star models, the shocks arise from instabilities in the WR wind itself. In binary models, the shocks form at the wind-wind interaction zone between a WR star and a massive companion. In Niemela et al. (1998), we used WFPC2 PC imaging to investigate the binary theory. For two WR stars we linked the non-thermal emission with the colliding wind region between binary companions. These observations support the conclusion that non-thermal radio emission arises in a wind collision zone between binary companions outside of the radio photosphere of the WR star.

However, not all WR stars exhibiting non-thermal radio emission are known to have a binary companion. To conclusively link non-thermal emission to binarity, we must demonstrate that all non-thermal emitters are binary, and that all thermal emitters are either single stars or binary systems with separations either too wide or too close to result in a wind-wind interaction that produces shocks.

In an effort to accomplish this goal we used the FGS1R interferometer on board the Hubble Space Telescope (HST) to observe 9 non-thermal emitting WR stars as well as a control sample of 9 thermal emitting WR stars. The FGS can resolve angular separations as small as 0".007 — within the currently accepted size of the radio photosphere for the relatively nearby stars in our sample. We found only one binary among the non-thermal emitting WR stars; WR 48 has a companion at a separation of 40 mas which is probably responsible for colliding winds non-thermal emission. However, the lack of evidence for companions among the other non-thermal emitters casts some doubt on the applicability of the colliding winds model for all non-thermal WR stars.

1. INTRODUCTION

As evolved descendants of massive O stars, most Wolf-Rayet (WR) stars have lost most or all of their H-rich outer envelopes via hot stellar winds. These winds, with terminal velocities in the range $v_\infty \approx$ 1000-3000 km/s (Crowther & Willis 1994), are responsible for the observed spectroscopic properties of WR stars. Their classification, into either WN (with strong nitrogen emission lines in their spectra) or WC (with strong carbon emission lines), is in reality a reflection of the WR wind composition (revealing N-enrichment from CNO burning in WN stars and C-enrichment from the triple-α process in WC stars).

Due to their fast and dense stellar winds, all OB and WR stars have excess thermal free-free IR and radio emission in addition to their photospheric radiation. However, many OB and WR stars have an additional, sometimes variable, non-thermal component of emission (with negative spectral index) that dominates at longer wavelengths (van der Hucht

& Williams 1993). For OB stars the incidence of non-thermal emission has been estimated at 25%, but for WR stars it could be as high as 50% (Chapman et al. 1999). This may be symptomatic of stronger WR winds and thus increased detectability of the non-thermal emission, but there may be other reasons as well.

Two separate models have been proposed to explain the non-thermal emission detected in some WR stars. In the binary star model the non-thermal radio emission arises from the WR wind colliding outside the WR radio photosphere with the wind of a companion O-type star (Williams et al. 1990). Not all WR non-thermal radio emitters have detected O-type star companions though, and a significant fraction of compact companions among WR stars is unlikely (Vanbeveren 1993). In the single WR star model, the emission is proposed to arise via synchrotron radiative processes in the outer (intrinsically unstable) WR wind (White 1985; Abbott et al. 1986).

In HST Cycle 6 (1996) and Cycle 7 (1997), we made WFPC2 observations which allowed us to confirm that the non-thermal emission of two WR stars arises from a region between the components of binary systems — as expected if the non-thermal emission is due to wind collision. However, other non-thermal WR sources that we observed were not resolved in our WFPC2 PC images. This might appear to support the single-star theory of the non-thermal emission, but the snapshot survey could not resolve binaries with projected separations less than about 150 mas (300 AU at a distance of 2 kpc).

2. OBSERVATIONS AND DATA REDUCTION

A Cycle 8 program using HST's FGS1R interferometer executed in 1999 August in an attempt to detect the companions, if any, responsible for the observed non-thermal radio emission seen in some WR stars. The FGS is capable of pushing the minimum detectable distance down to angular separations as small as 0".007. At still closer separations, WR binaries should be detectable by radial velocity observations so, in principle, the FGS1R observations close the gap of detectability. If the binary model for non-thermal emission is correct, we anticipated that most or all of our non-thermal emitters would have companions with separations in the range $0''.01 < d < 3''$. The upper limit comes from the observed offsets of < 3" between the optical and radio positions, while the lower limit comes from the projected size R_ν of the thermal radio photosphere (the companion must be outside the WR radio photosphere to produce observable non-thermal radio emission). Leitherer et al. (1997) derive $R_\nu \approx 10^{14-15}$ cm, or about 10 mas (20 AU) at a distance of 2 kpc (dependent on parameters such as stellar radius and mass loss rate). On the basis of the mass loss rate needed to generate the wind-wind interaction, the likelihood that the initial binary mass ratios are larger than 0.2 (Vanbeveren et al. 1997), stellar observations, and stellar evolution models, the companions are not expected to be fainter than the WR star by more than 4 optical magnitudes. They should therefore be, for the most part, detectable by the FGS (which can resolve favorable systems if the magnitude difference is less than about 3.5 magnitudes).

There are arguably 14 known non-thermal radio emitting WR stars. These include WR 11, WR 16, WR 40, WR 48, WR 90, WR 98, WR 105, WR 112, WR 125, WR 137, WR 140, WR 144, WR 146, and WR 147 (Abbott et al. 1986; Leitherer et al. 1997; Leitherer et al. 1995; Chapman et al. 1999; Dougherty & Williams 2000). WR 14, which we observed, was recently excluded as a non-thermal emitter (Dougherty & Williams 2000) as it was not detected at a 5σ level in the observations by Leitherer et al. (1997). WR 39, which we also observed, was also recently excluded (Dougherty & Williams 2000) as a non-thermal radio source based on uncertainties in the correspondance of the optical and radio positions.

In 1998 September, all of the known non-thermal radio emitting targets were selected (with a few exceptions) from the earlier works of Leitherer et al. (1997), Chapman et al. (1999), and Abbott et al. (1986). WR 112, WR 146, and WR 147 were not selected as they had been previously resolved into binary pairs with the WFPC2. WR 137 was a known long-period spectroscopic binary, and WR 11 was too bright to observe with the FGS. The observed targets, their classifications, their color, their distance, the projected separation for the resolving limit of the FGS1R at that distance, and preliminary conclusions are listed in Table 1.

In addition to the non-thermal radio emitting targets, a control group of thermal emitting WRs, listed in Table 2, was observed to determine if thermal WR stars also have companions. If the binary star model is correct, the absence of non-thermal emission means either that there is no companion, or that the companion is engulfed in the thermal radio envelope and would thus be too close to the WR star to produce observable non-thermal emission (and would not be resolved by the FGS).

All observations were taken in TRANSFER mode. In this configuration, the FGS1R performs a number of scans across the target along two mutually orthogonal directions to obtain the object's interferograms with 1 mas sampling steps. For each target, the individual scans were cross-correlated and co-added using standard STScI software. The high S/N co-added (x,y) interferograms were compared to those from point source

Table 1 Summary of HST FGS1R Observations of Non-Thermal Sources

Star	Filter	Spectral Type[a]	Color (b-v)[a]	Distance[a]	Resolution (projected)	FGS Resolved Companions
WR 14	F583W	WC7	0.15	2.00 kpc	> 20.0 AU	None
WR 39	F583W	WC7	1.54	1.61 kpc	> 16.1 AU	None
WR 48	F5ND	WC6 + O9.5 I	-0.11	2.40 kpc	> 24.0 AU	One
WR 90	F583W	WC7	-0.12	1.97 kpc	> 19.7 AU	None
WR 98	F583W	WN8o/WC7	1.08	5.09 kpc	> 50.9 AU	None
WR 105	F583W	WN9h	1.84	1.58 kpc	> 15.8 AU	None
WR 125	F583W	WC7 + O9 III	1.32	2.13 kpc	> 21.3 AU	None
WR 140	F5ND	WC7 + O4-5	0.27	1.34 kpc	> 13.4 AU	None
WR 144	F583W	WC4	2.01	0.79 kpc	> 7.9 AU	None

[a] values taken from van der Hucht (2000)

Table 2 Summary of HST FGS1R Observations of Thermal Sources

Star	Filter	Spectral Type[a]	Color (b-v)[a]	Distance[a]	Resolution (projected)	FGS Resolved Companions
WR 15	F583W	WC6	0.72	1.54 kpc	> 15.4 AU	None
WR 16	F550W	WN8h	0.30	2.37 kpc	> 23.7 AU	None
WR 22	F5ND	WN7h + O9 III-V	0.03	3.24 kpc	> 32.4 AU	None
WR 24	F5ND	WN6ha	-0.06	3.24 kpc	> 32.4 AU	None
WR 40	F550W	WN8h	0.11	2.26 kpc	> 22.6 AU	None
WR 78	F5ND	WN7h	0.20	1.99 kpc	> 19.9 AU	None
WR 89	F583W	WN8h + OB	1.22	2.88 kpc	> 28.8 AU	None
WR 134	F583W	WN6	0.20	1.74 kpc	> 17.4 AU	None
WR 136	F550W	WN6(h)	0.23	1.26 kpc	> 12.6 AU	None

[a] values taken from van der Hucht (2000)

calibration targets to facilitate the detection of companions. Note, however, that the appropriate calibration targets for some of these WR stars have yet to be observed. Therefore, we have derived preliminary results by using the observed interferograms from the thermal emitters as proxies for point source calibrators. This assumes that the thermal targets are single stars, an assumption which appears reasonable given that none of their interferograms indicated obvious duplicity, and they varied among themselves in a manner consistent with the effects of spectral color (the FGS1R has a mild sensitivity to color). The thermal targets were subdivided by the filter used in the observations, being either the F5ND attenuator or the wide bandpass F583W filter. For the analysis of a given target's interferograms, a calibration template was generated from the F5ND or F583W thermal emitter group by interpolation in $b - v$ color. The "calibration scan" was then compared to the non-thermal WR scan to search for companions.

3. DISCUSSION AND RESULTS

Down to a resolution limit of about 10 mas (a preliminary limit at which we are currently confident of detecting companions), only one Wolf-Rayet star, WR 48 (θ Muscae), was found to have a companion — confirming our previous speckle result (Hartkopf et al. 1999) in which we observed a companion at 46 mas, or a projected separation of about 110.4 AU. Although known to be a spectroscopic binary (Smith 1968) with an $\approx$ 18 day period, these observations confirmed a hypothesis by Moffat & Seggewiss (1977) who speculated that an additional long-period companion exists based on the exceptionally large mass ratios required to explain θ Mus as a binary system. The much higher quality FGS1R observations confirm that θ Mus has a wide component with a separation of 40.5 mas and a position angle of 100.9°. We also find the wide supergiant component to be 0.44 magnitudes (in the FGS filter system) brighter than the combined light of the short period binary.

No other binary systems were discovered, in either the non-thermal or the thermal emitting WR stars.

4. CONCLUSIONS

The results of the WFPC2 observations of WR 140, WR 146, and WR 147, and the ability of FGS1R to resolve stars with separations as small as the estimated size of the WR radio photosphere, led us to anticipate that the majority of the non-thermal emitting systems would show evidence of a companion. But only one non-thermal emitter, WR48, was found to be a binary. (The thermal systems were all found to be single as expected.) These results cast some doubt on the validity of the binary star model as the sole cause for non-thermal radio emission from WR stars. Can some non-thermal radio emitters be single?

The lack of binary detections in these observations does not necessarily establish singularity. For some of the WRs, the companion could be in an eccentric orbit near periastron and thus unresolvable with FGS1R (this could temporarily place the companion within the radio photosphere, causing the non-thermal emission to cease — possibly the case for WR 140 and WR 125; unfortunately, we do not have concurrent radio observations to assess this possibility). Also, the anticipated physical extent of the radio photosphere assumes a smooth and homogeneous wind structure (there is increasing evidence that neither is likely to be true; Moffat & Robert 1994; Grosdidier et al. 1998; Lépine & Moffat 1999), and a mass loss rate on the order of 10^{-5} $\dot{M}$ yr^{-1} (which may be too high). If so, a companion could remain undected outside a less extended radio photosphere. In principle though, these closer companions should be detectable by Doppler spectroscopy. Finally, for angular separations less than about 25 mas, binarity would be undetected by the FGS for $\Delta m > 2.5$ magnitudes, although wider systems would be detected provided $\Delta m < 3.5$ magnitudes.

Although no companion is currently known, WR 90's non-thermal emission is likely variable (Chapman et al. 1999) suggesting a possible close-period binary system. This could also be the case for WR 105 and WR 144. Of the stars not studied herein, WR 11 and WR 137 are known binaries which may account for the production of the observed non-thermal radio emission. Finally, with no companions detected, WR 16, WR 40, WR 90, WR 98, WR 140, and WR 144 have new upper limits on the potential companion separations. Having established these new upper limits, any companions, if present, should be detectable with Doppler spectroscopy.

References

Abbott, D. C., Bieging, J. H., Churchwell, E., & Torres, A. V. 1986, ApJ, 303, 239

Chapman, J. M., Leitherer, C., Koribalski B., Bouter, R., & Storey, M. 1999, ApJ, 518, 890

Crowther, P. A.,& Willis A. J. 1994, in *Evolution of Massive Stars: A Confrontation between Theory and Observations*, ed. D. Vanbeveren, W. van Rensbergen, and C. de Loore (Dordrecht: Kluwer), 85

Dougherty, S. M., & Williams, P. M. 2000, MNRAS, 319, 1005

Grosdidier, Y., Moffat, A. F. J., Joncas, G., & Acker, A. 1998, ApJ, 506, L127

Hartkopf, W. I., Mason, B. D., Gies, D. R., ten Brummelaar, T., McAlister, H. A., Moffat, A. F. J., Shara, M. M., & Wallace, D. J. 1999, AJ, 118, 509

Leitherer, C., Chapman, J. M., & Koribalski, B. 1995, ApJ, 450, 289

Leitherer, C., Chapman, J. M., & Koribalski, B. 1997, ApJ, 481, 898

Lépine, S., & Moffat, A. F. J. 1999, ApJ, 514, 909

Moffat, A. F. J., & Robert, C. 1994, ApJ, 421, 310

Moffat, A. F. J., & Seggewiss, W. 1977, A&A, 54, 607

Niemela, V. S., Shara, M. M., Wallace D. J., Zurek D. R., & Moffat A. F. J. 1998, AJ, 115, 2047

Smith, L. F. 1968, MNRAS, 138, 109

Vanbeveren, D. 1993, in *Evolution of Massive Stars: a Confrontation Between Theory and Observations*, ed. D. Vanbeveren, W. Van Rensbergen, and C. de Loore, Space Science Reviews, 66, 327

Vanbeveren, D., Van Bever, J., & De Donder, E. 1997, A&A, 317, 487

van der Hucht, K. A. 2000, New Astronomy Reviews, *in press*

van der Hucht, K. A., & Williams, P. M. 1993, Advances in Space Research, 13, 561

White, R. L. 1985, ApJ, 289, 698

Williams, P. M., van der Hucht, K. A., Pollock, A. M. T., Florkowski, D. R., van der Woerd, H., & Wamsteker, W. M. 1990, MNRAS, 243, 662

III

STARBURSTS, SUPERNOVAE, STELLAR WINDS

HIGH ENERGY EMISSION FROM STARBURST GALAXIES: THE ROLE OF X-RAY BINARIES

Ian R. Stevens
School of Physics & Astronomy, University of Birmingham, Edgbaston, Birmingham B15 2TT, UK
irs@star.sr.bham.ac.uk

Keywords: galaxies: starburst – galaxies: individual (M82, Hen 2-10, NGC 253, NGC 838) – X-rays: galaxies

Abstract The X-ray emission from starburst galaxies can potentially tell us a great deal about populations of X-ray binary systems, particularly the massive systems. With the recent launch of the *Chandra* and *XMM* X-ray satellites, with their enhanced capabilities compared to earlier generations of X-ray missions, new opportunities are opening up. In this paper I review recent X-ray results on a number of starburst galaxies, with an emphasis on the point source population, and review what they can tell us about the binary populations in starburst populations.

1. INTRODUCTION

High mass X-ray binary systems (HMXBs) are the product of the evolution of massive stars. Understanding the observed populations of such systems provides an important constraint on stellar and binary evolution. Although we can study binary populations in our own Galaxy and nearby objects such as the Magellanic Clouds, many X-ray binaries are sufficiently luminous to be detected (at X-ray wavelengths at least) in nearby galaxies. In this paper I shall concentrate on the evidence for populations of X-ray binary systems in starburst galaxies.

Starburst galaxies are key astrophysical objects for a whole host of reasons. They are the seats of rapid, ongoing star-formation and if the star-formation is of sufficient intensity (a star formation rate per unit area of $\Sigma_* \geq 0.1\, M_\odot\ \mathrm{yr}^{-1}\ \mathrm{kpc}^{-2}$; Heckman 2000) then the galaxy will be able to drive a superwind. Over a Hubble timescale these superwinds will heat the IGM and pollute it with with metals (Ponman et al. 1999; Nath & Trentham 1997). However, here we shall not concentrate on the

D. Vanbeveren (ed.), The Influence of Binaries on Stellar Population Studies, 183–198.

superwind emission but rather the stellar point-like emission from the disk of the galaxy. We shall discuss objects ranging from small dwarf starbursts up to large spiral disk starbursts. Observationally, starburst galaxies are prominent and interesting X-ray sources. Their X-ray emission comprises both extended superwind emission and point-like emission predominantly from the disk of the galaxy.

This year is a particularly propitious time to be considering the subject, as it is in the the process of being revolutionised by the two new major observatories, *XMM* and *Chandra*.

1.1. THE NEW X-RAY SATELLITES

Prior to 1999 the best view of the X-ray properties of starburst galaxies was provided by the *ROSAT* and *ASCA* satellites. These missions were quite complimentary, with *ROSAT* having good spatial resolution (but moderate spectral resolution) and *ASCA* having good spectral resolution (albeit with poor spatial resolution). These two missions have now been superseded by two much larger missions; the US led *Chandra* and the European led *XMM*. In this paper I will focus on the profound contributions that these two, also complimentary, missions are making on the subject of starbursts.

The US-led *Chandra* satellite, launched in July 1999, excels in spatial resolution, with $0.5'' - 1''$ resolution (Weisskopf et al. 1996). The *Chandra* satellite has several instruments, including the ACIS (Advanced CCD Imaging Spectrometer) cameras (ACIS-I and ACIS-S), which have good spatial resolution ($\sim 1''$) and reasonable spectral resolution. The HRC (High Resolution Camera), has limited spectral capability but the highest spatial resolution ($\sim 0.5''$). *Chandra* also has grating instruments, the LETG and HETG (low and high energy transmission gratings) which have the highest spectral resolution for point sources.

The European-led *XMM* satellite was launched in December 1999, and has unrivalled collecting area (up to 6500 cm^{-2}, compared to $\sim$ 1000 cm^{-2} for *ROSAT*), which combined with its broad energy range, reasonable spatial resolution and grating instruments offers excellent prospects in a range of fields (Jansen et al. 2001). The *XMM* satellite has three main instruments, the EPIC (European Photon Imaging Camera) instrument, a broad-band CCD instrument (which comes in two versions, the EPIC-PN and EPIC-MOS) which is capable of spectral-imaging, the RGS (Reflection Grating Spectrograph) which has excellent spectral resolution for point sources, and the Optical Monitor, which provides simultaneous optical images in a variety of filters.

This paper is organised as follows: in Section 2 I review the observational characteristics and origins of X-ray emission from starburst galaxies. In Section 3 I discuss case studies of four important starburst galaxies (M82, Henize 2-10, NGC 838 and NGC 253) that highlight recent results from both *XMM* and *Chandra*. In Section 4, I critically discuss the role of binaries, extrapolating from the binary population in our own Galaxy and in Section 5, I outline what I see as the way ahead in this subject.

2. X-RAY PROPERTIES OF STARBURSTS

The X-ray emission from starburst galaxies can be loosely categorised as either being due to diffuse emission or point-like emission. Of course at lower resolution there is the substantial possibility of confusion between multiple unresolved point sources and genuinely diffuse emission.

2.1. EXTENDED X-RAY EMISSION

Examples of extended emission extending well beyond the plane of the galaxy can be best seen in the following starburst galaxies: M82 (Strickland et al. 1997), NGC 253 (Pietsch et al. 2000), NGC 1569 (Heckman et al. 1995) and several others such as NGC 3079 (Pietsch et al 1998), NGC 3256 (Moran et al 1999). Many other galaxies have diffuse emission within the plane of the galaxy, for example M83 (Immler et al. 1999). This origin of the observed extended emission could be due to a range of physical processes and phenomena. Some possible origins include:

1 A superwind extending beyond the plane of the galaxy, driven by the combined energy and mass inputted by stellar winds and supernovae in the starburst (Strickland et al. 1997)

2 Superbubbles and HII regions. A localised burst of star-formation will inflate large superbubbles of X-ray emitting gas (Strickland & Stevens 1999). If the star-formation is strong enough, eventually these bubbles will merge to form an outflow.

3 Inverse Compton emission (Moran & Lehnert 1997). Starburst galaxies have a large population of IR photons and accelerated electrons due to the large population of supernova remnants (SNR, for instance, M82, Wills et al. 1997; Kronberg et al. 2000). The interaction of these could give rise to diffuse inverse Compton emission.

4 Unresolved stellar populations of both low and high mass stars (including binary systems). While both low and high mass single

stars are X-ray emitters and also occur in large numbers, the integrated X-ray luminosities of single stars will be dominated by X-ray binaries.

2.2. POINT SOURCE X-RAY EMISSION

Many starburst galaxies show multiple X-ray point sources in and around the galaxy disks. Perhaps the most striking study of a starburst galaxy in this regard, prior to *Chandra*, is the work on NGC 253 by Vogler & Pietsch (1999) and Pietsch et al. (2000). High spatial resolution observations with the *Chandra* satellite are currently rewriting our understanding of the X-ray binary populations in nearby starbursts and this shall be reported on in more detail later. The X-ray point sources in starburst galaxies are due to one of the following:

1 Massive black-holes ($M_{BH} \geq 10^6 M_\odot$) in the center of the galaxy. Many starbursts may harbour an AGN, being fuelled by starburst activity (one example is NGC 1365, a spiral galaxy with a central starburst, but where there is also X-ray evidence of an AGN; Stevens et al. 1999; Iyomoto et al. 1997). We shall also discuss the possibility of X-ray emission from intermediate mass black-holes ($M_{BH} \sim 10^2 - 10^3 M_\odot$), that do not necessarily lie at the center of the galaxy.

2 HII regions. Objects such as 30 Doradus will appear as moderately bright X-ray point sources in external galaxies. Examples are probably seen in NGC 5253 (Strickland & Stevens 1999). The matter is somewhat complicated in that HII regions are associated with massive stars, but so are Type II supernova and HMXBs.

3 Supernovae (SN) and SNR: here there are two possibilities, very young X-ray luminous SN, such as SN1988Z (Fabian & Terlevich 1996) or SN1986J (Houck et al. 1998) which can be very X-ray bright (with $L_X > 10^{40}$ erg s^{-1}) and older SN remnants which are often less luminous but could still be detected in external galaxies.

4 X-ray binaries: both HMXBs (associated with the starburst) and low mass systems (LMXBs, associated with the underlying stellar population). The distribution and characteristics of the LMXB population in galaxies is beginning to be investigated with *Chandra* observations of low-luminosity elliptical galaxies, such as NGC 4697 (Sarazin et al. 2000). Because of the lack of recent star-formation, in such systems we do not expect to see HMXBs. Interestingly the observations of Sarazin et al. (2000) do show a source population dominated by LMXBs, but also with evidence of systems

containing massive black-holes ($\sim 20M_\odot$), which are presumably descendants of massive stars and earlier star-bursting epochs.

2.3. X-RAY SPECTRAL CHARACTERISTICS

It has proved to be useful to divide up the X-ray spectra of starburst galaxies into soft and hard wavebands, with the dividing line at around 2 keV. However, we note that the fitting of what will be the intrinsically very complex X-ray spectra of starburst galaxies (with multiple thermal and non-thermal components) with relatively simple (two or three component) models can be a little problematic and lead to ambiguous results (Strickland & Stevens 1998; Dahlem et al. 2000).

The softer spectral component, with energies < 2 keV, will probably be dominated by thermal emission from hot gas associated with the starburst (which will have fitted X-ray temperatures of $kT \sim 0.3 - 1.0$ keV, for a single temperature Raymond-Smith plasma model, see for example Strickland et al. 1997). There will also be substantial contributions from point source emission (assuming we cannot spatially resolve the inner regions of the galaxy). Observationally this soft component is often extended.

The harder emission, photon energies > 2 keV on the other hand, should be dominated by non-thermal sources such as X-ray binaries. The harder X-ray emission is often more point like that the softer emission and is often variable. The spectra for the harder components can be fitted either by a powerlaw, with $\Gamma \sim 2$, or a thermal spectrum with $kT \sim$ a few keV. It has often been assumed that the hard component is due (in part at least) to a central AGN (Tsuru et al. 1997). As we shall see, this harder component is potentially very useful in understanding the binary population in starburst galaxies.

3. CASE STUDIES

3.1. M82: PROTOTYPICAL STARBURST

M82 is perhaps the prototypical starburst galaxy. It is a nearby (3.6Mpc), nearly edge on starburst galaxy, classified as a dwarf irregular galaxy (dIrr). The starburst in M82 was probably triggered by a past interaction with M81 around $\sim$ 200 Myr ago (Yun et al. 1993). M82 is an extremely bright IR source, with $L_{IR} \sim 3 \times 10^{10} L_\odot$. At radio wavelengths M82 is a complex source, with diffuse emission through the body of the galaxy and many radio point sources in the central starburst regions, which are believed to be a population of young SNR (Wills et al.

1997; Kronberg et al. 2000). The SNR lie in a region of $\sim$ 250pc of the galaxy center, and the inferred SN rate is of the order of 0.1 yr^{-1}.

Optical observations reveal the presence of more than 100 super star clusters visible (O'Connell et al 1995). Optical spectral analysis of the stellar populations in M82 suggests a complex starburst history, with some clusters having an age of $\sim$ 10Myr (Satyapal 1997), but some clusters being older $\sim$ 60Myr (Gallagher & Smith 1998).

X-ray observations of M82 with the *ROSAT* satellite showed that the galaxy has a very extended superwind, extending for $5 - 10$ kpc above the plane of the galaxy (Heckman et al. 1990; Strickland et al. 1997). Hα emission is also seen from the superwind and is also very extended, with a ridge of emission nearly 12kpc above the plane of the galaxy (see Lehnert, Heckman & Weaver 1999). Spectral fits, with single temperature thermal models, to the extended emission showed that the superwind had a temperature of $kT \sim 0.6 - 0.3\,\mathrm{keV}$, with a temperature gradient, with cooler material further away from the galaxy. Also seen in *ROSAT* and *ASCA* observations of M82 was variable, point-like X-ray emission from the general vicinity of the galaxy centre (Tsuru et al. 1997). The analysis of *ASCA* observations by Ptak & Griffiths 1999). The showed remarkable variability with one episode, lasting several weeks, that looked sinusoidal in form. The total X-ray luminosity of M82 is $L_X \sim 10^{41}\,\mathrm{erg\ s^{-1}}$, split between the extended superwind and the central point-like emission.

The new *Chandra* observations of M82 are painting a rather more complex picture, particularly of the point source emission in the central regions of M82. Results have been published by Matsumoto et al. (2000) and Kaaret et al. (2000). It should be noted that the 600s periodicity claimed for one of the sources in M82 by Kaaret et al. (2000) turned out to be spurious.

Chandra has found nine point sources in the central $1' \times 1'$ region of M82 (corresponding to 1kpc $\times$ 1kpc, see Fig.1). Around half of these sources show significant variability on a timescale of 3 months (which is the time between two *Chandra* pointings). The brightest source is very variable (by a factor of 7) and, very importantly, is **NOT** located at dynamical center of M82. The X-ray luminosity of this source is in the range of $L_X \sim 1.2 \times 10^{40} - 8.8 \times 10^{40}\,\mathrm{erg\ s^{-1}}$. Given this X-ray luminosity, this source is clearly too massive to be a normal neutron star X-ray binary (unless highly beamed). The variability probably rules out a young supernova origin. Consequently, these authors have concluded that this source is probably an intermediate mass black-hole with a mass $M \geq 400 M_{\odot}$. While on one hand the source could be a more massive black-hole radiating at well below the Eddington limit, the fact that it

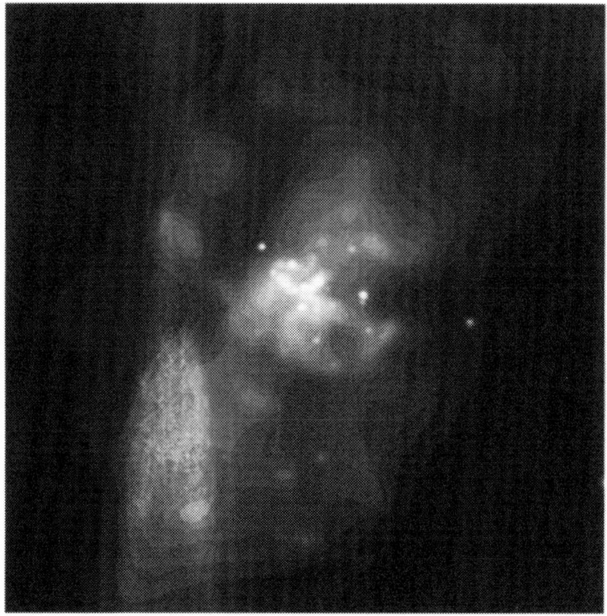

Figure 1 The *Chandra* X-ray image of M82, showing both the diffuse emission, associated with the starburst superwind, and the point source emission, mostly associated with HMXBs.

is not at the dynamical center means that it cannot be a dominant mass feature in the galaxy. This means that it is *NOT* an AGN.

For the other X-ray point sources seen with *Chandra* the X-ray luminosities are in the range of $L_X \sim 3 \times 10^{38} - 2.5 \times 10^{39}\,\mathrm{erg\ s^{-1}}$ and includes a transient source. Some of these X-ray sources are co-located with radio SNR, but most are not and are probably HMXBs.

3.2. HENIZE 2-10: A WOLF-RAYET GALAXY

Henize 2-10 is a nearby (9Mpc), dwarf starburst galaxy. It is also classified as a blue compact dwarf galaxy and a Wolf-Rayet galaxy. *HST* imaging of Hen 2-10 shows multiple super-star clusters, with masses in the range of $10^5 - 10^6 M_\odot$ (Conti & Vacca 1994; Johnson et al 2000). The number of Wolf-Rayet stars (of both varieties) in Hen 2-10 has been estimated to be $N_{WN} \sim 1000$ (for the WN subtype) and $N_{WC} > 250$ (for the WC subtype).

Hα observations of Hen 2-10 have revealed the presence of an outflow, with $V \sim 75 - 250\,\mathrm{km\,s^{-1}}$ (Méndez et al. 1999). Radio observations suggest a large number (~ 3500) of SNR (Méndez et al. 1999), but also the presence of optical thick emission from dense, highly obscured and very young ($<$ 1Myr) star clusters (Kobulnicky & Johnson 1999).

X-ray observations of Hen 2-10 with *ROSAT* show a soft, somewhat extended, X-ray source with $kT = 0.4\,\mathrm{keV}$ and $N_H = 3 \times 10^{21}\mathrm{cm}^{-2}$ (Stevens & Strickland 1998; Méndez et al. 1999). *ASCA* observations, with a higher bandwidth but with limited spatial resolution shows a hard tail to the X-ray emission, with the luminosity of this harder tail being $L_X(2.0 - 10\,\mathrm{keV}) = 1.2 \times 10^{39}\,\mathrm{erg\ s}^{-1}$ (Ohyama & Taniguchi 1999). The origin of this tail is that it is likely due to a population of X-ray binaries. Given the luminosity it could be due to several objects or one very luminous object. Importantly, Hen 2-10 has been accepted as a *Chandra* target, which should resolve (literally) the issue.

3.3. NGC 838: A STARBURST IN A GROUP

NGC 838 is a starburst galaxy within a Hickson Compact Group (HCG 16), and is an ongoing galaxy merger with strong starburst activity. As reported by Mendes de Oliveira et al. (1998) the galaxy shows a double optical core and multiple velocity components. The assumed distance to NGC 838 is $\sim$ 80Mpc, making it much more distant than the other galaxies considered in this paper (with a corresponding lack of spatial resolution). The *XMM* results for HCG 16 have now been published by Turner et al. (2001). While other galaxies in HCG 16 show spectral evidence of an obscured AGN, namely a highly obscured, power-law component, for NGC 838 the emission is believed to be thermal in nature. The disc of the galaxy is resolved in X-rays but there is no evidence of a point-like core indicative of an AGN. For NGC 838, the *XMM* data suggests a 2 component spectrum, with a soft component, with a fitted temperature of $kT = 0.6\,\mathrm{keV}$ and a luminosity of $L_{soft}(0.2 - 10\,\mathrm{keV}) = 4 \times 10^{40}\,\mathrm{erg\ s}^{-1}$ (see Fig. 2). In addition, for the soft component the metallicity is likely high, with $Z \sim 2 - 3Z_\odot$. The hard component in the spectrum can be fitted with a thermal model with $kT = 3.2\,\mathrm{keV}$, and has a luminosity of $L_{hard}(0.2 - 10\,\mathrm{keV}) = 8 \times 10^{40}\,\mathrm{erg\ s}^{-1}$. For this component, the fitted abundance is low, with $Z < 0.3Z_\odot$.

The soft component in the spectrum of NGC 838 is probably due to thermal emission associated with starburst and the hard component probably due to a population of HMXBs. As a general comment, the X-ray spectral characteristics of NGC 838 are similar to that for Hen 2-10, but the source is much more luminous.

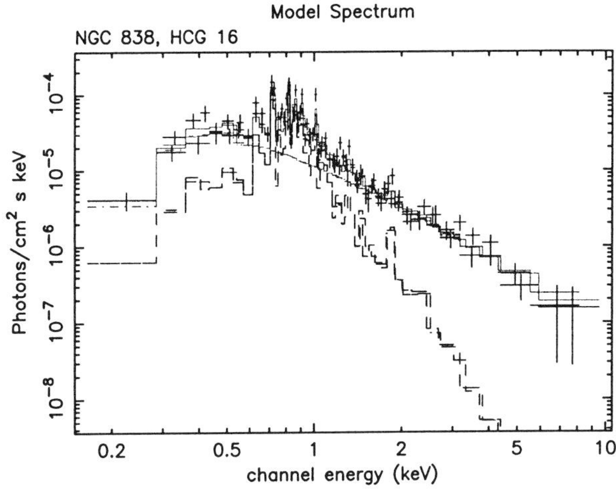

Figure 2 The X-ray spectrum of NGC 838 taken with the *XMM*-MOS instrument, showing the best-fit two component thermal model.

3.4. NGC 253: ANOTHER PROTOTYPICAL STARBURST

NGC 253 is a nearby (2.5Mpc), nearly edge-on spiral galaxy, of type Sc. It is a very vigorous nuclear starburst, with extended superwind. Along with M82 it is usually cited as the best example of a superwind, though in some respects NGC 253 represents a more typical starburst than M82, it being a normal L_* spiral galaxy, rather than a peculiar irregular galaxy.

ROSAT observations of NGC 253 have revealed a large number of point sources and an extended superwind (Read et al 1997; Vogler & Pietsch 1999, Pietsch et al. 2000). In total these authors found 73 X-ray sources in the NGC 253 field (as seen with *ROSAT*), of which 32 are probably associated with disk of galaxy. The central X-ray source has $L_X \sim 10^{39}\,\mathrm{erg\ s^{-1}}$ and is spatially extended. Other sources in NGC 253 have luminosities in the range of $7 \times 10^{36}\,\mathrm{erg\ s^{-1}} < L_X < 3 \times 10^{38}\,\mathrm{erg\ s^{-1}}$. A total of 13 sources are clearly variable, and of these all have $L_X > 3 \times 10^{37}\,\mathrm{erg\ s^{-1}}$ (though this is in part a selection effect). These sources are probably HMXBs close to Eddington limit. *ASCA* observations have revealed a number of emission lines (Ptak et al. 1997).

NGC 253 has been observed by both the *Chandra* (Strickland et al. 2000; Weaver et al. 2000) and *XMM* (Pietsch et al. 2001) satellites. The *Chandra* observations of Strickland et al. (2000) concentrate on the diffuse emission associated with the outflow from the starburst region.

Analysis of the point source characteristics will be presented in a later paper. The *XMM* observations detected a bright transient source, and monitored the variability of the brightest X-ray point source, which they conclude is a black-hole binary system.

4. THE EXPECTED NUMBER OF LUMINOUS HMXBS IN STARBURSTS

4.1. GALACTIC EXTRAPOLATIONS

Based on the massive X-ray binary population in our Galaxy we can make some very broad-brush extrapolations for the sort of HMXB populations we might expect in starburst galaxies, and then compare them with the observed X-ray properties of starburst galaxies.

The basic assumptions underlying these extrapolations are as follows.

1 Star-formation in the Galaxy is reasonably constant over a long period of time (see for instance Basu & Rana 1992)

2 The population of HMXBs in the Galaxy is reasonably constant

3 The local region ($r < 3.5$kpc) is representative of the whole Galaxy, and in turn is representative of a normal population of HMXBs.

Of course this approach is not with problems. For instance, one potential problem concerns the affect of metallicity on HMXB statistics. We know that in the LMC there does appear to be a different distribution of luminosities as compared to the galaxy, which in turn indicates that any Galactic population may not be very applicable to lower metallicity regions.

Based on the results of Miller & Scalo (1979, see also Dalton & Sarazin 1995) we estimate the following massive star-formation rate for the galaxy.

$$\mathrm{SFR}(M_i > 10M_\odot) \sim 0.1 - 0.2\, M_\odot\ \mathrm{yr}^{-1}$$

Now, if we consider only very luminous HMXBs, namely those with $L_X \geq 10^{37}\,\mathrm{erg\ s^{-1}}$, then we find that the Galactic HMXB population, for systems with $L_X \geq 10^{37}\,\mathrm{erg\ s^{-1}}$ is 27 ± 19 with a mean luminosity of $< L_X > \sim 5 \times 10^{37}\,\mathrm{erg\ s^{-1}}$ (Dalton & Sarazin 1995). This number is based on the number of such systems within a radios of 3.5kpc and extrapolated throughout the galaxy. Of course, the number extrapolated from is rather small (namely 4).

4.2. FOR STARBURST GALAXIES

If we assume that the timescale for ongoing star-formation in our starburst galaxies is greater than the formation timescale of HMXB then we can assume that a rough, quasi-steady state in the number of HMXB will hold. This assumption is questionable, in that star-formation in starbursts does not proceed in a simple way (even in such small systems as Hen 2-10; Conti & Vacca 1994), and is probably usually a series of sharp bursts of star-formation. From the work of Kennicutt (1998) we can estimate the star-formation rate in a starburst from:

$$\mathrm{SFR}(\,M_\odot\ \mathrm{yr}^{-1}\,) = 4.5 \times 10^{-44} L_{FIR}(\,\mathrm{erg\ s}^{-1}\,)$$

where the mass range for the star-formation rate is assumed to be $0.1 - 100M_\odot$. L_{FIR} is the far infrared luminosity. Then, assuming a Salpeter initial mass function (IMF), the high mass ($M_i > 10M_\odot$) star-formation rate is

$$\mathrm{SFR}(M_i > 10M_\odot) = 0.1 \times \mathrm{SFR}(0.1 - 100M_\odot)$$

Consequently, we can estimate the massive star-formation rate for our starburst galaxies, compare them with Galactic values to obtain an estimate of the number of HMXBs.

4.3. THE CASE STUDIES

Based on the previous section, we can convert the observed far infrared luminosities of the galaxy into star-formation rates and hence numbers of HMXBs and then compare the extrapolated numbers of HMXBs with the numbers of point sources actually observed with *Chandra* of the level of hard X-ray emission.

M82. From the FIR luminosity quoted in Read et al. (1997) we have a star-formation rate $\mathrm{SFR}(0.1 - 100M_\odot) \sim 3\,M_\odot\ \mathrm{yr}^{-1}$. Based on this we therefore expect a massive binary population of $N_{HMXB}(L_X \geq 10^{37}\,\mathrm{erg\ s}^{-1}\,) = 50 - 100$.

This number is probably rather more than is actually seen in M82 with *Chandra* (Matsumoto et al. 2000), but only by a factor of 2 or so. In addition, heavy absorption is a problem in the observations of M82 and this probably means that the observed population of X-ray sources is not complete down to a level of $10^{37}\,\mathrm{erg\ s}^{-1}$.

Hen 2-10:. Hen 2-10 has a FIR luminosity of $L_{FIR} \sim 10^{43}\,\mathrm{erg\ s}^{-1}$ (Stevens & Strickland 1998) implying a star-formation rate of $\mathrm{SFR}(0.1 - 100M_\odot) = 0.4\,M_\odot\ \mathrm{yr}^{-1}$. Based on this we expect a total population of

$N_{HMXB}(L_X \geq 10^{37}\,\mathrm{erg\ s^{-1}}) = 5 - 10$. This expected X-ray binary population is, in turn, of a reasonable level to explain the observed hard component of $L_X \sim 10^{39}\,\mathrm{erg\ s^{-1}}$. However, *Chandra* observations are needed to define the source population.

NGC 838:. For NGC 838 we have $L_{FIR} = 3.3\times10^{44}\,\mathrm{erg\ s^{-1}}$ (Turner et al. 2001) and so SFR$(0.1 - 100M_{\odot}) = 15\,M_{\odot}\ \mathrm{yr^{-1}}$. Therefore we expect $N_{HMXB}(L_X \geq 10^{37}\,\mathrm{erg\ s^{-1}}) = 250 - 500$. In a similar way to Hen 2-10, this represents a reasonable numbers of binaries to explain the observed hard component of $L_X \sim 10^{41}\,\mathrm{erg\ s^{-1}}$. Even with *Chandra* this galaxy is probably too distant to substantially resolve the source population.

NGC 253:. For NGC 253 we have $L_{FIR} = 6 \times 10^{43}\,\mathrm{erg\ s^{-1}}$ (Read et al. 1997), so that SFR$(0.1 - 100M_{\odot}) = 1.5\,M_{\odot}\ \mathrm{yr^{-1}}$. Therefore we expect $N_{HMXB}(L_X \geq 10^{37}\,\mathrm{erg\ s^{-1}}) = 25 - 50$. The observed number of X-ray point sources with $L_X \geq 10^{37}\,\mathrm{erg\ s^{-1}}$ is 25 (Vogler & Pietsch 1999; Strickland et al 2000), again implying reasonable agreement.

5. SUMMARY AND CONCLUSIONS

5.1. STATING THE OBVIOUS

From what has been shown in this paper it barely needs saying that X-ray binaries do contribute to the high energy emission from starburst galaxies, and so understanding binary evolution is important to understanding high energy emission from starburst galaxies. The converse is also true to some extent, namely that observations of the binary population in starburst galaxies will shed light on binary evolution.

Having stated that these facts, then it is perhaps worth highlighting some of the major outstanding problems in this area and a brief discussion of my view of the way ahead.

5.2. THE NEGATIVE VIEW

The problem with X-ray binaries are manifold - they are a particularly inhomogeneous group. As a class they have a very wide range of luminosities, ranging from a few $10^{33}\,\mathrm{erg\ s^{-1}}$ for some systems (and certainly lower than this, though whether they would still be called X-ray binaries is another matter) up to more than $10^{38}\,\mathrm{erg\ s^{-1}}$ for sources such as Sco X-1 (a low mass system) or Cen X-3 (a high mass system) and even higher if we include the transient sources or possible black-hole sources. They are extremely variable, and can change between luminosity states, vary within an orbit and indeed completely switch off in poorly understood ways. Their X-ray spectra while sharing some characteristics are also variable between systems, and the spectra of a given system will change between states. Their evolution is complex, the X-ray phase is relatively short-lived, with the duration depending very sensitively on system parameters, and so on. For instance, in starburst galaxies the star-formation can locally be very short-lived in particular star-clusters, but typically over the galaxy star-formation will have been occurring in bursts over a longer period. This can mean that there is no steady state at all in the number of X-ray binaries, let alone in the total luminosity. It would be hard to name another class of object that was less like a standard candle than X-ray binaries. In many cases, in an observation of a starburst galaxy it is impossible to tell whether a given apparent point source is a single X-ray binary, a SNR or a collection of sources.

A consequence of this is that a simple analysis of the spatially unresolved X-ray spectra of a given starburst galaxies does not really provide an easy method of estimating the binary contribution, in terms of number of systems or type (both high-mass and low-mass). The example of the hard X-ray tail seen in the dwarf starburst Hen 2-10 (Van Bever & Vanbeveren 2000) has been used as an example to show that binary evolution models can explain the observed X-ray properties of a starburst galaxy. While this is true, the uncertainties in the number of actual X-ray binaries is large. For instance, the integrated hard X-ray emission could come from a small number of fairly high luminosity systems (the hard component luminosity in Hen 2-10 is only $\sim 10^{39}$ erg s^{-1}), or could be just a single very luminous system, or a large population of less luminous systems. Consequently, this particular system provide only a mild constraint on the models (i.e. the models can explain the data and that is about it). In larger systems, such as NGC 253, the statistics of the number of systems is larger and perhaps less variable. However, the attraction of dwarf starbursts such as Hen 2-10 is that the massive star-formation tends to be a big more homogeneous. In the case of M82 the star-formation history is complex and spread over at least 60Myr.

5.3. THE POSITIVE VIEW

The previous section may have painted quite a bleak picture of the utility of X-ray observations of starburst galaxies in understanding or putting any constraints at all on binary populations. However, while that may be true if we were just to consider integrated luminosities or spectra, that is not the only view. The *Chandra* satellite offers a fantastic new opportunity in this field for the nearby starbursts, and for which observations are at a very early stage. The observations of M82 and NGC 253 by Matsumoto et al. (2000) and Strickland et al. (2000) are clear indicators of what will be possible.

Consequently, the way ahead in this field is through *CHANDRA* observations of nearby starburst galaxies and detailed follow-up observations at other wavelengths to investigate the likely origin of the X-ray point sources.

Potential science threads would include, looking at spatially resolved spectra of the galaxies, which sources contribute at higher energies. Also possible, for the brighter sources, is measuring the X-ray orbital lightcurves in order to determine binary periods. This will require multiple observations, but the rewards will be quite large. For the brighter sources it will be possible to determine (relatively crude) X-ray spectra of the sources, which will again be useful in determining the likely nature of the source. This is an area where *XMM* will be useful for those sources that are reasonably well separated.

In summary, X-ray observations of starburst galaxies are capable of being explained by binary models, with harder emission being a likely good tracer of HMXB populations. On the other hand, X-ray binaries are very poor standard candles, and it will require detailed observations and monitoring, particularly with the *Chandra* satellite to investigate the binary populations in starburst galaxies.

References

Basu S., Rana N.C., 1992, ApJ, 393, 373

Conti P.S., Vacca W.D., 1994, ApJ, 424, L97

Dahlem M., Parmar A., OosterbroekT., Orr A., Weaver K.A., Heckman T.M., 2000, ApJ, 538, 555

Dalton W.W., Sarazin C.L., 1995, 440, 280

Fabian A.C., Terlevich R.J., 1996, MNRAS, 280, L5

Gallagher J.S., Smith L.J., 1999, MNRAS, 304, 540

Heckman T.M., Armus L., Miley G.K., 1990, ApJS, 74, 833

Heckman T.M., Dahlem M., Lehnert M.D., Fabbiano G., Gilmore D., Waller W.H., 1995, ApJ, 448, 98

Heckman T.M., 2000, ASP Conf. Series "Gas and Galaxy Evolution" (in press)

Houck J.C., Bregman J.N., Chevalier R.A., Tomisaka K., 1998, ApJ, 493, 431

Immler S., Vogler A., Ehle M., Pietsch W., 1999, A&A, 352, 415

Iyomoto N., Makishima K., Fukazawa Y., Tashiro M., Ishisaki Y., 1997, PASJ, 49, 425

Jansen F., Lumb D., Altieri B. et al. 2001, A&A, (in press)

Johnson K.E., Leitherer C., Vacca W.D., Conti P.S., 2000, AJ, 120, 1273

Kaaret P., Prestwich A.H., Zezas A., Murray S.S., Kim D.-W., Kilgard R.E., Schlegel E.M., Ward M.J., 2000, ApJL, (in press)

Kennicutt R.C., ARA&A, 1998, 36, 189

Kobulnicky H.A., Johnson K.E., 1999, ApJ, 527, 154

Kronberg P.P., Sramek R.A., Birk G.T., Dufton Q.W., Clarke T.E., Allen M.L., 2000, ApJ, 535, 706

Lehnert M.D., Heckman T.M., Weaver K.A., 1999, ApJ, 523. 575

Matsumoto H., Tsuru T.G., Koyama K., Awaki H., Canizares C.R., Kawai N., Matsushita S., Prestwich A., Ward M., Zezas A.L., Kawabe R., 2000, ApJL, (in press)

Mendes de Oliveira C., Plana H., Amram P., Bolte M., Boulestix J., 1998, ApJ, 507, 691

Méndez D.I., Esteban C., Filipovic M.D., Ehle M., Haberl F., Pietsch W., Haynes R., 1999, A&A, 349, 801

Miller G.E., Scalo J.M., 1979, ApJS, 41, 513

Moran E.C., Lehnert M.D., 1997, ApJ, 478, 172

Moran E.C., Lehnert M.D., Helfand D.J., 1999, ApJ, 526, 649

Nath B., Trentham N., 1997, MNRAS, 291, 505

O'Connell R.W., Gallagher J.S., Hunter D.A., Colley W.N., 1995, ApJ, 446, L1

Ohyama Y., Taniguchi Y., 1999, In: Wolf-Rayet Phenomena in Massive Stars and Starburst Galaxies. Proc. IAU Symp. No. 193, eds. K.A. van der Hucht, G. Koenigsberger and P.R.J. Eenens. p.523

Pietsch W., Trinchieri G., Vogler A., 1998, A&A, 340, 351

Pietsch W., Vogler A., Zinnecker H., 2000, A&A, 342, 101

Pietsch W., et al. 2001, A&A, (in press)

Ponman T.J., Cannon D., Navarro J., 1999, Nature, 397, 135

Ptak A., Serlemitsos P., Yaqoob T., Mushotzky R. Tsuru T., 1997, AJ, 113, 1286

Ptak A., Griffiths R., 1999, ApJ, 517, L85

Read A.M., Ponman T.J., Strickland D.K., 1997, MNRAS, 286, 626

Sarazin C.L., Irwin J.A., Bregman J.N., 2000, ApJL, (in press)

Satyapal S., et al. 1997, ApJ, 483, 148

Stevens I.R., Strickland D.K., 1998, MNRAS, 294, 523

Stevens I.R., Forbes D.A., Norris R.P., 1999, MNRAS, 306, 479

Strickland D.K., Ponman, T.J., Stevens, I.R., 1997, A&A, 320, 378

Strickland D.K., Stevens I.R., 1998, MNRAS, 297, 747

Strickland D.K., Stevens I.R., 1999, MNRAS, 306, 43

Strickland D.K., et al. 2000, ApJ, (in press)

Tsuru T.G., Awaki H., Koyama K., Ptak A., 1997, PASJ, 49, 619

Turner M.J.L., et al. 2001, A&A, (in press)
Van Bever, J., Vanbeveren, D., 2000, A&A, 358, 462
Vogler A., Pietsch W., 1999, A&A, 342, 101
Weaver K.A., Heckman T.M., Strickland D.K., Dahlem M., 2000, ApJ, (in preparation)
Weisskopf M.C., O'Dell S.L., van Speybroeck L.P., 1996, Proc. SPIE, 2805, 2
Wills K.A., Pedlar A., Muxlow T.W.B., Wilkinson P.N., 1997, MNRAS, 291, 517
Yun M.S., Ho P.T.P. Lo K.Y., 1993, ApJ, 411, L17

SUPERNOVA TYPES AND RATES *

Enrico Cappellaro and Massimo Turatto
Osservatorio Astronomico di Padova
vicolo dell'Osservatorio, 5
I-35122, Padova (Italy)
cappellaro@pd.astro.it; turatto@pd.astro.it

Keywords: Supernova models, supernova rates, nucleosynthesis

Abstract We review the basic properties of the different supernova types identified in the current taxonomy, with emphasis on the more recent developments. To help orienting in the variegate zoo, the optical photometric and spectroscopic properties of the different supernova types are presented in a number of summary figures.
We also report the latest estimates of the supernova rates and stress the need for a dedicated effort to measure SN rates at high redshift.

1. INTRODUCTION

Supernovae (SNe) are at the confluence of many different streams of astronomical researches. As the final episode of the life of many kinds of stars, SNe allow crucial tests of stellar evolution theories. Yet, in many cases we are still entangled with the basic problem of finding a viable progenitor scenario for each SN type. During and immediately after the explosions, there are a variety of fundamental physical mechanisms which can be probed, such as neutrino and gravitational wave emissions, flame propagation and explosive nucleosynthesis, radioactive decays and shocks with circumstellar matter. Relics of the explosions are collapsed remnants, neutron stars or black holes, and expanding gas clouds which heat and pollute the interstellar medium. Indeed, SNe are a main agent in the chemical evolution of galaxies.

*Most of the material presented in this paper has been retrieved from the Padova Archive of SN observations. This is the product of a long standing effort for the monitoring of SNe conducted using ESO-La Silla and Asiago telescopes (Turatto et al. 1990).

D. Vanbeveren (ed.), The Influence of Binaries on Stellar Population Studies, 199–214.

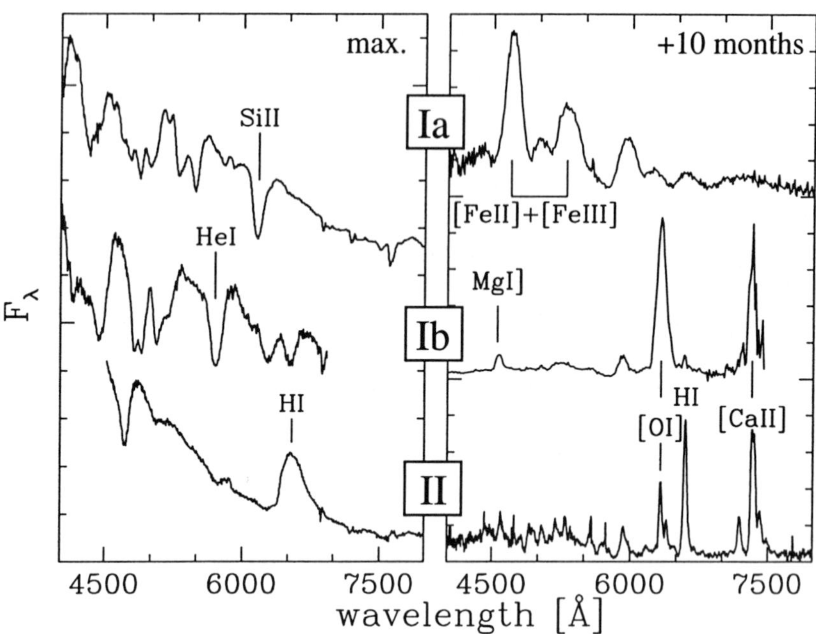

Figure 1 Basic SN types spectra. A SN which near maximum (left panel) shows clear signature of Hα is defined as type II, if it shows a strong SiII absorption at about 6150 Å is a type Ia, otherwise it is of type Ib/c (in the figure we show a SNIb characterized by strong He lines). Ten months later (right panel) SN Ia show strong emissions of [FeII] and [FeIII], SN Ib/c are dominated by [CaII] and [OI]. These same lines and strong Hα emission are typical of SNII.

Moreover, because of their huge luminosity and the fact that they can be accurately calibrated, type Ia SNe are the best distance indicators for measuring the geometry of the Universe.

For all these reasons, in the last decades there has been an increased interest in SN studies and a renewed effort in SN searches. As a consequence, the number of SN discoveries has increased from about 20/yr in the early '80, to about 200/yr in the last few years raising the number of SNe discovered to date to $\sim$ 1800 (Barbon et al. 1999; `http://merlino.pd.astro.it/~supern/` for a frequently updated version). It is not surprising that the enlarged sample combined with the better quality of the observations obtained by the new generation instruments is deeply changing our understanding of the SN phenomenon.

The aim of this paper is to give a snapshot of the SN zoo including the latest acquisitions. Other information can be found in Filippenko

(1997), Wheeler and Benetti (2000) and reference therein. We will also present the latest estimates of the SN rates which are fundamental to link the evolution of stars and stellar systems.

2. BASIC SN TYPES

The affair of SN classification began with Minkowskii (1941) who noticed that there are at least two different kinds of SNe, showing (type II) and not showing (type I) H in their spectra.

This subdivision was maintained until, in the '80s, it was recognized that a number of peculiar SN I missing in the spectrum near maximum the typical feature at $\sim$ 6150 Å, had a completely different nebular spectrum dominated by forbidden Ca and O emissions instead of Fe lines (Gaskell et al. 1986). Examining more carefully the near maximum spectra it was noticed that these SNe come in at least two flavors: those showing strong He lines were labeled Ib, the others Ic. We will argue later however that this subdivision is at a lower hierarchical level. The first branch of the SN classification is illustrated in Fig. 1.

In general, in the early phases the optical depth of the ejecta remains high and the emergent spectrum only probes the outermost layers. Therefore the early spectrum is most sensitive to variations in the density and composition of the progenitor envelope. With time the density and temperature decrease, the photosphere recedes in mass coordinates and eventually the ejecta becomes optically thin. At this point the innermost regions becomes exposed and the yields of the explosions can be probed. The fact that the nebular spectrum of SN Ib/c is similar to that of SN II, but for the H emission, prompts for a similar explosion mechanism. Instead, the weakness of intermediate element emissions, in particular O, in the nebular spectrum of SN Ia excludes that the progenitors of SN Ia are massive stars.

The SN taxonomy which originally was based on spectra near maximum may result somewhat confusing when one try to build a coherent progenitor scenario: indeed a more physical classification could be based on the distinction between SNe arising from the collapse of massive stars (SNII and SNIb/c) and SNe due to thermonuclear explosions of low mass stars (SN Ia). In the following we will give a brief description of the characteristics of each SN type describing the detailed taxonomy illustrated in Fig. 2.

3. TYPE IA SNE

SN Ia are quite homogeneous events with similar luminosity and spectral evolution. Indeed until the early '90s, it was commonly accepted

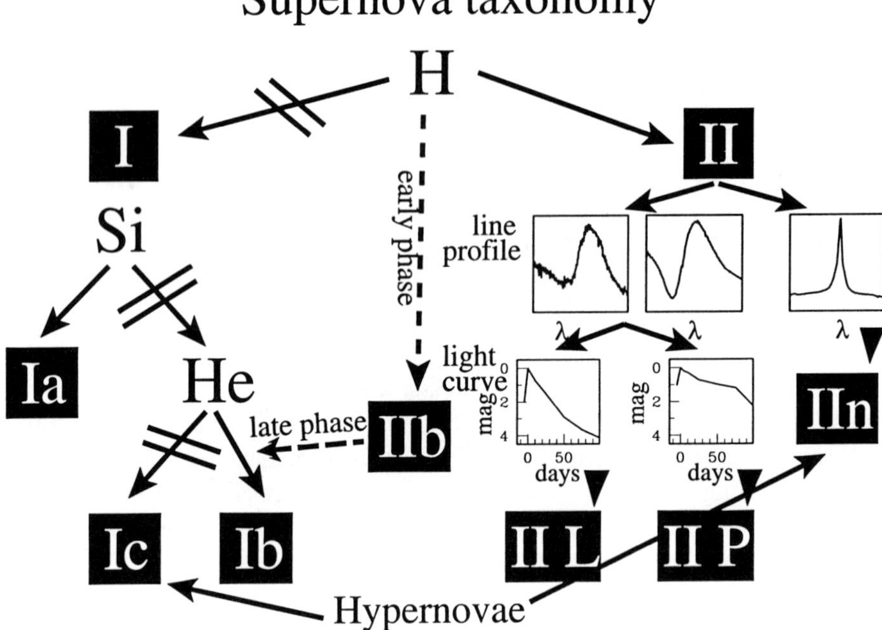

Figure 2 The detailed classification of SNe requires not only the identification of specific features in the early spectra, but also the analysis of the line profiles, luminosity and spectral evolutions

that SN Ia were identical and that observed differences were mainly due to observational errors (e.g Leibundgut & Tammann 1990). With the improvement of the signal to noise of the observations it was definitely demonstrated that differences do exist. In particular, though the light curve shapes remain similar, the absolute luminosities and the decline rates are different (Fig. 3). Actually, a relation has been found between the rate of the luminosity decline and the absolute magnitudes at maximum, with fast declining SNe being fainter (and redder). Because this is crucial for recovering SN Ia as distance indicators, a large efforts is being devoted to the accurate calibration of the different expressions of this relation (Phillips et al. 1999, Riess et al. 1998, Perlmutter et al. 1999). To some extent, the spread in luminosity correlates with variations in the spectral properties (Nugent et al. 1995). As shown in Fig. 4, very bright or very faint SNe show distinctive spectral features, whereas the differences among "normal" SN Ia stand out only after a careful analysis.

It is generally agreed that the difference in luminosity call for one order of magnitude variance in the mass of the radioactive ^{56}Ni which

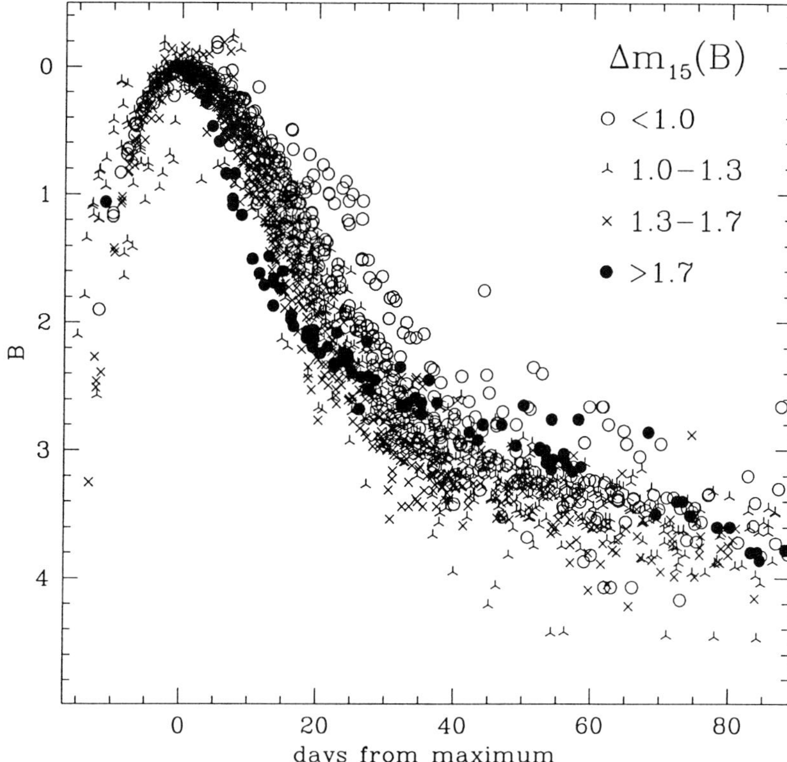

Figure 3 Light curves of 87 SN Ia relative to the phase and magnitude of maximum. The symbols identify objects with different $\Delta m_{15}(B)$, the magnitude decline in the early 15 days (Altavilla 2000).

is powering the early light curve. Controversial is the claim that light curves and spectral synthesis models require a factor two in the total progenitor mass (Cappellaro et al. 1997). The standard scenario for the SN Ia explosion consists of a binary system where one of the star is a white dwarf (WD) which accretes matter for the secondary star, reaches the Chandrasekhar limit, and undergoes a disruptive, thermonuclear explosion (Nomoto et al. 1984,Woosley & Weaver 1986). It is still unclear whether it is possible to reproduce the observed variance of SN Ia within the standard scenario (Mazzali et al. 2000).

An interesting finding in this respect is that on average SN Ia in early type galaxies show a faster decline rate (hence are less luminous) than SN Ia in late type galaxies (Fig. 5, cf. van den Bergh & Pazder(1992), Hamuy et al.(2000)). This suggests that (some of) the variance of SN Ia is related to different ages of the progenitor population. Until this is

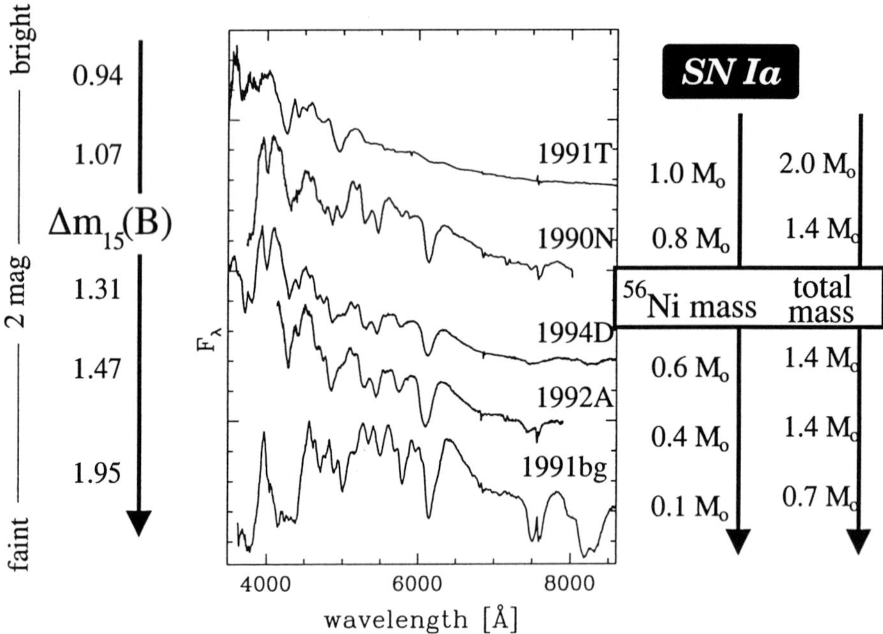

Figure 4 Comparison of the maximum light spectra of SN Ia with different decline rates (left), hence luminosities. On the right the ranges in ^{56}Ni and total masses suggested by light curve and spectral synthesis models are indicated.

not well understood, a severe caveat remains on the use of SN Ia as cosmological distance indicators.

4. TYPE II SNE

One might say that the only thing in common for SN II is the presence of H lines in the spectra. Indeed the maximum luminosities range over two orders of magnitudes and the spectral energy distributions and line profiles also shows dramatic differences. Historically, the first attempt to sort out some order was based on the light curve shapes (Barbon et al. 1979). In some cases, SN II after maximum remain on a plateau of almost constant luminosity for periods up to 2-3 months (II-P), in other cases they decline more or less linearly (II-L). The two classes are not separated and intermediate cases do exist (Clocchiatti et al. 1996). An extreme case of SN II-P is usually considered SN 1987A, the best observed SN ever, which after maximum showed a steadily increase of

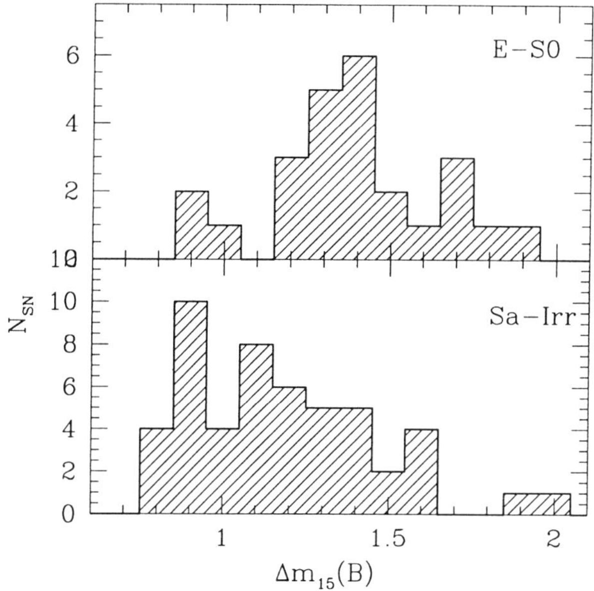

Figure 5 Histogram of $\Delta m_{15}(B)$ for SN Ia in early type galaxies (upper panel) and in late type galaxies (lower panel). A Kolmogorov-Smirnov test shows that the probability that the two distributions derive from the same population is only 0.004 (Altavilla 2000).

magnitude, lasting abound 3 months, until eventually the decline began (Fig. 6).

In general, it is assumed that SNII-L are brighter than SN IIP, but there are many exceptions (Patat et al. 1994), e.g. SN 1983K the bright II- P shown in Fig. 6. At late time (> 200 d), many SN II settle on a linear decline rate of about 1 mag/100d. This is powered by the second branch of the radioactive decay chain ^{56}Ni–^{56}Co–^{56}Fe. The fact that, despite the enormous difference at maximum, SN II at late time converge to a similar luminosity was taken as an indication that the ^{56}Ni mass produced in SN II was very similar, namely ~ 0.1 $M_\odot$ (Turatto et al. 1990). Actually, even in this respect there may be significant variations (Schmidt et al. 1994). An extreme case is that of SN 1997D (Fig. 6) which produced only 0.002 $M_\odot$ of ^{56}Ni and showed extremely low expansion velocities (Turatto et al. 1998).

It is generally accepted that SN II result from the core collapse of stars in the range of 10-30 $M_\odot$ (Woosley et al. 1988). Because of various mechanisms (eg. difference in metallicities, interaction in binary system) the size and the mass of the H envelope at the time of explosion can be very different even for progenitors of the same initial mass. In a typical

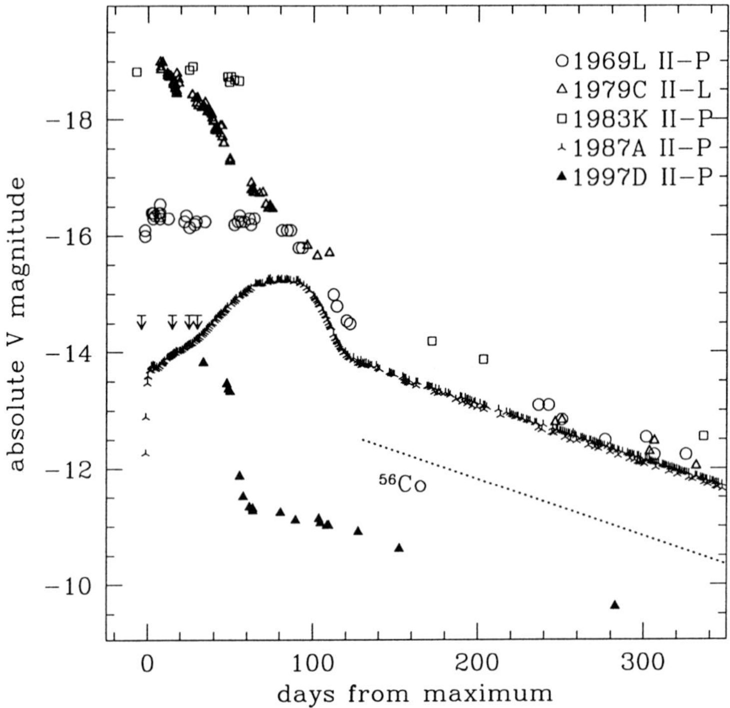

Figure 6 Representative light curves of SN II. The dotted line is the expected luminosity decline rate when the light curve is powered by the decay of ^{56}Co.

SN IIP the H envelope mass is $\sim 5-10$ $M_\odot$ and the radius is $\sim 10^{15}$ cm. The shock wave originated by the explosion rapidly reaches the surface completely ionizing the H. When the ejecta expands the energy from the recombination of H sustains the photosphere at an almost constant radius and temperature. The length of the plateau depends on the envelope mass and when this gets very small ($< 1 - 2$ $M_\odot$) the decline is faster (SN IIL). The different line profiles of II-P and II-L, with II-L missing the P-Cygni absorption of the Balmer lines (Fig. 7) is also due to their reduced envelope masses.

Instead the early luminosity depends on the radius of the progenitor star at the time of explosion. If the progenitor was compact, 10^{12} cm in the case of SN 1987A, much of the available energy is dissipated for the expansion and until the radioactive decay input becomes dominant, the luminosity remains lower than in the case of an extended progenitor.

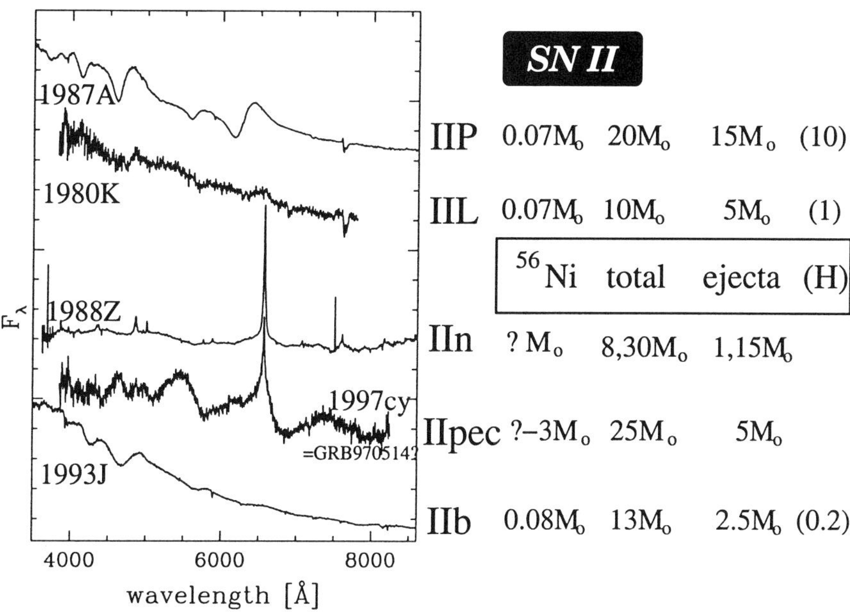

Figure 7 Representative spectra of SN II. On the right we report for each object the best estimate of the ^{56}Ni, total and ejecta masses (in parenthesis is the H mass in the ejecta).

In most cases, the remnant of the explosion is expected to be a neutron stars, but it has been argued that the unusual properties of SN 1997D may be understood if the collapsed remnant is a black hole (Turatto et al. 1998).

In general, because of the very high expansion velocity, the emission lines in the SN spectra are very broad but for possible contamination of interstellar gas emissions. A significant fraction of SN II (10 – 15 % of the SN II currently discovered) show narrow components on top of broader emissions (Fig. 7). These SNe are labeled IIn, where "n" stands for narrow–line (Schlegel 1990). It turns out that in many cases SN IIn are very bright and their luminosity evolution is much slower than other SN II (Fig. 8). Therefore, an additional source of energy powering the light curve in addition to the radioactive decay is required. It is currently believed that the kinetic energy of the ejecta is converted into radiation when the ejecta shock a dense circumstellar medium (CSM) (Aretxaga et al. 1999). This dense CSM is most likely a relic of a strong mass loss episodes occurring shortly before explosion. Differences in density and

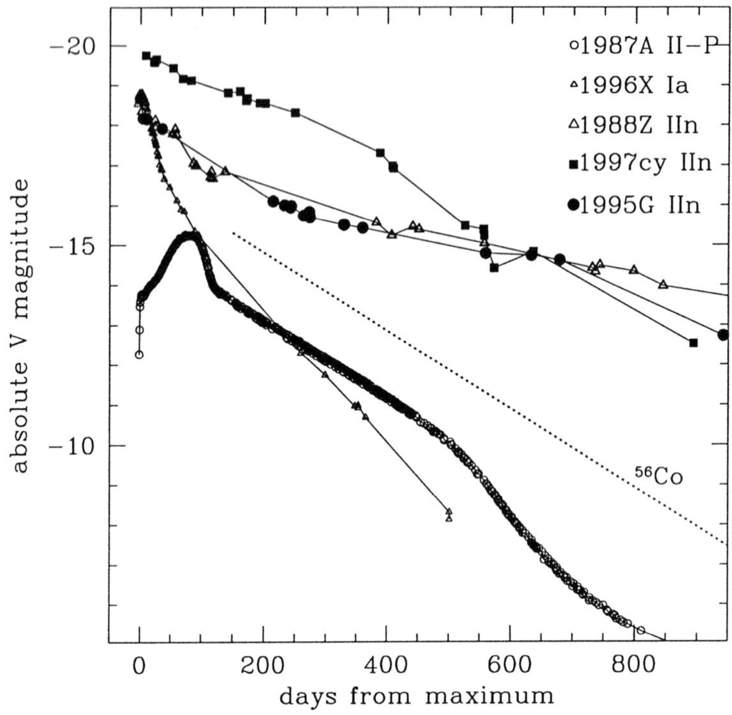

Figure 8 Representative light curves of SN IIn are compared with that of a SN II-P and of a SN Ia. The dotted line is as in Fig. 6

distribution of the CSM might explain the large variance of luminosity evolution between different SN IIn.

A most interesting case is that of SN 1997cy for its possible association with GRB970514. Because of the emission line profile (Fig. 7), we believe that also in this case the CSM-ejecta interaction is powering the light curve (Fig.8) and a huge Ni mass is not required (Turatto et al. 2000).

5. TYPE IB/C SNE

The explosion of SN 1993J in M81 was the missing link between SNe of type II and Ib/c. The spectrum of this SN evolved from that of a rather normal type II to that of a type Ib in few weeks. The object is therefore classified as the prototype of the intermediate class of SN IIb. Its observed properties have been explained with the explosion of a

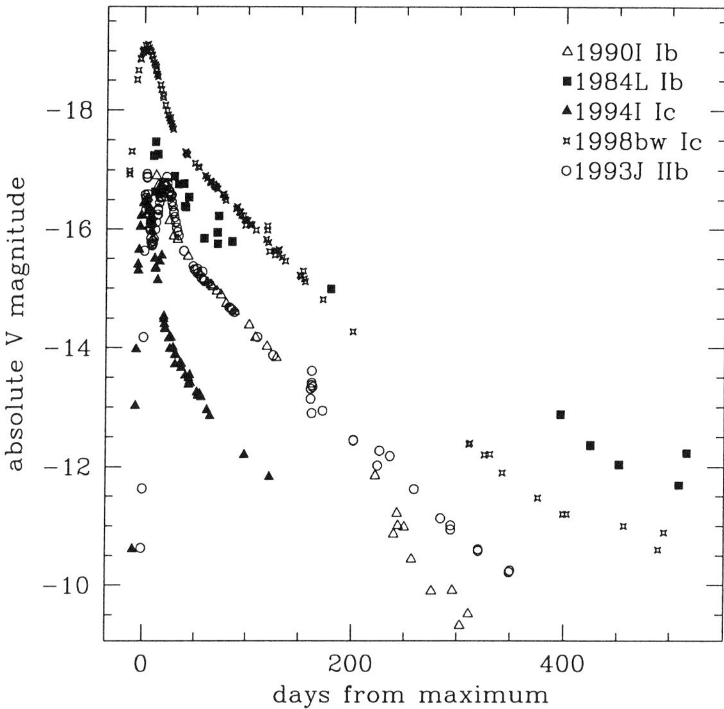

Figure 9 Representative light curves of SN Ib/c.

massive progenitor with a very small H envelope (0.2 $M_{\odot}$). It is believed that if the exploding star was left without the H envelope, the SN would appear as a typical SNIb. If even the He shell was removed the result will closely match the observational properties of type Ic. Other cases of SN Ic in which is detected a small amounts of H in the ejecta are reported in the literature (cf. Filippenko 1992; 2000H, Pastorello et al. 2000). There are also intermediate cases between type Ib and Ic in which residual amounts of He have been detected (eg. Filippenko et al. 1995). These considerations have motivated the introduction of a unifying scenario where the sequence Ic-Ib-IIb-IIL has mass loss as the driving parameter (Nomoto et al. 1996). At the moment the mechanism for this huge mass loss has not yet been identified, although interaction in close binary systems is the first candidate.

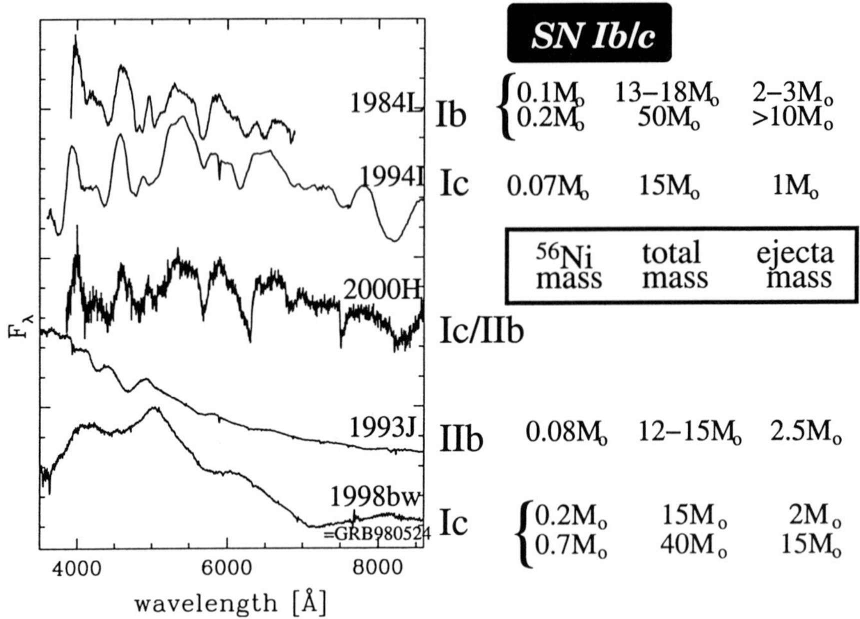

Figure 10 Representative spectra of SNIb/c. On the right the best estimates of the ^{56}Ni, total and ejecta masses are reported . In some case different modeling produces significantly different results.

One additional complication is that there are SN Ib/c in which the late light curves seem to track the radioactive decay of ^{56}Co (eg. SN 1984L) and other showing a luminosity decline similar or even faster than that of SN Ia (Fig. 9). Because in all cases it is believed that radioactive ^{56}Co is present, one must admit that the trapping of the γ-rays from ^{56}Co is different which in turn call for a much smaller ejecta mass in fast declining SNIb/c than in the slow declining ones. A summary of the ^{56}Ni and ejecta masses for some SNIb/c is reported in Fig. 10

More recently two unusual SNe were discovered (SN 1997ef and 1998bw) that, although formally classified as type Ic because they miss the 6150 Å dip of SNIa, the H line of type II and the He lines of type Ib, show indeed a very peculiar spectrum (Fig. 10). This is the result of an extremely high expansion velocities and in turn, of high explosion energy (Iwamoto et al. 1998, 2000). Most interesting is the case of SN 1998bw because of his close association with GRB250498 (Patat et al. 2000) Because of these extreme characteristics these SNe, along with SN 1997cy mentioned in the previous section, are often referred as **hypernovae** and may indeed require specific progenitor scenarios and explosion mechanism.

6. SN RATES

The most recent estimate of the rates of the different SN types has been published in Cappellaro et al. (1999) and are summarized in Tab. 1. In average, these values are in fair agreement with previous estimates (cf. van den Bergh & Tammann 1991).

Table 1 The local SN rates. Units are $SNu = \mathrm{SN}/10^{10}\mathrm{L}_{\odot,\mathrm{B}}/100\mathrm{yr}$. $\mathrm{h}=\mathrm{H}_0/100$.

galaxy	SN type			
type	Ia	Ib/c	II	All
E-S0	$0.32 \pm .11\ h^2$	$< 0.02\ h^2$	$< 0.04\ h^2$	$0.32 \pm .11\ h^2$
S0a-Sb	$0.32 \pm .12\ h^2$	$0.20 \pm .11\ h^2$	$0.75 \pm .34\ h^2$	$1.28 \pm .37\ h^2$
Sbc-Sd	$0.37 \pm .14\ h^2$	$0.25 \pm .12\ h^2$	$1.53 \pm .62\ h^2$	$2.15 \pm .66\ h^2$
All	$0.36 \pm .11\ h^2$	$0.14 \pm .07\ h^2$	$0.71 \pm .34\ h^2$	$1.21 \pm .36\ h^2$

One interesting result is that the rate of SN Ia **per unit B luminosity** is constant from ellipticals to late spirals. Even if one may argue about what is the stellar population which dominates the blue luminosity in spirals, a significant fraction must derive from young stars. This immediately implies that the average age of the SN Ia progenitors is shorter in spirals than in ellipticals (cf. Fig.5). Also worth noticing is that the SN Ib/c are only 15% of all core collapse events.

Based on the values of Tab. 1, we can derive an estimate of the SN rate in our own Galaxy making the basic assumption that the Galaxy has an average SN rate for his morphological type (here assumed Sb-Sbc) and luminosity ($2.3 \times 10^{10} L_{B,\odot}$). If we adopt $\mathrm{H}_0 = 65\,\mathrm{km\,s^{-1}\,Mpc^{-1}}$, we expect 0.4 ± 0.2 SN Ia and 1.5 ± 1.0 SNII+Ib/c per century, that is roughly one SN every 50 years. Formally, this is a factor two lower than derived from counts of historical SNe and SN remnants (eg. van den Bergh & Tammann 1991, Strom 1994). When considering the large uncertainties however, the two values are not in disagreement.

For what concerns the SN rate in the local Universe, we have now reached the point where the statistics is no longer limited by the SN discoveries but by the sample of galaxies for which the necessary information (distance, luminosity, morphological types and inclination) are available. In other words, to improve significantly the statistical accuracy of the results, one needs not only to continue SN searches but also to obtain more complete galaxy catalogues.

A major effort is still required, instead, in "un-biased" SN search at high redshift. In fact, the evolution of the SN rate with redshift brings unique information on the progenitor scenarios, galaxy star formation history and initial mass function (eg. Yungelson & Livio 2000, Dahlèn & Fransson, C. 1999).

Table 2 Estimate of SN rates at different redshift. $h = H/100$

$< z >$	Ia h^2 SNu	II h^2 SNu	N(SN)	source
0.01	0.36 ± 0.10	0.85 ± 0.41	137	Cappellaro et al. 1999
0.1	$0.44^{+.0.35}_{-0.21}$		4	hardin
0.38	$0.82^{+.0.65}_{-0.45}$		3	Pain et al. 1996
0.55	0.62 ± 0.19		38	Pain et al. 2000#

Not yet published. Reported in Ruiz-Lapuente & Canal 2000.

Current high-z SN searches are aimed to the discovery of SN Ia for their use as distance indicators. By construction, they are not well suited for estimating SN rates. This is obvious from Tab. 2 where we list all observational estimates of SN rates available at "high" redshift and compare them with the local estimates. In this table N(SN) gives the statistics of SNe on which the estimate is based. The very low statistics and the fact that there is not an estimate yet for core-collapse SN rate outside the local Universe is a clear indication of the work to be done.

References

Altavilla, G., 2000, Thesis Universita' di Padova.

Aretxaga I., Benetti S., Terlevich R. J., Fabian A. C., Cappellaro E., Turatto M., della Valle M., 1999, *MNRAS* , 309, 343

Barbon R., Ciatti F., Rosino L., 1979, *A&A* , 72, 287

Barbon R., Buondí V., Cappellaro E., Turatto M., 1999, *A&AS* , 139, 531

Cappellaro E., Mazzali P. A., Benetti S., Danziger I. J., Turatto M., della Valle M., Patat F., 1997, *A&A* , 328, 203

Cappellaro E., Evans R., Turatto M., 1999, *A&A* , 351, 459

Clocchiatti A., Benetti S., Wheeler J. C., et al., 1996, *AJ* , 111, 1286

Dahlèn, T., Fransson, C.: 1999 *A&A* 350 359.

Filippenko A. V., 1992, *ApJ* (Letters) , 384, L37

Fillipenko A. V., 1997, *ARAA* , 35, 309
Filippenko A. V., Barth A. J., Matheson T., et al., 1995, *ApJ* (Letters) , 450, L11
Gaskell C. M., Cappellaro E., Dinerstein H. L., Garnett D. R., Harkness R. P., Wheeler J. C., 1986, *ApJ* (Letters) , 306, L77
Hamuy, M., Trager, S. C., Pinto, P. A., Phillips, M. M., Schommer, R. A., Ivanov, V. & Suntzeff, N. B. 2000, *AJ* , 120, 1479
Hardin, D., Afonso, C., Alard, C. et al, 2000 *A&A* in press (astro-ph 0006424)
Iwamoto K., Mazzali P. A., Nomoto K., et al., 1998, *Nature* , 395, 672
Iwamoto K., Nakamura T., Nomoto K., et al., 2000, *ApJ* , 534, 660
Leibundgut B., Tammann G. A., 1990, *A&A* , 230, 81
Mazzali, P.A., Nomoto, K., Cappellaro, E., Nakamura, Y., Umeda, H., Iwamoto, K., 2000, *ApJ* in press (astro-ph/0009490)
Minkowski R., 1941, *PASP* , 53, 224
Nomoto K., Thielemann F. -., Yokoi K., 1984, *ApJ* , 286, 644
Nomoto K., Iwamoto K., Suzuki T., Pols O. R., Yamaoka H., Hashimoto M., Hoflich P., van den Heuvel E. P. J., 1996, IAU Symp. 165: Compact Stars in Binaries, 165, 119
Pain R., Hook I. M., Deustua S., et al., 1996, *ApJ* , 473, 356
Nugent P., Phillips M., Baron E., Branch D., Hauschildt P., 1995, *ApJ* (Letters) , 455, L147
Pastorello, A., Altavilla, G., Cappellaro, E., Turatto, M., Benetti S., 2000. IAU Circular 7367.
Patat F., Barbon R., Cappellaro E., Turatto M., 1994, *A&A* , 282, 731
Patat, F., Cappellaro, E., Mazzali, P.A., et al., 2000, *ApJ* in press
Perlmutter S., Aldering G., Goldhaber G., et al., 1999, *ApJ* , 517, 565
Phillips M. M., Lira P., Suntzeff N. B., Schommer R. A., Hamuy M., Maza J. ;, 1999, *AJ* , 118, 1766
Riess A. G., Filippenko A. V., Challis P., et al., 1998, *AJ* , 116, 1009
Ruiz-Lapuente, P., Canal, R., 2000, *ApJ* (Letters) submitted (astro-ph 0009312)
Schmidt B. P., Kirshner R. P., Eastman R. G., et al., 1994, *AJ* , 107, 1444
Schlegel E. M., 1990, *MNRAS* , 244, 269
Strom R. G., 1994, *A&A* , 288, L1
Turatto M., Bouchet P., Cappellaro E., Danziger, I. J., della Valle, M., Frabsson, C., Gouiffes, C., Lucy, L., Mazzali, P., Phillips, M., 1990, The Messenger, 60, 15
Turatto M., Cappellaro E., Barbon R., della Valle M., Ortolani S., Rosino L., 1990, *AJ* , 100, 771

Turatto M., Mazzali P. A., Young T. R., et al., 1998, *ApJ* (Letters) , 498, L129

Turatto M., Suzuki T., Mazzali P. A., et al., 2000, *ApJ* (Letters) , 534, L57

van den Bergh, S. & Pazder, J. 1992, *ApJ* , 390, 34

van den Bergh S., Tammann G. A., 1991, *ARAA* , 29, 363

Wheeler, J. C., Benetti, S., in *Allen's astrophysical quantities, 4th ed.* Publisher: New York: AIP Press; Springer, 2000.Ȧ. N. Cox.ed. p. 451.

Woosley S. E., Weaver T. A., 1986, *ARAA* , 24, 205

Woosley S. E., Pinto P. A., Ensman L., 1988, *ApJ* , 324, 466

Yungelson, L., Livio, M.: 2000 *ApJ* 528 108.

STELLAR WINDS FROM MASSIVE STARS

Paul A. Crowther
Department of Physics & Astronomy, University College London, Gower St., London WC1E 6BT, U.K.
pac@star.ucl.ac.uk

Keywords: Stars: mass-loss; early-type, supergiants; Wolf-Rayet

Abstract We review the various techniques through which wind properties of massive stars – O stars, AB supergiants, Luminous Blue Variables (LBVs), Wolf-Rayet (WR) stars and cool supergiants – are derived. The wind momentum-luminosity relation (e.g. Kudritzki et al. 1999) provides a method of predicting mass-loss rates of O stars and blue supergiants which is superior to previous parameterizations. Assuming the theoretical $Z^{0.5}$ metallicity dependence, Magellanic Cloud O star mass-loss rates are typically matched to within a factor of two for various calibrations. Stellar winds from LBVs are typically denser and slower than equivalent B supergiants, with exceptional mass-loss rates during giant eruptions $\dot{M} = 10^{-3} \ldots 10^{-1} M_{\odot}$ yr^{-1} (Drissen et al. 2001). Recent mass-loss rates for Galactic WR stars indicate a downward revision of 2–4 relative to previous calibrations due to clumping (e.g. Schmutz 1997), although evidence for a metallicity dependence remains inconclusive (Crowther 2000). Mass-loss properties of luminous ($\geq 10^5 L_{\odot}$) yellow and red supergiants from alternative techniques remain highly contradictory. Recent Galactic and LMC results for RSG reveal a large scatter such that typical mass-loss rates lie in the range $10^{-6} \ldots 10^{-4} M_{\odot}$ yr^{-1}, with a few cases exhibiting $\sim 10^{-3} M_{\odot}$ yr^{-1}.

1. INTRODUCTION

Winds are ubiquitous in massive stars, although the physical processes involved depend upon the location of the star within the H-R diagram. Mass-loss crucially affects the evolution and fate of a massive star (Meynet et al. 1994), while the momentum and energy expelled contribute to the dynamics and energetics of the ISM. The interested reader is referred to Lamers & Cassinelli (1999) for the topic of stellar winds in general, or Kudritzki & Puls (2000) for a detailed discussion of mass-loss from OB stars. This review will cover the broad topics of

D. Vanbeveren (ed.), The Influence of Binaries on Stellar Population Studies, 215–230.

mass-loss from early-type (OBA, Wolf-Rayet, Luminous Blue Variable) and late-type (yellow/red supergiants) stars. Theoretical predictions for hot star winds follow line-driven radiation pressure (Castor et al. 1975; Pauldrach et al. 1986). Pulsations are thought to initiate the winds of cool supergiants, together with other mechanisms (e.g. Alfven waves). Radiation pressure on dust grains helps to maintain outflows in their cool, outer envelopes (Wickramasinghe et al. 1966).

2. HOT STAR WIND DIAGNOSTICS

Global wind properties can be characterized by mass-loss and wind velocity. In hot stars, the former depends on application of varying complexity of theoretical interpretation, while the latter can fortunately be directly measured with minimal interpretation.

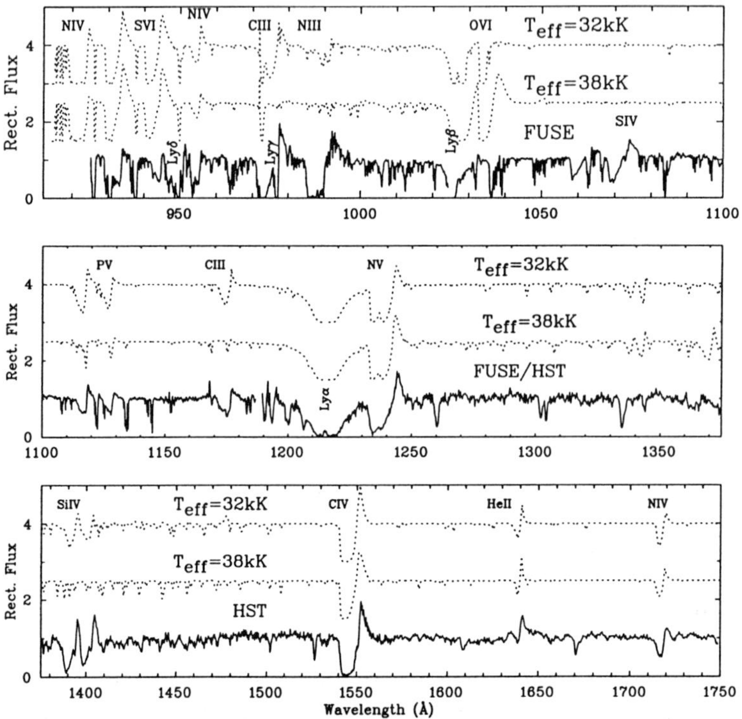

Figure 1 Comparison between HST UV and FUSE far-UV spectra of Sk 80 (O7 Iaf) with synthetic spectra (including shocks) at $T_{\rm eff}$=37,500Kand 32,000K, illustrating the improved agreement for the latter value(Fullerton et al. 2000).

2.1. UV AND FAR-UV SPECTROSCOPY

Ultraviolet P Cygni profiles, ubiquitous in O-type stars (Walborn et al. 1985) provide a direct indication of stellar winds. Such metal resonance lines are generally analysed via the Sobolev with exact integration (SEI) method (Lamers et al. 1987; Haser et al. 1998). Mass-loss determinations require knowledge of elemental abundances, the degree of ionization of the ion producing the line, and an assumed form of the velocity law that is predicted by radiation driven winds. In practice, accurate wind velocities can be readily obtained from saturated UV P Cygni profiles (Groenwegen et al. 1989; Prinja et al. 1990), while unsaturated P Cygni profiles provide a better means of determining mass-loss rates, albeit still subject to ionization results from non-LTE atmospheric models. Such models are subject to uncertain contributions of ionizing radiation from shocks in the stellar outflow. In view of these difficulties, Lamers et al. (1999) have combined column densities from unsaturated UV P Cygni profiles with independently determined mass-loss rates (from radio or Hα, see below) to derive empirical ionization fractions.

The Far-Ultraviolet Spectroscopic Explorer (FUSE) offers the possibility of determining mass-loss rates solely from UV data alone, since the far-UV significantly increases the number of stellar wind diagnostics, including C III, N III-IV, O VI, S IV-VI and P V. The additional ionization information provided by FUSE permits stellar temperature determinations. Fullerton et al. derived $T_{\rm eff}$=32–33kK for two Magellanic Cloud O7 supergiants from UV wind metal diagnostics, in contrast to $T_{\rm eff}$=38kK from the usual optical helium plane-parallel hydrostatic methods (see Fig. 1). Although previously derived mass-loss rates were confirmed, lower bolometric corrections implied reduced stellar luminosities, with consequences for current mass-loss calibrations.

2.2. OPTICAL AND NEAR-IR SPECTROSCOPY

Although Hα has long been recognized as the prime source of mass-loss information in early-type stars, accurate determinations rely upon a complex treatment of the lines and continua, i.e. non-LTE, spherically extended models which treat the sub- and supersonic atmospheric structure. Use of such codes also provide surface temperatures and gravity in OB stars via optical photospheric lines (e.g. He I-II in O stars). Problems with using fits to Hα to derive mass-loss rates include blending with He II λ6560 in O stars, together with nebular contamination from H II regions (e.g. NGC346; Walborn et al. 1995) and broadening of the central emission component by stellar rotation. In A supergiants, Hα

behaves as a scattering line, with a characteristic P Cygni profile, so they have the advantage over other hot stars that wind velocities and mass-loss rates may be simultaneously determined in external galaxies without the need for UV spectroscopy.

Although moderate to high spectral resolution observations of OB stars are rare at near-IR wavelengths (e.g. Fullerton & Najarro 1998), analogous lines to optical wind (and photospheric) diagnostics are available (Bohannan & Crowther 1999). This provides the prospect of the determination of mass-loss properties for hot stars obscured at optical wavelengths via IR diagnostics (Najarro et al. 1994).

2.3. IR–RADIO CONTINUA

Winds in hot stars can be readily observed at IR-mm-radio wavelengths via the free-free (Bremsstrahlung) 'thermal' excess caused by the stellar wind, under the assumption of homogeneity and spherical symmetry. Mass-loss rates can be determined via application of relatively simple analytical relations (Wright & Barlow 1975; Panagia & Felli 1975). Barlow & Cohen (1977) have used IR excess fluxes to determine mass-loss rates for a large sample of Galactic OBA supergiants, which generally compare favourably with recent Hα results for A and B supergiants (Kudritzki et al. 1999). Unfortunately, OB stars with relatively weak winds do not show a strong IR excess or radio flux, so mass-loss results have solely been for nearby hot stars with dense winds (Bieging et al. 1989; Drake & Linsky 1989). Observations at multiple frequencies are necessary to ensure against non-thermal (synchrotron) radio emission from colliding winds (e.g. Chapman et al. 1999). Hα and radio continuum mass-loss rates agree reasonably well (Lamers & Leitherer 1993) despite sampling quite different parts of the stellar wind – $\sim$1.5 stellar radii for Hα and $\sim$1000 stellar radii for the radio photosphere – placing constraints on relative clumping factors in these regions.

3. RESULTS FOR OBA STARS - ROLE OF METALLICITY?

Although the techniques used to study hot, massive stars are similar, we first discuss 'normal' early-type stars, and leave 'exotica' until subsequent sections. The mass-loss rate of a star is a fundamental ingredient in evolutionary models. Unfortunately, radiatively driven wind theory in its current form (Puls et al. 1996; Vink et al. 2000) is unable to reproduce the observed mass-loss properties for hot luminous stars sufficiently accurately (see below). Nevertheless, since their winds are driven by photon momentum transfer through metal line absorption, the stellar momenta and wind velocities are expected to be functions of stellar metallicity, Z, such that mass-loss rates scale with $Z^{0.5}$ for O stars and early B stars, $Z^{0.8}$ for mid-B supergiants, and $Z^{1.7}$ for A supergiants (Kudritzki & Puls 2000).

3.1. TERMINAL WIND VELOCITIES

Large compilations of wind velocities, v_∞, of O-stars and AB supergiants have been made by Prinja et al. (1990) and Lamers et al. (1995), revealing a spectral type and luminosity class dependence. Lamers et al. highlighted the so-called 'bi-stability' jump in wind properties around B 1.5 (~21,000), above which $v_\infty \simeq 2.65 v_{\rm esc}$ (escape velocity), and $v_\infty \simeq 1.4 v_{\rm esc}$ below, resulting from the change in ionization of the elements contributing to the line force (Vink et al. 1999). Wind velocities of LMC O stars differ little from Galactic counterparts (Garmany & Conti 1985), while a more prominent effect is observed in the SMC, particularly amongst early O stars (Walborn et al. 1995; Prinja & Crowther 1998).

Table 1 Coefficients for wind momentum luminosity relationship of Galactic O-stars and AB supergiants (Kudritzki & Puls 2000)

Sp Type	*log D_0*	x
O dwarf/giant	19.87±1.21	1.57±0.21
O supergiant	20.69±1.04	1.51±0.18
B0–1 supergiant	21.24±1.38	1.34±0.25
B1.5–3 supergiant	17.07±1.05	1.95±0.20
A0–3 supergiant	14.22±2.41	2.64±0.47

3.2. MASS-LOSS RATES

Observationally, estimates of mass-loss rates from Hα, UV wind lines or radio fluxes have been available for the past 20 years, although it is only recently that a reasonable degree of consistency has been reached between these different approaches via theoretical improvements. Puls et al. (1996) and Kudritzki et al. (1999) identified that O-stars and AB supergiants obey a wind momentum-luminosity relationship, as follows

$$\log \dot{M} v_\infty (R/R_\odot)^{0.5} = \log D_0 + x \log(L/L_\odot)$$

Stars of different spectral types have differing D_0 and x values, listed in Table 1, because of the change of lines driving the stellar wind with $T_{\rm eff}$.

Hot stars in the Magellanic Clouds provide us with the means to compare emirical mass-loss properties with predictions at lower Z (e.g. Garmany & Conti 1985). This has proved to be difficult given (i) the inability to measure radio fluxes for extra-galactic hot stars; (ii) the need to combine high quality UV and optical spectroscopy with sophisticated model techniques; (iii) the limited metallicity range spanned, 0.15–$1Z_\odot$. Although the number of stars analysed in detail for the Magellanic Clouds remains small (Puls et al. 1996), the mass-loss rates of LMC and SMC O stars tend to be lower than Galactic counterparts. As an example, let us consider Sk 80 (O7 Iaf+) in the SMC giant H II region NGC 346 which has a metallicity of $0.2Z_\odot$ (Haser et al. 1998; Peimbert et al. 2000). From the wind momentum-luminosity relationship we would expect an equivalent Galactic supergiant to have a mass loss rate four times higher than its observed value (Puls et al. 1996), suggesting a higher exponent of $Z^{0.8}$ in this case.

Table 2 Comparison ($\dot{M}(T_{\rm eff}, L)/\dot{M}_{\rm obs}$; σ in parenthesis) between predicted OBA mass-loss rates from various parameterizations with empirical data from Puls et al. (1996) and Kudritzki et al. (1999). Parameterizations are de Jager et al. 1988; Lamers & Cassinelli 1996; Vanbeveren et al. 1998; Kudritzki & Puls (2000); Vink et al. (2000) or Achmad et al. (1997) for A supergiants. Except for Lamers & Cassinelli, no metallicity dependence is accounted for, so we here assume a metallicity dependence of $Z^{0.5}$ for O stars, with $0.4Z_\odot$ for the LMC, $0.15Z_\odot$ for the SMC.

Spect. type (#)	*Galaxy*	*de Jager et al.*	*Lamers & Cassinelli*	*Vanbeveren et al.*	*Kudritzki & Puls*	*Vink et al.*
O (23)	Gal	0.9 (0.7)	1.1 (0.6)	0.9 (0.7)	1.3 (0.8)	1.8 (1.3)
O (6)	LMC	0.5 (0.5)	1.3 (0.7)	0.5 (0.3)	0.7 (0.3)	1.3 (0.4)
O (5)	SMC	1.3 (0.9)	2.5 (1.5)	1.3 (0.9)	2.3 (1.3)	6.5 (7.9)
B (14)	Gal	6.8 (4.6)	11.3 (7.7)		1.0 (0.3)	38.8 (29.4)
A (4)	Gal	5.0 (1.6)	3.7 (1.9)		1.0 (0.3)	2.0 (1.0)

Evolutionary calculations require $\dot{M}(T_{\rm eff}, L)$ parameterizations based on empirical data. Up until recently, the only compilation covering hot, luminous stars was that of de Jager et al. (1988). In an attempt to update the mass-loss calibration for evolutionary calculations, Vanbeveren et al. (1998) provided an updated relationship for Galactic OB stars, while Lamers & Cassinelli (1996) additionally considered LMC/SMC O stars. Most recently, Achmad et al. (1997) and Vink et al. (2000) have provided a set of predictions from current radiative driven winds for A, F and G supergiants and OB stars, respectively. How do these different parameterizations compare with observed mass-loss rates?

We have taken empirical data, $\dot{M}_{\rm obs}$, from Puls et al. (1996) for O stars and Kudritzki et al. (1999) for AB supergiants. In Table 2 we compare the mean ratio $\dot{M}(T_{\rm eff}, L)/\dot{M}_{\rm obs}$ for each calibration. Perhaps surprisingly, de Jager et al. (1988) fares as well as the recent Lamers & Cassinelli or Vanbeveren et al. (1998) parameterizations for O stars. In contrast with previous theoretical results, Vink et al. (2000) predicts sufficiently strong winds, albeit a factor of 2 too *strong* for Galactic O stars. Overall, the mass-loss rates of AB supergiants are typically overestimated (by 5–10) with the exception of Kudritzki et al. (1999) and Achmad et al. (1997). Predictions by Vink et al. (2000) show the poorest agreement for mid-B supergiants.

Overall, solely the wind momentum-luminosity prescription provides excellent agreement with all OBA observations. One note of caution is warranted, since the $T_{\rm eff}$-calibration for O stars, and hence the luminosity scale, relies almost universally on results from plane-parallel studies which neglect stellar winds (recall Sect 2.1).

3.3. ROTATION

O star rotational velocities ($v \sin i$) are typically 100 km s^{-1} (Howarth et al. 1997), with important consequences for their evolution (Meynet & Maeder 2000). Friend & Abbott (1986) first considered the effect of centrifugal acceleration on hot star winds, extended by Bjorkman & Cassinelli (1993) to allow for the azimuthal dependence, who found that $\dot{M}$ increases (and v_∞ decreases) towards the equator, leading to an oblate 'wind-compressed disk'.

Subsequently, Owocki et al. (1996) established that quite different predictions may result, by additionally allowing for non-radial components of the line force, and the increased polar radiation flux via gravity darkening (von Zeipel 1924). Due to a reduced escape velocity, the equatorial velocity law is slower than at the pole, inhibiting the equatorial disk and the original scaling of mass-loss can be reversed, such that a prolate geometry results. As an example, a mid-B dwarf rotating at 85% of its critical velocity was found by Petrenz & Puls (2000) to have a strongly prolate structure ($\rho_{\rm pole}/\rho_{\rm eq} \leq 15$) with a polar/equatorial terminal velocity of 1030/730 km s^{-1}. The global mass-loss rate was not found to deviate from its 1D value by greater than 10–20%, except for supergiants close to the Eddington limit.

3.4. STRUCTURE

Observationally, evidence for structure in O stars winds has taken various forms, including absorption variability observed from optical (Fullerton et al. 1996) and UV (Prinja 1992) spectroscopic intensive monitoring, and the presence of soft X-ray emission (Chlebowski et al. 1989), attributed to shocks in their winds (Lucy & White 1980). This view has recently been questioned by Chandra observations of ζ Ori (Waldron & Cassinelli 2000) which reveal that the X-ray emitting gas is exceptionally dense, with no evidence for expansion.

Theoretical studies of hot star winds reveal that line driven acceleration is subject to a strong instability, causing a variable and structured winds (Owocki et al. 1988; Feldmeier 1995). Since the region where Hα forms in OB stars ($\leq 1.5R_*$) is not anticipated to be heavily structured, mass-loss determinations may not be greatly affected by clumping/structure. Recent extensive monitoring of HD 152408 (WN9ha or O8 Iafpe) by Prinja et al. (2000) indicate variations of $\pm$10% in global mass-loss rate, although with a time averaged Hα profile that is remarkably stable over several years.

4. LUMINOUS BLUE VARIABLES

The stellar wind properties of LBVs are poorly constrained, although it is apparent that their winds are generally slower and denser than OBA supergiants (Crowther 1997), with little variation in mass-loss during their 'normal' irregular excursions across the H-R diagram (Leitherer 1997). To illustrate this, let us consider P Cygni, whose parameters have been determined by Najarro et al. (1997). Compared to normal B1.5 supergiants (780 km s^{-1}; Lamers et al. 1995), P Cyg has a very slow wind (185 km s^{-1}) and a mass-loss rate which exceeds the Galactic wind momentum relation (Sect. 2) by a factor of 15. Recall that from Petrenz & Puls (2000), non-rotating models of supergiants close to the Eddington limit may overestimate global mass-loss rates by a factor of $\sim$2, so that differences may be somewhat less severe. There are numerous examples of extreme supergiants which are not LBVs yet show remarkably high mass-loss rates and low wind velocities (e.g. Hillier et al. 1998).

In the case of giant LBV eruptions, information is very scarce and up until recently solely reliant on the 1840's η Car eruption causing the formation of the Homunculus, in which $\sim$3 $M_{\odot}$ was ejected over $\sim$30 years (Davidson & Humphreys 1997), corresponding to 0.1 $M_{\odot}$

yr^{-1}. Within the past few years, two extra-galactic LBVs have had sudden giant eruptions, HD 5980 in the SMC (Barba et al. 1995) and V1 in NGC 2363 (Drissen et al. 1997). These eruptions, in stars at low metallicity environments, appear to be quite different from that of η Car during its eruption, since peak mass-loss rates of each are $\sim 10^{-3} M_{\odot}\mathrm{yr}^{-1}$ (Drissen et al. 2001), equivalent to the current mass-loss for η Car (Hillier et al. 2000).

5. WOLF-RAYET STARS

The diagnostics of mass-loss for Wolf-Rayet stars mimic O stars except that their stronger winds mean that UV/optical wind diagnostics are not restricted to resonance lines or Hα. Wind velocities can be accurately determined either from the usual UV P Cygni profiles (Prinja et al. 1990), or from optical/IR He I P Cygni profiles (Eenens & Williams 1994). Radio and mm fluxes have been used to determine mass-loss rates of Galactic WR stars (Leitherer et al. 1997), while recent theoretical progress (Schmutz 1997; Hillier & Miller 1998) now permits detailed non-LTE analysis of UV and optical observations. WR winds are typically spherical (Harries et al. 1998), but clumped (Lepine et al. 2000). Spectroscopic tools make use of electron scattering wings in WR winds to estimate clumping factors since their strength depends linearly on density, versus the square of density for the underlying recombination line (Hillier 1991). Typical volume filling factors are 5–10%, such that mass-loss rates in a clumpy medium are 3–4 times lower than those from homogeneous models (e.g. Hillier & Miller 1999).

Although great progress has been made in recent years, detailed analysis requires labour intensive use of complex models, so the sample of stars studied in detail remains small. Recently, Nugis & Lamers (2000) combined empirical IR-radio observations with analytical models accounting for clumping to estimate mass-loss rates as a function of luminosity, He (Y) and metal (Z) mass fractions, i.e.

$$\log(\dot{M}_{\mathrm{WN}}) = -13.6 + 1.63 \log L/L_{\odot} + 2.22 \log Y$$

$$\log(\dot{M}_{\mathrm{WC}}) = -8.3 + 0.84 \log L/L_{\odot} + 2.04 \log Y + 1.04 \log Z$$

Recent evolutionary calculations for hydrogen-poor WR stars follow mass-loss calibrations of Langer (1989) or Vanbeveren et al. (1998). Table 3 compares predictions from these calibrations with results from detailed non-LTE line blanketed spectroscopic analyses, assuming the Schaerer & Maeder (1992) mass-luminosity relationship. The formulation of Langer (1989) overestimates mass-loss rates by factors of 2–4, while Vanbeveren et al. (1998) and especially Nugis & Lamers (2000) show much improved agreement with spectroscopic results.

Table 3 Comparison between WR mass-loss rates, in $10^{-5} M_{\odot} yr^{-1}$, predicted by Langer (1989, L89), Vanbeveren et al. (1998, VDV98), semi-empirical calibration of Nugis & Lamers (2000, NL2000) with detailed non-LTE spectroscopic results (nLTE) of hydrogen poor Galactic WN and WC stars.

Star	*Sp Type*	*L89*	*VDV98*	*NL2000*	*nLTE*	*Ref*
WR6	WN4	12.5	5.5	5.7	3.2	a
WR147	WN8(h)	9.5	4.5	3.2	2.5	b
WR111	WC5	2.8	2.0	1.3	1.5	c
WR90	WC7	4.8	3.2	2.6	2.5	d
WR135	WC8	2.0	1.6	1.6	1.2	d

(*a*) Schmutz (1997); (*b*) Morris et al. (2000); (*c*) Hillier & Miller (1999); (*d*) Dessart et al. (2000)

The question of a Z–dependence of mass-loss in WR stars remains open. There is a clear distinction amongst spectral subtypes in galaxies of differing metal content, which has been qualitatively explained from evolutionary and spectroscopic models (Smith & Maeder 1991; Crowther 2000). LMC WN stars (Crowther & Smith 1997; Hamann & Koesterke 2000) show negligible spectroscopic difference from Galactic counterparts. The situation is less clear for the SMC, since most WR stars are complicated by binarity. Sk41 is a possible exception, whose properties compare closely with Galactic or LMC counterparts (Crowther 2000). Further progress requires detailed study of apparently single WR stars beyond the Magellanic Clouds (e.g. Smartt et al. 2000), especially in nearby metal poor galaxies such as IC10 (Massey & Armandroff 1995).

6. MASS-LOSS DIAGNOSTICS IN YELLOW AND RED SUPERGIANTS

Various techniques have been applied to quantify mass-loss in yellow (FG) and red (KM) supergiants, whose wind driving mechanism remains uncertain. UV/optical observations of binaries with an early-type dwarf companion permit analysis of the red-supergiant chromosphere and wind in absorption (Bennett et al. 1996). Radio continua measurements have provided upper limits to mass-loss rates of F-supergiants (Drake & Linsky 1986) and M-supergiants (e.g. Harper et al. 2001). Alternatively, sub-mm CO emission or mid-IR dust emission permit mass-loss determinations in cool supergiants. CO emission is a reliable probe for AGB stars, although the UV radiation field from supergiant chromospheres could lead to a different CO emission per unit mass loss from star to star. Radial pulsations are generally assumed to provide the mechanism of initiating mass-loss in M supergiants, as evidenced from circumstellar dust shells (Bowers et al. 1983), with some notable exceptions – VY CMa (M5 Ia) and ρ Cas (F8 Ia$^+$). Outflow velocities of RSG are typically 10–40 km s^{-1} (Jura & Kleinmann 1990).

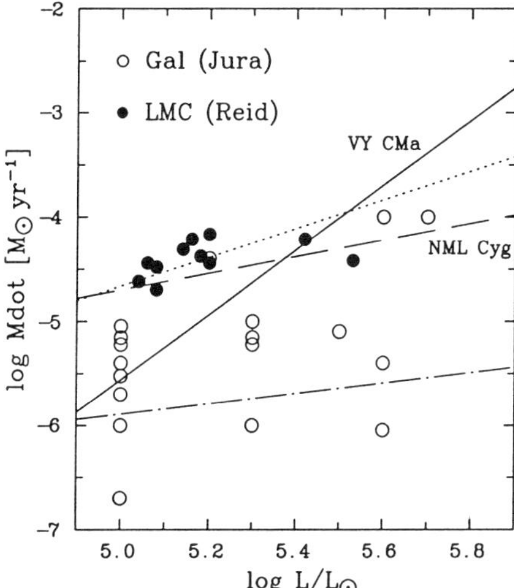

Figure 2 Mass-loss rates derived for Galactic (open circles) RSG by Jura & Kleinmann (1999) and LMC (filled circles) RSG by Reid et al.(1990) using the Jura (1986) formula, including predictions from de Jager et al. (1988, solid) and Reimers (1975, dot-dash) for $T_{\rm eff}$=3,500K. Fits to Reid et al. data are by Vanbeveren et al. (1998, dashed) and Salashnich et al. (1999, dotted).

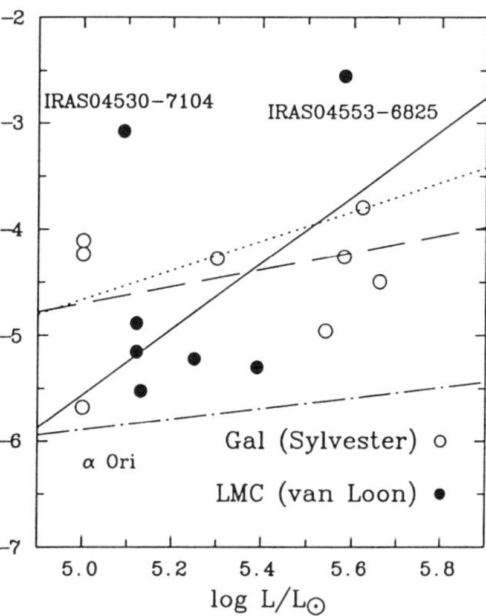

Figure 3 Mass-loss rates derived from modelling mid-IR spectroscopy of Galactic (open circles) and LMC (filledcircles) RSG by Sylvester et al. (1998) and van Loon et al. (1999). Distances to Galactic RSG follow Sylvester et al. except for W Per and μ Cep, which have been adjusted to distances derived by Garmany & Stencel (1992). Lines are as described in the adjacent caption.

Since we are concerned with massive stars in this review, we shall restrict our discussion to the most luminous 'hypergiants' (de Jager 1998), with log $(L/L_{\odot})$ >5. Studies have largely focussed on M supergiants, with only highly unusual FG supergiants studied in detail (e.g. ρ Cas, HR 8752, IRC+10420). The latter is widely interpreted as a RSG rapidly moving blueward (Jones et al. 1993), which has recently experienced huge variability in mass-loss rate, 10^{-2} to $10^{-4} M_{\odot}$ yr^{-1}, via analysis of its circumstellar dust shell (Blöcker et al. 1999).

Jura & Kleinmann (1990) applied the 'approximate' Jura (1986) formula for intermediate mass stars to IRAS 60μm photometry of nearby yellow and red supergiants to estimate mass-loss rates. Reid et al. (1990) followed the same approach for a sample of LMC RSG, which have been used by Vanbeveren et al. (1998) and Salasnich et al. (1999) to provide 'empirical' calibrations for evolutionary calculations, as illustrated in Fig. 2. Jura & Kleinmann (1990) data for Galactic RSG reveal systematically lower rates, except for VY CMa and NML Cyg. Predictions from the empirical calibration of de Jager et al. (1988, solid) and the formulation of Reimers (1975, dot-dash) are also illustrated, for an assumed $T_{\rm eff}$=3,500K (~M3 I). Although the scatter is large, most Galactic data lie between the Reimers (1975) and de Jager et al. (1988) calibrations.

Recently, van Loon et al. (1999) have modelled ISO spectroscopy and photometry of a similar LMC sample to Reid et al., and found rates which are 5–10 times *lower*. Meanwhile, Sylvester et al. (1998) have used the 10μm silicate emission feature for a sample of Galactic M supergiants, following the Skinner & Whitmore (1988) relation, which was calibrated against sub-mm CO determinations. Their determinations revealed *higher* mass-loss rates than Jura & Kleinmann (1990). From this combined dataset, presented in Fig. 3, there is little evidence of a luminosity dependence of mass-loss. The Vanbeveren et al. (1998) and Salasnich et al. (1999) calibrations match Galactic data reasonably well, except for α Ori (M2.2 Iab), which probably has the most secure determination. In contrast with Fig. 2, the majority of the LMC data indicate lower mass-loss rates than Galactic RSG, hinting at a metallicity dependence. However, van Loon et al. identified two LMC RSG with exceptionally high mass-loss rates of order $10^{-3} M_{\odot} {\rm yr}^{-1}$ (IRAS 04553-6825 and IRAS 04530-7104) which are unmatched in our Galaxy.

Overall, the level of consistency between different IR/sub-mm techniques remains embarrassingly poor. Sylvester et al. (1998), using the power in 10μm silicate emission (calibrated against earlier CO line data), derived a mass-loss rate for μ Cep (M2 Ia) which was 25 times *higher* than Jura & Kleinmann (1990) obtained from 60μm dust emission, or 250 times *higher* than Josselin et al. (2000) based on an alternative CO line calibration of Olofsson et al. (1993). Indeed, VY CMa has values ranging over a factor of twenty – $3.5\times10^{-4} M_{\odot} {\rm yr}^{-1}$ (Stanek et al. 1995) to $1.6\times10^{-5} M_{\odot}$ ${\rm yr}^{-1}$ (Josselin et al. 2000).

In summary, the mass-loss rates for RSG are by far the most controversial when recommending empirical calibrations for evolutionary calculations. The majority of mass-loss estimates lie between the Reimers (1975) and Vanbeveren et al. (1998) calibrations. This is illustrated in Table 4, where we compare rates assumed in the recent evolutionary model of Meynet & Maeder (2000) for an initially rapidly rotating $40M_\odot$ star with 'empirical' calibrations of Kudritzki & Puls (2000) for OBA stars, Vanbeveren et al. (1998) for RSG and Nugis & Lamers (2000) for WR stars. Agreement is very good, except for the brief, low mass-loss B supergiant phase and *highly uncertain* high mass-loss RSG phase. Continuing studies into the latter are highly desirable. Typical outflow velocities for each evolutionary phase are also indicated in Table 4.

Table 4 Summary of wind properties during various phases during the evolution of a rotating $\sim 40M_\odot$ star ($V_{\rm init}$ =300 km s^{-1}) at solar metallicity (following Meynet & Maeder 2000). Two sets of mass-loss rates are given - those adopted by Meynet & Maeder, $\dot{M}^{\rm MM}$, and those taken from recent empirical calibrations, $\dot{M}^{\rm emp}$.

Age *Myr*	*Mass* $M_\odot$	$\log L$ $L_\odot$	*Sp Type*	$\dot{M}^{MM}$	$\dot{M}^{emp}$ $10^{-6}M_\odot yr^{-1}$	v_∞ $km\,s^{-1}$
0.0	40	5.4	O4 V	1.2[a]	0.4[e]	3000[h]
3.7	35	5.5	O7 I	2.1[a]	2.1[e]	2100[h]
5.1	31	5.7	B3 I	7.6[a]	0.7[e]	490[i]
5.1	31	5.8	A0 I	22.[b]	14.[e]	170[i]
5.1	29	5.8	M0 I	400.[b]	86.[f]	30[j]
5.3	18	5.7	WNL	30.[c]	26.[g]	750[h]
5.3	16	5.6	WNE	25.[d]	31.[g]	2000[h]
5.4	13	5.5	WCE	15.[d]	28.[g]	2500[h]

(*a*) Lamers & Cassinelli (1996); (*b*) de Jager et al. (1988); (*c*) Nugis et al. (1988); (*d*) Langer (1989) corrected for clumping (Schmutz 1997); (*e*) Kudritzki & Puls (2000); (*f*) Vanbeveren et al. (1998); (*g*) Nugis & Lamers (2000); (*h*) Prinja et al. (1990); (*i*) Lamers et al. (1995); (*j*) Jura & Kleinmann (1990)

Acknowledgments

Many thanks to the hospitality of JILA, University of Colorado at Boulder where this review was written, especially Prof P.S. Conti. Financial support is provided by a Royal Society URF. Thanks to Prof M.J. Barlow and Dr G. Harper for useful discussions and to Dr G. Meynet for providing recent grids of evolutionary calculations.

References

Achmad, L., Lamers, H.J.G.L.M. & Pasquini, L., 1997, A&A **320**, 196

Barba, R.H., Niemela, V.S., Baume, G. & Vazquez, R.A., 1995, ApJ **446**, L23

Barlow, M.J. & Cohen, M., 1977, ApJ **213**, 737

Bennett, P.D., Harper, G.M., Brown, A. & Hummel, C.A., 1996, ApJ **471**, 454

Bieging, J.H., Abbott, D.C. & Churchwell, E.B., 1989, ApJ **340**, 518

Blöcker, T., Balega, Y., Hofmann, K.-H. et al. 1999, A&A **348**, 805

Bohannan, B. & Crowther, P.A., 1999, ApJ **511**, 374

Bowers, P.F., Johnston, K.J. & Spencer J.H., 1983, ApJ **274**, 733

Bjorkman, J.E. & Cassinelli, J.P., 1993, ApJ **409**, 429

Castor J.I., Abbott D.C. & Klein, R.I., 1975, ApJ **195**, 157

Chapman J.M., Leitherer, C., Koribalski, B., Bouter, R. & Storey, M, 1999, ApJ **518**, 890

Chlebowski, T., Harnden, F.R.Jr. & Sciortino, S., 1989, ApJ **341**, 427

Crowther, P.A., 1997, in Proc. LBVs: Massive Stars in Transition, ASP Conf. Ser. **120** (eds. Nota, A., Lamers, H.J.G.L.M.) p.51

Crowther, P.A., 2000, A&A **356**, 191

Crowther, P.A. & Smith, L.J., 1997, A&A **320**, 500

Davidson, K. & Humphreys, R. M., 1997, ARA&A **35**, 1

de Jager, C., 1998, A&A Rev **8**, 145

de Jager, C., Nieuwenhuijzen, H. & van der Hucht, K.A., 1988, A&AS **72**, 259

Dessart, L., Crowther, P.A., Hillier, D.J. et al. 2000, MNRAS **315**, 407

Drake, S.A. & Linsky, J.L., 1986, AJ **91**, 602

Drake, S.A. & Linsky, J.L., 1989, AJ **98**, 1831

Drissen, L., Roy, J.-R. & Robert, C., 1997, ApJ **474**, L35

Drissen, L., Crowther, P.A., Smith, L.J. et al. 2001, ApJ in press (astro-ph/0008221)

Eenens, P.R.J. & Williams, P.M., 1994, MNRAS **269**, 1082

Feldmeier, A., 1995, A&A **299**, 523

Friend, D.B. & Abbott, D.C., 1986, ApJ **311**, 701

Fullerton, A.W., Gies, D.R. & Bolton, C.T., 1996, ApJS **103**, 475

Fullerton, A.W. & Najarro, F., 1998, in Proc. Boulder-Munich II: Properties of hot, luminous stars, ASP Conf Ser. **131** (I.D. Howarth, ed.), p.47

Fullerton A.W., Crowther, P.A., De Marco, O. et al., 2000, ApJ **538**, L43

Garmany, C.D. & Conti, P.S., 1985, ApJ **293**, 407

Garmany, C.D. & Stencel, R.E., 1992, A&AS **94**, 211

Groenewegen, M.A.T., Lamers, H.J.G.L.M. & Pauldrach, A.W.A., 1989, A&A **221**, 78

Hamann, W-.R. & Koesterke, L., 2000, A&A **360**, 647

Harper, G.M., Brown, A. & Lim, J., 2001, ApJ submitted

Harries, T.J., Hillier, D.J. & Howarth, I.D., 1998, MNRAS **296**, 1072

Haser, S.M., Pauldrach, A.W.A., Lennon, D.J. et al., 1998, A&A **330**, 285

Hillier, D.J., 1991, A&A **247**, 455
Hillier, D.J., Crowther, P.A., Najarro, F. & Fullerton, A.W., 1998, A&A **340**, 483
Hillier, D.J., Davidson K., Ishibashi, K. & Gull T., 2000, ApJ in press
Hillier D.J. & Miller D., 1998, ApJ **496**, 407
Hillier D.J. & Miller D., 1999, ApJ **519**, 354
Howarth, I.D., Siebert, K.W., Hussain, G.A.J. & Prinja, R.K., 1997, MNRAS **284**, 265
Jones, T.J., Humphreys, R.M., Gehrz, R.D. et al. 1993, ApJ **411**, 323
Josselin, E., Blommaert J.A.D.L., Groenewegen, M.A.T., Omont, A. & Li, F.L., 2000, A&A **357**, 225
Jura, M. 1986, ApJ **303**, 327
Jura, M. & Kleinmann, S.G., 1990, ApJS **73**, 769
Kudritzki, R-.P., Puls, J., Lennon, D.J. et al. 1999, A&A **350**, 970
Kudritzki, R-.P. & Puls, J. 2000, ARA&A **38**, 613
Lamers, H.J.G.L.M., Cerruti-Sola, M. & Perinotto, M., 1987, ApJ **314**, 726
Lamers, H.J.G.L.M. & Leitherer, C., 1993, ApJ **412**, 771
Lamers, H.J.G.L.M., Snow, T.P. & Lindholm, D.M., 1995, ApJ **455**, 269
Lamers, H.G.J.L.M. & Cassinelli, J.P., 1996, in Proc. From Stars to Galaxies: The Impact of Stellar Physics on Galaxy Evolution, ASP Conf. Ser. **98** (eds. Leitherer C., Fritze-vob-Alvensleben, U., Huchra, J.), p.162
Lamers, H.G.J.L.M. & Cassinelli, J.P., 1999, Introduction to Stellar Winds, CUP, Cambridge
Lamers, H.G.J.L.M., Haser, S., de Koter, A. & Leitherer, C., 1999, ApJ **516**, 872
Langer, N. 1989, A&A **220**, 135
Leitherer, C., 1997, in Proc. LBVs: Massive Stars in Transition, ASP Conf. Ser. **120** (eds. Nota, A., Lamers, H.J.G.L.M.) p.58
Leitherer, C., Chapman, J.M. & Koribalski, B., 1997 ApJ **481**, 898
Lepine, S., Dalton, M.J., Moffat, A.F.J. et al. 2000, AJ in press
Lucy, L.B. & White, R.L. 1980, ApJ **241**, 300
Massey, P. & Armandroff, T.E., 1995, AJ **109**, 2470
Meynet, G., Maeder, A., Schaller, G., Schaerer, D. & Charbonnel, C., 1994, A&AS **103**, 97
Meynet, G., & Maeder, A., 2000, A&A **361**, 101
Morris P.W., van der Hucht, K.A., Crowther, P.A. et al. 2000, A&A **353**, 624
Najarro, F., Hillier, D.J., Kudritzki, R-.P. et al. 1994, A&A **285**, 573
Najarro, F., Hillier, D.J. & Stahl, O. 1997, A&A **326**, 1117
Nugis, T., Crowther, P.A. & Willis, A.J. 1998, A&A **333**, 956

Nugis, T. & Lamers, H.J.G.L.M. 2000, A&A **360**, 227
Panagia, N. & Felli, M. 1975, A&A **39**, 1
Pauldrach A.W., Puls J. & Kudritzki R-.P. 1986, A&A **164**, 86
Peimbert, M., Peimbert, A. & Ruiz, M.T. 2000, ApJ **541**, 688
Petrenz, P. & Puls, J. 2000, A&A **358**, 956
Prinja, R.K., 1992, in Proc. Nonisotropic and Variable Outflows from Stars, ASP Conf. Ser. **22**, (eds. Drissen, L., Leitherer, C., Nota, A.) p.167
Prinja, R.K., Barlow, M.J. & Howarth, I.D., 1990, ApJ **361**, 607
Prinja, R.K. & Crowther, P.A. 1998, MNRAS **300**, 828
Prinja, R.K., Stahl, O., Kaufer A. et al. 2000, A&A submitted
Puls, J., Kudritzki, R-.P., Herrero, A. et al. 1996, A&A **305**, 171
Reid, N., Tinney, C. & Mould, J., 1990, ApJ **348**, 98-119
Reimers, D., 1975, In: Problems in stellar atmospheres and envelopes, Springer-Verlag (New York), p.229
Olofsson, H., Eriksson, K., Gustafsson, B. & Carlstrom, U. 1993, ApJS **87**, 267
Owocki, S.P., Castor, J.I. & Rybicki, G.B. 1988, ApJ **335**, 914
Owocki, S.P., Cranmer, S.R. & Gayley, K.G. 1996, ApJ **472**, L115
Salasnich, B., Bressan, A. & Choisi, C. 1999, A&A **342**, 131
Schaerer, D. & Maeder, A. 1992, A&A **263**, 129
Schmutz, W. 1997, A&A **321**, 268
Skinner, C.J. & Whitmore, B. 1988, MNRAS **231**, 169
Smartt, S.J., Crowther, P.A., Dufton, P.L. et al. 2000 MNRAS submitted (astro-ph/0009156)
Smith, L.F. & Maeder, A., 1991, A&A **241**, 77
Sylvester, R., Skinner, C.J. & Barlow, M.J. 1998, MNRAS **301**, 1083
Stanek, K.Z., Knapp, G.R., Young, K. & Phillips, R.G. 1995, ApJS **100**, 169
van Loon, J.Th., Groenewegen, M.A.T., de Koter, A. et al. 1999, A&A **351**, 559
Vanbeveren D., De Loore, C. & Van Rensbergen, W. 1998, A&A Rev. **9**, 63
Vink, J.S., de Koter, A. & Lamers, H.J.G.L.M. 1999, A&A **350**, 181
Vink, J.S., de Koter, A. & Lamers, H.J.G.L.M. 2000, A&A in press
von Zeipel, H., 1924, MNRAS **84**, 665
Walborn, N.R., Nichols-Bohlin, J. and Panek, R.J., 1985, IUE Atlas of O-type spectra from 1200 to 1900Å. NASA RP 1155.
Walborn, N.R., Lennon, D.J., Haser, S., Kudritzki, R-.P. & Voels, S.A., 1995, PASP **107**, 104
Waldron, W.L. & Cassinelli, J.P. 2000, ApJ Letters submitted
Wickramasinghe, N.C., Donn, B.D. & Stecher T.P., 1966, ApJ **146**, 590
Wright, A.E. & Barlow, M.J. 1975, MNRAS **170**, 41

IV

STELLAR EVOLUTION

NEWS ON THE EVOLUTION OF MASSIVE STARS

Andre Maeder
Geneva Observatory
CH - 1290 Sauverny, Switzerland
andre.maeder@obs.unige.ch

Keywords: Stellar evolution, rotation, massive stars, Wolf–Rayet stars

Abstract The evolution of massive stars strongly depends on both mass loss and rotation. We show that the account of rotation modifies all the output of stellar evolution: tracks in the HRD, lifetimes, age determinations, chemical abundances, balance between blue and red supergiants as well as all the properties of WR stars.

1. INTRODUCTION

Models with mass loss were generally very successful to explain many observational properties of massive stars evolution. However, several significant discrepancies between model predictions and observations have been found for massive stars, and also for red giants of lower masses. These difficulties have been listed and examined (cf. Maeder 1995). Both rotation and binary interactions appear as possible mechanisms to account for the observed discrepancies. The question immediately arises, which one of these two mechanisms (rotation or binarity) is the dominant one for explaining the various abundance anomalies in massive stars. We shall consider this point with a particular attention in Section 2 below.

Recently, large excesses of [N/H] have been found in A–type supergiants, particularly in the SMC (Venn et al. 1998; Venn 1999), where the excesses may reach up to an order of magnitude with respect to the average [N/H] local value in the SMC. These excesses, not predicted by current models, are the clear signatures of mixing processes over the entire star. This extensive mixing changes the size of the reservoir of nuclear fuel available for evolution, and thus the lifetimes, the tracks and all the model outputs (cf. also Langer 1992 ; Heger et al. 2000;

D. Vanbeveren (ed.), The Influence of Binaries on Stellar Population Studies, 233–247.

Maeder & Meynet 2000a). Also the chemical abundances, the yields and the final stages are modified. Thus, it is an essential question to understand the origin of the mixing present in massive stars. We show below that rotation is likely the leading parameter responsible for the observed discrepancies, which does not preclude, however, binary effects to also play a role in some cases.

The physical effects of stellar rotation are numerous (cf. Maeder & Meynet 2000a). The basic equations of stellar structure need to be modified. The meridional circulation and its interaction with the horizontal turbulence, the diffusion effects produced by shear turbulence, the inhibition of the transport mechanisms by the μ–gradients, the transport of the angular momentum and of the chemical elements, the loss of mass and angular momentum at the surface, the enhancement of mass loss by rotation (cf. Maeder and Meynet 2000b), etc... are some of the effects to be included in realistic models. In addition to the model physics, rotation also brings a number of new numerical problems in the stellar code, such as the 4th order equation for the transport of the angular momentum, which has to be coupled to the equations of stellar structure.

2. ROTATION VERSUS BINARY EFFECTS

Both rotation and binarity produce an upwards extension of the top of the Main Sequence (Maeder 1971), or may modify the balance between blue and red supergiants, or may produce an earlier entry in the WR phase. Thus, disentangling the two effects from the HR diagram is not easy. However, the study of the chemical abundances may provide a solution. The chemical abundances of Helium, CNO elements and Na are modified by both rotation and binarity and both effects produce some mixing of CNO elements with original cosmic abundances and CNO equilibrium abundances. Fortunately, we can identify a few effects which behave differently in the case of rotation or in the case of binarity.

In addition, it is worth to emphasize that several discrepancies between models and observations (cf. Maeder 1995) are consistent with rotation. In particular, as shown by Herrero et al. (1999; 2000) there are no fast rotating OB stars with normal He– abundances.

Boron versus CNO abundances:

Boron depletion has been found in some main sequence B stars (cf. Venn et al 1996). In some cases, the B-depletion goes along with a N-excess, while there are also stars with a B-depletion and a normal N-content.

Rotation can produce both Boron depletions and N–excesses, if the rotation velocity is high. Rotation can also produce a Boron depletion and no N–excess, if the rotation velocity is low, (Fliegner et al. 1996). However, binary mass transfer only leads to a destruction of Boron and at the same time to N-excesses (Wellstein & Langer 2000). Thus, the above mentioned observation of Boron–depletions together with normal N–abundances is only consistent with a rotational origin. This is a first indication in favour of rotational mixing rather than binary mass transfer as an effect responsible for the abundance anomalies in massive stars.

He- and N-excesses as a function of luminosity:
All supergiant stars appear to have He– and N–excesses. However, larger He– and N–excesses have been found for supergiants of types A and B with higher luminosities (McEarlan et al 1999). This relation between evidences of CNO processed elements and luminosity is quite consistent with the properties of rotating stellar models (Heger et al. 2000; Meynet and Maeder 2000), while a relation between He–, N–excesses and stellar luminosities is not predicted by binary models (Wellstein & Langer 2000). It may thus be concluded that the contribution of the binary scenario does not seem to be the dominant one for explaining the abundance anomalies in A– and B–type supergiants.

He– and N–excesses as a function of metallicity:
The observations by Venn et al. (1998) and Venn (1999) also show that the the He– and N—excesses in A–type supergiants are higher for stars with lower metallicity Z, like in the SMC, as compared to the excesses for supergiants in the Galaxy. This feature is quite consistent with the predictions of models of rotating stars at low Z (Maeder & Meynet 2000a), which show higher He– and N–enrichments in A–type supergiants of lower metallicities, for reasons given just below. As to binary stars, there is no obvious reason why the effects would be larger at lower Z, and there is no appropriate model in agreement with this observation.

The above relation between evidences of CNO processed elements and metallicity Z is quite consistent with the fact that the relative fraction of Be–stars is larger at lower metallicities, which suggests that the average rotational velocities of OB stars near the end of the Main Sequence are generally larger and would thus indicate larger rotational effects at lower Z. Another effect contributing to the above relation is that, at lower Z, models of supergiants stay longer in the B–supergiant phase than at solar composition, which enables the surface enrichments to proceed further (Maeder & Meynet 2000a).

There are probably additional discriminating tests than the three above ones. However, this first basis supports the view that rotational

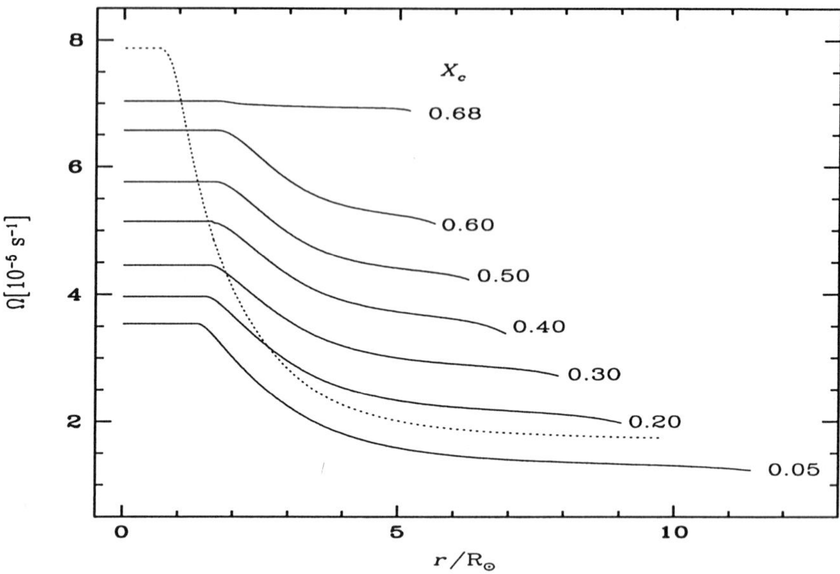

Figure 1 Evolution of the angular velocity Ω as a function of the distance to the center in a 20 $M_{\odot}$ star with $v_{\rm ini}$ = 300 km s^{-1}. X_c is the hydrogen mass fraction at the center. The dotted line shows the profile when the He–core contracts at the end of the H–burning phase.

effects, rather than binary effects, are responsible for the many discrepancies found in massive stars.

3. EVOLUTION OF ROTATION

Even in some recent works the assumption of solid body rotation is considered. It is, however, much more physical to examine the evolution of $\Omega(r)$ resulting from the transport of angular momentum by shears, meridional circulation and convection, also from the effects of central contraction and envelope expansion and from the losses of angular momentum by stellar winds. In particular, the large losses of angular momentum at the surface lead to a redistribution of the angular momentum in the interior by meridional circulation, shear turbulent diffusion and convection. The whole problem is self–consistent, since the meridional circulation and shear transport depend in turn on the degree of differential rotation. Thus, the full system of equations has to be solved in order to provide the internal evolution of Ω (cf. Maeder & Meynet 2000a).

The evolution of $\Omega(r)$ during the MS evolution of a 20 $M_\odot$ star is shown in Fig. 1 from Meynet & Maeder (2000). We notice that initially (i.e. for high values of the central H–content X_c), the degree of differential rotation is small, then, it grows during the evolution. The ratio between the central and surface values of Ω remains however small during MS evolution. The rotation rate does not vary by more than about a factor two throughout the star. *Although this degree of differential rotation is not large, it plays an essential role in the shear effects and the related transports of angular momentum and chemical elements.* The same remark also applies to the effects of differential rotation on the transport by meridional circulation, since $U(r)$ critically depends on the derivatives of Ω in the star (cf. Zahn 1992; Maeder & Zahn 1998).

We notice a general decrease of $\Omega(r)$, even in the convective core which is contracting. The main reason is the mass loss at the stellar surface, which removes a substantial fraction of the total angular momentum. This makes $\Omega(r)$ to decrease with time everywhere, because of the internal transport mechanisms, which ensure some coupling of rotation. The same behaviour was obtained by Langer (1998) in the case of rigid rotation (*i.e.* in the case of maximum coupling). In the present models, the shear transports the angular momentum outwards, which reduces the core rotation, while circulation makes an inward transport in the deep interior and an outwards transport in the external region (Gratton–Öpik term in the velocity of meridional circulation). This makes the progressive flattening of the curves above $r = 4R_\odot$ in Fig. 1.

From the end of the MS evolution onwards, i.e. when $X_c \leq 0.05$, central contraction becomes faster and starts dominating the evolution of the angular momentum in the center. The central Ω grows quickly (cf. dotted curve in Fig. 1) and this will in principle continue in the further evolutionary phases until core collapse. The evolution of the angular momentum after the H–burning phase is mainly dominated by the local conservation of the angular momentum, since the secular transport mechanisms have little time to operate. This is the assumption we are making in the present models. During these phases, the chemical elements continue to be rotationally mixed in the radiative layers mainly through the shear effect. In the further evolutionary phases, the strong central contraction will lead to very large central rotation, unless some other processes as fast dynamical instabilities or magnetic braking occur in these stages.

Fig. 2 shows the evolution of $\frac{\Omega}{\Omega_{\rm crit}}$ as a function of the age for the present models. Firstly, we notice that for models without mass loss, as shown for the 20 $M_\odot$ with $\dot{M} = 0$, v and $\frac{\Omega}{\Omega_{\rm crit}}$ go up fastly so that the critical velocity would be reached near the end of the MS phase. The

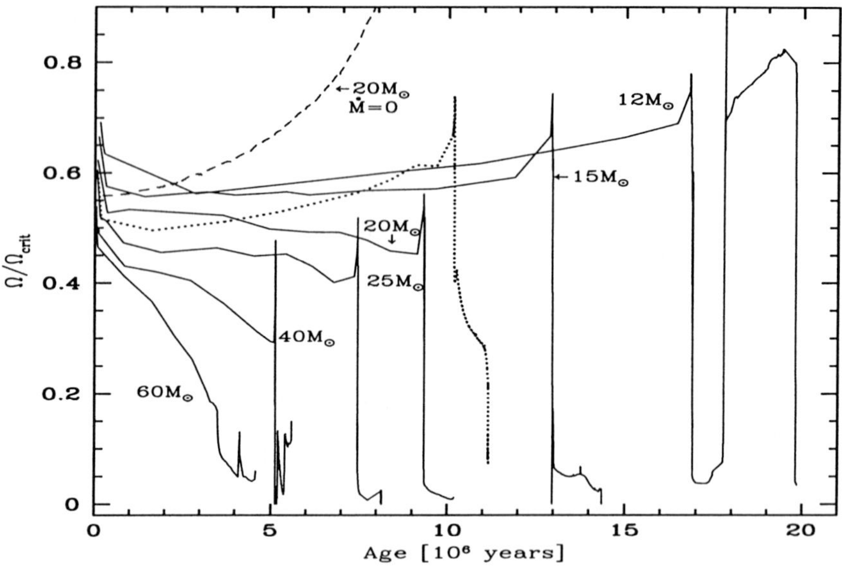

Figure 2 Evolution of the ratio $\Omega/\Omega_{\rm crit}$ of the angular velocity to the break–up velocity at the stellar surface. as a function of time for stars of different initial masses with $v_{\rm ini}$ = 300 km s^{-1}. The continuous lines refer to solar metallicity models, the dotted line corresponds to a 20 M$_\odot$ star with $Z = 0.004$. The dashed line corresponds to a 20 M$_\odot$ star without mass loss.

current model of 20 M$_\odot$ with mass loss shows a significant decrease of v, while the critical ratio remains almost constant during most of the MS phase. Fig. 2 shows how fastly rotation decreases at the surface of the most massive stars, which lose a lot of mass. Consistently we see that the reduction of the surface rotation is much larger for the more massive stars. This is of course a consequence of the removal of large amounts of angular momentum by the stellar winds. The effect is amplified by the increase of the mass loss in fast rotators. We see that the decrease of $\frac{\Omega}{\Omega_{\rm crit}}$ is so strong that it will prevent a massive star to reach the critical velocity near the end of the MS phase.

For initial stellar masses of 12 or 15 M$_\odot$, a peak is reached during the overall contraction phase at the end of the MS phase. Then, v and the critical ratio go down as the star moves to the red supergiant phase. During such a fast evolution, we may say that the rotation evolves almost like in the simple case, where the angular momentum is conserved locally. The large growth of the radius implies a decrease of v and of $\frac{\Omega}{\Omega_{\rm crit}}$. Later during the blue loops, where the Cepheid instability strip is crossed, the

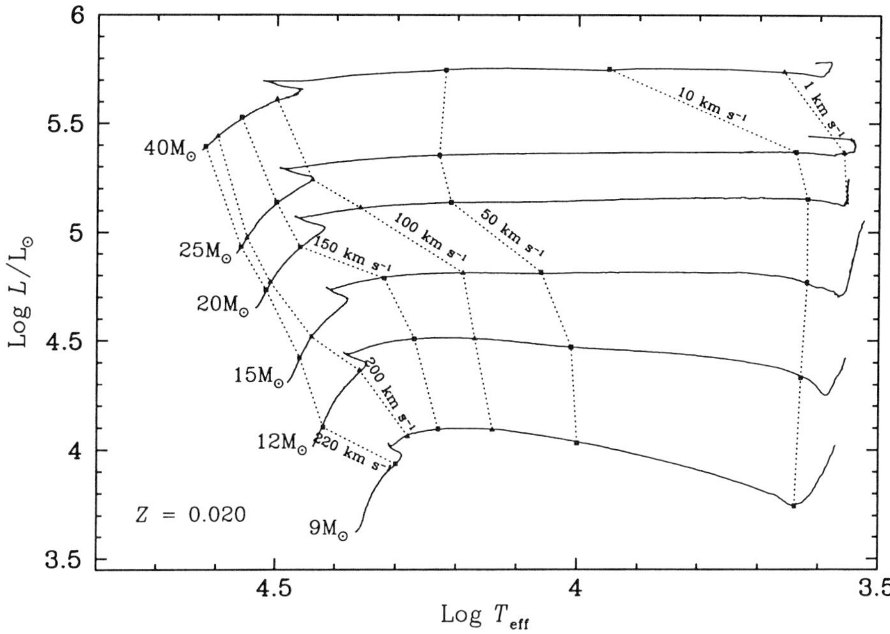

Figure 3 Evolution of the equatorial surface velocities along the evolutionary tracks in the HR diagram starting from $v_{\rm ini}$ = 300 km s^{-1}. For purpose of clarity, only the first part of the 40 M$_\odot$ track is shown.

rotation velocity becomes very large again and could easily become close to critical. This behaviour, also found by Heger & Langer (1998) results from the stellar contraction which concentrates a large fraction of the angular momentum of the star (previously contained in the extended convective envelope of the RSG) in the outer few hundredths of a solar mass. This result suggests that rotation may also somehow influence the Cepheid properties.

Amazingly, we notice that the stars with initially low mass loss during the MS phase, like for stars with M $\leq$ 12 M$_\odot$, have more chance to reach the break–up velocities near the end of the MS and thus to experience huge mass loss than the more massive stars which lose a lot of mass in the early MS phase. It is somehow surprising that little mass loss during most of the MS may favour large mass loss rates at the end of the MS phase. This may explain why stars close to break–up, like the Be stars, do not form among the O–type stars, but mainly among the B–type stars, where we see that the ratio $\frac{\Omega}{\Omega_{\rm crit}}$ may increase during the MS phase. Another related observation is the fact that the relative number

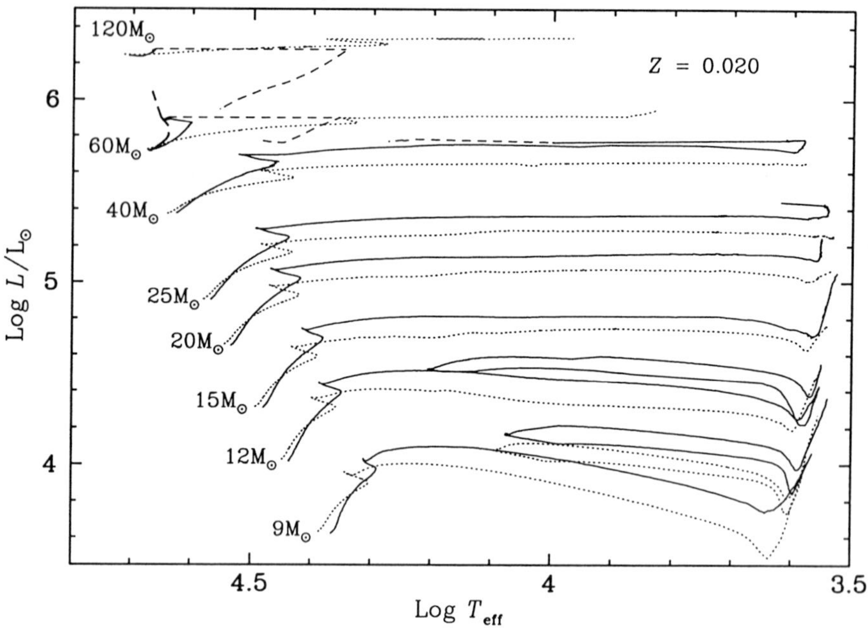

Figure 4 Evolutionary tracks for non–rotating (dotted lines) and rotating (continuous lines) models with solar metallicity. The rotating models have an initial velocity $v_{\rm ini}$ of 300 km s^{-1}. For purpose of clarity, only the first part of the tracks for the most massive stars (M $\geq$ 40 M$_\odot$) is shown. Portions of the evolution during the W–R phase for the rotating massive stars are indicated by short–dashed lines. The long–dashed track for the 60 M$_\odot$ model corresponds to a very fast rotating star ($v_{\rm ini}$ $\sim$ 400 km s^{-1}), which follows a nearly homogeneous evolution.

of Be–stars with respect to B–type stars is much higher in the SMC and LMC than in the Galaxy (cf. Maeder et al. 1999). In the SMC and LMC, due to the lower metallicity, the average mass loss rates are lower and thus these stars may keep higher rotation in general and thus form more Be stars. This explanation does not exclude differences in the initial $v_{\rm ini}$ as well.

The evolution of v in the HR diagram is shown in Fig. 3. Starting with models having a velocity of 300 km s^{-1} on the zero–age main sequence, we give some lines of constant v over the HR diagram. On the MS, we notice in particular that the decrease is much faster for the most massive stars than for stars with M $\leq$ 15 M$_\odot$. This difference remains also present in the domain of B– and A–supergiants. During the crossing of the HR diagram, the rotational velocities decrease fastly, to become very small, i.e. of the order of a few km s^{-1}, in the red supergiant phase.

4. HR DIAGRAM, LIFETIMES AND AGES

Figure 4 from Meynet & Maeder (2000) shows the evolutionary tracks of non–rotating and rotating stellar models for initial masses between 9 and 120 $M_\odot$. For initial masses $M \leq 40\ M_\odot$, the MS tracks with rotation become more luminous than for no rotation. On the whole, the tracks with rotation for masses below 40 $M_\odot$ mimic the tracks with overshooting with more extended tracks, (the effects of atmospheric distorsion for an average orientation is also accounted for in Fig. 4). This results from essentially two effects. On one side, rotational mixing brings fresh H–fuel into the convective core, slowing down its decrease in mass during the MS phase. For a given value of the central H–content, the mass of the convective core in the rotating model is therefore larger than in the non–rotating case. On the other side, rotational mixing transports helium into the radiative envelope. The He–enrichment lowers the opacity. This contributes to the enhancement of the stellar luminosity and favours a blueward track.

Rotation reduces the MS width in the high mass range ($M \geq 40\ M_\odot$). Let us recall that when the considered stellar mass increases, the ratio of the diffusion timescale for the chemical elements to the MS lifetime decreases (Maeder 1998). As a consequence, starting with the same $v_{\rm ini}$ on the ZAMS, massive stars will be more mixed than low mass stars at an identical stage of their evolution. This reduces the MS width since greater chemical homogeneity makes the star bluer. Moreover, due to both rotational mixing and mass loss, their surface will be rapidly enriched in helium.

For $Z = 0.020$, the lifetimes are increased by about 20–30% when the mean rotational velocity on the MS increases from 0 to ~200 km s^{-1}. The He–burning lifetimes are less affected by rotation than the MS lifetimes. The changes are less than 10%. Thus, the ratios t_{He}/t_H of the He to H–burning lifetimes are only slightly decreased by rotation and remain around 10–15%.

On Fig. 5 isochrones for ages between about 8 and $20 \cdot 10^6$ yr. computed from non–rotating and rotating stellar models are presented. An isochrone with rotation is almost identical to an isochrone without rotation with log age smaller by 0.1 dex. This has for consequence that an average rotation with $v_{\rm ini} = 200$ km s^{-1} leads to an increase of the age estimated for a given cluster by about 25%. We may wonder whether this effect explains the difference between the age estimates based on the upper MS and the estimates based on the lithium content of the very low mass stars (see e.g. Martin et al. 1998; Barrado y Navascués et al. 1999).

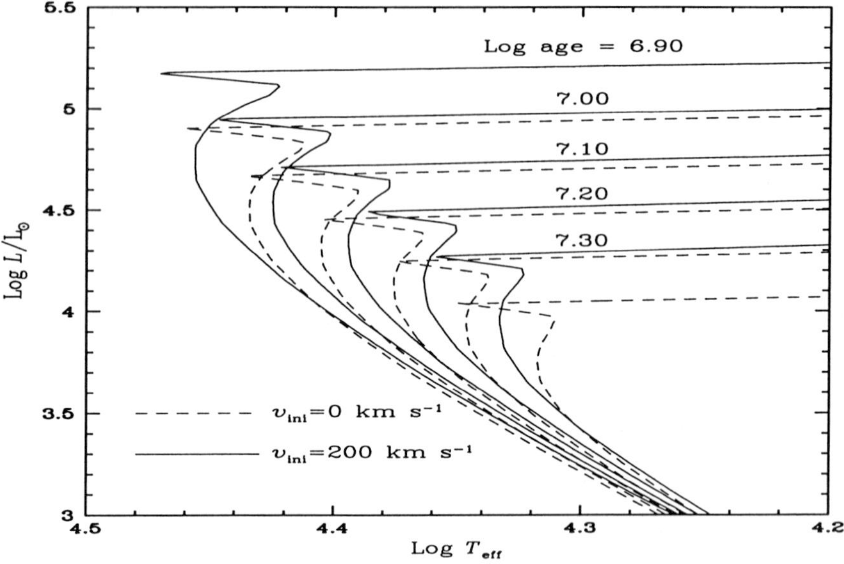

Figure 5 Isochrones computed from stellar evolutionary tracks for the solar metallicity. The dashed and continuous lines correspond to the case of non–rotating and rotating stellar models respectively. In this last case, models have an initial velocity $v_{\rm ini}$ of 200 km s^{-1}. The logarithms of the ages (in years) label the isochrones computed from the models with rotation.

Rotation leads to deviations from the current mass–luminosity relations. High rotation produces larger He–cores and He–rich envelopes, which implies overluminous stars with lower gravity as evolution proceeds. This can occur even for the slow rotating objects because they could have had a sufficiently high initial rotational velocity. We estimate that differences of ∼30% in mass at a given luminosity are quite possible.

5. CHEMICAL ABUNDANCES

The rotational mixing produces by shear diffusion and meridional circulation produces He–, N–enrichments and related C– and O–depletions at the stellar surface of rotating stars. These changes at the surface are the consequence of a general modification of the internal distribution of the elements. Fig. 6 from Meynet & Maeder (2000) shows the changes of the surface ratio N/C in stellar models of rotating and non rotating stars evolving from the ZAMS to the red supergiant stage. The N/C ratio appears as the most sensitive observable parameter. For

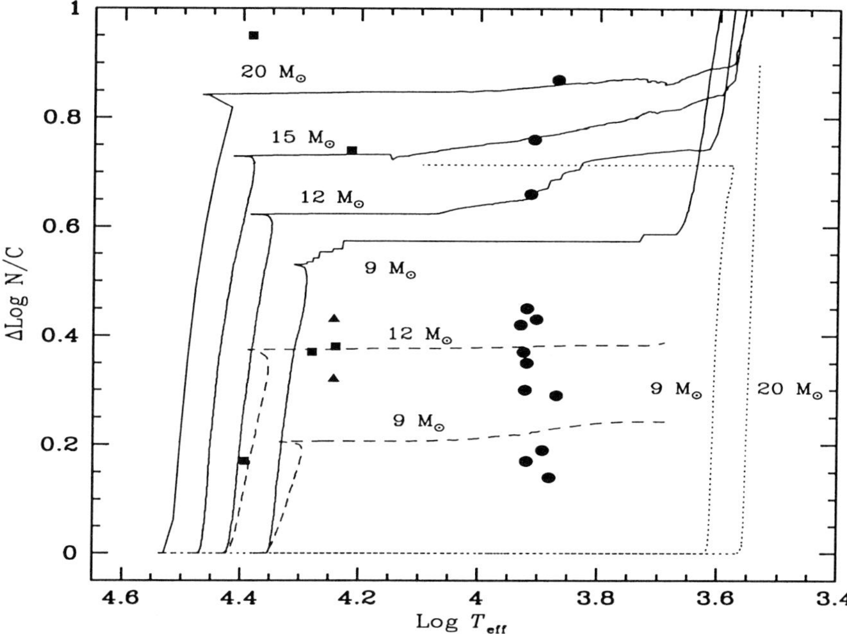

Figure 6 Evolution as a function of log $T_{\rm eff}$ of $\Delta \log \frac{\rm N}{\rm C} = \log({\rm N/C}) - \log({\rm N/C})_i$, where N and C are the surface abundances (in number) of nitrogen and carbon respectively, the index i indicates initial values. The continuous tracks correspond to rotating stellar models with $v_{\rm ini}$ = 300 km s^{-1}. The dashed and dotted lines are respectively for initial $v_{\rm ini}$ = 200 and 0 km s^{-1}. Only parts of the tracks with $v_{\rm ini}$ = 200 km s^{-1} are shown. The initial masses are indicated. The solid squares and triangles correspond to the observations of blue supergiants by Gies & Lambert (1992) and Lennon (1994) as reported by Venn (1995b). The solid circles show the positions of galactic A–type supergiants observed by Venn (1995a, b).

non–rotating stars, the surface enrichment in nitrogen only occurs when the star reaches the red supergiant phase; there, CNO elements are dredged–up by deep convection. The behaviour is the same as for the He–enrichments. For rotating stars, N–excesses already occur during the MS phase and they are larger for higher rotation and initial stellar masses.

Comparisons with observations, as indicated in the figure caption, are also made in Fig. 6. We notice first that at the $T_{\rm eff}$ of the observed supergiants, non–rotating stellar models predict no enrichment at all unless there is a blue loop. However, all the observed points are likely not at the tip of a blue loop. Indeed some of the stars have initial

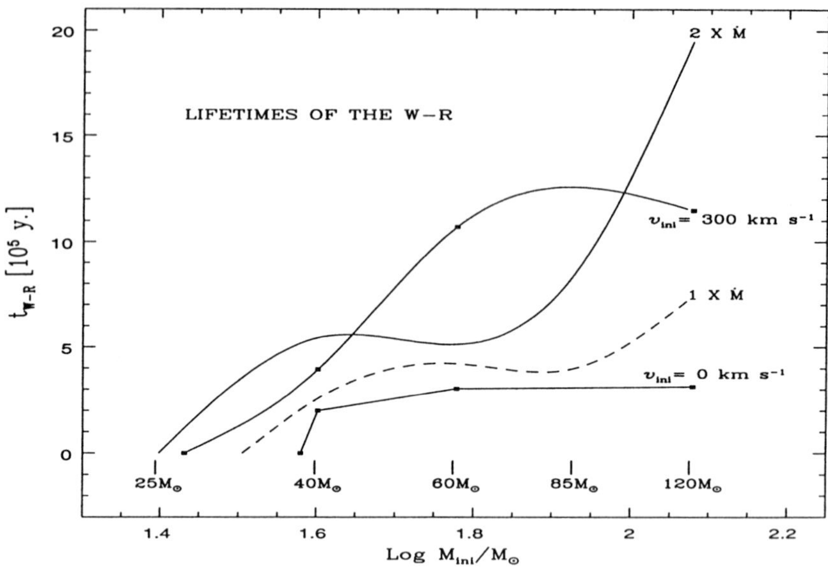

Figure 7 Duration of the Wolf–Rayet phase in units of 10^5 years as a function of the initial mass expressed in solar units. The relations labeled by 1 x $\dot{M}$ and 2 x $\dot{M}$ correspond to the solar metallicity models computed with standard and high mass loss rates respectively (Schaller et al. 1992; Meynet et al. 1994). The other two relations correspond to the present models computed with and without rotation.

masses above 15 $M_\odot$ and, at solar metallicity, current grids of models only predict blue loops for masses equal or lower than 15 $M_\odot$. Also, many of the observed stars have N/C ratios too low to result from a first dredge–up episode . Therefore Venn (1995b) suggested that the stars presenting the lowest N/C ratios are on their way from the MS to the red giant branch and have undergone some mixing in the early stage of their evolution. Such stars are not at all accounted for by standard evolutionary tracks, while rotating models well reproduce their observed abundances.

6. WOLF–RAYET STARS

A difficulty of the previous grids of models was the need of mass loss rates twice the observed values, in order to well account for the observed number ratios of WR stars to O-type stars. Rotational mixing as well as the increase of the mass loss rates by rotation contribute also to the formation of WR stars. Rotation, by making smoother internal composition gradients, also modifies the evolution of the surface abundances, particularly in the WC stage.

Fig. 7 compares the lifetimes of models with zero rotation and with an initial velocity of 300 km/s to the previous results (Schaller et al. 1992; Meynet et al. 1994). The interesting result is that rotation lowers the mass threshold for forming WR stars from about 38 $M_{\odot}$ to about 27 $M_{\odot}$ when an initial velocity of 300 km/s is considered. In that respect, we recall that according to Fig. 3 the average corresponding velocity during the MS is much lower. Also, rotation considerably increases the WR lifetimes, thus there is no need of an arbitrary increase of the mass loss rates to well account for the observations.

Construction of models at various metallicities Z will enable us to examine the effects of rotation on the number ratios of WR stars, blue, red supergiants, O–tyope stars in galaxies of the Local Group. In this respect, we emphasize that models now in progress at low Z show that when rotation is included the massive stars spend a much longer time in the red supergiant stage and this is a very favourable result in view of the ususal difficulty (cf. Langer & Maeder 1995) to account for the large number of red supergiants at low Z as in the SMC.

Rotation also brings interesting results for the chemical abundances in WR stars. Firstly, rotational mixing, like mass loss, favours the appearance at the stellar surface of He– and N–abundances typical of the the WN stars. Thus, we cannot exclude, if surface abundances are the leading factor for WR stars, that some WR stars are still in the H–burning phase.

The CNO abundances in WN star models without rotation are usually just the equilibrium CNO abundances. The inclusion of rotation makes the evolution of the surface abundances much more progressive (Maeder & Meynet 2000a), so that the CNO abundances in WN stars may significantly deviate from equilibrium values, in the sense that the equilibrium values are not yet reached. Rotation also enables us to better understand the occurence of the so–called transition WN/WC stars. Rotational mixing brings some new ^{12}C and ^{22}Ne in regions where ^{14}N is not yet depleted. Therefore even some Neon excesses are possible in WN stars. Changes of the ^{22}Ne/^{14}N ratios should even be noticeable on human timescale in some transition WN/WC stars !

The abundances in WC stars also show quite interesting results. Without rotation, the abundances of He, C and O are almost the same in general, with some dominance of ^{12}C. With rotation, the rise of ^{12}C and ^{16}O at the beginning of the WC phase is very much slower. Thus, during nearly all the WC phase, the abundance of ^{12}C is much lower than the abundance of helium (Maeder & Meynet 2000a). The difference is even larger for ^{16}O. These results look quite promising in view of recent abundance determinations in WC stars.

7. CONCLUSIONS

Nowadays rotation appears as an essential ingredient of massive star evolution, as is the mass loss by stellar winds. In the case of binary evolution, it is likely that some mixing effects similar to those we have considered here for rotation will also intervene and need to be considered, in order to make consistent models of binary evolution.

Acknowledgments

I express my best thanks to Dr. Georges Meynet for the most active collaboration we have on these matters.

References

Barrado y Navascués D., Stauffer J.R., Patten B.M., 1999, ApJ 522, L53

Fliegner J., Langer N., Venn K.A., 1996, A&A 308, L13

Gies D.R., Lambert D.L., 1992, ApJ 387, 673

Heger A., Langer N., 1998, A&A 324, 210

Heger A., Langer N., Woosley S.E., 2000, ApJ 528, 368

Herrero A., Corral L.J., Villamariz M.R., Martin E.L., 1999, A&A 348, 542

Herrero A., Kudritzki R.P., Vilchez J.M., Kunze D., Butler K., Haser S., 1992, A&A 261, 209

Herrero A., Puls J., Villamariz M.R., 2000, A&A 354, 193

Langer N., 1992, A&A 265, L17

Langer N., 1997, The Eddington Limit in Rotating Massive Stars. In: Nota A., Lamers H. (eds.) Luminous Blue Variables: Massive Stars in Transition. ASP Conf. Series 120, p. 83

Langer N., 1998, A&A 329, 551

Langer N., Maeder A., 1995, A&A 295, 685

Lennon D.J., 1994, Space Sci. Rev. 66, 127

Maeder A., 1971, A&A 10, 354

Maeder A., 1995, Population I Stellar Structure and Evolution: facing the lingering difficulties to make a step forward. In: Stobie R.S., Whitelock P.A. (eds) Astrophysical applications of stellar pulsation. ASP Conf. Series 83, p. 1

Maeder A., 1998, The Physics of Rotational Mixing in Massive Stars. In: Howarth I. (ed.) Boulder-Munich II: Properties of Hot, Luminous Stars. ASP Conf. Ser. 131, p. 85

Maeder A., Meynet G., 2000a, ARAA, vol 38, in press.
Maeder A., Meynet G., 2000b, A&A 361, 159(Paper VI)
Maeder A., Zahn J.P., 1998, A&A 334, 1000 (Paper III)
Maeder A., Grebel E., Mermilliod J.C., 1999, A&A 346, 459
Martin E.L., Basri G., Gallegos J.E., Rebolo R., Zapatero-Osorio M.R., Bejar V.J.S., 1998, ApJ 499, L61
McErlean N.D., Lennon D.J., Dufton P.L., 1999, A&A 349, 553
Meynet G., Maeder A., Schaller G., Schaerer D., Charbonnel C., 1994, A&AS 103, 97
Meynet G., Maeder A., 2000, A&A 361, 101 (Paper V)
Schaller G., Schaerer D., Meynet G., Maeder A., 1992, A&AS 96, 269
Venn K.A., 1995a, ApJ 449, 839
Venn K.A., 1995b, ApJS 99, 659
Venn K.A., 1996, ApJ 308, L13
Venn K.A., McCarthy J.H., Lennon D.J., Kudritzki R.P., 1998, A-Supergiant Abundances in Local-Group Galaxies. In: Howarth I. (ed.) Boulder-Munich II: Properties of Hot, Luminous Stars. ASP Conf. Ser. 131, p. 177
Venn K.A., 1999, ApJ 518, 405
Wellstein S., Langer N., 2000, A&A350, 148
Zahn J.P., 1992, A&A 265, 115

MASSIVE CLOSE BINARIES

Dany Vanbeveren

Astrophysical Institute, Vrije Universiteit Brussel, Pleinlaan 2, 1050 Brussels, Belgium

dvbevere@vub.ac.be

Keywords: Massive Stars, stellar winds, populations

Abstract The evolution of massive stars in general, massive close binaries in particular depend on processes where, despite many efforts, the physics are still uncertain. Here we discuss the effects of stellar wind as function of metallicity during different evolutionary phases and the effects of rotation and we highlight the importance for population (number) synthesis. Models are proposed for the X-ray binaries with a black hole component. We then present an overall scheme for a massive star PNS code. This code, in combination with an appropriate set of single star and close binary evolutionary computations predicts the O-type star and Wolf-Rayet star population, the population of double compact star binaries and the supernova rates, in regions where star formation is continuous in time.

1. INTRODUCTION

Population number synthesis (PNS) of massive stars relies on the evolution of massive stars and, therefore, uncertainties in stellar evolution imply uncertainties in PNS. The mass transfer and the accretion process during Roche lobe overflow (RLOF) in a binary, the merger process and common envelope evolution have been discussed frequently in the past by many authors (for extended reviews see e.g. van den Heuvel, 1993, Vanbeveren et al., 1998a, b). In the present paper we will focus on the effects of stellar wind mass loss during the various phases of stellar evolution and the effect of the metallicity Z. The second part of the paper summarises the Brussels overall massive star PNS with a realistic fraction of binaries. We compare theoretical prediction with observations of O-type stars, Wolf-Rayet (WR) stars, supernova (SN) rates and binaries with two relativistic components [neutron star (NS) or black hole (BH)], in regions where star formation is continuous in time. In the last part of the paper, we will discuss overall evolutionary effects of rotation and

D. Vanbeveren (ed.), The Influence of Binaries on Stellar Population Studies, 249–271.

the consequences for PNS. As usual, we will classify binaries as Case A, Case B_r, Case B_c and Case C depending on the onset of RLOF.

2. STELLAR WIND FORMALISMS

We distinct four stellar wind (SW) phases: the core hydrogen burning (the OB-type phase) prior to the Luminous Blue Variable (LBV) phase, the LBV phase, the Red Supergiant (RSG) phase, the Wolf-Rayet (WR) phase. To study the effect on evolution one relies on observed SW mass loss rates and how these rates vary as function of stellar parameters (luminosity, temperature, chemical abundances, rotation). In the following, $\dot{M}$ is in $M_{\odot}$/yr and the luminosity L is in $L_{\odot}$.

The OB-type and LBV phase. SW mass loss formalisms of OB-type stars have been reviewed by Kudritzki and Puls (2000) (see also P. Crowther, in these proceedings). Note that all formalisms published since 1988 give very similar results when they are implemented into a stellar evolutionary code.

LBVs are hot, unstable and very luminous OB-type supergiants. Those brighter than $M_{\rm bol}$ = -9,5 (corresponding to stars with initial mass $\geq$ 40 $M_{\odot}$) are particularly interesting as far as evolution is concerned. They are losing mass by a more or less steady stellar wind ($\dot{M}$ = 10^{-7}-10^{-4} $M_{\odot}$/yr) and by eruptions where the star may lose mass at a rate of 10^{-3}-10^{-2} $M_{\odot}$/yr. The outflow velocities are typically a few 100 km/s. The observed SW-rates are highly uncertain, but due to the fact that no RSGs are observed brighter than $M_{\rm bol}$ = -9,5 an evolutionary working hypothesis could be: *the $\dot{M}$ during the LBV + RSG phase of a star with $M_{\rm bol} \leq$ -9,5 must be sufficiently large to assure a RSG phase which is short enough to explain the lack of observed RSGs with $M_{\rm bol} \leq$ -9,5.*

The RSG phase. Using the formalism proposed by Jura (1987), Reid et al. (1990) determined the SW mass loss rate for 16 luminous RSGs in the Large Magellanic Cloud. The following equation relates these $\dot{M}$ values to the luminosity

$$Log(-\dot{M}) = 0.8 Log L - C \quad (19.1)$$

with C = 8.7. However, RSG mass loss rates are subject of possible large uncertainties and C = 9 ($\dot{M}$ is a factor 2 smaller than predicted with C = 8.7) can not be excluded.

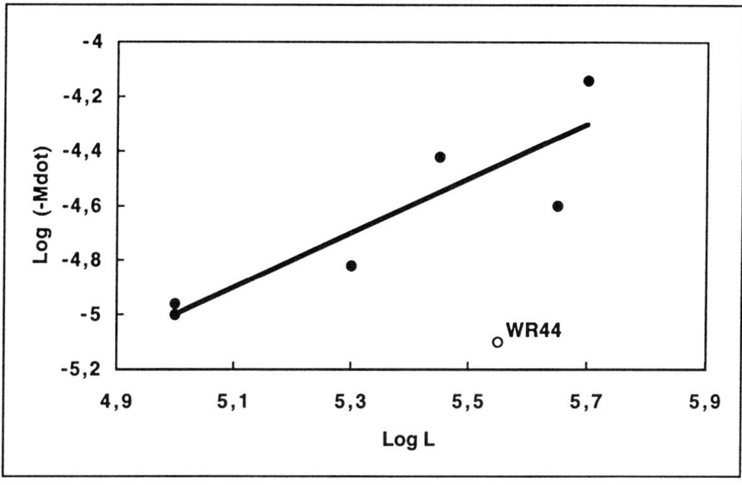

Figure 1 The Log $\dot{M}$-Log L diagram of galactic WR stars

The WR phase. There is increasing evidence that the outflowing atmospheres of WR stars are inhomogeneous and consist of clumps (Crowther, the present proceedings). SW mass loss rates of WR stars determined by methods that include the effects of clumps are a factor 2-4 times smaller than those obtained with smooth wind models.

Table 1 Luminosity and mass loss rates of galactic WR stars. [1] = De Marco et al., 2000, [2] = Hillier and Miller, 1999, [3] = Morris et al., 2000, [4] = Hamann and Koesterke, 1998, [5] = Nugis et al., 1998

WR Star	*Log L*	*Log (-$\dot{M}$)*	*ref*
γ^2 Vel	5	-5	[1]
HD 165763	5.3	-4.8	[2]
WR 147	5.65	-4.59	[3]
HD 50896	5.45	-4.42	[4]
V444 Cyg	5	-5.04	[5]
WR 44	5.55	-5.1	[4]
WR 123	5.7	-4.14	[4]

Table 1 lists a number of galactic WR stars where the $\dot{M}$ and the luminosity is determined by detailed spectroscopic analysis.

Notice that the spectroscopic $\dot{M}$-value of the WR component of the binary V444 Cyg corresponds to independent determinations which are based on the binary period variation (Khaliulin et al., 1984) and on polarimetry (St.-Louis et al., 1988; Nugis et al., 1998), whereas γ^2 Vel is a nearby WR binary and its distance (based on Hipparcos measurements) is very well known, thus also its luminosity. Figure 1 illustrates a possible $\dot{M}$-L relation. We have given large statistical weight at the observations of V444 Cyg and of γ^2 Vel. WR 44 is a peak which seems to be unrealistic from statistical point of view. Either the luminosity is too high or the mass loss rate is too small. Omitting WR 44, the following formula links $\dot{M}$ and L

$$Log(-\dot{M}) = LogL - 10 \tag{19.2}$$

Remark that including WR 44 gives a relation which predicts even smaller $\dot{M}$-values for Log L $\geq$ 5 compared to Equation 19.2.

The effect of metallicity Z on $\dot{M}$. SW rates resulting from spectral analysis are uncertain by at least a factor 2. It is therefore difficult to deduce any kind of metallicity dependence from observations. Theory predicts that for a radiation driven SW, $\dot{M} \propto Z^{\zeta}$ with $0.5 \leq \zeta \leq 1$. We expect that OB-type mass loss rates obey this rule.

We do not know how the SW rates of LBVs and RSGs depend on Z. We like to remind that the $\dot{M}$-L relation (Equation 19.1) holds for RSGs in the LMC. If the $\dot{M}$ of RSGs depends on Z as well, the effect on the evolution of galactic stars could be even larger.

Hamann and Koesterke (2000) investigated 18 WN stars in the LMC. Figure 2 illustrates the Log $\dot{M}$-Log L diagram. There seems to be no obvious relation however, as an exercise, we used a linear relation similar to (Equation 19.2) and shifted the line until the sum of the distances between observed values and predicted ones was smallest (a linear regression where the slope of the line is fixed). This resulted in a shift of -0.2 which, accounting for the Z value of the Solar neighbourhood and of the LMC corresponds to $\zeta = 0.5$. To be more conclusive, many more observations are needed where detailed analysis may give reliable stellar parameters, but from this exercise it looks as if *WR mass loss rates are Z-dependent.*

3. MASSIVE STAR EVOLUTION

When the most recent SW mass loss formalisms of OB-type stars are implemented into a stellar evolutionary code, one concludes that *stellar*

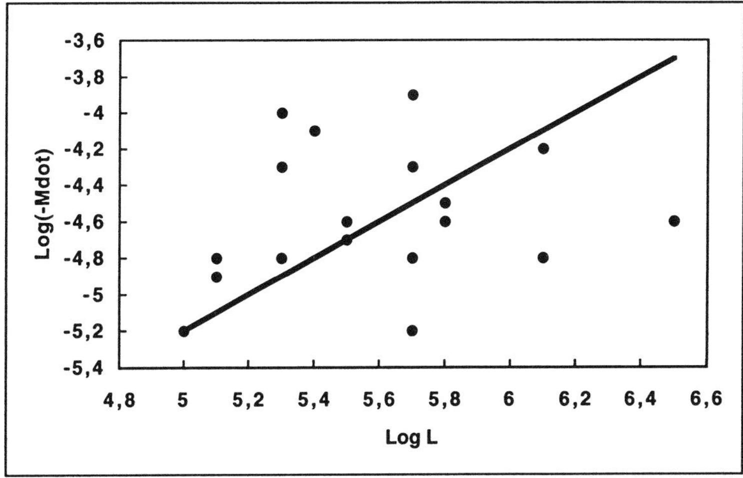

Figure 2 The Log $\dot{M}$-Log L diagram for WR stars in the LMC

wind mass loss during core hydrogen burning (prior to the LBV phase) is not important for the evolution of galactic stars with initial mass smaller than 40 $M_\odot$.

A star with initial mass larger than 40 $M_\odot$ loses most of its hydrogen rich layers by OB+LBV-type stellar wind. When it is a close binary component and when the binary period P is large enough so that RLOF does not start before the onset of the LBV phase (roughly P $\geq$ 10 days), mass loss by RLOF may be largely suppressed (or may not occur at all), i.e. *due to OB-type + LBV-type stellar wind, the importance of RLOF in a binary with orbital period* $\geq$ *10 days and with primary mass larger than 40* $M_\odot$ *may be significantly reduced; RLOF may even be avoided.*

Of particular importance is the evolution of the binary period due to stellar wind. The specific angular momentum ($\dot{J}/\dot{M}$) of a spherically symmetric wind can be expressed as

$$\frac{\dot{J}}{\dot{M}} = \lambda A^2 \Omega \tag{19.3}$$

with A the distance between the two binary components and $\Omega = \frac{2\pi}{P}$. When the wind velocity of the mass loser (with mass M_1 and mass loss $\dot{M}_1$) is large compared to the orbital velocity ΩA of the companion (with mass M_2), this specific angular momentum can be approximated by the momentum of matter at the moment it leaves the loser, i.e.

$$\lambda = \left(\frac{M_2}{M_1 + M_2}\right)^2 \tag{19.4}$$

resulting in an overall binary period increase.

However, when the wind velocity is comparable to the orbital velocity, the outflowing matter can gain orbital angular momentum by torque and the value of λ may be significantly larger than predicted by Equation 19.4. This is particularly important for the evolution of binaries with very small mass ratio, possible progenitors of the Low Mass X-ray Binaries (LMXBs) with a BH component (see Sect. 4).

3.1. THE EFFECT OF RSG MASS LOSS

The influence of RSG mass loss rates (Equation 19.1) on single star evolution has been described in extenso by Vanbeveren et al., (1998a, b) and Salasnich et al. (1999). One of the important consequences of the larger RSG rates is that *Galactic single stars with initial mass as low as 20(15)* $M_\odot$ *become hydrogen deficient core helium burning stars = WR stars.*

Hamann and Koesterke (1998, 2000) determined the luminosity and temperature of a large sample of WR stars in the Galaxy and in the LMC. When one considers the WN stars without hydrogen and if at least some of the WR stars with the lowest luminosity are single, their position in the HR-diagram is indirect evidence for the RSG-wind evolution of single stars as discussed in the papers cited above.

When a star with initial mass $\geq$ 15-20 $M_\odot$ is a binary component and when the period is large enough so that RSG wind starts before the star fills its Roche lobe (Case B_c and Case C binaries), it is clear that RSG wind can significantly reduce the importance of the RLOF mass loss (and the common envelope process). Even more, case C will not happen. Let us remark that an RSG wind is very slow. This means that λ in Equation 19.3 may be substantially larger than predicted by Equation 19.4.

If RSG mass loss rates are Z-dependent as predicted by the radiatively driven wind theory, it can readily be understood that the minimum mass of single WR star progenitors in small Z-regions will be larger than in the Galaxy.

3.2. THE EFFECT OF WR MASS LOSS

Due to LBV and/or RSG stellar wind mass loss or due to RLOF, a massive star may become a hydrogen deficient core helium burning star. The further evolution is governed by a WR-type stellar wind mass loss.

The effect on massive star evolution of present day WR-type mass loss rates (Equation 19.2) has been studied by Vanbeveren et al.(1998 a, b, c). Recently, Wellstein and Langer (1999) discussed the core helium burning (CHeB) evolution of massive stars by adopting a stellar wind mass loss rate relation

$$Log(-\dot{M}) = 1.5 Log L - 12.25 \tag{19.5}$$

for hydrogen deficient CHeB stars with log L $\geq$ 4.45. Compared to Equation 19.2, two differences are important. Equation 19.5 assumes a stronger luminosity dependence whereas it predicts $\dot{M}$-values for the stars of table 1 which are still significantly larger than the observed values. We will compare our CHeB results with results calculated with Equation 19.5.

The Galactic WR lifetime. To deduce the WR-lifetime one obviously needs a definition of a WR star that is useful for evolution. Most of the single WR stars have Log L $\geq$ 5 (Hamann and Koesterke, 1998, 2000) whereas most of the WR components in WR+OB binaries have a mass $\geq 8\ M_{\odot}$ (Vanbeveren et al., 1998b). Accounting for the theoretical M-L relation of WR stars discussed later, it can be concluded that both conditions are similar. We will use the Log L $\geq$ 5 criterion. Figure 3 shows the galactic WN and WC lifetimes as a function of initial progenitor mass. We only consider the core helium burning WR stars. The most massive stars may become WNL stars (WN stars with significant hydrogen in their atmosphere) already during core hydrogen burning burning (see Andre Maeder, present proceedings) and this obviously increases the total WN lifetime compared to the lifetimes shown in Figure 3. Remark that most of the WNE (WN stars without hydrogen) stars originate from stars with initial mass between 20 $M_{\odot}$ and 40 $M_{\odot}$ whereas most of the WC stars have M $\geq$ 40 $M_{\odot}$ progenitors. The figures illustrate that a PNS study of WR stars will not critically depend on whether evolutionary results are used where the RSG stellar wind mass loss is computed with Equation 19.1 with C = 8.7 or with C = 9. The different wind formalisms during the WR phase have a larger effect (typically a factor 2).

The M-L relation of WR stars. A mass luminosity relation of WR stars has been proposed by Vanbeveren and Packet (1979), Maeder (1983), Langer (1989), Vanbeveren (1991), Schaerer and Maeder (1992). All relations closely match, although different assumptions were made

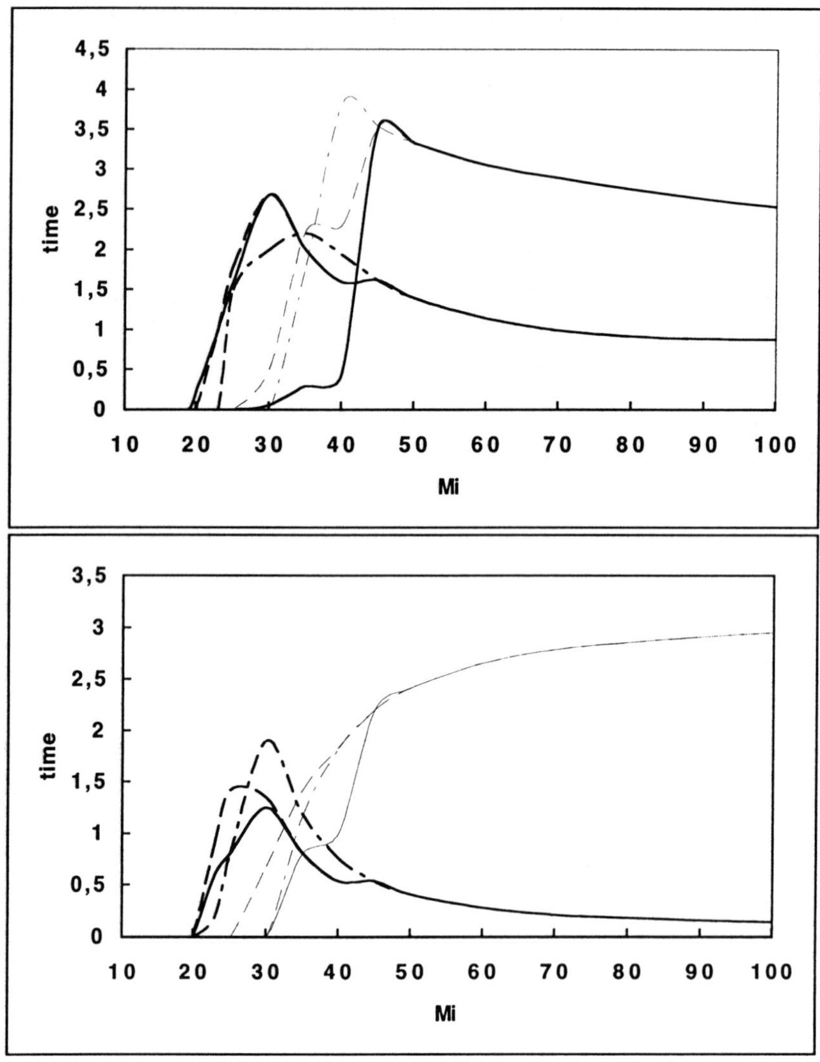

Figure 3 The WN (thick lines) and WC (thin lines) lifetime (in 10^5 yr) as function of initial progenitor mass (in $M_\odot$). Dashed-dotted (respectively dashed lines and full lines) lines apply to case B binaries (respectively single stars where the RSG mass loss is calculated with Equation 19.1 with C = 8,7 and with C = 9. Above 40 $M_\odot$ all stars lose their hydrogen rich layers very fast due to the LBV-type formalism discussed in Sect. 2. The top-figure (respectively bottom-figure) corresponds to our preferred WR mass loss formalism Equation 19.2 (respectively Equation 19.5).

concerning a number of uncertainties during CHeB. This relation can therefore be considered as a very robust one.

The final mass prior to core collapse. Figure 4 gives the final mass of the star and of the CO-core prior to core collapse, for single stars and for Case B binary components. Remind that Case B is the more frequent class of interacting binaries. Compared to Case B, it is trivial to understand that Case A components have smaller final masses whereas it is obvious that Case C (and non-interacting) binaries are similar to single stars. Equation 19.1 decides upon the mass loss during the RSG phase. The top-figure holds for the WR rates predicted by Equation 19.2 whereas the bottom-figure corresponds to the formalism adopted by Wellstein and Langer (2000). Notice the much larger final masses when Equation 19.2 holds, i.e. *with our preferred WR mass loss formalism, stars with initial mass between 40 $M_\odot$ and 100 $M_\odot$ end their life with a mass between 10 $M_\odot$ and 20 $M_\odot$ corresponding to CO-cores with a mass between 5 $M_\odot$ and 15 $M_\odot$.*

It must be clear that PNS of binaries with relativistic components, NSs or BHs, depends critically on the adopted WR mass loss formalism.

Figure 4 can be used to link the minimum mass of BH formation of single stars with the one of primaries of case B-type binaries. Suppose that the latter is 40 $M_\odot$. Based on the value of the CO-core mass, the corresponding minimum mass for single stars is ~30 $M_\odot$ (RSG stellar wind according to Equation 19.1 with C = 8.7) or ~20 $M_\odot$ (Equation 19.1 with C = 9). Obviously, these mass limits also apply for Case C binaries and for Case B_c ones where the RSG wind phase begins before the onset of the RLOF (Sect. 3.1).

Case BB versus stellar wind. Habets (1985, 1986) computed the evolution of helium stars with $2 \leq M/M_\odot \leq 4$ up to neon ignition, assuming that the SW mass loss during CHeB is small in this mass range. He concluded that CHeB stars with $2 \leq M/M_\odot \leq 2.9$ expand significantly during the He shell burning phase, after CHeB. If a star like that is a close binary component, it can be expected that it will overflow the critical Roche lobe for a second time: the process is known as case BB RLOF. However, when we extrapolate the stellar wind mass loss formalism (Equation 19.2) and we calculate the evolution of CHeB stars with $2 \leq M/M_\odot \leq 2.9$, it follows that case BB is largely suppressed (or does not happen at all). It can readily be checked that the difference between case BB and SW mainly concerns the period evolution of binaries.

4. APPLICATIONS

ν Sgr. The binary has a period P = 138 days, a hydrogen deficient A-type supergiant with minimum mass = 2.5 $M_\odot$ and a B-type companion with minimum mass = 4 $M_\odot$ (Dudley and Jefferey, 1990). A 2.5

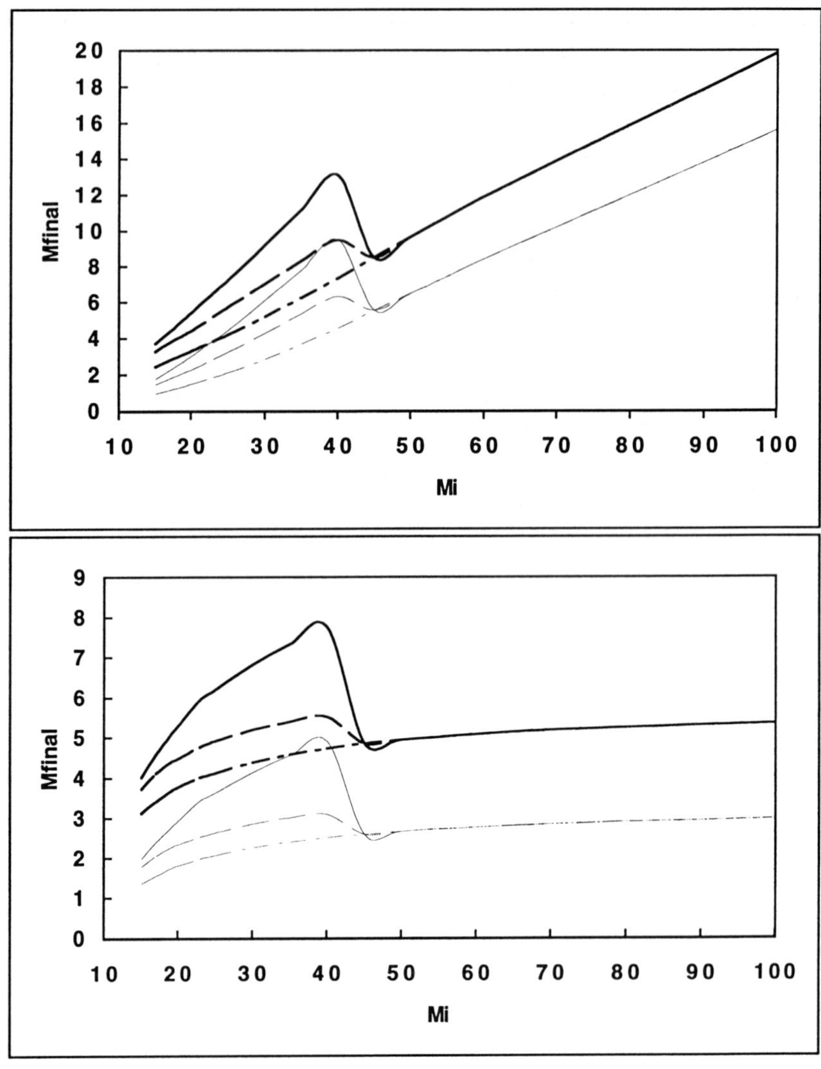

Figure 4 The final mass (in $M_\odot$) of the star (thick lines) and of the CO core (thin lines) as a function of initial mass (in $M_\odot$) of the star. The line-types have the same meaning as in Figure 3. The top-figure (respectively bottom-figure) corresponds to our preferred WR mass loss formalism Equation 19.2 (respectively Equation 19.5).

$M_\odot$ helium core corresponds to a primary with initial mass ~12 $M_\odot$ and we propose a 12 $M_\odot$ + 4 $M_\odot$ progenitor binary. The present period indicates that the initial period was very large, corresponding to Case B_c or Case C binaries, and this means that the binary experienced common envelope evolution. When it is assumed that all the energy that is needed to remove the hydrogen rich layers of the 12 $M_\odot$ star comes from the orbit, it is impossible to explain the present binary period. We propose an alternative model: when we use the RSG stellar wind mass loss formalism (Sect. 2), the star may lose ~5-6 $M_\odot$ by stellar wind before the onset of the common envelope. Orbital energy is then responsible for the removal of the remaining 3-4 $M_\odot$ and for the reduction of the binary period to 138 days. We conclude that the binary ν Sgr gives indirect evidence for the importance of RSG stellar wind mass loss in massive stars, possibly down to 10-12 $M_\odot$.

Cyg X-1. The compact star in the high mass X-ray binary Cyg X-1 is most likely a BH with mass ~10 $M_\odot$ (Gies and Bolton, 1986; Herrero et al., 1995). Nelemans et al. (1999) demonstrated that the space velocity of the binary (~50 km/s) can be explained if ~4 $M_\odot$ were ejected during a previous supernova explosion. This means that the final mass of the BH progenitor was at least 14 $M_\odot$. Figure 4 (top-figure) illustrates that the initial mass of the BH progenitor was larger than 60-70 $M_\odot$. With the WR mass loss rates proposed by Wellstein and Langer, it is not possible to explain the BH mass. The initial mass indicates that LBV+RSG stellar wind mass loss has played a fundamental role in the evolution of the progenitor system (Sect. 3 and Sect. 3.1).

The LMXBs with a BH component. Bailyn et al. (1998) list a number of LMXBs with a (probable) BH component. The BH masses range between 2 $M_\odot$ and 18 $M_\odot$ (7 $M_\odot$ is the average), the binary period between 0.2 days and 6 days, the mass of the optical component between 0.3 $M_\odot$ and 2.6 $M_\odot$. To explain the space velocity of Nova Sco 1994 ($\geq$ 100 km/s), Nelemans et al. (1999) propose that ~4 $M_\odot$ of the BH progenitor was lost during the supernova explosion. If this value can be considered as some canonical one, the average final mass of the BH progenitors was ~11 $M_\odot$. Figure 4 (top-figure) then indicates that the initial mass of these progenitors was $\geq$ 30-40 $M_\odot$. In particular, XN Mon 75 and XN Cyg 89 contain a very massive BH component (10-15 $M_\odot$). Accounting for the mass that was lost during the SN explosion, it follows from Figure 4 (top-figure) that their progenitors had an initial mass in excess of 40 $M_\odot$ (possibly as large as 60-70 $M_\odot$). This means that the evolution of these progenitors has been heavily influenced by stellar wind mass loss during an LBV and RSG phase, similar as the progenitor of the BH in Cyg X-1. Brown et al. (1999) propose a case C type of evolution to explain these systems. However they deny the effects of stellar wind mass loss during the LBV and/or RSG phase of a massive star (Sect. 3 and Sect. 3.1). As a consequence of these SW phases, Case C does not happen in binaries with primary mass ≥ 20 $M_{\odot}$ and we therefore do not support the idea of Brown and Bethe.

Portegies Zwart et al. (1997) explored models which are based on the classical spiral-in formalism and concluded that it is extremely difficult to explain the observed numbers of LMXBs with BH components.

Using the stellar wind mass loss rate formalisms during the LBV phase and during the WR phase discussed above (Equation 19.2) and accounting for the final masses of massive stars which depend on these formalisms, Vanbeveren et al. (1998a, b) proposed the following model for the progenitors of LMXBs with a BH component: LMXBs with a massive BH component originate from binaries with a primary mass ≥ 40 $M_{\odot}$ and with a period which is large enough (of the order of a few dozen days) so that LBV stellar wind mass loss starts before the RLOF (and thus before classical spiral-in) happens (Sect. 3). The effect of RLOF and of the spiral-in is largely reduced.

Here, we propose a second channel: LMXBs with (not too massive) BH components originate from Case B_c binaries with extreme mass ratio, with primary mass between 20(30) $M_{\odot}$ and 40 $M_{\odot}$ (Sect. 3.2), with a period which is large enough that the RSG stellar wind phase starts before the RLOF-spiral-in phase reducing the importance of the latter (Sect. 3.1).

Preliminary calculations reveal that most of the binaries survive this phase, i.e. merging is avoided. This may explain the (relatively) large number of observed LMXBs with a BH component. The periods of LMXBs range between 0.2 days and a few days, which at first glance contradict a stellar wind scenario. However, an LBV or RSG wind is dense and the outflow velocity may be comparable to the orbital velocity of the low mass component. The orbital angular momentum of the outflowing matter can therefore be significantly larger than the momentum of matter at the moment it leaves the loser (Sect. 3) implying a large reduction of the binary period. Moreover, BH formation may be preceded by a SN explosion. If the latter is asymmetrical, even a binary with a pre-SN period of the order of several days can become a post-SN binary with very small period.

5. POPULATION NUMBER SYNTHESIS (PNS) OF MASSIVE STARS

The skeletons of most PNS codes with a realistic fraction of interacting binaries are very similar. Binaries with mass ratio q ≤ 0.2 (low mass component is a normal core hydrogen burning star) evolve through a spiral-in phase. Many of them merge.

Case B_c and Case C systems pass through a common envelope phase where one has to account for the effect of RSG stellar wind mass loss, Case A and Case B_r with q $\geq$ 0.2 experience RLOF (whether RLOF is conservative or not is still a matter of debate; mass loss from the system is parameterised).

Post-supernova binaries with a NS or a BH component evolve through a common envelope phase. In the case of merging, a so called Thorne-Zhytkow object is formed.

Differences in the results of massive star PNS performed by different research teams are due to

- differences in the evolutionary calculations of single stars and of close binaries that are used to perform PNS
- differences in the mass and angular momentum loss formalism of a non-conservative RLOF, common envelope and spiral-in process
- differences in the physics of the SN explosion and how the SN explosion of one component affects the orbital parameters of a binary
- differences in the criteria used to decide when a binary will merge or not.

The PNS code of the Brussels team has been described in Vanbeveren et al. (1998a,b,c). Summarising:

- we use evolutionary computations of single stars and of binary components with moderate amount of convective core overshooting as proposed by Schaller et al. (1992), no rotation, with the stellar wind mass loss rate formalisms during the different phases as described above (OB-phase, LBV phase, RSG phase, WR phase).
- the effect of mass accretion on the evolution of the mass gainer is followed in detail. We have a library of over 1000 evolutionary calculations of massive close binaries where the evolution of the mass gainer is followed up to the end of its CHeB phase
- our evolutionary library of single stars and of close binaries contains calculations for different Z values (0.002 $\leq$ Z $\leq$ 0.02)
- when mass is leaving the binary during RLOF, we assume that this happens through the second Lagrangian point (matter forms a ring around the system). The angular momentum loss therefore

can be computed from first physical principles and is parameter free

- BHs form out of primaries in Case B_r binaries with initial mass larger than 40 $M_\odot$, single stars and primaries of Case B_c and Case C systems with initial mass larger than ~25 $M_\odot$, mass gainers who became single after the first SN explosion and who's mass after mass transfer is larger than ~25 $M_\odot$

- the observed space velocities of pulsars indicate that the SN explosion is asymmetrical. This asymmetry can be translated in terms of the kick velocity of the compact remnant. We study effects of kicks on PNS in full 3-D. The adopted kick velocity distribution is a χ^2-distribution (with a tail of large velocities) and we investigate the consequences for different values of the average velocity. BH formation may be preceded by a SN explosion (Israelian et al, 1999; Nelemans et al., 1999) but our code does not account for the effect.

- **merging conditions**: interacting binaries experience either a spiral-in process, a common envelope process, RLOF and mass transfer. All these processes stop when the mass loser has lost almost all its hydrogen rich layers or when both components merge before. After the removal of most of the hydrogen, a helium burning remnant restores its thermal equilibrium and shrinks rapidly. From our evolutionary calculations, we deduce the following relation for the equilibrium radius R_e of the helium remnant (R_e in $R_\odot$) as function of mass M (in $M_\odot$):

$$R_e = 0.000062M^3 - 0.0049M^2 + 0.18M + 0.17 \qquad (19.6)$$

When a star does not accrete mass, its radius remains constant. When accretion happens, our evolutionary library gives us the radius after accretion when the star restored its equilibrium. *Merging is avoided when the equilibrium radii of both components are smaller than their Roche radii.* Binaries with $q \leq 0.2$ at birth or with a NS or BH component evolve through a spiral-in phase. For these binaries we compare the equilibrium radius R_e of the He remnant with the orbital radius of the low mass star, NS or BH. Merging happens when the latter is smaller than R_e. In our PNS code we assume that the mergers of two normal stars evolve in a similar way as single stars with mass = the sum of the masses of both components before the mass loss process. However, in most of the cases, merging happens when the primary is a hydrogen shell burning star, i.e. when the helium core is fixed. If the process of merging is similar to the process of accretion of mass (= the mass of the lower mass component), a new star is formed with a helium core that is smaller than the one of a normal single star with mass = the sum of the masses of both components and the evolution of such stars may be different from the evolution of normal single stars.

5.1. PNS RESULTS FOR O-TYPE STARS AND WR STARS

In this paper we consider regions where star formation is continuous in time. Starbursts will be considered by Van Bever and Vanbeveren later on. To account for the properties of the observed population of O and early B type binaries, and to account for statistical biases, one is forced to conclude that at least 60% of all massive stars are born as the primary of an interacting binary with orbital period between ~1 day and 10 years (see also Mason and Van Rensbergen, these proceedings). A significant fraction is expected to have mass ratio q $\leq$ 0.2 where the evolution will be governed by the spiral-in and merger process.

We predict that 15- 25% of the O-type stars are post-SN mass gainers of massive close binaries and it was shown by De Donder et al. (1997) that $\sim$ 40% of the latter class are runaway candidates (i.e. 5-10% of the O-type stars are expected to be runaways due to a previous SN explosion in a binary). Compared to the observed number of runaways, we conclude that at least 50% of them are formed by the SN mechanism in binaries. If the average SN kick velocity = 500 km/s (resp. 150 km/s), less than 20% (resp. more than 50%) of these runaways have a compact companion (NS or BH). These percentages should be very similar in the Galaxy and in the Magellanic Clouds.

The observed overall WR population of the Solar neighbourhood is well reproduced if we start with an initial massive interacting close binary frequency $\geq$ 60%. We predict a WR+OB binary frequency $\sim$ 40% and a WN/WC number ratio around 1 corresponding to the observed values. It is interesting to remark that about 40-50% of our predicted WR+OB sample did not experience a classical RLOF with significant mass transfer. They became WR+OB through the LBV and/or RSG stellar wind scenario discussed above. At least 50% of the single WR stars have a binary past (originate from merged binaries or from binary mass gainers where the SN explosion of the primary disrupted the binary) and only few WR stars are expected to have a compact companion (cc) (most of these WR+cc are WR+BHs with a period of the order of days to decades). This explains why very few WR-type (hard) X-ray binaries are observed.

We repeated these computations for low Z regions (Z = 0.002, the metal value of the SMC) assuming a similar OB-type binary frequency as in the Galaxy and the RSG stellar wind mass loss depends on Z. We first used evolutionary calculations where the WR stellar wind mass loss rates are the same as in the Galaxy. PNS predicts that 80% of all WR stars have an OB-type companion (corresponding to observations in the SMC, see C. Foelmi these proceedings), and about 2/3 have experienced RLOF and mass transfer (i.e. only 1/3 were formed by LBV and/or RSG stellar wind mass loss). However, the predicted WN/WC ratio ~1-1.5 whereas the observed SMC ratio ~9. We conclude that a Z-dependence of the RSG mass loss can not be the only reason for the observed difference between the WN/WC number ratio in the Galaxy and the SMC. We then used evolutionary computations where the WR wind is Z-dependent and we obtain a theoretical WN/WC number ratio ~5-10 which corresponds to the observed value. Although SMC statistics of WR stars is small number statistics, we are tempted to conclude that *the variation of the WN/WC number ratio as function of Z is indirect evidence that the WR stellar wind mass loss rates are Z-dependent as predicted by the radiatively driven wind theory.*

Table 2 compares the observed and the theoretically predicted WR/O number ratio for the Solar neighbourhood (SN), the outer Milky Way, the LMC and the SMC. In all cases we assume an interacting close binary frequency $\geq$ 60%.

Table 2 Comparison between the observed and theoretically predicted WR/O number ratio for the Solar neighbourhood (SN), the outer Milky Way, the LMC and the SMC

Z	$\frac{WR}{O}_{observed}$	$\frac{WR}{O}_{predicted}$
0.02 (SN)	~ 0.1	0.03-0.05
0.013 (outer MW)	0.03	0.03
0.008 (LMC)	0.04	0.03-0.04
0.002 (SMC)	0.02	0.016-0.03

It has been noted frequently in literature that in the Solar neighbourhood the WR/O number ratio ~0.1. However, Hipparcos made clear that all O-type distance calibrations are very poor and due to the severe reddening complications in the Milky Way, the number of O-type stars is highly uncertain. PNS predicts a WR/O number ratio ~ 0.03-0.05, however this value may increase significantly when one accounts for the formation process of O-type stars (Bernasconi and Maeder, 1996; Norberg and Maeder, 2000). Therefore, a comparison between observations and predictions for the Solar neighbourhood seems to be difficult at present. Remark that the correspondence between theory and observation is much better for the outer Milky Way.

The distance-problem is obviously less severe for the Magellanic Clouds and it is interesting to notice that the observed WR/O number ratio in the LMC agrees with the PNS prediction.

The effect of semi-convection on the evolution of massive stars with low Z deserves some attention. During the hydrogen shell burning phase, massive stars expand and may become RSGs. The expansion time scale depends on the treatment of semi-convection and on the effect of metallicity. When semi-convection is treated as a very slow diffusion process (the Ledoux criterion), the expansion is always very rapid and a massive single star becomes a RSG almost at the onset of the CHeB phase. Furthermore, the RLOF of a primary with initial mass $\leq 40\ M_{\odot}$ in a Case B_r binary is always very fast (independent from Z) and stops at the beginning of CHeB when the primary has lost most of its hydrogen rich layers (resembling a WR star). However, when semi-convection is treated as a very fast diffusion process (Schwarzschild criterion), the expansion timescale is much longer. The models of Meynet et al. (1994) for small Z show that massive single stars ($\leq 40\ M_{\odot}$) straggle around $logTeff = 4.2 - 4$ and become RSG only at the end of CHeB. As a consequence RSG stellar wind mass loss plays a very restricted role and single WR stars rarely form out of this mass range. Moreover, since the timescale of the RLOF in Case B_r binaries depends on the expansion timescale during hydrogen shell burning, it follows that in low Z-regions the RLOF of a primary with initial mass $\leq 40\ M_{\odot}$ in a Case B_r binary lasts for a considerable fraction of the CHeB timescale. The remaining post-RLOF CHeB lifetime is small and this means that only few WR+OB binaries are expected from this mass range. The observed WR/O number value for the SMC ~ 0.02 whereas PNS predicts 0.016 (semi-convection is fast) or 0.03 (semi-convection is slow). We conclude that the correspondence is satisfactory for the SMC as well.

5.2. PNS RESULTS FOR SUPERNOVA RATES

Stellar evolution predicts whether or not a star ends its life with hydrogen rich outer layers. If one knows which stars explode as a supernova, it is therefore possible to use PNS codes to determine the rates of SN type I (no hydrogen) and of SN type II (with hydrogen). This has been done by De Donder and Vanbeveren (1998) with the code described above. One distinguishes type Ia from the other type I (=type Ibc). A type Ia SN may be the result of a White Dwarf in a binary that accretes mass from a main sequence star or red giant that fills its Roche lobe (Ken Nomoto, present proceedings). Type Ibc supernovae progenitors are massive stars that lost their hydrogen by stellar wind mass loss or by RLOF (hypernovae may be a subclass of these type I's) whereas type II progenitors are massive single stars where stellar wind mass loss was not large enough to remove all hydrogen rich layers. It is clear that the relative rates of the different types depend on the binary population (binary frequency, mass ratio and period distribution). We summarise the following three results that were discussed in De Donder and Vanbeveren.

- To determine the *observed* SN II/SN I SN number ratio of one particular galaxy, it is customary to collect data of a large sample of galaxies of the same type. The ratio holding for the whole data set is then taken as representative for all galaxies of this type (Cappellaro et al., 1993a, b; see also Cappellaro present proceedings). However, since the ratio depends critically on the population of binaries, it is clear that this is meaningful only if the binary population is similar in all the galaxies of the data set. Whether this is true or not is a matter of faith.

- Early type (respectively late type) spiral galaxies have an average SN II/Ibc ratio ~2.3 (respectively 5.5). If this difference is due to a difference in the binary population, we conclude that the massive close binary formation rate in late type spirals may be a factor 2 smaller than in early type spirals.

- Consider the data of all galaxies with any type. By comparing the SN II/Ibc number ratio with PNS prediction, we conclude that the average cosmological massive interacting close binary formation rate ~50%.

5.3. PNS RESULTS FOR DOUBLE COMPACT STAR BINARIES

PNS predicts the post-massive star collapse population. To estimate the double compact star binary population, let us consider the OB+cc (cc=NS or BH) class binaries. Their further evolution is governed by the spiral-in process of the cc in the OB star and by the various stellar wind mass loss phases of the OB star discussed in Sect. 2. When the binary survives the (spiral-in)+(stellar wind mass loss) phase, the OB star will be a hydrogen deficient CHeB star, resembling a WR star. The further evolution will be governed by WR stellar wind mass loss (Sect. 2) and to predict the population of double compact star binaries it is essential to use CHeB evolutionary computations with the most up to date WR stellar wind rates. Remark that beside the fact that these WR wind rates critically determine whether a massive star ends its life as a NS or as a BH, they also determine the binary period evolution.

The **spiral-in survival condition** has been discussed at the beginning of Sect. 5. In Brussels, we use this survival condition in our PNS code since 1997.

Results. PNS computations with the Brussels code have been published by De Donder and Vanbeveren (1998). Summarizing:

- PNS predicts that the birth rate of Galactic double NS systems is about $10^{-6} - 10^{-5}$/yr, corresponding with observations (van den Heuvel, 1994). Remark however that the observed value is still very uncertain

- mixed systems consisting of a BH and a NS are (2-5) times more frequent than NS+NS binaries

- the formation rate of double BH systems is at least a factor 100 higher than the formation rate of double NS systems

- most of the double NS systems have a period ≤ 2 days which means that most of them will merge within the Hubble time. All double BH binaries and at least 50% of the mixed systems have a period ≥ 10 days. It is questionable whether these systems will merge within the Hubble time.

6. ROTATION

The influence of rotation on massive single star evolution has been studied in detail by Meynet and Maeder (2000) (see also A. Maeder, present proceedings). They showed that in order to understand the evolutionary history and the chemistry of the outer layers of individual stars, the effects of rotation are very important.

For PNS research and to investigate the chemical evolution of galaxies, an important question is obviously in how far rotation affects average evolutionary properties.

From the study of Meynet and Maeder, we restrain the following important evolutionary effects.

The extend of the convective core. The larger the rotation the larger the convective core during the core hydrogen burning phase and this means that convective core overshooting mimics the effect of rotation on stellar cores. The average equatorial velocity of OB-type stars ~100-150 km/sec (Penny, 1996; Vanbeveren et al., 1998a, b). When this property is linked to the calculations of Meynet and Maeder, we conclude that *the effect of rotation on the extend of the convective core during the core hydrogen phase of massive single stars is on the average similar to the effect of mild convective core overshooting as described in Schaller et al. (1992)*

Due to RLOF a binary component loses most of its hydrogen rich layers and the remnant corresponds more or less to the helium core left behind after core hydrogen burning. Rotation makes larger helium cores so that also here convective core overshooting mimics rotation.

Mixing of the layers outside the convective core. Rotation induces mild mixing in all mass layers outside the convective core (rotational diffusion or meridional circulation).

It is a general property that as nuclear burning proceeds, the mass extend of the convective core during core hydrogen burning decreases. In this way a composition gradient is left behind in the stellar interior. Especially in massive stars a situation occurs where these chemically inhomogeneous layers are dynamically stable but vibrationally unstable and this may initiate mixing, a process commonly known as semi-convection (Kato, 1966). Whether this process is fast or slow is still a matter of debate (Langer and Maeder, 1995) but semi-convection and rotational diffusion imply similar effets. Remark however that semi-convection operates only in the inhomogeneous layers whereas rotational diffusion also happens in the outer homogeneous layers. The latter is then responsible for the appearence of CNO elements in the stellar atmosphere but is has little overall evolutionary consequences.

Semi-convection and rotational diffusion decide about the occurence of blue loops in post-core-hydrogen-burning phases, i.e. the blue and red supergiant timescales during core helium burning.

Rotation and stellar wind mass loss. The evolution of massive stars is significantly affected by stellar wind mass loss. Question: how does rotation affects the $\dot{M}$? When a theoretical hydrodynamic atmosphere model is used to investigate the effect of $\dot{M}$ on stellar evolution, it is clear that stellar rotation should be included although, at least for stars which are not too close to the Eddington-limit, the effect is at most of the order of 10-20% (Petrenz and Puls, 2000). We like to emphasise that *a study of the effect of rotation on $\dot{M}$ (and thus on stellar evolution) is meaningful only if $\dot{M}$ is determined using a theoretical model.*

However, most massive star evolutionary studies use (semi)-empirical SW rate formalisms like the ones discussed in Sect. 2. Empirical rates are determined by combining observations and atmosphere models. It is clear that these atmosphere models must account for all physical processes, thus also rotation. In this context, it is interesting to notice that empirical SW rates resulting from the analysis of the H_α spectral feature, tend to be too large when 1-dimensional atmosphere models are used (i.e. when rotation is not taken into account) although, again the effect is small for stars that do not rotate close to the break up velocity (Petrenz and Puls, 1996).

Overall conclusion

As far as overall massive single star evolution is concerned, old-fashion semi-convection + mild convective core overshooting mimic the effect of new-fashion rotation.

Rotation and PNS. To study the population of massive stars, present day PNS uses evolutionary calculations with a moderate amount of convective core overshooting (the one proposed by Schaller et al., 1992). I made some trial PNS computations for the Galaxy where evolutionary models are used with varying overshooting (simulating the situation where stars have different rotational properties) and it can be concluded that *rotation hardly changes the conclusions resulting from massive star PNS for the Galaxy where evolutionary calculations are used with moderate amount of overshooting.* The population of blue and red supergiants may be critically affected by rotation but due to the uncertainty in the treatment of semi-convection, it is difficult to draw any firm conclusion.

References

Bailyn, C.D., Jain, R.K., Coppi, P. & Orosz, J.A.: 1998, Ap.J. 499, 367.

Bernasconi, P.A. & Maeder, A.: 1996, A.&A. 307, 829.

Brown, G. E., Lee, C.-H.& Bethe, H. A.: 1999, NewA 4, 313.

Cappellaro, E., Turatto, M., Benetti, S., Tsvetkov, D. Y., Bartunov, O. S. & Makarova, I. N.: 1993a, A.&A. 268, 472.

Cappellaro, E., Turatto, M., Benetti, S., Tsvetkov, D. Y., Bartunov, O. S. & Makarova, I. N.: 1993b, A.&A. 273, 383.

De Donder, E. & Vanbeveren, D.: 1998, A.&A. 333, 557.

De Donder, E., Vanbeveren, D. & Van Bever, J.: 1997, A.&A. 318, 812.

De Jager, C., Nieuwenhuyzen, H. & van der Hucht, K.A.: 1988, A.&A. Suppl. 72, 259.

De Marco, O., Schmutz, W., Crowther, P. A., Hillier, D. J., Dessart, L., de Koter, A. & Schweickhardt, J.: 2000, A.&A. 358, 187.

Dudley, R.E. & Jeffery, C.S.: 1990, MNRAS 247, 400.

Gies, D.R. & Bolton, C.T.: 1986, Ap.J. 304, 371.

Hamann, W.-R. & Koesterke, L.: 2000, A.&A. 360, 647.

Hamann, W.-R. &Koesterke, L.: 1998, A.&A. 333, 251.

Herrero, A., Kudritzki, R.P., Vilchez, J.M., Kunze, D., Butler, K. & Haser, S.: 1992, A.&A. 261, 209.

Hillier, D.J. & Miller, D.L.: 1998, Ap.J. 496, 407.

Israelian, G¿, Rebolo. R¿, Basri, G., Casares, J. & Martin, E.L.: 1999, Nature 401, 142.

Jura, M.: 1987, Ap.J. 313, 743.

Kato, S.: 1966, Publ. Astron. Soc. Japan 18, 374.

Khaliullin, K.F., Khaliullina, A.I. & Cherepashchuk, A.M.: 1984, Soviet Astron. Letters 10, 250.

Kudritzki, R-.P. & Puls, J.: 2000, ARA&A 38, 613.

Langer, A. & Maeder, A.: 1995, A.&A. 295, 685.

Langer, N.: 1989, A.&A. 210, 93.

Maeder, A.: 1983, A.&A. 120, 113.

Meynet, G. & Maeder, A.: 2000, A.&A. 361, 101.

Meynet, G., Maeder, A., Schaller, G., Schaerer, D. & Charbonnel, C.: 1994, A.&A. Suppl. 103, 97.

Morris, P. W., van der Hucht, K. A., Crowther, P. A., Hillier, D. J., Dessart, L., Williams, P. M. & Willis, A. J.: 2000, A.&A. 353, 624.

Nelemans, G., Tauris, T. M. & van den Heuvel, E. P. J.: 1999, A.&A. 352, 87.

Norberg, P. & Maeder, A.: 2000, A.&A. 359, 1025.

Penny, L.R.: 1996, Ap.J. 463, 737.

Petrenz, P. & Puls, J.: 1996, A.&A. 312, 195.

Petrenz, P. & Puls, J.: 2000, A.&A. 358, 956.
Portegies Zwart, S.F., Verbunt, F., Ergma, E.: 1997, A.&A. 321, 207.
Reid, N., Tinney, C. & Mould, J.: 1990, Ap.J. 348, 98.
Salasnich, B., Bressan, A. & Chiosi, C.: 1999, A.&A. 342, 131.
Schaerer, D. & Maeder, A.: 1992, A.A. 263, 129.
Schaller, G., Schaerer, D., Meynet, G. & Maeder, A.: 1992, A.&A. Suppl. 96, 268.
St.-Louis, N., Moffat, A.F.J., Drissen, L., Bastien, P. & Robert, C.: 1988, Ap.J. 330, 289.
Van den Heuvel, E.P.J.: 1993, in Interacting Binaries, Saas-Fee Advanced Course 22, eds. H. Nussbaumer & A. Orr, Springer-Verlag: p. 263.
Vanbeveren, D.: 1991, A.&A. 252, 159.
Vanbeveren, D. & Packet, W.: 1979, A.&A. 80, 242.
Vanbeveren, D., Van Rensbergen, W. & De Loore, C.: 1998a, The Astron. Astrophys. Rev. 9, 63.
Vanbeveren, D., Van Rensbergen, W. & De Loore, C.: 1998b, The Brightest Binaries, ed. Kluwer: Dordrecht.
Vanbeveren, D., De Donder, E., Van Bever, J., Van Rensbergen, W., & De Loore, C.: 1998c, NewA 3, 443.
Wellstein, S. & Langer, N.: 1999, A.&A. 350, 148.

MASSIVE SINGLE AND BINARY STAR MODELS: A COMPARISON

Norbert Langer
Astronomical Institute, Utrecht University, The Netherlands
n.langer@astro.uu.nl

Stephan Wellstein
Potsdam University, Potsdam, Germany
stephan@astro.physik.uni-potsdam.de

Alexander Heger
Lick Observatory, UC Santa Cruz, U.S.A.
alex@ucolick.org

Keywords: Massive stars: surface abundances, remnant masses, remnant spins

Abstract We compare the predictions of recent single and binary star models with respect to several observable features. First, we find that rotationally induced mixing and accretion from a close binary component may lead to almost identical surface abundance patterns — with the boron to nitrogen ratio as a marked exception. We then compare the formation of Wolf-Rayet and helium stars in isolation and binary systems and relate them to Type Ib and Type Ic supernovae. We furthermore discuss the influence of a binary companion on the critical initial masses for the white dwarf–neutron star and neutron star–black hole transitions. E.g., assuming an upper initial mass limit for the formation of white dwarfs of $8\,M_{\odot}$ in single stars, we argue that this limit may be as high as $16\,M_{\odot}$ in binary systems. Finally, we discuss trends for the spins of compact objects emerging from the most recent single star and binary models.

1. INTRODUCTION

When a distinct feature is observed in a massive star, then it is often not immediately clear whether this feature occurs in this star because it occurs in all stars of the same initial mass and metallicity at a cer-

D. Vanbeveren (ed.), The Influence of Binaries on Stellar Population Studies, 273–286.

tain point in their evolution. Alternatively, this feature could have its origin in a peculiarity of the observed star, the most frequent one may be that it has, or had, a close binary companion. Examples for such still badly understood observed features are numerous: peculiar surface abundances in main sequence stars (e.g., McErlean et al. 1998, 1999; Smith & Howarth 1998; Herrero et al. 1999; Korn et al. 2000) or evolved stars (e.g., Venn 1999), eruptions as those occurring in Luminous Blue Variables (e.g., Drissen et al. 2001; and references therein), (aspherical) circumstellar shell (Nota et al. 1995), the B[e] supergiant phenomenon (Langer & Heger 1998), the Wolf-Rayet phenomenon (van der Hucht 2001), supernovae of Types Ib and Ic (cf., Podsiadlowski 1996, Langer & Woosley 1996), the rings around Supernova 1987A (Maran et al. 2000), young neutron stars with rapid or slow spins (cf. discussion in Heger et al. 2000), or peculiar supernova remnants, e.g. the Crab nebula (e.g., Fesen et al. 1997).

Even though for some of the mentioned features there are indications that binarity plays a role — e.g., some Wolf-Rayet stars are known to have very close companions; however, how many of the more than 200 Galactic Wolf-Rayet stars (van der Hucht 2001) would, at this time, not be a Wolf-Rayet star had they not had a close companion, is unknown — the contribution of binarity to these features is generally unknown for most of them.

As our group has, in the recent past, computed a large number of evolutionary models for massive single stars and for close binary systems (cf. references below), we present here a comparison of both types of models, i.e. of single star models with those of either close binary primaries (the mass losing, initially more massive components) and secondaries (the possibly accreting, initially less massive components), whatever is adequate. Such a comparison turns out to be helpful in evaluating the relative importance of binary models for an explanation for some of the observational features mentioned above.

A second goal of this comparison is to provide some insight to groups doing population synthesis studies for massive stars and binaries (for which there are several examples presented in these proceedings) into which stellar properties remain unaffected by the presence of a companion star, and which features are affected — may be even to such extent that they do not occur in single stars at all.

Due to the limited space, we consider in the following only the most important form of binary interaction, namely Roche lobe overflow.

2. MAIN SEQUENCE STARS

When a massive binary starts its first Roche lobe overflow phase, mass is being transfered from the initially more massive star (the primary) to the companion star (the secondary), where the latter is in almost all possible cases a main sequence star. Depending on the initial orbital separation and mass ratio, the secondary star does accrete a large or small fraction of the overflowing matter, with higher accretion efficiencies for initially tighter orbits and mass ratios closer to one (Wellstein 2001).

In this section, we do not discuss the primary stars — those become helium or Wolf-Rayet stars in most cases (cf. Section 3). We will also not discuss those secondary stars which accrete nothing or very little, as those might remain indistinguishable from single stars (except for having a companion star for some time). However, it is generally believed that a significant fraction of all secondaries accretes more than, say, 50% of the overflowing matter. Are any properties of these stars expected to stick out compared to those of single massive main sequence stars?

In principle, there may be three ways to identify those stars. Firstly, they may still have a binary companion, e.g., a helium or Wolf-Rayet star, or a compact object. And if they lost their companion when it exploded as supernova, they may have received an anomalously high space velocity. However, exactly for those cases where the secondary accreted substantially the corresponding effects are smallest: The accretion makes the secondary gain mass and it becomes very luminous, likely outshining the companion. Further, it makes the mass ratio much smaller than one, i.e. the orbital motion of the (bright) secondary will be small. And for the same reason, the explosion of the primary removes only a small fraction of the total mass from the system, and the imposed space velocity will be smaller than in binaries without accretion. Even though there are few cases where accretion onto the secondary has been shown to exist (e.g., in the massive X-ray binary Wray 977, cf. Wellstein & Langer 1999; or the massive Algol system V729 Cygni, cf. Wellstein 2001), those "proven cases" are exceptions.

Secondly, accretion stars may be recognised by their extremely rapid rotation: The accretion of matter is connected with the accretion of large quantities of angular momentum, which can quickly spin the accretion star up to critical rotation (Packet 1981). Indeed, one interpretation of the Be/X-ray binaries is that the Be star is spun up by accreting matter from the progenitor of the compact companion which now produces the X-ray emission (Pols et al. 1991). The problem to identify X-ray quiet binaries with spun-up accretion stars is that massive stars are rapid rotators in general, and as the inclination angle of the rotation axis is

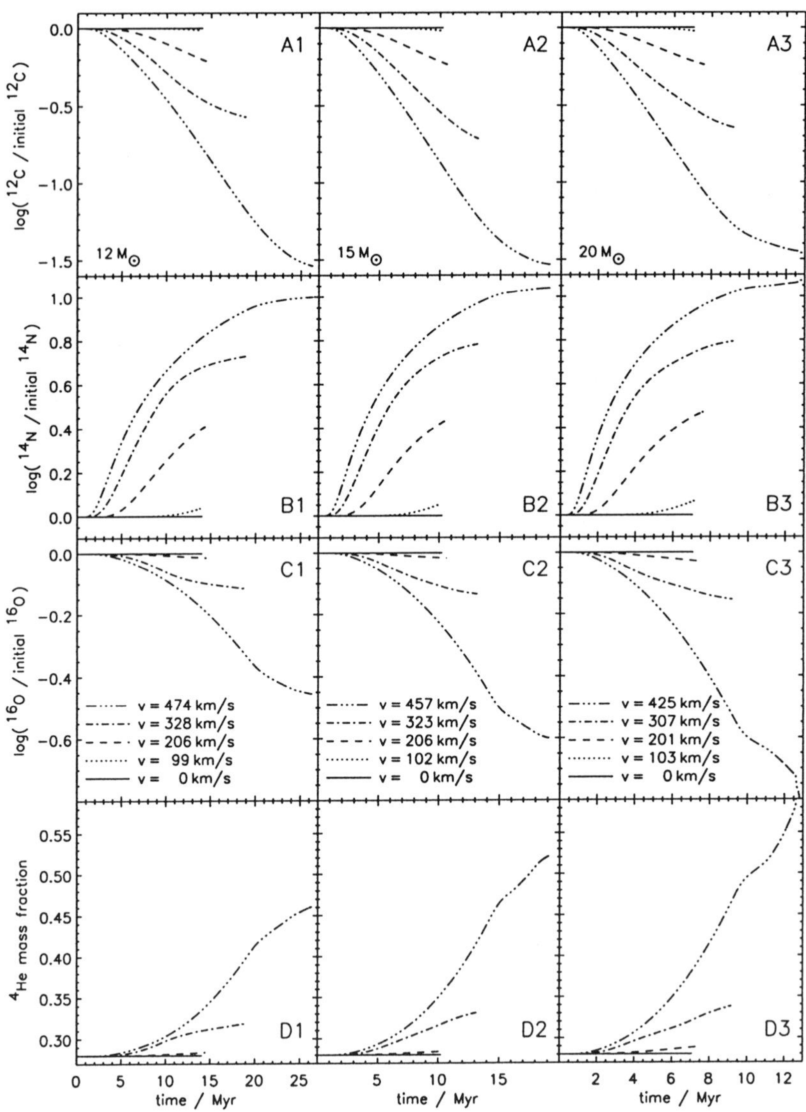

Figure 1 Evolution of the surface abundances during core hydrogen burning, for stars of 12 $M_\odot$ (right column), 15 $M_\odot$ (middle column) and 20 $M_\odot$ (left column), for various initial rotation rates, as indicated (cf. Heger & Langer 2000).

mostly unknown it is hard to pick extreme cases. Furthermore, stars above 15...20 $M_\odot$ have sufficient angular momentum loss due to their radiation driven winds to quickly lose any extreme rotation, even if they were spun-up to critical rotation in the past (cf. Langer 1998, and Langer & Heger 1998).

Table 1 Surface mass fractions, or enhancements for the surface abundances relative to the initial abundances, for secondary stars after the major mass transfer phase, for various Case A and Case B systems. Systems No. 1 to 74 do not include rotational physics (spin up due to accretion and rotationally induced mixing) and are described in more detail in Wellstein et al. (2001). Sequence R1 does include rotational physics (see Wellstein 2001); here, the abundances are given before (bA) and after the Case A mass transfer.

Nr.	$M_{1,i}$ $M_\odot$	$M_{2,i}$	$\dot{M}$	$M_{2,f}$ $M_\odot$	He $[\frac{X}{X_i}]$	Li	Be $[\log\frac{X}{X_i}]$	B	C $[\frac{X}{X_i}]$	N $[\frac{X}{X_i}]$	O $[\frac{X}{X_i}]$	Na $[\frac{X}{X_i}]$
1	12	11.5	A	21.5	1.23	-8.0	-22	-13	0.56	3.9	0.85	1.7
3	12	11	B	20.4	1.14	-8.8	-11	-10	0.61	3.4	0.88	n.b.
27	12	8	B	17.5	1.12	-10	-22	-17	0.45	3.7	0.89	n.b.
31	12	7.5	A	17.8	1.26	-10	-29	-27	0.23	5.4	0.81	2.0
34	16	15.7	A	27.8	1.20	-9.8	-22	-11	0.16	5.4	0.85	1.7
35	16	15.7	B	27.5	1.19	-9.2	-18	-8.3	0.58	3.8	0.84	1.8
51	16	13	B	24.8	1.13	-9.8	-19	-11	0.28	4.6	0.88	1.6
60	16	11	A	23.8	1.26	-10	-22	-25	0.10	6.0	0.81	2.0
65	25	24	A	40.6	1.28	-5.0	-13	-14	0.21	5.5	0.80	1.8
68	25	24	B	39.2	1.12	-9.4	-22	-8.7	0.49	3.7	0.88	1.5
74	25	16	A	33.3	1.28	-10	-19	-23	0.04	6.5	0.79	1.9
R1	16	15	bA	15.8	1.00	-2.9	-1.5	-0.3	0.97	1.1	1.00	1.0
R1	16	15	A	23.8	1.12	-8.6	-13	-4.5	0.47	3.7	0.92	1.5

Then, thirdly, one may try to identify accretion stars by enrichments of the surface in hydrogen burning products, which they might have acquired from their companion star. Indeed, the models of Wellstein et al. (2001), which were computed assuming maximum accretion efficiency, show that the N/C-ratio may increase by a factor of 10...100, and the N/O-ratio by a factor 3...10 (see Table. 1). Enrichments of this kind have in fact been observed in several OB stars (e.g., Schönberner et al. 1988). However, their interpretation remains ambiguous in most cases since such enrichments are also predicted by the most recent single star models for massive stars (cf. Fig. 1), where they are caused by rotationally induced internal mixing processes. Even an observed correlation of the surface enrichment with the stellar rotation rate does not break this degeneracy, as it is naturally predicted by both, the single star and the binary models.

The perhaps clearest way to distinguish both evolutionary channels has been proposed by Fliegner et al. (1996): a simultaneous determination of nitrogen and *boron* surface abundances. As boron is predicted to be destroyed in massive rotating stars very early (cf. Heger & Langer 2000), a boron depletion combined with a normal nitrogen abundance clearly points towards the action of rotational mixing and can not be reproduced by binary mass transfer (cf. Table 1). However, a boron depletion combined with a nitrogen enhancement — which seems to be the more frequent case; cf. Fliegner et al. (1996), Venn et al. (1996) — can be due to either rotation or binarity. This behaviour is confirmed by the most recent binary models of Wellstein (2001), which include rotation for both components of the binary models. Unfortunately, we do not yet know the boron abundances for a large sample of OB stars, as its determination requires high-resolution UV spectroscopy. Accordingly, we do not yet know which fraction of the enriched OB stars are accretion stars.

3. TO FORM AND EXPLODE WOLF-RAYET STARS

Current single star evolutionary models predict that stars with initial masses above 20...30 $M_\odot$ can lose all of their hydrogen due to stellar winds and become Wolf-Rayet stars — the uncertainty in the limiting mass coming mostly from poorly known red supergiant mass loss rates (cf. Vanbeveren et al. 1998). In case of very rapid rotation, this mass limit may be lower (Fliegner & Langer 1994, Meynet 2000, Heger et al. 2000).

A mass limit above $\sim 20\,M_\odot$ implies that Wolf-Rayet stars formed by single stars have masses of $M_{WR} \gtrsim 5\,M_\odot$ (Schaller et al. 1992, Woosley et al. 1993, Heger et al. 2000, Maeder & Meynet 2000). In a close binary, however, there is no such mass limit. Primary components of any initial mass lose their hydrogen envelopes completely in case of a Roche lobe overflow evolution. Therefore, one expects from binary models that hydrogen-free stars of all masses are formed. Those hydrogen-free stars which start out with a low mass show pure helium (plus trace elements) at their surface throughout their evolution, and never develop high carbon or oxygen surface abundances, since the helium star winds become weaker for lower masses (cf. Langer 1989, Woosley et al. 1995).

The cores of helium stars which are born with a mass above $\sim 2\,M_\odot$ collapse to neutron stars at the end of their lives and explode as supernovae (Woosley et al. 1995). The same holds for most of the more massive Wolf-Rayet stars formed from single or binary stars — whether or not the most massive Wolf-Rayet stars form black holes rather than supernovae is currently still debated (Wellstein & Langer 1999; Brown et al. 2001; see also Section 4).

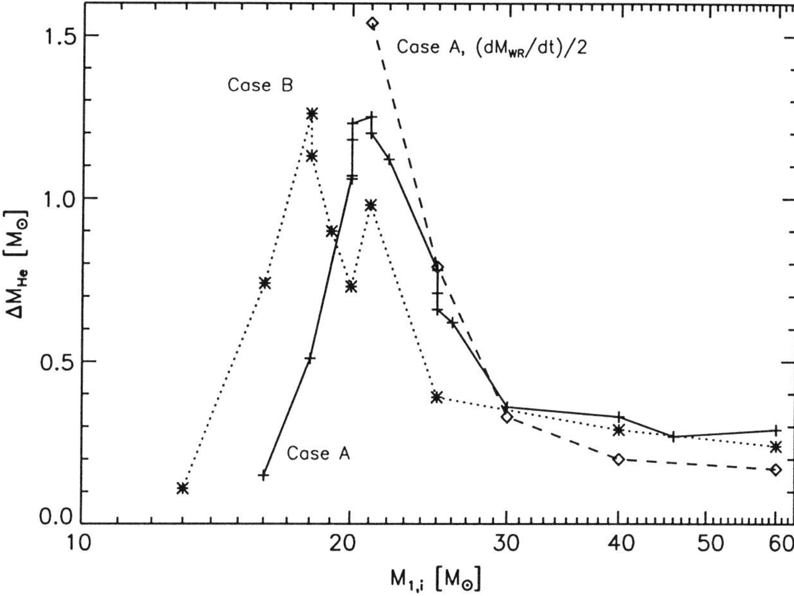

Figure 2 The amount of helium present in the primary components at their immediate pre-supernova stage, as function of their initial mass, for Case A (solid line) and Case B (dotted line) models. The dashed line corresponds to Case A systems computed with a WR wind mass loss rate reduced by 50 % (cf. Wellstein & Langer 1999.)

Clearly, supernovae originating from initially massive stars which end their lives without hydrogen envelope correspond to observed supernovae of Type Ib and Type Ic (Nomoto et al. 1996), of which the Ibs show helium lines in their spectra, and the Ics don't. As the observed rate of Type Ib/c supernovae seems to be higher than what is expected from single stars only (Capellaro et al. 1999), it seems clear that the Wolf-Rayet and helium stars formed due to binary mass transfer make up for a major fraction of them.

Wellstein & Langer (1999) have investigated the amount of helium left in the envelopes of helium and Wolf-Rayet star models which formed in binaries, at the time of their supernova explosion (cf. Fig. 2). They found two initial mass ranges with less than a few tenths of a solar mass of helium remaining: One is 12...15 $M_{\odot}$ (note that less massive primaries evolve into white dwarfs; cf. Section 4), and the other comprises of initial masses above $\sim 30\,M_{\odot}$ (which overlaps with WR stars from single stars which end up with little helium). Attributing the latter to Type Ic supernovae, Type Ib supernovae would be produced by primaries of massive close binaries in the intermediate mass range.

4. MASS LIMITS FOR THE FORMATION OF COMPACT OBJECTS

The evolution of stars, and more specifically their fate, is determined by their masses. Unfortunately, the *initial* mass of a star is *not* always sufficient to predict its evolution and fate, because the star may lose mass on their way. In particular, mass loss alters the evolution of the core when the core is not surrounded by a nuclear shell source. E.g., any mass loss during the core hydrogen burning phase of evolution may affect the fate of the star — which causes the predictions for the end of the most massive stars ($M_{\rm initial} \gtrsim 30\,{\rm M}_{\odot}$) to be a function of their metallicity and initial rotation rate (Schaller et al. 1992, Maeder & Meynet 2000, Heger et al. 2000).

One implication is that for the primary components of binaries, the initial mass limits for the formation of compact objects (white dwarf, neutron star, or black hole) are significantly increased for most binary evolution channels. This is evident if mass transfer (i.e. heavy mass loss from the primary) commences already during the primary's core hydrogen burning phase (so called Case A mass transfer). However, it remains true for systems which start their mass transfer due to the expansion of the primary star towards the red supergiant stage (Case B mass transfer). Only if the mass transfer starts after the primary has gone through most of its core helium burning would the mass limits not be affected by the mass transfer.

Table 2 illustrates this at the example of a $16\,{\rm M}_{\odot}$ star, evaluating the final helium core masses assuming various evolutionary histories. Without significant mass loss as a single star, it evolves a $4.3\,{\rm M}_{\odot}$ helium core by the end of its life. While this value may depend on the rotation rate of the star or on convective core overshooting, let us disregard these effects here and only consider the consequences of mass loss. For Case A primaries, this effect is drastic (cf. Fig. 3); depending on the exact time during the core hydrogen burning evolution of the primary when the mass transfer starts, it may reduce the final helium core mass by more than 50%. But also for Case B primaries the effect of mass loss is significant (about 20% in this example). Here, it is due to the indirect effect of the envelope loss. In single stars, the mass of the helium core grows substantially during the core helium burning phase, as the hydrogen burning shell source adds mass to it. As Case B primaries lose their

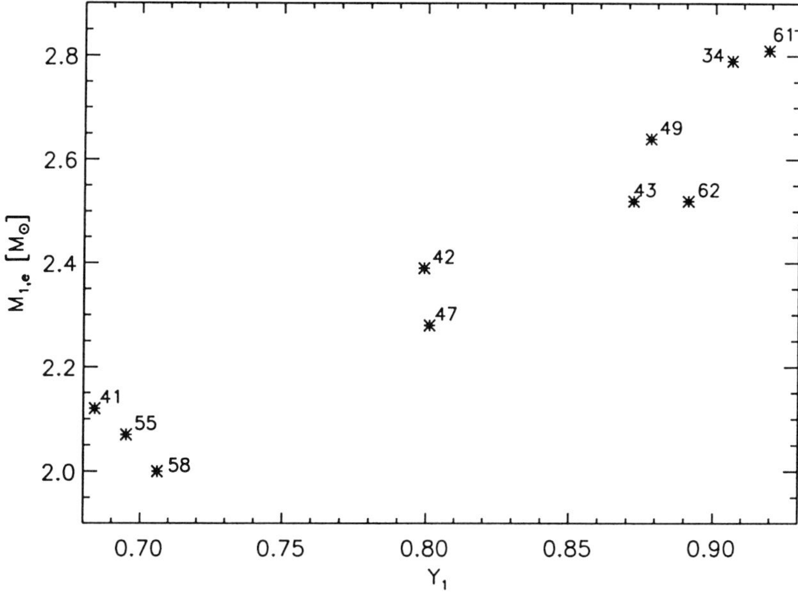

Figure 3 Masses during core helium burning for primaries of Case A systems with an initial mass of $16\,M_\odot$, as function of their central helium abundance at the onset of Case A mass transfer. See Wellstein et al. (2001), for more details.

entire hydrogen envelope at the beginning of core helium burning, the

Table 2 Final helium core masses for a $16\,M_\odot$ star as function of its evolutionary history, i.e. assuming it evolves as a single star, as a primary of a Case B binary, or of a Case A binary.

evolutionary path	final He-core mass
single star	$4.3\,M_\odot$
binary: Case B	$3.6\,M_\odot$
binary: Case A	$2.0...2.8\,M_\odot$

Table 3 Critical initial masses for the white dwarf–neutron star and neutron star–black hole transition, for various evolutionary paths (see text).

evolutionary path	WD/NS-limit	NS/BH-limit
single stars	$\sim 8\,M_\odot$	$\sim 25\,M_\odot$ (?) except 40...80 $M_\odot$ (?)
binary: Case B	10...12 $M_\odot$	∞ (?)
binary: Case A	12...16 $M_\odot$	∞ (?)

helium core can not grow in this case. Therefore, Case B primaries end up with smaller cores than single stars of the same initial mass.

We point out that this is an important effect in (binary) population synthesis models. Table 3 gives an overview of the expected shifts of the critical mass limits for compact object formation for close binary primaries of solar metallicity. The white dwarf/neutron star limit is based on models of Wellstein et al. (2001). Due to persisting uncertainties in the mass loss rates of Wolf-Rayet stars, the neutron star/black hole limit is significantly more uncertain. Current models predict an initial mass limit for black hole formation in single stars of the order of $25\,M_\odot$ (Fryer 1999), which is consistent with the SN 1987A being the explosion of a $20\,M_\odot$ star (Arnett et al. 1989). However, due to a turnover in the final mass–initial mass relationship in single star models (Schaller et al. 1992, Maeder & Meynet 2000), stars born in the mass range 40...80 $M_\odot$ may perhaps end up as neutron stars, but massive black holes may be formed for the highest stellar masses (Brown et al. 2001). There are two major uncertainties involved in this prediction, the mass loss rates of Wolf-Rayet stars, and the physics of convection in the stellar interior (cf. Fig. 5 in Wellstein & Langer 1999) For Case A and Case B binaries, the prediction of Wellstein & Langer (1999) and Wellstein et al. (2001) is much simpler: primaries of all initial masses above the white dwarf limit are expected to form neutron stars. Note that a reduction of the Wolf-Rayet mass loss rate as suggested by recent studies of Wolf-Rayet winds (Hammann & Koesterke 1998) did not alter these conclusions (Wellstein & Langer 1999).

The Wolf-Rayet mass loss rates continue to be uncertain, and it is conceivable that they are presently still overestimated (although the opposite is equally likely). The prediction of neutron star remnants from

Case A and B primaries poses a problem to explain the half a dozen or so known low mass black hole binaries, i.e. black holes of typically $5...7\,M_\odot$ with a low mass companion (Bailyn et al. 1998). However, it is questionable whether even lower Wolf-Rayet mass loss rates can alleviate this problem. The reason is that if Wolf-Rayet stars lose mass at any rate, the helium burning convective core can not grow in mass. As this reduces the rate of mixing of α-particles into the core during late core helium burning, it leads to a rather high final C/O-ratio at core helium exhaustion. High C/O-ratios, however, imply smaller iron core masses and thus an increased probability of forming a neutron star rather than a black hole when the iron core collapses (Woosley & Weaver 1993, 1995; Brown et al. 2001). Therefore, a Case C scenario may be the most likely explanation for the low mass black hole binaries at this time (Brown et al. 1999). Unfortunately, due to the uncertain post-main sequence radius evolution of massive stars (cf. Langer & Maeder 1995), current stellar evolution models can not strongly constrain the number of expected Case C events in massive binaries.

5. SPINS OF COMPACT OBJECTS

During recent years, evolutionary calculations for rotating single stars have been carried out through all evolutionary stages, which leads to predictions for the initial spin rates of compact remnants. For stars in the initial mass range $10...20\,M_\odot$, Heger ct al. (2000) found iron core specific angular momenta of the order of $10^{16}\,cm^2\,s^{-1}$, with lower values for higher masses (e.g., $3\,10^{15}\,cm^2\,s^{-1}$ for $25\,M_\odot$) due to angular momentum loss associated with main sequence stellar wind mass loss (cf. Langer 1998). The collapsar model for γ-ray burster by MacFadyen & Woosley (1999) seems to require higher angular momenta than those obtained in single star models for stars massive enough to produce black holes. As the mass loss rate of main sequence stars is lower for stars of lower metallicity, the situation may be more optimistic in the early universe.

A recent preliminary investigation of the so far neglected effects of magnetic fields (Spruit, in preparation) leads to specific angular momenta which are up to one order of magnitude lower (Heger, Spruit and Woosley, in preparation). This may help to bring anticipated pulsar rotation rates close to observed ones (cf. discussion in Heger et al. 2000).

Models for a rotating white dwarf progenitor ($M_{\rm initial} = 3\,M_\odot$) — without B-fields — have been computed by Langer et al. (1999), predicting a specific angular momentum of $2\,10^{15}\,cm^2\,s^{-1}$ for the emerging white dwarf. The corresponding rotational velocity of $30\,km\,s^{-1}$ is not in contradiction with upper limits derived by Heber et al. (1997) and Koester et al. (1998) for local white dwarf samples. Clearly slower rotating individual white dwarfs could be either braked by internal magnetic fields (this scenario has not been modelled yet) or stem from stars below $M_{\rm initial} \simeq 1.5\,{\rm M}_\odot$: Those stars, like the sun, lose most of their angular momentum during their main sequence evolution by magnetic braking and are thus expected to lead to much slower spinning white dwarfs as $3\,{\rm M}_\odot$ stars — which may actually have a reflection in planetary nebulae morphologies; cf. García-Segura et al. (1999, 2001).

How does binary evolution affect the remnant spins? Qualitatively, there are two important effects. First, as mass loss is coupled to angular momentum loss even if magnetic fields play no role (Langer 1998), the mass loss due to the Roche lobe overflow is expected to spin the primary down, at least for Case A and Case B mass transfer where the mass loss time scale is of the order of the thermal time scale or longer. And second, any accretion onto the secondary star is expected to rapidly spin it up to critical rotation (Packet 1981).

Recently, Wellstein (2001) has computed the first massive binary models which include these effects in detail, using the rotational physics of Heger et al. (2000) for both stellar components, and including tidal synchronisation for a complete description of spin-orbit coupling. Primaries in Case A and B models for $16\,{\rm M}_\odot + 15\,{\rm M}_\odot$ systems in which both stars start out with an equatorial rotational velocity of $200\,{\rm km\,s^{-1}}$ evolve into low mass Wolf-Rayet or helium star stars with spin periods like that of our sun (they also happen to have a radius of about $1\,{\rm R}_\odot$). From those models, we expect Case A primaries to give birth to slowly rotating neutron stars.

On the other hand, the accretion of matter in an advanced stage of core hydrogen burning for close binary secondaries opens the possibility that the effect of a reduced specific angular momentum due to main sequence stellar wind mass loss, as it occurs in massive single stars (see above) is damped, and that they retain enough angular momentum to produce a γ-ray burst. Similar arguments may hold for the case that both star of a binary merge in an advanced evolutionary phase. The mentioned $15\,{\rm M}_\odot$ secondaries grow in mass up to $\sim 23\,{\rm M}_\odot$, and obtain rotation rates of up to $500\,{\rm km\,s^{-1}}$.

Acknowledgments

This work has been supported by the Deutsche Forschungsgemeinschaft through grant La 587/15, and by the Alexander von Humboldt-Stiftung (FLF-1065004).

References

Arnett, W. D., Bahcall, J. N., Kirshner, R.P., Woosley, S.E. 1989, ARAA 27, 629

Bailyn, C. D., Jain, R. K., Coppi, P., Orosz, J. A., 1998, ApJ 499, 367

Brown, G. E.; Lee, C.-H.; Bethe, H. A., New Astr. 4, 313

Brown G.E., Heger A., Langer N., Lee C.-H., Wellstein S., Bethe H.A., New Astr., submitted

Capellaro, E., Evans, R., Turatto, M. 1999, A&A 351, 459

Dissen L., Crowther, P. A., Smith, L. J., Robert C. Roy, J.-R., Hillier, D. J. 2001, ApJ 546, 484

Fesen, R., Shull, J. M., Hurford, A.P., 1997, AJ 113, 354

Fliegner J., Langer N., 1994, in: *Wolf-Rayet Stars: Binaries, Colliding Winds, Evolution*, Proc. IAU-Symp. No. 163, K.A. van der Hucht et al., ed., Kluwer, p. 326

Fliegner J., Langer N., Venn K., 1996, A&A 308, L13

Fryer C.L., 1999, ApJ 522, 413

García-Segura G., Langer N., Różyczka N., Franco J., 1999, ApJ 517, 767

García-Segura G., Franco, J., Lopez, J. A., Langer N., Różyczka N., 2001, in: *Ionized Gaseous Nebulae*, RMxAA, in press

Hamann, W.-R., Koesterke, L. 1998, A&A 335, 1003

Heber U., Napiwotzki R., Reid I.N., 1997, A&A 323, 819

Heger A., Langer N. 2000, ApJ 544, 1016

Heger A., Langer N., Woosley S.E., 2000, ApJ 528, 368

Herrero, A., Corral, L.J., Villamariz, M.R., Martin, E.L. 1999, A&A 348, 542

van der Hucht, K. A., 2001, New Astron. Rev 4, 123 Koester D., Dreizler S., Weidemann V., Allard N.F., 1998, A&A 338, 617

Korn, A.J., Becker, S.R., Gummersbach, C.A., Wolf, B. 2000, A&A 353, 655

Langer N., 1989, A&A 220, 135

Langer N., 1998, A&A 329, 551

Langer N., Woosley S.E., 1996, in: *From Stars to Galaxies — The Impact of Stellar Physics on Galaxy Evolution*, eds. C. Leitherer, U. Fritze-von Alvensleben, & J. Huchra, ASP Conf. Series Vol. 98, p. 220

Langer N., Heger A., 1998, in: Proceeding of Workshop on B[e] stars, C. Jaschek, A. M. Hubert, (eds.), Kluwer, p. 235

Langer N., Maeder, A., 1995, A&A 295, 685

Langer N., Heger A., Wellstein S., Herwig F., 1999, A&A 346, L37

MacFadyen, A. I., Woosley, S. E., 1999, ApJ 524, 262

Maeder, A., Meynet, G., 2000, A&A 361, 159

Maran, S.P., Sonneborn, G., Pun, C.S.J., Lundqvist, P., Iping, R.C., Gull, T.R., 2000, ApJ 545, 390

McErlean, N.D., Lennon, D.J., Dufton, P.L. 1998, A&A 329, 613

McErlean, N.D., Lennon, D.J., Dufton, P.L. 1999, A&A 349, 553

Meynet, G., 2000, in: *Massive Stellar Clusters*, eds, A. Lancon, C. Boily, ASP Conf. Ser., p. 105

Nomoto, K., Iwamoto, K., Suzuki, T., Pols, O. R., Yamaoka, H., Hashimoto, M., Hoflich, P., van den Heuvel, E. P. J. 1996, in *Compact stars in binaries*, proc. IAU-Symp. No. 165, J. van Paradijs et al., eds., Kluwer, Dordrecht, p.119

Nota, A., Livio, M., Clampin, M., Schulte-Ladbeck, R. 1995, ApJ 448, 788

Packet W., 1981, A&A 102, 17

Podsiadlowski, P. 1996, in *Hydrogen deficient stars*, ASP Conference Series Vol. 96, ed.C. S. Jeffery and U. Heber, p. 419

Pols O.R., Cote, J., Waters, L. B. F. M., Heise, J. 1991, A&A 241, 419

Schaller, G., Schaerer, D., Meynet, G., Maeder, A., A&AS 96, 269

Schönberner, D., Herrero, A., Becker, S., Eber, F., Butler, K., Kudritzki, R. P., Simon, K. P, A&A 197, 209

Smith, K.C., Howarth, I.D. 1998, MNRAS 299, 1146

Vanbeveren, D., de Donder, E., van Bever, J., van Rensbergen, W., de Loore, C., 1998, New Astr. 3, 443

Venn, K.A. 1999, ApJ 518, 405

Venn, K.A., Lambert, D. L., Lemke, M. 1996, A&A 307, 849

Weaver T.A., Woosley S.E., 1993, Phys. Rep. 227, 65

Wellstein S. 2001, PhD Thesis, Potsdam University

Wellstein S., Langer N. 1999, A&A 350, 148

Wellstein S., Langer N., Braun H., 2001, A&A, in press

Woosley S.E., Langer N., Weaver T.A.: 1993, ApJ 411, 823

Woosley S.E., Langer N., Weaver T.A., 1995, ApJ 448, 315

Woosley S.E., Weaver T.A., 1995, ApJS 101, 181

CHEMICAL EVOLUTION OF LOW- AND INTERMEDIATE-MASS STARS

Nami Mowlavi
INTEGRAL Science Data Center, Geneva Observatory, CH-1290 Versoix, Switzerland

Keywords: single stars, stellar evolution, AGB phase

Abstract The chemical evolution of M $\lesssim$ 8 $M_\odot$ stars during their red giant and asymptotic giant branch phases is reviewed. Open issues in abundance predictions are discussed, in particular those related to the modeling of the AGB phase.

1. INTRODUCTION

Low- and intermediate-mass ($M \lesssim 8$) stars represent about 80% of stars in the Galaxy. They burn H and He in their core, and end their life as white dwarfs. Their lifetimes vary from less than 50 million years to more than the age of the Universe (Table 1).

During their evolution, they are characterized by two main phases of expansion. The first is the red giant branch (RGB) phase which occurs after core H exhaustion. The structure of the star consists of a H-burning shell surrounding a He core (degenerate in the electrons for low-mass stars M $\lesssim$ 1.8 $M_\odot$). The second expansion phase is the asymptotic giant branch (AGB) after core He exhaustion. The star consists of two burning shells, one of H and one of He, surrounding a C-O core (degenerate in the electrons for all stellar masses considered here). Both the RGB and the AGB phases are characterized by a deep convective envelope extending from the surface down to the H-burning shell.

The stellar radius significantly increases during each of those two expansion phases. For a 1.5 $M_\odot$ star of solar metallicity, for example, the radius reaches 200 $R_\odot$ at the tip of the RGB phase (it is 2 $R_\odot$ on the main sequence), and several hundred solar radii during the AGB phase. Heavy mass losses also characterize those giant phases. This is particularly true for AGB stars in which mass loss rates up to 10^{-4} $M_\odot$/y are

D. Vanbeveren (ed.), The Influence of Binaries on Stellar Population Studies, 287–297.

observed. Those stars eventually eject all their envelope in the interstellar medium, possibly as a planetary nebula, leaving a white dwarf remnant.

Table 1 Main sequence lifetimes (in 10^9 years) for a set of low- and intermediate-mass stars at two different metallicities Z. From Schaller et al. 1992.

M ($M_\odot$)	t_{MS} ($\times 10^9$ y)	
	Z=0.02	Z=0.001
0.8	25	15
1	9.8	6.2
2	1.1	0.85
4	0.17	0.14
7	0.04	0.04

Observations reveal that both RGB and AGB phases are characterized by peculiar surface chemical abundances when compared to solar-scaled abundances, and that the abundances at the surface of AGB stars are very peculiar compared to those at the surface of RGB stars. The presence of a giant star in a binary system can thus significantly impact on the chemical evolution of the system when mass is transferred from the giant star to the other one (see, for example, the review of Jorissen, 1999 and references therein). Barium and Tc-poor S stars are examples of the outcome of such interactions.

In this review, I concentrate on the chemical evolution of single red giant stars, with particular attention to AGB stars. Section 2 gives an overview of the chemical properties of those stars. Section 3 is devoted to current open issues in the modeling of AGB stars. Conclusions are drawn in Sect. 4.

2. CHEMICAL PROPERTIES OF RED GIANT STARS

I restrict this section to a qualitative description of chemical production by red giant stars. Elsewhere (Mowlavi 1998), I have reviewed the chemical properties of red giant stars element by element, where the interested reader can find a more extensive list of references. The reader is also referred to the review article by Busso et al. (1999), and to the

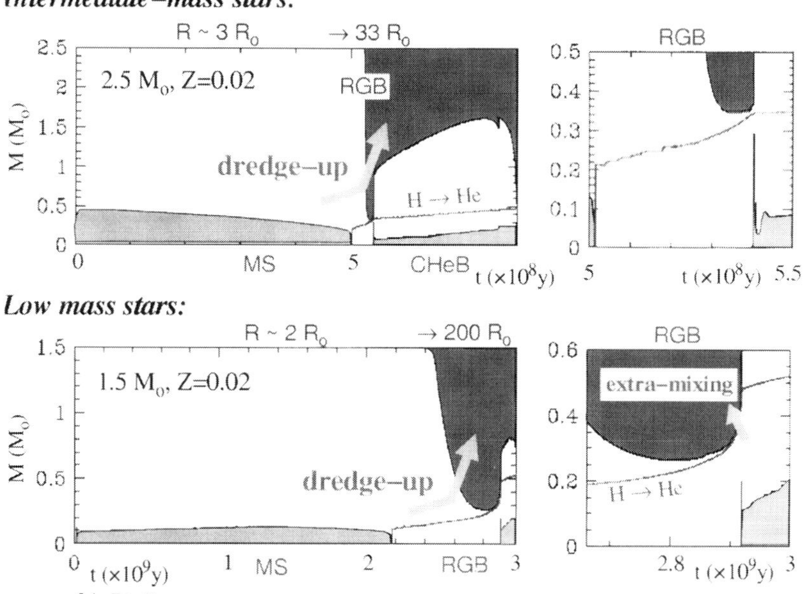

Figure 1 Structural evolution of an intermediate-mass (upper figures) and a low-mass (lower figures) star during the main sequence (MS), the red giant branch (RGB) phase and part of the core helium burning (CHeB) phase. Filled areas indicate convective regions (in orange when nucleosynthesis occurs, in brown for nuclearly inactive regions). The H burning shell is also shown in red line. The maximum radius reached during the red giant phase is shown on top of the left figures. The figures on the right are zooms on the RGB phase. (a colored version of the figure is available at *http://obswww.unige.ch/~mowlavi/publications*).

proceedings of the IAU Symp. 191 edited by Leberte et al. (1999) and entirely devoted to AGB stars.

2.1. RED GIANT BRANCH STARS

The deep convective envelope in RGB stars is responsible for the transport of the ashes from H burning from the deep interior to the surface. Nuclei affected by H-burning nucleosynthesis are, besides light nuclei (^{3}He, B, Li), those involved in the well known CNO, NeNa and MgAl chains (e.g. Arnould et al. 1999). They are synthesized during the main sequence, left over on top of the H-burning core, and mixed to the surface during the RGB phase by the so-called first dredge-up. This picture holds well for intermediate-mass stars (El Eid 1994 and references therein). We note that the dredged-up nuclei are 'secondary' in the sense that they require the presence of elements heavier than helium in the initial stellar composition[1].

Observations of red giants in star clusters show that the surface abundances further vary along the red giant branch for low-mass stars (e.g. Gilroy and Brown 1991, Smith and Tout 1992). Those stars are characterized by the fact that their H-burning shell reaches the H-He discontinuity left over by the convective envelope (Fig. 1). The bottom of the convective envelope thereafter remains very close above the H-burning shell as the latter advances in mass. It is believed that extra-mixing processes take place at this stage and transport freshly produced ashes from the H-burning shell to the convective envelope. We refer to the recent work of Weiss et al. 2000 and references therein) who apply a parameterized deep mixing procedure to study the abundance anomalies observed in low-metallicity red giant stars.

2.2. ASYMPTOTIC GIANT BRANCH STARS

The structure of AGB stars is more complex than that of RGB stars. The presence of two thin burning shells on top of the C-O core is known since Schwarzschild and Härm (1967) to lead to the occurrence of thermal instabilities in the He-burning shell. Those instabilities liberate, periodically and on a short time-scale (several tens of years), 10^2 to 10^6 times the energy provided by the H-burning shell. They lead to the development of a convective zone (pulse) in the He-burning shell which extends up to the H-burning shell and engulfs the ashes left over by the latter.

An important structural response is predicted to occur after pulse extinction. The convective envelope penetrates through the now extinct H-burning shell down to the H-depleted regions. It eventually sinks into the carbon-rich layers and mixes to the surface material processed by the pulse. This process is called 'third dredge-up'. As a result of this process, products of both H- and He-burning nucleosynthesis are brought to the surface of AGB stars.

The rich nucleosynthesis occurring in AGB stars is summarized in Fig. 2. Six processes contribute tp the chemical evolution of those stars.

1) The first site for nucleosynthesis is the H-burning shell during the interpulse period. All three of the CNO, NeNa and MgAl chains of hydrogen burning are in operation. Among the recent H-burning related

[1]'Primary' elements are those that can be synthesized in stars of the first generation consisting of only H and He at the time of their formation

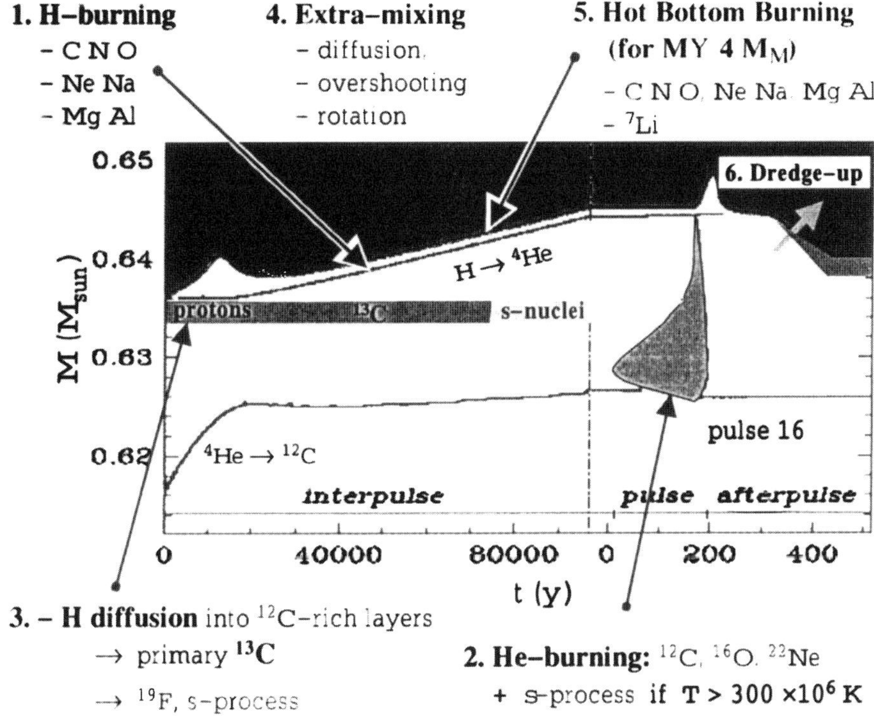

Figure 2 Overview of the processes occurring in AGB stars which impact their chemical evolution. The figure displays an interpulse (during which H burns in a shell), pulse (He burns in the pulse convective tongue) and afterpulse (dredge-up mixes burning ashes to the surface) sequence. Brown and orange filled areas have the same meaning as in Fig. 1. Regions filled with H-burning ashes are shown in light blue. Layers where protons have partially diffused from the envelope to the C-rich layers are shown in green. The subsequent synthesis of ^{13}C and, later during the interpulse, of s-nuclei is also shown in the figure (a colored version of the figure is available at *http://obswww.unige.ch/~mowlavi/publications*).

studies in AGB stars, I mention the production of ^{26}Al (Mowlavi & Meynet 2000).

2) The He-burning shell constitutes the second obvious nucleosynthesis site. It contributes mainly to the production of ^{12}C. Successive occurrences of third dredge-ups lead to a gradual surface carbon enrichment, eventually turning a M star with C/O<1 to a C star with C/O>1 (e.g. Marigo et al. 1999).

Another important contribution from the He-burning shell is provided by the nucleosynthesis resulting from the injection into the pulse of the H-burning shell ashes. For example, The burning of ^{14}N in the pulses leads to the production of ^{22}Ne by two consecutive α captures. If the temperatures at the base of the pulse reach 300×10^6 K, ^{22}Ne also burns and releases neutrons through $^{22}\mathrm{Ne}\,(\alpha\,,\mathrm{n})\,^{25}\mathrm{Mg}$. This leads to some s-process nucleosynthesis in massive AGB stars and/or at a late stage on the AGB (pulse temperatures increase with pulse number for a given star or with stellar mass at a given pulse number).

3) A third site for nucleosynthesis resides just below the convective envelope at the completion of each third dredge-up. The penetration of the H-rich envelope into the C-rich layers leads to a small transition region with steep abundance profiles. Partial mixing is expected to occur across the H-C discontinuity, either from the shears induced by differential rotation between the slowly rotating envelope and the fast core rotator, or simply from overshooting (Sect. 3). This partial mixing brings some protons into the C-rich material. After reheating, an incomplete CN cycle occurs, the result of which is the production of ^{13}C. When the intershell temperatures reach 90×10^9 K, ^{13}C burns through $^{13}\mathrm{C}\,(\alpha\,,\mathrm{n})\,^{16}\mathrm{O}$, leading to an efficient s-process and ^{19}F nucleosynthesis during the interpulse (Straniero et al. 1995). We note that the ^{13}C produced in this way is primary.

4) The fourth process which can alter surface abundances during the AGB phase is a slow mixing between the H-burning shell and the convective envelope during the interpulse. The process is called cool bottom processing and is similar to that occurring along the RGB in low-mass stars. Such a transport mechanism is required to explain peculiar oxygen isotopic ratios (Boothroyd et al. 1995).

5) H-burning can also operate directly *within* the envelope of an AGB star if the temperatures at the bottom of the envelope are higher than about 50×10^6 K. This phenomenon is predicted to occur in stars more massive than $\sim 4\,\mathrm{M}_{\odot}$, and is called hot bottom burning (HBB). It alters the surface composition of those stars without the need to invoke a dredge-up scenario. Among the signatures of HBB, the production of ^{7}Li is quite interesting since this light nuclei is easily destroyed at

temperatures as low as 3 million K. The mechanism at work to produce ^{7}Li in massive AGB stars, explained by Cameron and Fowler (1971), combines the production of ^{7}Be at the bottom of the convective envelope by $^3\mathrm{He}(\alpha,\gamma)^7\mathrm{Be}$ and its convective transport to the surface where it is transformed to ^{7}Li by electron capture (Sackmann and Boothroyd 1992). A recent work in relation with ^{7}Li production in AGB stars can be found in Ventura et al. (2000; see also references therein).

6) Finally, I mention the third dredge-up itself as a process which impacts nucleosynthesis in AGB stars. We have seen (point 3) that it induces the production of primary ^{13}C in the intershell layers. In addition, it can have a feedback on the H-burning shell nucleosynthesis during the interpulse phase. This results from that fact that the material which is processed in the H-burning shell comes from the envelope, the chemical composition of which is modified by the third dredge-up. An example of such a feedback is the possible production of primary ^{23}Na (Mowlavi 1999b). The sequence of operations is dredge-up of primary ^{12}C, $^{12}\mathrm{C} \rightarrow {}^{14}\mathrm{N}$ in the H-burning shell, $^{14}\mathrm{N} \rightarrow {}^{22}\mathrm{Ne}$ in the pulse, dredge-up of ^{22}Ne, $^{22}\mathrm{Ne} \rightarrow {}^{23}\mathrm{Na}$ in the H-burning shell, and finally dredge-up of ^{23}Na. The efficiency of ^{23}Na production through this scenario is very dependent on the amount of carbon dredged-up and on the number of dredge-ups experienced by the AGB star. But it is remarkable to see how dredge-ups may impact on the nucleosynthesis occurring in AGB stars.

3. CURRENT ISSUES IN MODELING ASYMPTOTIC GIANT BRANCH STARS

Given the complex structure of AGB stars and the rich nucleosynthesis occurring in them, only complete evolutionary model calculations including the relevant nucleosynthesis can provide reliable chemical yields. No evolutionary model calculation yet exist which consistently integrate all required processes in a consistent way, as I will show in the next sections. There are at least three difficulties that AGB modelers do face. They concern the modeling of the third dredge-up process, the handling of mixing in the intershell layers, and the adoption of a realistic mass loss prescription. Those are addressed in turn.

3.1. MODELING THIRD DREDGE-UP

An AGB star is basically composed of a H-depleted core and a H-rich envelope. The core is very dense and compact with a radius of about $0.10 - 0.15\,\mathrm{R}_\odot$ and a mean density of $10^5 - 10^7$ g cm^{-3}. In contrast, the envelope is rather dilute with a mean density of about 10^{-6} g cm^{-3}, and extends up to the photosphere several hundreds of solar radii away from the center of the star. Convection is its dominant mode of energy transport.

As dredge-up proceeds, H-depleted matter is added from the core into the envelope. That matter is lifted to higher radius by convection, and its potential energy increases. The characteristic time-scale of that process is determined by the thermal relaxation time-scale of those layers engulfed in the envelope and which must expand to fit the envelope structure. This time-scale imposes a maximum rate at which dredge-up proceeds. Model calculations indeed confirm that basic physical picture, with a dredge-up rate of few $10^{-5}\,\mathrm{M}_{\odot}/\mathrm{yr}$ (Mowlavi, 1999a).

The difficulty in modeling third dredge-up comes from the steep composition gradients which develop at the interface between the envelope and the core. In that region, the opacity is dominated by Thompson opacity [$\kappa_{th} = 0.2(1 + X)$ where X is the hydrogen mass fraction]. H-rich material is opaque to radiation under the thermodynamic conditions characterizing the intershell layers, while He-rich material is not. This property is at the origin of many difficulties in the modeling of third dredge-up when a local theory of convection is used. With such a local prescription, convection settles only in layers containing hydrogen, and a H-He discontinuity develops. In reality, the discontinuity is unphysical and some H-rich material, whatever small the amount, is always mixed across the H-He interface. As soon as hydrogen is injected in the He-layers, the latter becomes opaque to radiation and dredge-up proceeds inward (at a maximum rate as described above).

Some extra-mixing process must thus be included in AGB models to correctly follow the third dredge-up process[2]. It is interesting to note that the speed of convective bubbles crossing the H-He discontinuity is estimated to be much higher (by about four orders of magnitude) than the penetration rate of the envelope during the dredge-up (Mowlavi, 1999c). This basically ensures that dredge-up predictions are not dependent on the adopted extra-mixing parameters.

Two extra-mixing processes are used so far in AGB modeling. The first one considers diffusive overshooting at the base of the envelope (Herwig et al. 1997). The second mechanism invokes differential rotation between a slowly rotating envelope and a fast rotating core (Langer et al. 1999). The impact of diffusive overshooting on the structural and

[2]Some evolutionary models available in the literature obtain third dredge-up without apparently any extra-mixing prescription. We believe this is the result of uncontrolled numerical diffusion. Dredge-up predictions would in this case be dependent on the spatial and time resolution of the models.

chemical evolution of AGB stars has recently been studied in more details by Herwig (2000). Further studies, and by other AGB modelers, should be performed. The impact of rotation on AGB evolution, on the other hand, has not yet been studied in detail. It certainly constitutes one of the main issues on which the efforts of stellar modelers will concentrate during the next decade.

3.2. S-PROCESS NUCLEOSYNTHESIS

Predicting consistently the ^{13}C pocket required to explain the observed abundances of s-process nuclei (cf. 2.2) is another difficulty which gathers the efforts of AGB modelers.

The diffusive extra-mixing process required for the operation of third dredge-up, whether overshooting or rotationally induced mixing (Sect. 3.1), provides a natural mechanism to explain the formation of a ^{13}C pocket in the intershell layers (e.g. Herwig and Langer 2000 for a recent discussion). Contrary to the dredge-up case, however, the amount of protons diffused in the C-rich layers is sensitive to the partial mixing parameters such as the extent and efficiency of overshooting or the amplitude of rotation. Consequently, the ^{13}C abundance profile after H burning is also sensitive to those partial mixing parameters.

Therefore, s-process nucleosynthesis in AGB stars has so far only be studied by *assuming* a parameterized initial abundance profiles in the intershell layers. Two approaches are found in the literature. The first one parameterizes the ^{13}C abundance profile (Gallino et al. 1998), the parameters being chosen in order for abundance predictions to match observations. Successes of this approach are reviewed in Busso et al. (1999) to which the reader is referred for further information. The second approach parameterizes the initial H abundance profile (Goriely & Mowlavi, 2000), the ^{13}C pocket being then obtained consistently after H ignition in the intershell layers. It has the advantage to enable the study of s-process nuclei synthesis together with that of other nuclei produced at H-ignition in the partial mixing region (such as F or Na). It, however, also suffers from the parameterized approach. Future works in this field should reduce the uncertainties in the H abundance profiles predicted by extra mixing mechanisms.

3.3. MASS LOSS

Mass loss has always been and remains a difficult issue in AGB modeling. Its importance cannot be underestimated. Mass loss determines the duration of the AGB phase, the number of third dredge-up events that the star experiences as well as their efficiencies, and, in massive AGB stars, the efficiency of hot bottom burning. It is thus a key ingredient for chemical yields.

Several mass loss prescriptions have been used in AGB model calculations, varying from the use of an enhanced Reimers formula (most widely used) to attempts to correlate with observational data or theoretical calculations (e.g. Vassiliadis & Wood 1993, Blöcker 1995, Schröder et al. 1999). The subject is far outside the scope of this paper, and I will not discuss it further. An interesting recent attempt to obtain further observational hints on mass loss in AGB stars can be found in Alard et al. (2001).

4. CONCLUSIONS

The open issues in AGB modeling are not restricted to the three topics developed in Sect. 3. Other issues are equally important, such as the treatment of the atmosphere, the prescription used to model convection, nuclear reaction rate uncertainties, or multi-dimensional considerations. All of them influence the structural and chemical evolution of AGB stars.

Significant progress has been recorded in the last decade in the field of AGB modeling, but many challenges are still facing us. In view of the increasing power of computers, and thus of model generation, it is important to assess the physical consistency of the models and to evaluate the uncertainties affecting calculations. Comparison between model predictions from independent groups should be encouraged, and differences analyzed. Hopefully, consistent and reproducible predictions will soon become available.

References

Alard C., Blommaert J., Cesarky C., et al., 2001, ApJ, in press
Arnould M., Goriely S., Jorissen A., 1999, A&A 347, 572
Blöcker T.: 1995, A&A 297, 727
Boothroyd A., Sackmann I.J., Wasserburg G.J., 1995, ApJ 442, L21
Busso M., Gallino R., Wasserburg G.J., 1999, ARAA 37, 239
Cameron A.G.W., Fowler W.A., 1971, ApJ 164, 111
El Eid M., 1994, A&A 404, 271
Gallino R., Arlandini C., Busso M., et al. 1998, ApJ 497, 388
Gilroy K.K., Brown J.A.: 1991, ApJ 371, 578
Goriely S., Mowlavi N., 2000, A&A 362, 75
Herwig F., 2000, A&A 360, 952
Herwig F., Blöcker T., Schönberner D., El Eid M., 1997, A&A 324, L81

Herwig F., Langer N., 2000, Mem. Soc. Astron. It., in press (astro-ph/0007390) Torino-Melbourne-Pasadena Workshop in honour of G.J. Wasserburg

Jorissen A., 1999, in 'Asymptotic Giant Branch Stars' (IAU Symp. 191), eds. Le Bertre T., Lèbre A. and Waelkens C., ASP, p. 437

Langer N., Heger A., Wellstein S., Herwig F., 1999, A&A 346, L37

Le Bertre T., Lèbre A., Waelkens, 1999, editors of the IAU Symp. 191 (Asymptotic Giant Branch Stars), PASP

Marigo P., Girardi L., Bressan A., 1999, A&A 344, 123

Mowlavi N., 1998, in: Cosmic Chemical Evolution, IAU Symp. 187, in press[3]

Mowlavi N., 1999, A&A 344, 617

Mowlavi N., 1999, A&A 350, 73

Mowlavi N., 1999, in: Nuclei in the Cosmos V, eds. N. Prantzos and S. Harissopulos, Editions Frontières, p 170[3]

Mowlavi N., Meynet G., 2000, A&A 361, 959

Sackmann I.-J., Boothroyd A.I., 1992, ApJ 392, L71

Schaller G., Schaerer D., Meynet G., Maeder A., 1992, A&AS 96, 269

Schröder K.-P., Winters J.M., Sedlmayr E., A&A 349, 898

Schwarzschild M., Härm R., 1967, ApJ 150, 961

Smith G.H., Tout C.A.: 1992, MNRAS 256, 449

Straniero O., Gallino R., Busso M., et al., 1995, ApJ 440, L85

Vassiliadis E., Wood P.R.: 1993, ApJ 413, 641

Ventura P., D'Antona F., Mazzitelli I., 2000 A&A 363, 605

Weiss A., Denissenkov P.A., Charbonnel C., 2000, A&A 356, 181

[3]Available at *http://obswww.unige.ch/~mowlavi/publications*

THE EVOLUTION OF INTERMEDIATE-MASS BINARIES

Peter P. Eggleton

Institute of Geophysics and Planetary Physics, Lawrence Livermore Laboratory, L-413, 7000 East Ave, Livermore CA 94550, USA

ppe@ast.cam.ac.uk

Keywords: Binary processes, Binary evolution

Abstract I discuss three aspects of binary stellar evolution, based on (1) an exploration of a large number (5550) of conservative Case A evolutionary models by Nelson & Eggleton (2001), (2) a rather smaller number (150) of conservative Case B evolutionary evolutionary models by Han, Tout & Eggleton (2000), and (3) an attempt by myself to understand the circumstances in which very non-conservative Case C evolution can lead, through a common-envelope phase, to short-period highly evolved systems containing one or two white dwarfs and/or hot subdwarfs.

1. A LARGE GRID OF CONSERVATIVE CASE A MODELS

Nelson & Eggleton (2001) computed a large grid of Case A binaries, following the evolution of gainer as well as loser and using the assumption of conservative evolution. Conservation, of angular momentum as well as mass, seems a reasonable assumption for Case A systems in the spectral range early B to mid-F or early G ('hot Algols'). At earlier spectral types (O – B1) stellar winds assisted by radiation pressure may be a problem, and stars later than mid-F often show evidence of magnetic activity that is probably related to magnetic braking, and which we try to model in this paper. Our work suggests that indeed several semidetached binaries where both components are in the range B – F are well approximated by conservative evolution, and also confirms that many such binaries at later spectral types ('cool Algols') are badly approximated by conservative models. There are several cool Algols that must have lost substantial angular momentum, and some must have lost considerable mass as well.

D. Vanbeveren (ed.), The Influence of Binaries on Stellar Population Studies, 299–309.

The grid of models computed had 37×10×15 elements, for 3 independent variables M_{10}, q_0 anp P_0. These are respectively the initial mass of $*1$, defined as the component with the larger *initial* mass, the initial mass ratio, defined as $M_{10}/M_{20} \geq 1$, and the initial period. We took

$$\log M_{10} = -0.10, -0.05, \ldots, 1.7 \ , \tag{22.1}$$

$$\log q_0 = 0.05, 0.10, \ldots, 0.5 \ , \tag{22.2}$$

$$\log(P_0/P_{ZAMS}) = 0.05, 0.1, \ldots, 0.75 \ . \tag{22.3}$$

where P_{ZAMS}, a function of M_{10}, is the period at which the initially more massive component would just fill its Roche lobe on the zero-age main sequence. We used the approximation

$$P_{ZAMS} \approx \frac{0.19M_{10} + 0.47M_{10}^{2.33}}{1 + 1.18M_{10}^2} \ . \tag{22.4}$$

We computed the evolution on the Compaq Teracluster of LLNL, a massively parallel array of 512 processors. This was a relatively unsophisticated procedure, since each system evolves independently of all other systems; but in the future we hope to combine such stellar evolution with the dynamical evolution of a cluster.

Because the longest orbital periods considered were only about 5.5 times the period at which $*1$ would fill its Roche lobe while still on the ZAMS, the majority of computed cases were Case A, but with a substantial minority of Case B systems. The main assumptions of our RLOF model were:
(i) Orbital angular momentum is kept constant.
(ii) $*1$ loses mass at a rate which is proportional to the cube of the radius excess – stellar radius minus Roche-lobe radius. The constant is sufficiently large that the radius excess is $\sim$ 1% if RLOF is on a thermal timescale.
(iii) $*2$ gains mass at the same rate that $*2$ loses it.
(iv) The mass accreted by $*2$ was assumed to be at the same entropy and composition as the material already at the surface of $*2$. This is not a particularly good assumption, especially once $*1$ is stripped down to its helium-rich core, but we do not believe that this is a major effect.
(v) A degree of enhanced mixing was included, as calibrated by Schröder et al. (1997) – see also Pols *et al.* (1998) – in order to account for the properties of certain accurately-determined ζ Aur binaries, where one component is evolved to core helium-burning near the red supergiant region. Our model of enhaced mixing gives uniform mixing over a region substantially larger than the basic Schwarzschild-criterion core, but no

extra heat transport, positive or negative, was included near the core boundary.

The numerical models were terminated when one of the following criteria was satified:
(a) RLOF from $*1$ rose rapidly above the thermal timescale, and appeared to be heading for a dynamical timescale
(b) $*2$ came to fill its Roche lobe, sometimes when $*1$ was still doing so and therefore leading to a contact configuration, and sometimes after $*1$ had shrunk within its Roche lobe, leading to reversae RLOF
(c) the age of the system exceede 20 Gyr.
A few other criteria were also used, including that the computation broke down for some indeterminate reason.

We found that Case A evolution can be divided into about 8 principal sub-cases. Many of these evolve into contact, and many more into a drastic (dynamical timescale) RLOF from $*1$ to $*2$. Those that avoid both of these evolve to a reverse RLOF phase, once $*2$ evolves to fill its own Roche lobe after $*1$ has shrunk to a hot remnant (or neutron star) well inside its Roche lobe. In this third group we expect that the RLOF stage is also, as in the second group, on a dynamical timescale: not because the loser ($*2$) has a convective envelope but because the mass ratio is very severe (~ 0.1) at this stage.

The 8 subcases are:
(i) AD – Drastic RLOF: the mass-transfer rate becomes very rapid very quickly, but it is unclear whether the result will be a contact binary or a common-envelope phase
(ii) AR – Rapid (thermal timescale) evolution into contact, after only a small amount of mass is transferred
(iii) AS – Slow (nuclear timescale) evolution into contact while both stars are still within the MS band, but after substantial mass is transferred
(iv) AE – Early overtaking: $*2$, its evolution much accelerated by mass gain, evolves into contact while crossing the Hertzsprung gap
(v) AG – (sub)Giant contact: contact reached when $*2$ evolves to the base of the giant branch, $*1$ being already a giant (or rather subgiant)
(vi) AL – Late overtaking: $*2$ evolves to reverse RLOF while $*1$ is in a compact core-helium-burning phase
(vii) AB – the classic Case AB: $*1$ after becoming a small He burning star evolves back to large radius and refills its Roche lobe for a while
(viii) AN – No overtaking: $*1$ reaches a degenerate state (WD or NS) before $*2$ reaches reverse RLOF.
The first two cases happen for all initial mass ratios greater than a critical value which depends primarily on initial period: $q_0 \gtrsim 1.4$ if P_0 is only slightly greater than $P_{\rm ZAMS}$, increasing to ~ 1.6 if, near the

TMS, $P_0 \sim 4P_{\rm ZAMS}$. The remaining cases happen if q_0 is less than this critical value, and are ordered above roughly according to $P_0/P_{\rm ZAMS}$. However there is significant M_{10}-dependence, and not all cases occur at all masses.

There are several observed semidetached binaries which are at least arguably on one or other of the above routes. From the frequency of such systems it is usually concluded that the timescale of mass transfer is nuclear rather than thermal, or *a fortiori* dynamical. They are therefore not likely to be products of the first two routes. Those with observed $q \lesssim 0.3 - 0.4$ are unlikely to be Cases AE or AS, since these generally reach contact while $q \gtrsim 0.3 - 0.4$. Nelson & Eggleton (2000) attempted to fit the observed systems with models from their grid, using a χ^2 test. Observation can in favourable cases determine all of M_1, q, P, R_2, T_1, T_2; they also determine R_1, L_1, L_2, but as these variables are functionally dependent on the first 6 variables (on the assumption that $*1$ fills its Roche lobe), we excluded them from our test. The theoretical models have 4 parameters (M_{10}, q_0, P_0 and age) and so there are two degrees of freedom: if the models are right, and if the observational errors are distributed normally, we would expect a mean χ^2 of 2. For the 8 hot-Algol systems that we selected, somewhat subjectively on the basis that we believed them to have the most accurate data, the mean χ^2 was 3 – somewhat above the desired value but not far above it. The worst system (AF Gem) was in fact at the cool edge of our definition of 'hot Algol', with spectral type (for the cooler component) of G0. It could be, according to our hypothesis of the first paragraph, that stars at G0 can still be subject to dynamo activity and angular-momentum loss.

The two next worse cases were λ Tau and DM Per. Interestingly, both of these systems have a close third body, with orbital periods of respectively 33 d (Fekel & Tomkin 1982) and 100 d (Hilditch, Hill & Khalesseh 1992). It is quite possible, as argued by Kiseleva, Eggleton & Mikkola (1998), that this third body can drain appreciable angular momentum from the semidetached pair. The mechanism is that the third body injects a fluctuating eccentricity into the close pair, which tidal friction then tries to dissipate. As commonly happens in self-gravitating systems, dissipation tends to take angular momentum from the inner regions (the close pair) while depositing it in the outer regions (the third-body orbit). Our comparison with the grid does indeed suggest that the discrepancy is in the sense of a loss of angular momentum from the semidetached pair: both have a rather small mass ratio, suggestive of Case AL, while having a rather short period suggestive of Case AE/AS. Thus these two cases perhaps should be excluded from the comparison. If we reject all three worst cases, we are left with a mean χ^2 of 1.3, somewhat smaller than we expect. However the number of examples left is rather small (5). They appear to be consistent with the expectation that 'hot Algols' evolve rather conservatively.

The situation regarding cool Algols is however strongly against the hypothesis of conservation. We considered 10 cool Algols using the same χ^2 test and found a mean χ^2 of 11. Most of these systems have quite small mass ratios ($q \sim 0.1 - 0.3$), indicative of Case AL/AN, while having low angular momenta indicative of Case AE/AS. This discrepancy is in the right sense to have been caused by angular momentum loss, which can plausibly be attributed to magnetic braking after $*1$ became a cool subgiant or giant during slow (nuclear-timescale) mass transfer. One of the cool Algols we included, R CMa, has such low *total* mass (Sarma, Vivekananda Rao & Abhyankar 1996) that it is reasonable to suppose that significant mass loss, as well as angular momentum loss, has taken place.

Although the hypothesis of angular momentum loss in cool Algols is hardly radical, it appears to be less well known that there is strong evidence for significant *mass* loss in cool binaries, even before RLOF begins. The system Z Her, of RS CVn type, has the more evolved component significantly less massive than the companion (by $\sim 20\%$, Popper 1988), despite the fact that neither component yet fills its Roche lobe. This suggests that the timescale for mass loss is much enhanced relative to what it would be in an isolated red subgiant, and this in turn suggests a phenomenon of 'binary-enhanced stellar wind' (Tout & Eggleton 1988). Eggleton (2000) has suggested a rather simplistic model of mass and angular momentum loss, due to rotation and dynamo activity enhanced by tidal friction, which can increase the mass-loss rate of a red subgiant or giant by a factor of $\sim 10^3$ in close binaries containing a dynamo-active component, relative to what it would be in a single star. If this can happen in pre-RLOF systems like Z Her, it can *a fortiori* happen in semidetached cool Algols. Algol itself is a very active star – the brightest radio source in the sky – and it may be that it started with a significantly greater total mass than its present value.

2. AN INITIAL-FINAL MASS RELATION FOR EARLY CASE B SYSTEMS

Han et al. (2000) considered a grid of models with RLOF in Case B. They computed 10 initial primary masses, $\log M_{10} = 0.0, 0.1, \ldots, 0.9$, 5 initial mass ratios, $q_0 = 1.1$, 1.5, 2,0, 3.0 and 4.0, and 3 initial periods. The choice of the latter was based on the fraction of the distance (in $\log R$) across the Hertzsprung gap (HG) that $*1$ had to evolve before starting RLOF: 10%, 50% or 90%. This means that we cover early Case B, and not late Case B, where $*1$ is already on the giant branch before RLOF begins (but not yet at helium ignition).

Unlike the Case A calculations of Section 1, these calculations did not pursue the evolution of the gainer. This is because in these wider binaries there is less scope for $*2$ to swell up and fill its own Roche lobe during the initial thermal-timescale mass transfer (Case AR). In the shortest-period Case B systems considered here, the Case BR analogue can occur, particularly at $q_0 \sim 3, 4$. This can also happen at longer period if M_{10} is above the range considered here, but is not expected to be common for the intermediate and longer period in the present mass range. There is also no scope for Case B analogues of subCases AE/AS, at least within the framework of the conservative assumption. Most systems are likely to be BL/BN analogues of subCases AL/AN.

Although it is commonly assumed that the remnant of $*1$ in Case B is just the core mass at the terminal main sequence (TMS), and so depends mainly on M_{10}, we found a fairly significant dependence on both q_0 and P_0. At the lowest M_{10} the final mass M_{1f} can vary by about 10% ($\sim 0.32 - 0.35\,M_\odot$) depending on the other parameters, and at the highest mass ($8\,M_\odot$) by nearly 30% ($0.90 - 1.17\,M_\odot$). For $q_0 = 1.5$, and the mid-HG P_0, $M_{1f}(M_{10})$ can be approximated to about 1% by

$$M_{1f} = \frac{174.6 - 36M_{10}^2 + 19M_{10}^{2.5}}{1000 - 369M_{10} + 52M_{10}^2}$$

Han et al. (2000) give interpolation formulae for the more detailed function $M_{1f}(M_{10}, q_0, P_0)$ that they obtain. Reading the result in reverse, knowledge of the WD mass leads to an estimate of the precursor which can be rather uncertain: an M_{1f} of $0.36\,M_\odot$ can be reached with $M_{10} \sim 1 - 2\,M_\odot$, with an even greater range if one cannot exclude Case A, or late Case B.

In several models the remnant WD, though of low mass, was a CO WD rather than an He WD, because the He core of stars with $M_{10} \sim 2\text{–}4\,M_\odot$ can contract to He ignition before becoming entirely degenerate and yet with core masses in the range $0.32 - 0.45\,M_\odot$, i.e. below the limit for the helium core flash in lower-mass stars.

3. CASE C EVOLUTION OF LOW-MASS BINARIES: SPIRAL-IN OR NOT?

The 'Common Envelope' (CE) scenario (Paczyński 1976) was invented largely to account for the possibility that cataclysmic variables – very close binaries containing a WD – should evolve from initially rather wide binaries in which *1 would have enough room to evolve to a red (super)giant with a substantial WD core. This important scenario also has relevance to a great number of other situations, arguably leading to double-neutron star binaries, LMXBs, Thorne-Żytkow objects, and double-WD objects, not to mention merged objects of a variety of types.

A modest number of short-period binaries is known which are almost as short-period as CVs, but which are apparently detached. Several of them are planetary-nebula nuclei (PNNi), and appear to be the just-born remnants of the CE process. Presumably some at least of them will evolve further by a combination of magnetic braking and tidal friction (although in a few cases by nuclear evolution) to become CVs. They are potentially much better than CVs in allowing us to elucidate the CE mechanism, because they have not yet started RLOF, and therefore their masses and periods are presumably just what was left over by the CE process. Current CVs have transferred a largely indeterminate amount of mass, and lost an indeterminate amount of angular momentum, so that their immediately post-CE parameters are not known.

But as well as the short-period post-CE binaries that are seen, there are also a number of binaries that appear to have similar components (WD/SDOB + MS/RG), that are *not* short-period, and yet can arguably only be the products of the same sort of original systems as CVs. These objects are not commonly discussed, at least in the same context as the short-period systems, but it seems to me that they have as much to say: they may represent 'failed-CE' evolution.

Table 1 lists, with period P and eccentricity e, a number of binaries, divided loosely into 4 categories: (i) PNNi, (ii) SDOB binaries (which probably were PNNi not long ago), (iii) WD binaries where the companion is non-compact (MS, RG), and (iv) WD/SDOB binaries where the companion is also compact (probably also WD). I have selected those for which good data exists on both masses (M_1, M_2), but also added some other systems where there is less good data, but enough (I hope) to be able to infer the necessary characteristics (M_{10}, P_0) of the precursor binary, which I presume in all cases to have contained a red giant or supergiant. Table 1 also lists my estimates of what the precursor M_{10} and P_0 could have been.

Most of the M_1, M_2 entries come straight from the literature cited, or a reference in that literature. A few, and all the M_{10}, P_0 entries, have some guesswork by myself. Space does not permit me to justify these in detail: a discussion will be given elsewhere (Eggleton *et al.* 2002). In many cases I have used the measured or conjectured mass of the WD/SDOB component to estimate how large the precusor RG/RSG's

Table 1 Some detached WD/SDOB binaries related to Common Envelope evolution

Name	P	e	M_1	M_2	M_{10}	P_0	Reference
KV Vel	.357		.55	.24	2.0	800	Hilditch et al. 1996
UU Sge	.465		.63	.29	2.5	1000	Bell et al. 1994
V477 Lyr	.472		.51	.15	1.5	300	Pollacco & Bell 1994
V651 Mon	16.0	.07	.6	1.8	3.0	200	Smalley 1997
HW Vir	.1167		.48	.14	2.0	100	Wood & Saffer 1999
AA Dor	.262		.25	.05	1.2	10	Włodarczyk 1984
BE UMa	2.29		.7	.36	3.5	900	Ferguson et al. 1999
FF Aqr	9.2		.6	2.5	3.0	100	Etzel et al. 1977
HD 137569	530	.11	.45	.7	1.1	800	Bolton & Thomson 1980
GD448	.103		.41	.10	1.5	350	Maxted et al. 1998
NN Ser	.130		.57	.12	2.0	1200	Catalán et al. 1994
0308-096	.284		.39	.18	1.5	250	Saffer et al.1993
GK Vir	.344		.51	.10	2.0	300	Fulbright et al. 1993
V471 Tau	.521		.72	.72	3.5	800	Young & Lanning 1975
G203-47	14.7	.07	.56	.28	1.2	800	Delfosse et al. 1999
IK Peg	21.7		1.1	1.7	5.5	2000	Landsman et al. 1993
AY Cet	56.8	.09	.55	2.1	2.5	200	Simon et al. 1985
G77-61	245.5		.55	.33	1.1	1000	Dearborn et al. 1986
0957-666	.061		.37	.32	2.5	200	Moran et al. 1997
0422-54	.090		.60	.60	3.5	200	Koen et al. 1998
HD 49798	1.55		.80	1.00	4.0	500	Bisscheroux et al. 1997

orbit should have been, assuming either late Case C or in some instances late Case B RLOF.

A property that is striking about many of the short-period post-CE systems is how small is the mass of the companion. This runs rather counter to the general expectation that most CVs started with an MS component that was, say, a G/K dwarf, but has lost mass to become the currently-observed M dwarf. Many of the observed immediately-post-CE systems already have M dwarf components, often of very small mass. By comparison, several of the failed-CE systems have a higher-mass companion. This leads me to speculate that in a pre-CE binary a rather extreme mass ratio is necessary for substantial spiral-in, and that if the mass ratio (in the pre-CE) is less than ~ 4 (in the sense RSG/companion) then there is little or even no spiral-in.

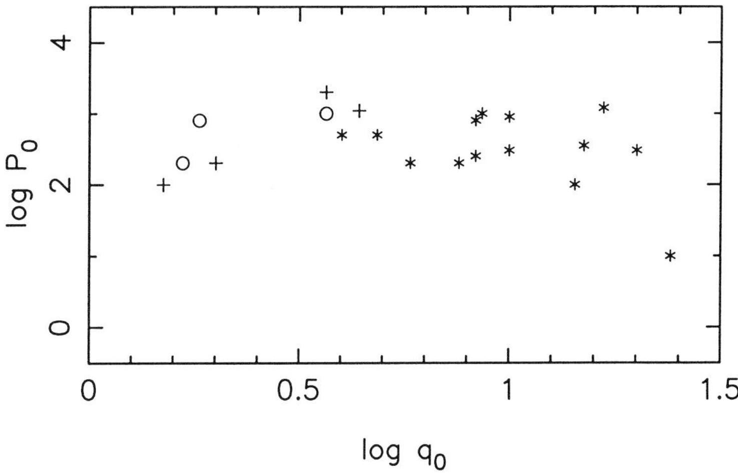

Figure 1 Estimated $\log P_0$ versus estimated $\log q_0$, for the systems of Table 1. These are the period and mass rati PRIOR to the most recent episode of RLOF. Systems which were still quite wide (P larger than 50d) AFTER RLOF are indicated with a circle; those which now have very short period (P smaller than 1d) by an asterisk; and those with an intermediate final period by a plus.

In Fig 1 I plot the estimated precursor (log) period against the estimated precursor (log) mass ratio $q_0 = M_{10}/M_2$. Asterisks represent those systems whose current periods are very short ('full CE'), circles where it is rather long ('failed CE') and plusses for an intermediate period ('partial CE'). In the diagram there is some overlap between the various categories, but there seems to be something of a division at about $q_0 \sim 4$ between the full CE cases and the others.

Of course there exists a large category of totally-failed-CE systems, the Barium stars (Jorissen et al. 1998). These probably are located near the top left corner of Fig 1, coming from systems with $q_0 \lesssim 1.5$ and $P_0 \gtrsim 400\,$d. They reinforce the view that a rather high mass ratio is necessary for full CE evolution. Another category of failed-CE stars is the Algol class, at least those widish Algols (like RX Cas, RZ Oph) that must have come from late Case B or late Case C evolution. They probably should be on the left-hand edge of the diagram, perhaps also with $q_0 \sim 1.5$ but with $P_0 \sim 5 - 400\,$d. They, and perhaps also a proportion of the Ba stars, seem to be an indication of 'binary-enhanced stellar wind', i.e. systems where the more evolved component, on first coming within a factor of ~ 2 of its Roche-lobe radius, is spun up to corotation by tidal friction and therefore has much greater magnetic activity at its surface than a single (sub)giant of the same size. Such a star may lose mass up to $\sim$ 1000 times as fast as a single counterpart, and may therefore reduce its mass ratio from say 1.3 to $\lesssim 0.7$ in the process. This could prevent the RLOF from leading to CE, and instead just lead to a moderate, nuclear-timescale, RLOF.

To sum up this section, although there is a common expectation that late Case B and late Case C RLOF should lead to full CE evolution, some of the post-CE objects that we see appear to tell a different story: only if the mass ratio is fairly extreme ($q_0 \gtrsim 4$) does full CE seem to be the preferred option. However, one would like to see observational determinations of the parameters of many more post-CE systems, including failed-CE and partial-CE as well as full-CE, before being confident of this.

Acknowledgments

This work was undertaken in relation to the DJEHUTY project, a project to model single and binary stellar structure in 3-D. Work performed at LLNL is supported by the DOE under contract W7405-ENG-48. The above discussion illustrates the well-known fact that most, indeed probably all, evolution of binaries leads, either in the short or the long term, to a situation in which some fully 3-D work has got to be done on a hydrodynamical timescale: whether drastic RLOF, probably leading to common-envelope evolution and either a merger or a short-period binary, or a slower RLOF that leads to contact. In either case, simplistic analysis is unlikely to lead to a clear indication of where the system goes next, and a fully hydrodynamic 3-D modelling of the situation looks like the best way forward. We believe that DJEHUTY should be able to achieve that.

References

Bell, S. A., Pollacco, D. L. & Hilditch, R. W. (1994) MN, 270,449

Bisscheroux, B. C., Pols, O. R., Kahabka, P, Belloni, T. & van den Heuvel, E. P. J. (1997) A&A, 317, 815

Bolton, C. T. & Thomson, J. R. (1980) ApJ, 241, 1045

Catalán, M. S., Davey, S. C., Sarna, M. J., Smith, R. C. & Wood, J. H. (1994) MN, 269, 879

Dearborn, D. S. P., Liebert, J., Aaronson, M., Dahn, C. C., Harrington, R., Mould, J. & Greenstein, J. L. (1986) ApJ, 300, 314

Delfosse, X., Forveillet, T., Beuzit, J.-L., Udry, S., Mayor, M. & Perrier, C. (1999) A&A, 344, 897

Eggleton 2000 Bormio conf.

Etzel, P. B., Lanning, H. H., Patenaude, D. J. & Dworetsky, M. M. (1977) PASP, 89, 616

Fekel, F. C. Jr., & Tomkin, J. (1982) ApJ, 263, 289

Ferguson, D. R., Liebert, J., Haas, J., Napiwotzki, R. & James, T. A. (1999) ApJ, 518, 866

Fulbright, M. S., Liebert, J., Bergeron, P. & Green, R. (1993) ApJ, 406, 240
Han, Z., Tout, C. A. & Eggleton, P. P. (2000) MN, 319, 215
Hilditch, R. W., Harries, T. J. & Hill, G. (1996) MN, 279, 1380
Hilditch, R. W., Hill, G. & Khalesseh, B. (1992) MN, 254, 82
Jorissen, A., Van Eck, S., Mayor, M. & Udry, S. (1998) A&A, 332, 877 [Ba star orbits]
Kiseleva, L. G., Eggleton, P. P. & Mikkola, S. (1998) MN, 300, 292
Koen, C., Orosz, J. A. & Wade, R. A. (1998) MN, 300, 695
Landsman, W., Simon, T. & Bergeron, P. (1993) PASP, 105, 841
Maxted, P. F. L., Marsh, T. R., Moran, C., Dhillon, V. S. & Hilditch, R. W. (1998) MN, 300, 1225
Moran, C., Marsh, T. R. & Bragaglia, A. (1997) MN, 288, 538
Nelson, C. A. & Eggleton, P. P. (2001) ApJ, accepted
Paczyński, B. (1976) in *Structure and Evolution of Close Binary Systems*, IAU Symp. No. 73, eds Eggleton P.P., Mitton S., Whelan J., Reidel, Dordrecht, p75
Pollacco, D. L. & Bell, S. A. (1994) MN, 267, 452
Pols, O. R., Schröder, K.-P., Hurley, J. R., Tout, C. A. & Eggleton, P. P. (1998) MN, 298, 525
Popper, D. M. (1988) AJ, 96, 1040
Saffer, R. A., Wade, R.A., Liebert, J., Green, R. F., Sion, E. M., Bechtold, J., Foss, D. & Kidder, K. (1993) AJ, 105, 1945
Sarma, M. B. K., Vivekananda Rao, P. & Abhyankar, K. D. (1996) ApJ, 458, 371
Schröder, K.-P., Pols, O. R. & Eggleton, P. P. (1997) MN, 285, 696
Simon, T., Fekel, F. C. & Gibson, G. M. (1985) ApJ, 295, 153
Smalley, B. (1997) Obs, 117, 338
Tout, C. A. & Eggleton, P. P. (1988) ApJ, 334, 357
Wood, J. H. & Saffer, R. A. (1999) MN, 305, 820
Włodarczyk, K. (1984) AA, 34, 381
Young, A. & Lanning, H. H. (1975) PASP, 87, 461

CLOSE BINARIES IN TRIPLE SYSTEMS

Ludmila Kiseleva-Eggleton
UC Davis, Physics Department, Davis CA 95616, USA & Institute of Geophysics and Planetary Physics, Lawrence Livermore National Laboratory, L-413, 7000 East Ave, Livermore CA 94550, USA; e-mail: lkisseleva@igpp.ucllnl.org

Peter P. Eggleton
Institute of Geophysics and Planetary Physics, Lawrence Livermore National Laboratory, L-413, 7000 East Ave, Livermore CA 94550, USA; email: ppe@igpp.ucllnl.org

Abstract The presence of of the third companion, even a very distant and/or low-mass one, may significantly affect the orbit of a close binary. The ex-eclipsing binary SS Lac, which has reccently been proved to have a companion star, is an interesting example of such an influence. We apply a model in which both the orbit and the intrinsic spins of the stars in a close binary evolve subject to a number of perturbing forces, including the effect of a third body in a possibly inclined wider orbit to this system as well as to hypothetical young Algol (β Per) models with a detached binary of two ZAMS stars. We find that our model of SS Lac is quite constrained by the relatively good observational data of this system, and leads to a specific inclination (29°) of the outer orbit relative to the inner orbit at epoch zero (1912). Although the intrinsic spins of the stars have little effect on the orbit, the converse is not true: the spin axes can vary their orientation relative to the close binary by up to 120° on a timescale of about a century. For Algol we find that although at present the semi-detached binary orbit is practically unaffected by perturbations, the initial binary if it had a period of $P \gtrsim 3$ days would have shortened it to under 3 days. This sets an upper limit to the amount of angular momentum loss that the binary may have subsequently experienced, when one component became a magnetically-active red subgiant.

1. INTRODUCTION

Most of the observed triple systems in the field as well as in open clusters are hierarchical, i.e. a close binary has a distant component. This is not surprising because non-hierarchical triples (except a very few special cases like, for example, the Eulerian or Lagrangian configurations) are

D. Vanbeveren (ed.), The Influence of Binaries on Stellar Population Studies, 311–318.

dynamically unstable, i.e. they eventually (usually, within several crossing times) disintegrate into a bound binary and a detached single body which can escape to infinity. In this work we model the effect on a short-period binary-star orbit, and also on the spins of the two components, of the following perturbations:

(a) a third body (treated as a point mass) in a longer-period orbit
(b) the quadrupolar distortion of the stars due to their intrinsic spin
(c) the further quadrupolar distortion due to their mutual gravity
(d) tidal friction, in the equilibrium-tide approximation
(e) General Relativity.

The third body's effect is treated only at the quadrupole level of approximation, although in principle it should not be difficult to go to a higher order if necessary. The stellar distortion terms are also treated only at the quadrupole level. Each of the perturbing forces has been averaged over an approximately Keplerian orbit. We follow the analysis of Eggleton, Kiseleva & Hut (1998) for perturbuations due to quadrupolar distortion and tidal friction. The detailed mathematical description of the model used here is given by Eggleton & Kiseleva-Eggleton (2000).

We then apply the model to the interesting 14.4d binary SS Lac in the young open cluster NGC 7209, which eclipsed for the first 50 years of the 20th century, and stopped eclipsing later (Zakirov & Azimov 1990). Recently Torres & Stefanik (2000) – hereinafter TS00 – have demonstrated the existence of a third body in a $\sim$ 700d orbit, which if it is inclined at a suitable angle to the inner orbit could well account for the necessary change in the orientation of the orbit relative to the observer. We find that in order to accommodate the data given by TS00, we require the outer orbit to be inclined at about 29° (or 151°) to the inner orbit. We also require a specific value (37°) for the longitude of the outer orbit's axis relative to the inner orbital frame in 1912 (epoch zero). We find a modest variation with time in the eccentricity, and in the rate of change of inclination of the inner orbit to the line of sight. These cause us to revise slightly the masses found by TS00. Our model is fully constrained by the known data on this system, but it is not over-constrained, so that unfortunately we are not in a position to confirm that the model is correct. We can however make predictions for changes that should be capable of confirmation or refutation in about 20 years.

Let us note that there is another remarkable binary system V907 Sco (a possible member of the open cluster M7) whose eclipses have turned on and off twice within modern history. This binary which has an orbital period $\approx$ 3.8d is actually a triple with a faint (late K or white dwarf) third component which period $P_{out} \approx$ 99.3d (Lacy, Helt & Vaz 1999). There are other eclipsing binaries in triple systems (for example, RW Per,

LY Aur, IU Aur, CW Ser) where the light curves (depths of eclipses) are perturbed by a third star.

In hierarchical triple systems where two orbital planes have high (39° or more) relative inclination, Kozai cycles (Kozai 1962), i.e. cyclic large-amplitude variation of the eccentricity of the inner binary binary , in combination with tidal friction can make the inner orbit shrink even if it is initially too wide for tidal friction to be important. There is only a rather small number of known triples where the inclination of the outer to the inner orbit is directly measured, and even fewer in which it is clearly established that this inclination is large enough to cause Kozai cycles. In fact, the system β Per (Lestrade et al. 1993) whose relative orbital inclination $i_{\rm H}$= 100°, is the only example we know. We therefore consider how the orbit of this system might have been modified at an early stage in its life, when it was a detached near-ZAMS system.

2. THE CASE OF SS LAC

SS Lac is a binary which eclipsed before about 1950, but not subsequently. A likely explanation was the presence of a third body, in a non-coplanar orbit, and this was confirmed by TS00, who found long-period orbital motion in the CG of the short-period pair. By coincidence, the longer period in SS Lac is exactly the same as that in Algol (679d). Because the inner binary is rather wide ($P \sim 14$ d), provided its eccentricity (or more specifically its perihelion separation) does not vary significantly, tidal friction as well as quadrupolar effects should not be important for the orbital evolution. We modeled the evolution of the orbit according to a model (Eggleton & Kiseleva 2001; hereinafter EK01) in which the effect of the third body was approximated as a quadrupole perturbation. At this level of approximation, the period and semimajor axis of the inner orbit are constant (tidal friction being found to be negligible), but not the eccentricity or inclination. The outer orbit is entirely constant, in this approximation.

If quadrupole distortion as well as tidal friction are unimportant in SS Lac, then the evolution of the inner orbit and of eclipses depends on just 13 quantities. Following TS00, we use suffices A for the inner orbit, AB for the outer orbit, and Aa, Ab and B for the 3 components. The 13 quantities are the 3 masses $M_{\rm Aa}$, $M_{\rm Ab}$, $M_{\rm B}$, the two periods $P_{\rm A}$, $P_{\rm AB}$, the two eccentricities $e_{\rm A}$, $e_{\rm AB}$, the two angles $i_{\rm A}$, $\omega_{\rm A}$ (latitude and longitude) giving the direction of the observer relative to the (inner) orbital frame, the latitude and longitude $i_{\rm H}$, $\omega_{\rm H}$ of the axis of the outer orbit relative to the inner orbital frame, and the stellar radii $R_{\rm Aa}$, $R_{\rm Ab}$. The last two quantities do not directly affect the evolution, but they

Table 1 Parameters for SS Lac

Parameter	Epoch	Observed	Guessed	Computed	Source
P_{AB}		679 d			rv
e_{AB}		.159			rv
P_A		14.416 d			rv
i_A	1912	87.6°			lc
$\boldsymbol{i_A}$	1998		**73°**	**72.9°**	
e_A	1912		.115		
$\boldsymbol{e_A}$	1998	**.136**		**.138**	rv
$\Delta\phi_{II}$	1912	-.072			lc
ω_A	1912			168.8°	$\Delta\phi_{II}$, e_A
$\boldsymbol{\omega_A}$	1998	**178.3°**		**178.4°**	rv
i_H	1912		29°		
ω_H	1912		233°		
i_{AB}				75.7°	i_H, i_A, ω_H-ω_A
f_{Aa}	1998	$2.56M_\odot$			rv
f_{Ab}	1998	$2.49M_\odot$			rv
f_B	1998	$0.22M_\odot$			rv
M_{Aa}				$2.93M_\odot$	
M_{Ab}				$2.85M_\odot$	
M_B				$.798M_\odot$	
$R_{Aa}/\ a_A$		.0741			lc
$R_{Ab}/\ a_A$		.0715			lc
a_A				$44.7R_\odot$	
R_{Aa}				$3.36R_\odot$	
R_{Ab}				$3.20R_\odot$	
$\boldsymbol{\Delta T_E}$	1950	**38**yr		**37.7**yr	

[1] a_A is the semi-major axis of the inner binary.

[2] 'Computed' refers to values calculated using standard formulae as well as to values obtained in numerical simulations.

affect the size and shape of eclipses, if and when these occur. We would like to know all these quantities at epoch zero, which we take to be 1912, the mean date of the eclipse light curve from historic data, as determined by TS00. However, some are only known at epoch 1998, the date of the radial-velocity curve analysed by TS00. For example, e_A was measured as 0.136 in 1998, but the equations governing the orbital evolution (EK01) predict that e_A will change on much the same timescale as the inclination i_A. We therefore have to guess a number of values for

1912, but these guesses are constrained by the fact that the solution has to give certain observed values in 1998.

Table 1 lists a number of parameters, under five headings: epoch, observed, guessed, computed, and source. For observed data, 'source' is either the 1912 light curve (lc) or the 1998 radial velocity curve (rv), as analysed by TS00. The Table also lists the relevant date, 1912 or 1998 (except for $\Delta T_{\rm E}$, see below). If no date is given, the datum was constant over the interval, in our approximation. We must for example guess the inclination in 1998 in order to obtain masses from the observed mass functions. Obviously the masses will have been the same in 1912, but our other guesses for 1912 and 1998 are constrained by the condition (among others) that the 1998 inclination computed from the model should be the same as the 1998 value that was guessed.

The parameter $\Delta\phi_{\rm II}$ listed in the Table 1 is the displacement in phase of the secondary eclipse relative to the primary eclipse (in 1912). This determines the longitude $\omega_{\rm A}$ of periastron of the orbit in 1912, in terms of the eccentricity $e_{\rm A}$guessed for that epoch. Note that TS00 *assumed* that $e_{\rm A}$was constant; but we find it to be variable. We therefore obtain a different value for $\omega_{\rm A}$ in 1912 from the value TS00 determined. The further parameter $\Delta T_{\rm E}$ in the Table is the approximate time interval between epoch 1912 and cessation of eclipses around 1950.

We found that it was necessary to guess 4 quantities, some in 1912 and some in 1998, in order to determine all 13 starting conditions. But there turned out to be exactly 4 observational constraints, independent of those already used, which can in principle lead to an exact solution. These can be identified in the Table as four quantities (boldfaced in Table 1) which have both a numerically 'computed' value and an 'observed' or 'guessed' value. We found this solution by an eyeball search of parameter space, rather than by an iterative procedure, and so we cannot guarantee the uniqueness of the solution; but our search suggested strongly that all other areas of parameter space are rather far from a solution. Even changes of $\sim$ 1% from the guessed values listed gave a quite unacceptable difference between 'computed' and 'observed/guessed'. For the relative inclination between two orbital planes after a brief search found one quite accurate solution with a fairly modest inclination between the orbits (29°), and a rather more extended search suggested that there were unlikely to be any other solutions, except possibly at high inclination where the behaviour becomes rather chaotic due to Kozai effects. This solution is also in a good agreement with the upper limit for $i_{\rm H}(30°)$ suggested by TS00.

Although it is gratifying that there were as many as 4 constraints on the 4 guesses, so that a solution could be determined, it is a pity that there were no more than 4 constraints, since these might have provided a check on the validity of the model.

From our model we found that the period of precession of the inner orbit is about 1000 yr, and the inclination to the line of sight i_A oscillates between about 47°and 105°. The inclination of the inner orbit relative to the outer oscillates by only about 1°. However, the rotation axis of components stars can turn by more than 90°in the course of $\sim$ 500 yr.

Our model also predicts that the next series of eclipses is not to be expected until about 2500. Epochs when eclipses take place appear to be separated alternately by a long interval and a shorter interval, and unfortunately we seem to be entering a long interval. Each period of eclipses lasts about a century. Of course, these predictions cannot be tested in a human lifetime. However, we can make some other predictions which are testable: for example the inclination i_A should decrease to 70° in 2011, and to 65° in 2039. This should produce a measurable change in the mass-function. The eccentricity should be currently approaching its peak of $\sim$ 0.138, and so may not change significantly for about 30 yr, but should drop to 0.132 by 2040. It should reach a minimum of 0.09 in 2160.

3. APPLICATION TO A YOUNG ALGOL-LIKE SYSTEM

In Algol(β Per) ($m_1 = 0.8\,M_\odot$, $m_2 = 3.7\,M_\odot$, $P = 2.87\,$d; $m_3 = 1.7\,M_\odot$, $P_{\rm out} = 1.86\,$yr, $e_{\rm out} = 0.23$; $i_{\rm H} = 100^\circ$, $P_{\rm out}/P = 237$) two components of the inner semi-detached binary are sufficiently close to each other to be subjects to effects of tidal friction and qudrupolar distortion. Because of the high relative orbital inclination $i_{\rm H} = 100°$ this system could be dramatically affected by Kozai cycles of $e_{\rm A}$ (changes between 0 and 0.975) and $i_{\rm H}$ (between 100°and 140°). However, such Kozai cycles do not actually occur in this semidetached system. This is because two components of the inner binary are sufficiently close to each other to be subject to effects of tidal friction and quadrupolar distortion which kill Kozai cycles (Eggleton & Kiseleva-Eggleton 2000, Kiseleva, Eggleton & Mikkola 1998).

Although in its present configuration the inner pair in Algol is *not* subject to Kozai cycles, at an early stage in its life β Per must have been a detached binary of two main-sequence stars, with radii, and hence quadrupole moments, substantially smaller than at present. It may therefore have been subject to substantial Kozai cycles. If we believe that β Per has evolved conservatively, i.e. without mass loss (ML) or angular momentum loss (AML), we would be able to infer the period

Table 2 Proto-Algol binaries with different initial periods

P_o (d)	P_{end} (d)	T_{circ} (yrs)	i_o (deg)
3	2.8	3×10^7	99.8
5	1.9	1×10^7	97.3
10	1.7	8×10^5	95.0
15	0.8	9×10^4	94.0

at any mass ratio, from $P \propto (1+q)^6/q^3$, where $q = M_1/M_2$. Taking an illustrative $q_0 = 1.25$, the present period $P = 2.87$d and $q = 0.216$ imply that $P_0 \sim 0.6$d. But Algol is an active binary, with ongoing mass loss and magnetic activity that probably means it has lost angular momentum since its formation. Thus we can only say that its initial period was likely to have been greater. The combination of Kozai cycles and tidal friction set an *upper* limit to the initial period.

In our numerical simulations we take the inner masses to be 2.5 and $2\,M_\odot$ with ZAMS radii. However, we also tested other reasonable mass pairs, for example 3.0 and $1.5\,M_\odot$, and the results were very similar. We consider a variety of initial values of P, but otherwise take the same parameters for the outer orbit as for present-day Algol.

Even if the inner binary, when it is nearly circular, is too wide for tidal friction to play a role, the increase in eccentricity may make tidal friction important when the two stars come close together at some point in the Kozai cycle. We find that the initial period P_o of a 'proto-Algol', had it been longer than $\sim$ 3d, would have shrunk, by a combination of Kozai cycles and tidal friction, to a value under 3d in a fairly short interval of time ($\sim 10^4 - 10^7$yr depending on P_o).

Table 2 presents final periods P_{end} for proto-Algol systems with different initial binary periods P_o. The last two columns give circularisation time-scales T_{circ} and the initial relative orbital inclination i_o leading to a final value of $i_H \approx 100°$. One can see that as P_o increases values in all other columns decreases, especially P_{end} and T_{circ}, which places a rather strict upper limit on a possible value of P_o. However, both upper and lower limits of P_o (by the lower limit we mean the shortest binary period which still allows a shrinkage of the orbit due to the combination of Kozai cycles and tidal dissipation) depend on a number of system parameters, the most important of which are the relative inclination between inner and outer orbits the outer orbital period.

References

Eggleton, P. P., Kiseleva, L. & Hut, P. 1998, Ap.J, 499,853

Eggleton, P. P. & Kiseleva-Eggleton, L. 2001, Ap.J, submitted

Kiseleva, L., Eggleton P. P. & Mikkola, S. 1998, MNRAS, 300, 292

Kozai, Y. 1962, AJ, 67, 691

Lacy, C. H. S., Helt, B. E. & Vaz, L. P. R. 1999, AJ, 117, 541

Lestrade, J.-F., Phillips, R. B., Hodges, M. W. & Preston, R.A. 1993, Ap.J, 410, 808

Torres, G, & Stefanik, R. P. 2000, Ap.J., 119, 1914

V

POPULATION SYNTHESIS

THE POPULATION OF CATACLYSMIC VARIABLE SYSTEMS

Ulrich Kolb
Department of Physics and Astronomy, The Open University, Walton Hall, Milton Keynes, MK6 7AA, UK
U.C.Kolb@open.ac.uk

Keywords: cataclysmic variables, novae

Abstract I discuss studies involving cataclysmic variable (CV) population models. I summarize the main elements of CV formation and emphasize the principal difficulties of testing population synthesis calculations against observations. The continuity equation approach to describe the CV orbital period distribution serves as an illustration. I present an update on work studying the ^{26}Al production by the Galactic nova population, the age of the CV population, an alternative period gap model, and the effect of nova outbursts on the mass transfer rate spectrum. A discussion of the impact on CV research of the new generation of models of low–mass stars and brown dwarfs concludes this review.

1. INTRODUCTION

In cataclysmic variables (CVs) a white dwarf accretes hydrogen–rich matter from a low–mass Roche–lobe filling companion star. Orbital periods have been measured for about 400 systems (Ritter & Kolb 1998; updated). Most periods fall in the observationally "convenient" range of 1-10 hrs, making CVs ideal laboratories for the study of accretion physics (see e.g. Marsh 2000). The fairly well–defined orbital period distribution has prompted numerous statistical studies, aimed at deducing the evolutionary history of CVs from the shape of this distribution. A major obstacle is the need to disentangle genuine evolutionary effects from selection effects. To first order CVs represent a magnitude–limited sample, and most systems reside within 100-200 pc from the Sun. Yet it is often the time–variable flux accompanying accretion phenomena which draws attention to CVs even at much larger distances. Although this

D. Vanbeveren (ed.), The Influence of Binaries on Stellar Population Studies, 321–337.

limits the use of population synthesis techniques I shall show examples where their application is meaningful and helpful.

2. PREHISTORY - A SUMMARY

I start by summarising determining factors in the evolutionary history of CVs.

The white dwarf (WD) primary component is the remnant of the degenerate core region of a giant star which has dimensions much larger than the typical orbital separations of CVs. The progenitor binary must have been much wider and hence must have experienced substantial orbital decay by a factor $10^2 - 10^3$. This is usually attributed to the formation of a common envelope (see Taam & Sandquist 2000, Iben & Livio 1993, for reviews) as a result of runaway mass transfer when the giant progenitor overflows its Roche lobe. The core and the low–mass secondary star effectively orbit inside this common envelope (CE). The envelope gains energy and angular momentum through dynamical friction, and is eventually expelled from the system. It is usually assumed that a fraction $\alpha_{\rm CE}$ (of order unity) of the released orbital energy is used to unbind the envelope,

$$\frac{G(M_c + M_{\rm env})M_{\rm env}}{\lambda R} = \alpha_{\rm CE}\left(\frac{GM_cM_2}{2a_f} - \frac{G(M_c + M_{\rm env})M_2}{2a_i}\right). \quad (24.1)$$

Here a_i is the initial orbital separation, M_c and $M_{\rm env}$ the core and envelope mass of the giant, R the Roche–lobe radius of the giant at the time of first contact, and M_2 the mass of the secondary. The parameter λ takes account of the detailed structure of the giant and relates the phenomenological expression on the left hand side of (25.1) to the actual binding energy. For WD–producing intermediate–mass giants λ typically varies between 0.2 and 0.8 (Dewi & Tauris 2000a), while for higher–mass stars λ is generally smaller (Dewi & Tauris 2000b). If (25.1) cannot be satisfied for a positive a_f that would accommodate the secondary inside its own Roche lobe the system is assumed to merge. Although the product $\alpha_{CE}\lambda$ is the only free parameter in the formulation (25.1), there are other, hidden uncertainties. Without knowledge of the detailed physics of the ejection process it remains unclear to what extent the internal energy contributes to unbind the envelope.

The WD/red dwarf pair emerging from the CE is close but still detached. Subsequent evolution into contact is due to the loss of orbital angular momentum, either by the emission of gravitational waves, or by some other mechanism such as magnetic stellar wind braking (e.g. Verbunt & Zwaan 1981). Commonly used braking models represent an extrapolation of some form of the Skumanich law (Skumanich 1972), the empirical relation describing the rotational braking of single stars. The increasingly complex picture that has emerged through more recent observations (see e.g. Krishnamurthi et al. 1997; Sills et al. 2000; and references therein) still awaits assessment in the context of CVs.

A number of authors have used population models with standard assumptions about the distribution of newly forming binaries in orbital period and component mass space to study the formation of CVs (e.g. de Kool 1992; Han et al. 1995; Politano 1996; Hurley et al. 2000, Howell et al. 2000). These models show that with a standard Skumanich-type magnetic braking law only about 1% of ZAMS binaries with primary mass in the range $1 \lesssim M_p/\mathcal{M}_\odot \lesssim 9$ and orbital separation $\lesssim 10^6 R_\odot$ form a CV within a Hubble time.

Once a CV forms successfully it belongs to one of three distinct groups, depending on the nature of the white dwarf. CVs with a low-mass helium WD (with mass $\lesssim 0.45\mathcal{M}_\odot$) descend from binaries where the former giant star filled its Roche-lobe before the ignition of core helium burning. Helium WDs form only through a binary evolutionary channel. The progenitors of CVs with an intermediate-mass or high-mass C/O WD ($\gtrsim 0.55\mathcal{M}_\odot$) have been wide enough to allow the giant to evolve to the AGB stage before it filled its Roche lobe. The mass gap between helium and C/O WDs is due to the core mass growth during core helium burning. Progenitor binaries with primary stars evolving through core C/O burning but avoiding a supernova result in CVs with massive O/Ne/Mg WDs. The minimum mass of O/Ne/Mg WDs is uncertain but probably close to $1.2\mathcal{M}_\odot$ (e.g. Nomoto 1984).

The shape of the initial secondary mass spectrum is dominated by the effect of mass transfer stability (e.g. Ritter 1996), which requires $M_2/M_1 < q_{\rm crit} \simeq 1$ and therefore removes systems with high-mass donor stars. Although it is generally accepted that systems that are unstable against dynamical-timescale mass transfer (unstable systems with secondary mass $\lesssim 0.7\mathcal{M}_\odot$) disappear from the population, i.e. merge, the fate of systems undergoing thermal-timescale mass transfer (initial mass $\gtrsim 0.7\mathcal{M}_\odot$) is less clear (see below). Most population models tend to generate a spike in the formation rate distribution at $M_2 \simeq 0.4\mathcal{M}_\odot$, which is the upper limit for dynamical stability of mass transfer onto a $0.6\mathcal{M}_\odot$ WD, the dominant mass in the population. If thermal-timescale mass transfer systems are included and assumed to "appear" as CVs once their mass ratio is smaller than unity, they produce a large spike of systems forming between 6 and 10 hrs orbital period. Except for these features the initial secondary mass spectrum is roughly flat in $\log M_2$.

As a rule one finds that in order to successfully form a CV the initial mass ratio of the progenitor binary has to be in excess of 3 – 4 (Politano 1996), to ensure the formation of a common envelope. The initial separations are between $10^2 - 10^3 R_\odot$ (i.e. orbital periods of months to years), so as to establish first contact only after the formation of a helium or carbon/oxygen core. A moderate variation of the parameter $\alpha_{CE}\lambda$, perhaps in the range 0.1 – 1, simply shifts the parameter space of CV forming binaries from larger to smaller giant radii at first contact, without significantly altering the expected initial WD mass and secondary mass distribution (see e.g. de Kool 1992). The predicted total number of CVs in the Galaxy — assuming that CVs once born never die — is between 10^6 and a few times 10^7, translating into local space densities between $10^{-5} - 10^{-4}$ pc^{-3}. The space density is smaller if the initial mass ratio distribution is more sharply peaked to unity, as in this case fewer systems obey the initial mass ratio limit.

3. THE PERIOD HISTOGRAM - CONTINUITY EQUATION

Given a CV formation rate the continuity equation provides a first estimate of the shape of the CV orbital period distribution, if stationarity and an ideal convergence of evolutionary tracks holds. The latter property refers to the observation that evolutionary tracks of CVs forming with different initial secondary masses converge rather quickly to an essentially unique track in e.g. orbital period P – mass transfer rate $\dot{M}$ space. This behaviour is a consequence of the fact that the angular momentum losses driving mass transfer depend on the current system parameters only, and that low–mass main–sequence stars subject to mass loss rapidly reach a new equilibrium state (Stehle et al. 1996).

If this is the case the continuity equation in orbital period space reads

$$\frac{\mathrm{d}}{\mathrm{d}P}\left\{n(P)\,\dot{P}\right\} = \dot{n}(P), \tag{24.2}$$

where $n(P)$ is the true (intrinsic) number density of CVs in period space, $\dot{P}$ the local speed in period space due to secular evolution, and $\dot{n}(P)$ denotes the CV formation rate at period P (number density of systems forming per unit time).

The dominant observational selection effect is from the brightness of the system, which in turn is essentially given by the accretion luminosity $L_{\rm acc}$. With $L_{\rm acc} \propto \dot{M}$ the observed number density scales to lowest order as

$$n_{obs} \;\propto\; n\,\dot{M}^{\alpha}$$

$$= n\,\dot{P}^{\alpha}\left(\frac{\dot{M}}{\dot{P}}\right)^{\alpha}$$

$$= n\,\dot{P}^{\alpha}\left(\frac{\mathrm{d}M_2}{\mathrm{d}P}\right)^{\alpha}, \tag{24.3}$$

where the selection index α is of order unity (see e.g. Kolb 1996). The derivative $\mathrm{d}P/\mathrm{d}M_2$ is related to the mass–radius index $\zeta = \mathrm{d}\ln R/\mathrm{d}\ln M_2$ of the donor via

$$\frac{\mathrm{d}P}{\mathrm{d}M_2} = \frac{1}{2}\,(3\zeta - 1)\,\frac{P}{M_2}. \tag{24.4}$$

Given the track convergence, $\mathrm{d}P/\mathrm{d}M_2$ is a known function of P. Although weakly dependent on the mass transfer timescale $t_M = M_2/\dot{M}$, $\mathrm{d}P/\mathrm{d}M_2$ mainly reflects features of the mass radius relation of low–mass stars (cf. Kolb 1993).

Using (24.3) and integrating (24.2) gives

$$n_{\mathrm{obs}}(P) \propto \dot{P}^{\alpha-1}\left(\frac{\mathrm{d}P}{\mathrm{d}M_2}\right)^{\alpha}(P)\;\int_{\infty}^{P}\dot{n}(P')\mathrm{d}P' \tag{24.5}$$

while for $\alpha = 1$ this reduces to

$$n_{\mathrm{obs}}(P) \propto \left(\frac{\mathrm{d}P}{\mathrm{d}M_2}\right)(P)\;\int_{\infty}^{P}\dot{n}(P')\mathrm{d}P' \tag{24.6}$$

The last equation shows that if the selection factor is close to unity the shape of the period distribution is governed by two factors, the birth rate integral $\int_{\infty}^{P}\dot{n}(P')\mathrm{d}P'$ and the stellar response term $\mathrm{d}P/\mathrm{d}M_2$. As the formation rate $\dot{n}(P)$ appears only in integrated from it is not surprising that the observed period distribution is insensitive to detailed features of the birth rate distribution. Rather, via the response term the observed distribution might hold clues on stellar structure (Kolb 1993; but see Sect. 4.3 below). If α is significantly different from unity, an additional term involving $\dot{P}$ appears. This term depends on the uncertain angular momentum loss timescale. The main task in interpreting the observed period distribution is therefore to disentangle features due to the birth rate from effects due to the orbital evolution timescale.

As long as observational selection cannot be modelled with greater confidence, the diagnostic value of the observed orbital period distribution, and of population synthesis models trying to reproduce it, is limited.

4. APPLICATIONS

Despite this rather pessimistic note there are a wealth of applications where CV population synthesis models are vitally important tools. I give four examples.

4.1. THE AGE OF CVS

From the above discussion of the track convergence it is clear that the age of a CV does not affect its mass transfer rate at a given orbital period P. Hence the age distribution of CVs is not affected by brightness–selection. Different ages at the same P could come about due to different initial secondary masses, and due to the time spent evolving from different post–CE orbital separations into contact.

Population synthesis models predict the age variation of CVs as a function of P for a given set of population parameters. As an example, Kolb & Stehle (1996) used de Kool's standard model with $\alpha_{rmCE}\lambda = 0.5$, and the magnetic braking law from Verbunt & Zwaan (1981) with the calibration parameter set to unity. They found that systems above the CV period gap ($P \gtrsim 3$ hr) have an average age of 1 Gyr, with most of them being younger than 1.5 Gyr, while systems below the gap ($P \lesssim 2$ hr) display a wide range of all ages above about 1 Gyr, with a mean of 3-4 Gyr. Observational confirmation of such an age difference would be strong evidence for the suspected disparity of the angular momentum loss timescales in systems with low–mass ($\lesssim 0.3\mathcal{M}_{\odot}$) and high–mass secondaries.

In principle, the WD luminosity (or temperature) and the dispersion of CV space velocities could be used as an age indicator.

Assuming standard core cooling, the above CV ages translate into a WD temperature $\simeq 10 \times 10^3$ K and $\simeq 7 \times 10^3$ K for systems above and below the period gap, respectively (for a $0.8\mathcal{M}_{\odot}$ C/O WD, cf. Chabrier et al. 2000; a He WD with mass $0.4\mathcal{M}_{\odot}$ would have a similar temperature, cf. Hansen 1999). The observed WD temperatures are much higher, typically between 10 000 and 20 000 K below the gap, and 20 000 K up to 50 000 K above the gap (e.g. Gänsicke 1999). This suggests that accretional heating dominates over secular core cooling. The observed trend of cooler WDs in CVs with shorter periods could be a simple consequence of the higher mass accretion rate and hence more effective accretional heating above the period gap. This in itself can be regarded as independent evidence for the corresponding difference in angular momentum loss timescales.

A more promising test of the predicted age difference makes use of the empirical relation between age t and space velocity dispersion $\sigma(t)$ of a coeval group of stars. From field stars in the solar neighbourhood Wielen et al. (1977) derive $\sigma(t) = k_1 + k_2 t^{1/2}$ (k_1, k_2 are constants). Using this Kolb & Stehle (1996) obtained an intrinsic dispersion of the line of sight velocity of the centre of mass (the γ velocity) $\sigma_\gamma \lesssim 20$ km/s above the period gap, while $\sigma_\gamma \simeq 30 - 35$ km/s below the gap. The difference is mainly due to the time spent evolving from the post–CE orbit into contact. If magnetic braking does not operate in this detached phase, then the above values are $\sigma_\gamma \simeq 27$ km/s above the gap versus $\sigma_\gamma \simeq 32$ km/s below the gap.

Measurement errors introduce an additional spread of γ–velocities and could entirely wipe out the signal. This is the most likely reason why the data compiled by van Paradijs et al. (1995) from the literature do not show any systematic difference between short–period and long–period CVs. Other reasons why the calculated difference might overestimate the actual difference could be that the magnetic braking law assumed by Kolb & Stehle (1996) is a fairly efficient one for systems with secondary mass above $1\mathcal{M}_\odot$ (cf. the discussion in Kalogera et al. 1998), and that the underlying $\sigma(t)$ relation might be less steep as claimed by Wielen et al. (1977). Recent work on stars in the solar neighbourhood by Fuchs et al. (2000) largely confirms the Wielen et al. relation, while Binney et al. (2000) find a somewhat shallower slope $\sigma(t) \propto t^0.33$ from HIPPARCOS stars. In contrast, Quillen et al. (2000) claim that there is no age–dependent dispersion for stars that formed within the last 10 Gyrs.

Despite all these uncertainties a dedicated observational programme to improve on γ velocity measurements (North et al. 2000) has already resulted in improved values for four CVs. So far the data are consistent with a small velocity dispersion ($\lesssim 15$ km/s) above the period gap.

4.2. ALUMINIUM-26

During a classical nova outburst CVs eject a large fraction of the material that has previously accreted onto the WD into the interstellar medium. The outburst is a result of explosive hydrogen burning, a thermonuclear runaway (TNR), on the WD surface. Although nova ejecta are rich in heavy elements (see e.g. Starrfield 1999) the total contribution of novae to Galactic nucleosynthesis is probably small. Novae are, however, potentially important sources of e.g. ^{7}Li and the radioactive isotopes ^{22}Ne and ^{26}Al (see Gehrz et al. 1998 for a review).

CV population models are a natural tool to translate nucleosynthesis yields of individual TNR models into the Galactic contribution of the whole nova population. Adding a description of nova outbursts to population models inevitably introduces additional little constrained parameters due to our incomplete understanding of TNR physics. The main problems are the degree of mixing between WD core material and accreted material, and the mechanism of mass ejection. Estimates of the nova shell mass ejected per outburst are therefore rather uncertain.

Here I discuss the production of ^{26}Al which is detectable via its radioactive decay line at 1.8 MeV. A single TNR on an O/Ne/Mg WDs produces only a small amount of ^{26}Al. Nonetheless the accumulated contribution by a large number of sources over 10^6 yrs, the mean lifetime of ^{26}Al, generates a potentially detectable smooth background emission at 1.8 MeV centred on the Galactic equator. Early COMPTEL all–sky 1.8 MeV maps revealed an unexpected clumpiness (Diehl et al. 1995) which seemed to limit the contribution in a smooth background to less than $1\mathcal{M}_\odot$ ^{26}Al in the Galaxy. This prompted Kolb & Politano (1997) to reevaluate the ^{26}Al output from a nova population to see if this tighter upper limit constrains CV population models. Their result is expressed in terms of general parameters that allow a rescaling to match a particular dependence of the ^{26}Al production in an individual outburst on the mass M_1 of the white dwarf, on the mass accretion rate $\dot{M}$, and on the degree of mixing. The scaling functions are expressed relative to a base model (a $M_1 = 1.25\mathcal{M}_\odot$ WD with $\dot{M} = 10^{17}$ g/s and 50% mixing) which was assumed to give an ^{26}Al ejecta mass fraction of 9.5×10^{-2}, i.e. $6 \times 10^{-7}\mathcal{M}_\odot$ ^{26}Al per outburst. The preferred population model produced $0.15\mathcal{M}_\odot$ ^{26}Al in the Galaxy, while limits on mixing and the minimum mass of O/Ne/Mg WDs could be established from the observed upper limit of $1\mathcal{M}_\odot$.

A more recent revision of proton capture rates (see the summary in Starrfield et al. 1998) led to TNR models with significantly reduced ^{26}Al output (Starrfield et al. 1998) compared to earlier calculations (Politano et al. 1995). The ^{26}Al mass fraction in the base model decreased by a factor of 20; the $\dot{M}$ and M_1 dependence has not been calculated. With a reduction of the total ^{26}Al output by this order the currently observed upper limit is too large to place strong constraints on the more interesting CV population parameters. The most recent maps (Plüschke et al. 2000; see also Knödelseder et al. 1999) largely confirm the early limits on ^{26}Al in a smooth background. However, TNR models by José et al. (1999) do not reproduce this dramatic decrease of ^{26}Al prodcution when new reaction rates are used. Moreover, José & Hernanz (1998) find a steeper dependence on WD mass mass ($\beta = \mathrm{d}\ln X(^{26}\mathrm{Al})/\mathrm{d}M_1 \simeq 6$ rather than $\beta = 3$, the value preferred by Kolb & Politano 1997) which would increase the total ^{26}Al output by a factor of a few. They also find a much steeper dependence on the envelope mixing, while $\dot{M}$ has not been varied in their calculations.

Progress may have to await a convergence of predictions from TNR model calculations.

4.3. ALTERNATIVE PERIOD GAP MODELS

The disrupted orbital braking model is still the most consistent and convincing explanation for the CV period gap (Spruit & Ritter 1983; Rappaport et al. 1983), the dearth of CVs with orbital periods between 2 and 3 hrs. The model postulates that due to a sudden reduction of the orbital angular momentum loss rate the systems are detached when they evolve through the period range of the gap (see e.g. King 1988 for a review).

The lack of independent observational evidence for a key ingredient — magnetic stellar wind braking ceases to be effective when the donor star becomes fully convective — continues to provoke alternative gap models. A population synthesis procedure provides the means to test these against observations.

One of the latest alternative gap models is by Clemens et al. (1998) who link the gap to a particular property of the mass–radius relation of low–mass main–sequence stars. They derive this relation empirically from a colour magnitude diagram of nearby stars, using bolometric corrections, a temperature–colour scale, and an empirical mass–magnitude relation. The resulting $\log M - \log R$ relation has an unusually small slope $\zeta \simeq 1/3$ in the vicinity of masses $M = 0.15\mathcal{M}_\odot$ and $M = 0.4\mathcal{M}_\odot$ (ζ is usually between 0.5 and 1). As a consequence, CVs would pile up at the corresponding periods 3.5 hr and 2.0 hr (cf. eq. (24.4)), causing an apparent dearth of CVs in between, even if mass transfer is continuous.

Kolb et al. (1998) used the mass radius relation by Clemens et al. as an input for a population synthesis model with a continuous orbital angular momentum loss rate (gravitational radiation). The calculated discovery probabilty of systems as a function of orbital period is flat, with two pronounced but rather narrow spikes at the periods quoted above. The shape is insensitive to the selection parameter α, cf. (24.3). This is in contrast to the observed distribution where the number density of systems is much lower in the gap than outside the gap. Hence even if the lower main sequence were as claimed by Clemens et al. (1998) it would not by itself cause the CV period gap.

4.4. NOVA OUTBURSTS AND MASS TRANSFER SPECTRA

A long–standing problem in the theory of the secular evolution of CVs is the apparent contradiction between the track convergenge (cf. Sect. 3) and the apparent scatter of observationally deduced mass transfer rates (e.g. Warner 1995, his Fig. 9.8). The track convergence implies that any variation of $\dot{M}$ at a fixed P should largely reflect the dependence of the angular momentum loss rate $\dot{J}$ on the white dwarf mass M_1. Since $\dot{M}/M_2 \propto \dot{J}/J$ (e.g. Ritter 1996), and $\dot{J}$ is independent of M_1 for mechanisms like magnetic braking which are rooted in the secondary star, the expected dependence is $\dot{M} \propto J^{-1} \propto \sqrt{M_1 + M_2}/M_1$. This is too weak in view of the observed scatter of at least an order of magnitude.

Attempts to reconcile this with the successful period gap model sought to identify physical mechanisms that superimpose short–term fluctuations on the long–term mean mass transfer rate. The fluctuations should not modify the mean transfer rate, otherwise the coherence of evolutionary tracks is lost and collective features such as the period gap would become blurred.

One mechanism discussed extensively in the literature (see Ritter et al. 2000, and refereces therein) is the mass transfer instability induced by anisotropic, weak irradiation of the secondary star. This cannot, by itself, explain the full range of $\dot{M}$ fluctuations, as the instability disappears when the secondary star is predominantly convective. Recently Spruit & Taam (2000) suggested that a circumbinary disc acts as a sink of orbital angular momentum and increases the transfer rate by an amount that depends on the disc surface density distribution, i.e. on the age of the disc. This random element would however destroy the track coherence.

Another mechanism that has been studied before is the effect of nova outbursts on the transfer rate. All CVs are expected to experience some form of nuclear burning of the accreted hydrogen–rich material on the surface of the WD. In most cases this should be explosive, in the form of a classical nova outburst, and hence accompanied by the ejection of the burning layers, the nova shell. The ejection occurs on a very short timecsale ($\lesssim$ years) and effectively leads to an instantaneous change of the orbital separation. The magnitude and sign of the change is determined by the amount of angular momentum the shell removes from the orbit. As the transfer rate varies exponentially with the difference between the donor's stellar and Roche lobe radius R_L (e.g. Ritter 1988), the consequent instantaneous change of R_L causes a discontinuous change of $\dot{M}$. The transfer rate recovers on a timescale set by the thermal relaxation term induced in the donor star by its mass loss history. Previous studies (e.g. Schenker et al. 1998) concluded that the effect on the width of $\dot{M}$ distributions at fixed P are negligible as the transfer rate was found to stay close to the secular mean for most of the time between outbursts, despite amplitudes in excess of 1 dex.

Revisiting the problem, Kolb et al. (2000b) discovered that the mass transfer rate spectrum at given P changes shape when the system is close to dynamical instability and the accreted mass needed to trigger the outburst is in excess of a critical limit of a few $10^{-4}\mathcal{M}_\odot$. This critical value is a function of WD mass and depends on the photospheric density scaleheight of the donor star. Population synthesis models show that if trigger masses of that order are realised, then the width of transfer rate spectra reach easily 1 dex above 3 hr orbital period. This is largely due to CVs with WDs of mass $0.55 - 0.8\mathcal{M}_\odot$ and secondaries with mass close to the stability limit $M_2 \lesssim 0.7M_1$. Below the period gap the effect is negligible, whatever the trigger mass, as none of the systems is close enough to instability (Kolb et al. 2000b; Kolb 2000). The effect is maximal if there are no frictional angular momentum losses (see Schenker et al. 1998). The crucial question is whether trigger and hence ejection masses are as large as $10^{-4}\mathcal{M}_\odot$. TNR calculations usually give values smaller by a factor of 10, while observational estimates are indeed of this order (see e.g. Starrfield 1999).

5. THE IMPACT OF NEW STELLAR MODELS

Recent years have seen significant progress in the interpretation of secondary star properties from a new generation of models of low–mass stars and brown dwarfs (see Chabrier & Baraffe 2000 for a review). The models employ an equation of state dedicated to describe the dominant non–ideal plasma effects, and real atmosphere models (Allard & Hauschildt 1998) as an outer boundary condition. This allows one to assign colours and a spectral type to a given stellar model in a self–consistent way.

5.1. SPECTRAL TYPE VS ORBITAL PERIOD/DONOR MASS

It has long been known that the spectral type (SpT) of many CV secondaries is considerably later than the spectral type of a main–seqenence star that would fill its Roche lobe at the same orbital period. The new stellar models allow one to quantify this deviation. Baraffe & Kolb (2000) found that the ZAMS forms an upper envelope for the observed location of CV secondaries in the SpT vs P diagram (henceforth SP diagram). Principal causes for a deviation from the ZAMS are (a) thermal disequilibrium due to mass loss, and (b) evolution off the ZAMS prior to the CV stage. In case (a) the SpT is earlier than for equilibrium models if $P \gtrsim 6$ hr, and later if $P \lesssim 6$ hr. This reflects the fact that predominantly radiative stars contract on rapid mass loss and predominantly convective stars expand, while there is little change in effective temperature. In case (b), i.e. for somewhat nuclearly evolved stars, the deviation is in the opposite direction. For $P \gtrsim 5$ hr the SpT is later, for short periods the SpT much earlier than for corresponding ZAMS stars. In both cases the behaviour can be understood by noting that the SpT provides a rough indicator for the donor mass (Kolb et al. 2000a). Any offset from the ZAMS in the SP diagram is largely due to a shift along the P axis to longer P when the star is larger than a ZAMS star of the same mass, or to shorter P when it is smaller. Uncertainties in the theory of convection, made explicit by the difficulty in constraining the value of the mixing length in the standard mixing length theory of convection, make it impractical at present to use the SpT as a quantitative measure of stellar mass (Kolb et al. 2000a).

A comparison of the location of observed CV secondaries in the SP diagram with evolutionary tracks by Baraffe & Kolb (2000) shows that for $3 \lesssim P/\mathrm{hr} \lesssim 6$ the generally late SpT could arise from the effect of mass loss on unevolved stars, but transfer rates in excess of the canonical $2 \times 10^{-9} M_{\odot}\ \mathrm{yr}^{-1}$ (e.g. Kolb 1996) are needed if the whole range of SpTs were to be explained. For $P \gtrsim 6$ hr the data seem to imply that a large fraction of CVs have an evolved secondary star. A track proceeding with small transfer rate ($10^{-9} M_{\odot}\ \mathrm{yr}^{-1}$) that starts from a terminal age main–sequence star provides a lower envelope to the data points, lending credibility to this interpretation. Overall, a mass transfer rate that increases with decreasing period, and which is a function of central hydrogen content (as a measure of the age of the secondary) seems most consistent with the overall distribution of systems in the SP diagram, and with the track coherence required by the period gap.

In all these calculations a systematic error may arise from the way a one–dimensional stellar model with a certain mass and radius is translated into a Roche–lobe filling star. Most studies apply the concept of volume–equivalent radii. Here the radius coordinate r of the one–dimensional model is taken to represent the Roche equipotential that encloses the same volume as a sphere with radius r. Recent SPH calculations of three–dimensional polytropic stars (Baraffe et al. 2000) have shown that the volume–equivalent radius of this star when it fills its Roche–lobe is larger by a few percent than the equilibrium radius of the star when it is single. The radius increase is slowly increasing with decreasing mass ratio $q = M_2/M_1$, and depends on the polytropic index. For $n = 1.5$ it is $4 - 6\%$, (consistent with results by Uryu & Eriguchi 1999), while for $n = 3$ early indications suggest an even larger effect.

The $n = 1.5$ case is relevant for systems with fully convective secondaries, i.e. for CVs with periods below the period gap, and is discussed in the next section. The $n = 3$ case approximates predominantly radiative stars and would apply to the long–period regime. In the SP diagram it tends to shift tracks to longer periods, alleviating the problem of an uncomfortably large fraction of CV donors that otherwise appeared to be very close to the termination of core hydrogen burning. More detailed work to assess the impact of this radius increase, and the consequent change of the donor star's thermal time, is in progress.

CVs with nuclearly evolved donor stars probably descend from systems with a secondary that is up to three times more massive than the white dwarf. Such systems evolve through a phase of thermal–timescale mass transfer, probably appearing as supersoft X–ray sources. For the interpretation of the SP diagram it is vital to know the outcome of the thermal–timescale mass transfer phase. Baraffe & Kolb (2000) considered initial donor masses up to $1.5\mathcal{M}_\odot$ and found little dependence of the position of systems in the SP diagram below $P \lesssim 10$ hr on the detailed mass loss history at longer periods. A systematic study involving single stars and a bimodal mass transfer rate (constant high rate switches to a constant low rate) shows that larger initial masses give rise to slightly larger deviations from the ZAMS once they emerge from fast mass transfer, while detailed calculations of the thermal–timescale phase (King et al. 2000; Langer et al. 2000) reveal an even more complex picture. A more systematic study involving a self–consistent treatment of the rapid phase is needed.

5.2. PERIOD MINIMUM

Theoretical models have consistently failed to reproduce the observed shape of the CV period distribution at the short–period cut–off at 78 min. Pacyński (1971) was the first to relate the cut–off to the minimum orbital period which systems reach when mass transfer is driven by the emission of gravitational waves. The effect is due to the increasing dominance of electron degeneracy pressure slowing down the secondary's radius contraction.

The problem is twofold (see Barker & Kolb 2000): (a) calculations with one–dimensional stellar models place the period minimum at a value that is too short by at least 10%, and (b) a slow–down and reversal of the orbital period derivative implies a "period spike", an accumulation of systems close to the minimum period. There is no evidence for such a feature in the observed period distribution (e.g. Ritter & Kolb 1998).

Using up–to–date brown dwarf models (e.g. Chabrier & Baraffe 2000, and references therein) Kolb & Baraffe (1999) confirmed the small value of $P_{\rm min} \simeq 67$ min (if the white dwarf has mass $0.6\mathcal{M}_\odot$) found in earlier calculations with other codes. Use of rotational and tidal corrections following Chan & Chau (1979) had a negligible effect on $P_{\rm min}$. Adding a contribution of CVs which form with a brown dwarf donor star rather than evolving from non–degenerate higher–mass stars to become a brown dwarf neither improves on the value of the period minimum nor on the expected shape of the distribution. To remove the spike by selection effects requires a selection index $\alpha \simeq 66$, cf. (24.3), i.e. effectively a discontinuous change of the detectability. As the distribution of magnetic and non–magnetic CVs is statistically indistinguishable, with both showing the cut–off at the same period, the physical cause for such a selection effect is unlikely to be rooted in the accretion disc physics. There is a non–negligible probability that a period spike is not seen in the sample of magnetic CVs due to small number statistics.

At least three potentially promising avenues for progress arise. Both the missing period spike and the apparently too long $P_{\rm min}$ could be explained if CVs avoid the evolutionary phase with $\dot{P} = 0$ altogether.

This would be the case if the systems are detached when the sign of $\dot{P}$ reverses, e.g. as a consequence of disruption of remnant magnetic braking. This could be linked to the observed sharp drop in magnetic activity of low–mass stars at the boundary between spectral type M and L (e.g. Rutledge et al. 2000, and references therein).

Similarly, the phase $\dot{P} = 0$ could be avoided if CVs are much younger than we think. If the 78 min. edge represents the oldest systems in the population which are still on their way down to the true period minimum they would not reach it for another 0.9 Gyrs. In this case essentially no CV should have a donor with mass much less than $0.1\mathcal{M}_\odot$. The best candidate CV which appears to be in conflict with this limit is WZ Sge with a claimed mass $M_2 \simeq 0.02\mathcal{M}_\odot$ (e.g. Patterson 1998).

The SPH experiments by Baraffe et al. (2000) mentioned above point towards a $4-6\%$ underestimate of the secondary star radius for mass ratios and masses appropriate for minimum period CVs. The consequent period increase is smaller than the naive expectation of $6-9\%$ due to knock–on effects on the angular momentum loss timescale and thermal time of the donor. Neglecting the latter leads to a decrease of $\dot{M}$ by a factor 1.3 for these tidally expanded stars, and gives $P_{\rm min} \simeq 72$ min. The change of the thermal time is more difficult to account for. If the radius increase is accompanied by a decrease in luminosity as in the case of anisotropically irradiated stars (Ritter et al. 2000), then the effect of the smaller luminosity dominates over the larger radius and gives an overall shorter thermal time $\propto 1/RL$. At the time of writing it is not clear how big this effect is. A period spike would occur in all cases.

6. CONCLUSIONS

After a phase of consolidation lacking major new discoveries, independent progress in the understanding of the structure of low–mass stars and brown dwarfs, and in the study of X–ray binaries, has revived interest in the evolutionary history of CVs. Tantalising discoveries show that we are far from a clear understanding of even the main concepts, and further investigations are certain to reveal more surprises.

Acknowledgments

I thank Isabelle Baraffe, John Barker, Andrew King, Tom Marsh, Klaus Schenker, Saul Rappaport and Hans Ritter for useful discussions, and Andy Norton for improving the language of the manuscript.

References

Baraffe I., Kolb U. 2000, MNRAS, 318, 354
Baraffe I., Renvoizé V., Kolb U., Ritter H. 2000, in preparation
Barker J., Kolb U. 2000, this volume
Binney J., Dehnen W., Bertelli G., 2000, MNRAS, 318, 658
Chabrier G., Baraffe I. 2000, ARA&A, 38, in press (astro-ph/0006383)
Chabrier G., Brassard P., Fontaine G., Saumon D. 2000, ApJ, 543, 216
Chan K.L., Chau W.Y. 1979, ApJ, 233, 950
Clemens J.C., Reid I.N., Gizis J.E., O'Brien M.S. 1998, ApJ, 496, 352
De Kool M. 1992, A&A, 261, 188
Dewi J.D.M., Tauris T.M. 2000a, A&A, 360, 1043
Dewi J.D.M., Tauris T.M. 2000b, Bormio Proceedings
Diehl R., Dupraz C., Bennett K. et al. 1995, A&A, 298, 445
Fuchs B., Dettbar C., Jahreiss H., Wielen R., in *Dynamics of Star Clusters and the Milky Way*, S. Deiters, B. Fuchs, A. Just, R. Spurzem (eds.), ASP Conf. Series, in press (astro-ph/0009059)
Gänsicke B. 1999, in *11th European Workshop on White Dwarfs*, J.-E. Solheim, E.G. Meistas (eds.), ASP Conf. Series, Vol. 169, 315
Gehrz R.D., Truran J.W., Williams R.E., Starrfield S. 1998, PASP, 110, 3
Han Z., Podsiadlowski P., Eggleton P.P. 1995, MNRAS, 272, 800
Hansen B.M.S. 1999, ApJ, 520, 680
Howell S.B., Nelson L.A., Rappaport S. 2000, ApJ, submitted
Hurley J.R., Tout C.A., Pols O.R. 2000, MNRAS, submitted
Iben I., Livio M. 1993, PASP, 105, 1373
José J., Hernanz M. 1998, ApJ, 494, 680
José J., Coq A., Hernanz M. 1999, ApJ, 520, 347

Kalogera V., Kolb U., King A.R. 1998, ApJ, 504, 967
King A.R. 1988, QJRAS, 29, 1
King A.R., Schenker K., Kolb U., Davies M.B. 2000, MNRAS, in press (astro-ph/0009062)
Knödelseder J., Dixon D., Bennett K. et al. 1999, A&A, 345, 813
Kolb U. 1993, A&A, 271, 149
Kolb U. 1996, in *Cataclysmic Variables and Related Objects*, A. Evans, J.H. Wood (Eds.), Kluwer Academic Publishers, Dordrecht, IAU Coll. 158, p. 433
Kolb U. 2000, in *Evolution of binary and multiple star systems*, P. Podsiadlowski et al. (eds.), PASP, in press
Kolb U., Stehle R. 1996, MNRAS, 282, 1454
Kolb U., Politano M. 1997, A&A, 319, 909
Kolb U., Baraffe I. 1999, MNRAS, 309, 1034
Kolb U., King A.R., Ritter H. 1998, MNRAS, 298, L29
Kolb U., King A.R., Baraffe I., 2000a, MNRAS, in press (astro-ph/0009458)
Kolb U., Rappport S.A., Schenker K., Howell S.B. 2000b, ApJ, in preparation
Krishnamurthi A., Pinsonneault M.H., Barnes S., Sofia S. 1997, ApJ, 480, 303
Langer N., Deutschmann A., Wellstein S., Höflich P. 2000, A&A, in press (astro-ph/0008444)
Marsh T. 2000, in *Astro-tomography workshop*, H. Boffin, D. Steeghs (eds.), Springer
Nomoto K. 1984, ApJ, 277, 791
North R., Marsh T., et al. 2000, in preparation
Pacyński B. 1971, ARA&A, 9, 183
Patterson J. 1998, PASP, 110, 752
Plueschke S., Diehl R., Schönfelder V., et al. 2000, in Proceedings 4th INTEGRAL Workshop, Alicante, ESA-SP, in press
Politano M. 1996, ApJ, 465, 338
Politano M., Starrfield S., Truran J. W., Weiss A., Sparks W. M. 1995, ApJ, 448, 807
Quillen A.C., Garnett D.R. 2000, ApJ, submitted (astro-ph/0004210)
Rappaport S., Verbunt F., Joss P.C. 1983, ApJ,275, 713
Ritter H. 1988, A&A, 202, 93
Ritter H. 1996, in *Evolutionary Processes in Binary Stars*, R.A.M.J. Wijers, M.B. Davies, C.A. Tout (eds.), Kluwer, Dordrecht, p. 223
Ritter H. 2000, NewAR, 44, 105
Ritter H., Kolb U., 1998, A&AS, 129, 83
Ritter H., Zhang Z., Kolb U. 2000, A&A, 360, 959
Rutledge R.E., Basri G., Martín E.L., Bildsten L. 2000, ApJ, 538, L141

Schenker K., Kolb U., Ritter H. 1998, MNRAS, 297, 633
Sills A., Pinsonneault M.H., Terndrup D.M. 2000, 534, 335
Skumanich A. 1972, ApJ, 171, 565
Spruit H.C., Ritter H. 1993, A&A, 124, 267
Spruit H.C., Taam R.E. 2000, ApJ, in press (astro-ph/0010194)
Starrfield S. 1999, Physics Reports, 311, 371
Starrfield S., Truran J.W., Wiescher M.C., Sparks W.M. 1998, MNRAS, 296, 502
Stehle R., Ritter H., Kolb U. 1996, MNRAS, 279, 581
Taam R. Sandquist E. 2000, ARA&A 38, 111
Uryu K., Eriguchi Y. 1999, MNRAS, 303, 329
Van Paradijs J., Augusteijn T., Stehle R. 1996, A&A, 312, 93
Verbunt F., Zwaan C. 1981, ApJ, 100, L7
Wielen R. 1977, A&A, 60, 263

THE POPULATION OF CLOSE DOUBLE WHITE DWARFS IN THE GALAXY

Lev R. Yungelson
Institute of Astronomy of the Russian Academy of Sciences, 48 Pyatnitskaya Str., 109017, Moscow, Russia – E-mail: lry@inasan.rssi.ru

Gijs Nelemans
Astronomical Institute "Anton Pannekoek", Kruislaan 403, NL-1098 SJ Amsterdam, the Netherlands – E-mail: gijsn@astro.uva.nl

Simon F. Portegies Zwart
Massachusetts Institute of Technology, 725 Commonwealth Avenue, Boston, MA 01581, Cambridge, USA – E-mail: spz@komodo.bu.edu

Frank Verbunt
Astronomical Institute, Utrecht University, P.O. Box 80000, NL-3508 TA Utrecht, the Netherlands – E-mail: F.W.M.Verbunt@astro.uu.nl

Keywords: stars: white dwarfs – stars: statistics – binaries: close

Abstract We present a new model for the Galactic population of close double white dwarfs. The model accounts for the suggestion of the avoidance of a substantial spiral-in during mass transfer between a giant and a main-sequence star of comparable mass and for detailed cooling models. It agrees well with the observations of the local sample of white dwarfs if the initial binary fraction is $\sim 50\,\%$ and an *ad hoc* assumption is made that white dwarfs with $M \lesssim 0.3\,\mathrm{M}_\odot$ cool faster than the models suggest. About 1000 white dwarfs with $V \lesssim 15$ have to be surveyed for detection of a pair which has $M_1 + M_2 \gtrsim M_{Ch}$ and will merge within 10 Gyr.

D. Vanbeveren (ed.), The Influence of Binaries on Stellar Population Studies, 339–354.

1. INTRODUCTION

The interest in the close double white dwarfs (hereafter CDWD) stems from several reasons: (i) white dwarfs are the endpoints of stellar evolution; (ii) CDWD experienced at least two stages of mass exchange and thus provide an important tool for testing the evolution of binaries; (iii) merging double CO white dwarfs are a model for SNe Ia; (iv) CDWD may be the most important contributors to the gravitational wave noise at $\nu \lesssim 0.01$ Hz, possibly burying signals of other sources.

We review the data on the currently known CDWD and present results of the modelling of the population of white dwarfs, which involves several new aspects. The most important are the treatment of the first stage of mass loss without significant spiral-in Nelemans et al., 2000a, the use of detailed models for the cooling of white dwarfs and the consideration of different star formation histories for the Galactic disc.

2. OBSERVED CLOSE DOUBLE WHITE DWARFS

Table 1 lists the currently known CDWD for which the orbital period and the mass of at least one component are measured (Maxted and Marsh, 1999; Maxted et al., 2000a). The accuracy of the mass determinations is certainly not better than $\pm 0.05\,M_\odot$. With the exception of WD 0135−052 and 1204+450, the brighter components of the pairs are likely to be He white dwarfs, since their mass is lower than $0.46\,M_\odot$, the limiting mass for the helium ignition in a degenerate stellar core (Sweigart et al., 1990). In the range of $M \simeq (0.35-0.45)\,M_\odot$ white dwarfs could have CO cores and thick He envelopes (Iben and Tutukov, 1985). However, the probability of the formation of these so called "hybrid" WD is 4 – 5 times lower than for helium WD with the same mass (Tutukov and Yungelson, 1993; Nelemans et al., 2000b). The binary dwarfs WD 0957−666, 1101+364, 1704−481A, and 2331+290 are expected to merge because of angular momentum loss via gravitational wave radiation. If both components are He dwarfs, the merger may result in the formation of a nondegenerate star or a supernova-scale explosion Nomoto and Sugimoto, 1977. In the case of CO companions, the formation of an R CrB star is expected (Webbink, 1984; Iben et al., 1996).

We included in Tab. 1 also data on two sdB stars with white dwarf companions (Orosz and Wade, 1999; Maxted et al., 2000b). Subdwarfs are supposed to be helium-burning stars. In these particular systems central helium burning will be completed before components merge. This makes these binaries candidates for future merging CO+CO white

Table 1 Known close double white dwarfs and subdwarfs with WD companions

N	*Name*	$P(d)$	$M_1/M_\odot$	$M_2/M_\odot$
1	WD 0135−052	1.556	0.47	0.52
2	WD 0136+768	1.407	0.44	0.34
3	WD 0957−666	0.061	0.37	0.32
4	WD 1101+364	0.145	0.31	0.36
5	WD 1204+450	1.603	0.51	0.51
6	WD 1704+481A	0.145	0.39	0.56
7	WD 1022+050	1.157	0.35	
8	WD 1202+608	1.493	0.40	
9	WD 1241−010	3.347	0.31	
10	WD 1317+453	4.872	0.33	
11	WD 1713+332	1.123	0.38	
12	WD 1824+040	6.266	0.39	
13	WD 2032+188	5.084	0.36	
14	WD 2331+290	0.167	0.39	
15	KPD 0422+5421	0.09	0.51	0.53
16	KPD 1930+2752	0.095	0.5	0.97

Objects 1 - 14 are CDWD, objects 15 and 16 are sdB stars with suspected white dwarf companions. M_1 is the mass of the brighter component and M_2 is the mass of the companion. See the text for references.

dwarfs. Remarkably, for KPD 1930+2752 the total mass of the system exceeds M_{Ch}, making it a possible SN Ia candidate!

About 10 more WD and sdB stars with suspected close WD companions are known (Marsh, 2000; Maxted et al., 2001). However, for these systems the orbital periods or the masses of the components are not yet determined.

3. FORMATION OF HELIUM WHITE DWARFS

Three CDWD are of special interest – WD 0136+768, 0957−666, and 1101+364. In these systems the masses of both components suggest that they are helium dwarfs and, thus, descend from degenerate cores of low-mass (sub)giants. If we designate main-sequence stars as MS, red (sub)giants as RG, white dwarfs as WD, the stage of mass exchange, either stable or unstable, as RLOF, and refer by subscripts 1 and 2 to the initial primary and secondary star, the evolutionary sequence which results in the formation of a double helium WD may be described as follows: $MS_1 + MS_2 \rightarrow RG_1{+}MS_2 \rightarrow RLOF_1 \rightarrow WD_1{+}MS_2 \rightarrow WD_1{+}RG_2 \rightarrow RLOF_2 \rightarrow WD_1{+}WD_2$. If the mass transfer is unstable, the change of component separation a is usually calculated by balancing the binding energy of the envelope of the mass-losing star with the change of the orbital energy (Webbink, 1984):

$$\frac{M_1\,(M_1 - M_c)}{\lambda\, r_1} = \alpha\left[\frac{M_c\, M_2}{2\, a_{\rm f}} - \frac{M_1\, M_2}{2\, a_{\rm i}}\right]. \tag{25.1}$$

Here M_c is the mass of the core of the mass-losing star, r_1 is its radius, subscripts i and f refer to the initial and final values of the orbital separation, α is the efficiency of the deposition of orbital energy into the common envelope and λ is a parameter which depends on the density distribution in the stellar envelope; the usual assumption is $\lambda = 0.5$ (de Kool et al., 1987). It has to be noticed, however, that for stars more massive than $3\,M_\odot$ the values of λ for red giants may be significantly lower (Dewi and Tauris, 2000). It would be worthwhile to investigate also lower mass stars, which form most of the observed CDWD.

Giants with degenerate helium cores obey a unique core-mass – radius relation (Refsdal and Weigert, 1970). Neglecting a slight dependence on the total mass of the star, this dependence (for solar chemical composition objects) may expressed as (Iben and Tutukov, 1985):

$$R \approx 10^{3.5}\, M_{\rm c}^4, \tag{25.2}$$

where radius of the giant R and mass of the core $M_{\rm c}$ are in solar units. At the instant when the star fills its Roche lobe, the radius given by Eq. (25.2) is equal to the radius of the Roche lobe. If the mass loss is unstable, the mass of the core of the mass-losing star doesn't grow during this stage and the mass of the new-born white dwarf is equal to $M_{\rm c}$.

Applying Eq. (25.2) together with restrictions on the masses of the progenitors of helium white dwarfs ($0.8 - 2.3\ M_\odot$), Nelemans et al. (2000a) reconstructed the evolution of WD 0136+768, 0957−666, and 1101+364. Their conclusions may be summarized as follows: (i) in the first episode of mass loss, when the companion of the Roche lobe filling star was a main-sequence star of a comparable mass, no substantial spiral-in occurred, see Fig. 1; (ii) in the second mass loss episode, which resulted in the formation of the currently brighter white dwarf, the separation of the components was strongly reduced. The deposition of the energy into common envelope in this episode was highly efficient: in Eq. (25.1) the product $\alpha\lambda \lesssim 3$. A note has to be made in relation with the latter statement: since Eq. (25.1) provides nothing more than an order of magnitude estimate, a formal solution which gives $\alpha \geq 1$ indicates that energy deposited into common envelope has to be comparable to the orbital energy of the binary.

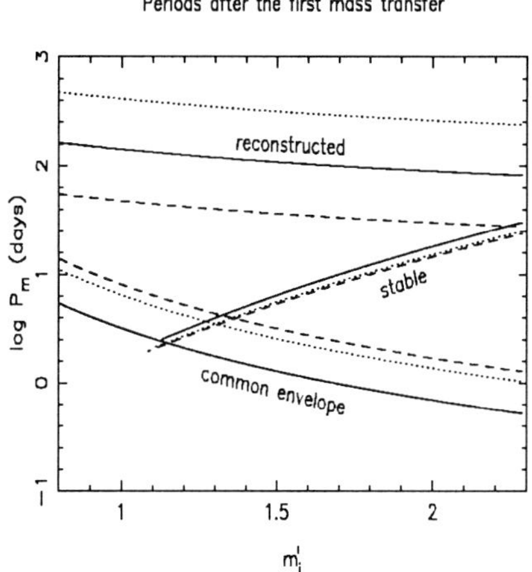

Figure 1 Periods after the first phase of mass loss as a function of the mass of the secondary component at this time. Solid, dashed, and dotted lines are for WD 0957-666, 1101+364, and 0136+768 respectively. The top three lines are periods needed to explain the mass of the last formed white dwarf. The middle three lines give the maximum period if the first white dwarf would be formed by conservative mass exchange. The lower three lines give the periods for the case when the formation of the dwarf was accompanied by a spiral-in (Eq. (25.1) with $\alpha\lambda = 2$).

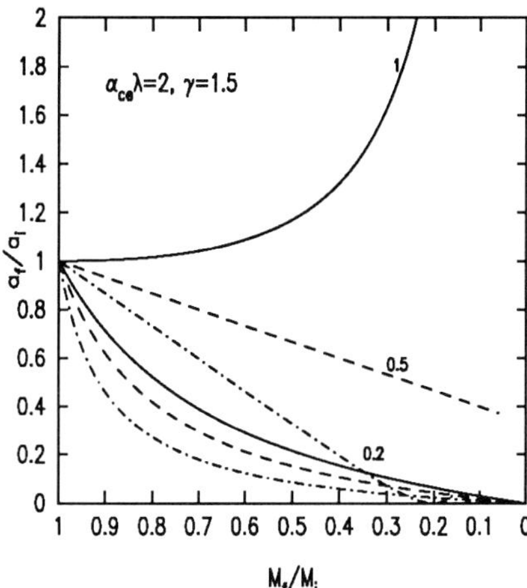

Figure 2 Dependence of the relative variation of the separation of the components a_f/a_i on the fraction of the mass lost by the star for the cases of spiral-in in a common envelope and AML regulated variation of a. Pairs of solid, dashed, and dash-dotted curves correspond to the initial mass ratio of 1.0, 0.5, and 0.2, respectively. The upper curve of every pair is for the "AML formalism" [Eq. (25.3)].

Since the first mass loss episode is neither stable mass transfer nor a spiral-in in a common envelope and the physical picture of the process is absent, Nelemans et al. (2000a) suggested to use in the population synthesis studies a simple *linear* equation for the angular momentum balance:

$$J_\mathrm{i} - J_\mathrm{f} = \gamma J_\mathrm{i} \frac{\Delta M}{M}, \tag{25.3}$$

where J_i and J_f are the angular momenta of the pre- and post-mass-transfer binary, respectively, ΔM is the amount of mass lost by the binary and M is the total initial mass of the system. The parameter γ is adjusted by fitting the orbital periods and masses of the three above mentioned helium CDWD. It follows that $1.4 \lesssim \gamma \lesssim 1.8$. As Fig. 2

illustrates, when Eq. (25.3) is applied, the separation between the components changes much less drastically, compared to the case when the "standard" common envelope formalism [Eq. (25.1)] is applied. For typical values of the fractional mass of the core M_f/M_i both formalisms give similar a_f/a_i when q decreases to $\simeq 0.2$. In the actual calculations, for a given M_f/M_i, the larger of two values of a_f/a_i was used.

4. COOLING OF WHITE DWARFS AND OBSERVATIONAL SELECTION

The observed sample of CDWD is biased. Some objects were selected for study since their low mass already suggested binarity. White dwarfs must be sufficiently bright for the mass determination and the measurement of the radial velocities. This suggests to compare a magnitude limited model sample of dwarfs with the observations. Hitherto, following Iben and Tutukov (1985), it was assumed in all population synthesis studies that all WD can be observed for 10^8 yr, unless close pairs merge in a shorter time. The actual cooling curves were never applied. However, recent studies (Driebe et al., 1998; Hansen, 1999; Benvenuto and Althaus, 1999; Sarna et al., 2000) show, that helium WD cool more slowly than carbon-oxygen ones. This is due to the higher heat content of the helium WD (Hansen, 1999) and residual nuclear burning in the relatively thick hydrogen envelopes (first noticed by Webbink (1975)). In our basic model, we take cooling curves for CO dwarfs from Blöcker (1995) and for He ones from Driebe et al. (1998, hereafter DSBH), see Fig. 3. The initial models for these curves are obtained by mimicking the mass loss by stars and therefore may be considered as the most realistic. However, as we show below, cooling times given by them probably can not be taken at face value and need revision downward. Mass loss in stellar wind and thermal flashes, which extinguishes hydrogen burning (Iben and Tutukov, 1986; Sarna et al., 2000; Althaus et al., 2000), may provide the necessary mechanism.

In addition to the sufficient brightness of the components, CDWD must have such orbital periods that radial velocities would be large enough to be measured, but small enough not to be smeared out during the integration. Following estimates of Maxted and Marsh (1999), we model this selection effect assuming that CDWD with $0.15\,\mathrm{hr} \leq P_{orb} \leq 8.5\,\mathrm{day}$ will be detected with 100% probability and that above 8.5 day the detection probability decreases linearly to 0 at $P_{orb} = 35\,\mathrm{day}$.

Our other basic assumptions may be listed as follows: we use Miller and Scalo (1979) IMF, flat distributions over initial mass ratios of the

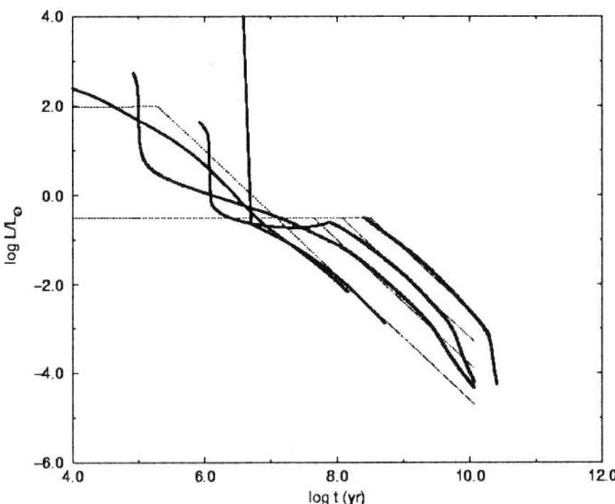

Figure 3 Cooling curves for 0.179, 0.300, 0.414 $M_\odot$ white dwarfs from Driebe et al. (1998) and for 0.6 and 0.8 $M_\odot$ ones from Blöcker (1995), from right to left. Straight lines are the fits to the curves used in the simulations.

components and the logarithm of the orbital separation, and a thermal distribution of orbital eccentricities.

5. AN OVERVIEW OF THE MODELS

For the modelling of the population of double white dwarfs we used the numerical code SeBa (Portegies Zwart and Verbunt, 1996) with modifications for low- and intermediate mass stars described by Nelemans et al. (2000b). Some of the important points are discussed below.

Since most progenitors of white dwarfs are low-mass stars, the Galactic star formation history influences the current birthrate of CDWD and the properties of their population. This factor was not studied before. For the present study we take a time-dependent star formation rate in the Galactic disk as

$$\mathrm{SFR}(t) = 15\ \exp(-t/\tau) \quad \mathrm{M}_\odot\ \mathrm{yr}^{-1}, \tag{25.4}$$

where $\tau = 7$ Gyr. Equation (25.4) gives current SFR of $3.6\,\mathrm{M}_\odot\ \mathrm{yr}^{-1}$, which is compatible with the observational estimates (Rana, 1991; van den Hoek and de Jong, 1997). With this equation, the amount of matter that has been turned into stars over the lifetime of the Galactic disk (10 Gyr in our model) is $\sim 8 \times 10^{10}\,\mathcal{M}_\odot$. It is higher than the current mass of the disk, since a part of this matter is returned to the ISM by supernovae and stellar winds. We also compute several models with constant SFR, to allow comparison with previous work (see Tab. 2).

The distribution of stars in the Galactic disk is taken as

$$\rho(R, z) = \rho_0 \, e^{-R/H} \, \mathrm{sech}(z/h)^2 \quad \mathrm{pc}^{-3}, \tag{25.5}$$

where we use $H = 2.5$ kpc (Sackett, 1997) and $h = 200$ pc. The age and mass dependence of h is neglected.

Table 2 gives an overview of our assumptions and model results and a comparison with some models of other authors. Model A is our basic model with an exponential star formation rate in the disk [Eq. (25.4)], initial fraction of binaries equal to 50% (i.e. with 2/3 of all stars in binaries) and cooling of white dwarfs according to DSBH, but modified as described below, in Sec. 6. The 50% fraction of binaries is suggested as a lower limit to their actual occurrence by the studies of normal main-sequence stars (Abt, 1983; Duquennoy and Mayor, 1991). This model, as well as all our models presented in the Table, were calculated with $\alpha\lambda = 2$ in Eq. (25.1) and $\gamma = 1.75$ in Eq. (25.3). Model A′ is similar to the model A, but assumes that the first stage of mass loss is a common envelope described by Eq. (25.1) instead of Eq. (25.3). Model B is similar to the model A, but has initially all stars in binaries, while model C has a constant star formation rate and 50% binaries. Model D has a constant star formation rate and 100% binaries. Models C and D were normalised to SFR of $4\,\mathrm{M}_\odot\ \mathrm{yr}^{-1}$, to allow comparison with the models of Iben et al. (1997, ITY) and Han (1998). The former model was recalculated by LRY for the disk age of 10 Gyr, the age assumed in this study. Note, that ITY assume that unstable mass loss always results in the formation of a common envelope, and their formulation of the equation for the energy balance is different from Eq. (25.1).

Briefly, the comparison of models shows the following. Models with an exponential SFR compared to the models with a constant SFR (mod. A vs. mod. C) have a higher number of old stars and a higher mass of the disk. This gives higher birthrates of CDWD and AM CVn stars. The rate of mergers giving SNe Ia is similar, since it is determined mainly by the SFR in the past ~ 300 Myr (Tutukov and Yungelson, 1994).

Model A with Eq. (25.3) for the first mass loss episode gives less mergers than model A' with Eq. (25.1), since it produces wider pairs. For the same reason the occurrence of SNe Ia is lower in model A'. Formation rate of interacting systems (AM CVn's) is higher in model A', because, if two common envelope phases occur, the second-formed WD is typically less massive than companion; if such a pair is brought into contact due to angular momentum loss via gravitational wave radiation, unequal masses of components favor stable mass transfer (Nelemans et al., 2001).

Model D has an IMF rather similar to Han's model, but treats the first mass loss episode differently, does not have companion reinforced

Table 2 Summary of models

Mod.	*SFR*	*bin* %	*WD* 10^9	ν 10^{-2}	$\nu_{\rm m}$ 10^{-2}	*SN Ia* 10^{-3}	*IWD* 10^{-3}	*CWD* 10^8	$\rho_{wd,\odot}$ 10^{-3}	$\nu_{\rm PN,\odot}$ 10^{-12}
A	Exp	50	9.2	4.8	2.2	3.2	4.6	2.5	19	2.3
A′	Exp	50		3.0	2.4	2.2	11.0	1.0		
B	Exp	100		8.1	3.6	5.4	8.8	4.1		
C	Cns	50	4.0	3.2	1.6	3.4	3.6	1.2	8.5	1.9
D	Cns	100		5.3	2.8	5.8	6.1	1.9		
HAN	Cns	100		2.9	2.8	2.6	23.0	0.9		
ITY	Cns	100	4.0	8.7	1.7	1.9	8.5	3.5	8.3	1.5
OBS						4 ± 2			$4 \div 20$	3

The columns list the identifiers of the models, type of star formation rate assumed for the model (exponential or constant), initial fraction of binaries, total Galactic number of WD, rates of formation and merger of CDWD per 100 yr, rate of merger of CDWD with $M_1 + M_2 \geq M_{Ch}$ per 1000 yr (SNe Ia), rate of formation of Interacting double WD (AM CVn type stars) per 1000 yr, total number of Close double WD in the Galaxy, local density of all WD per pc^3, local rate of formation of planetary nebulae per pc^3 per yr. HAN and ITY denote (Iben et al., 1997) and (Han, 1998) models, OBS - observational data.

stellar wind, and has a higher $\alpha\lambda$ value. This results in relatively less mergers in the first stage of mass loss and in higher birthrate and total number of CDWD and higher occurrence rate of SNe Ia.

The Iben et al. model differs from all other models by applying a different equation for the evolution in the common envelope, which is, in practice, equivalent to the usage of much higher value of $\alpha\lambda$. This results in less frequent mergers in both stages of mass loss, hence, a higher total number of CDWD. The total number of the Galactic WD in this model is lower, compared to our models. This is due to the ITY assumption that close and wide binaries obey different distributions over the mass ratios of the components: flat for close systems and $\propto q^{-2.5}$ for wide systems (see Tutukov & Yungelson (1993) for details).

Our models share with the Han and ITY models the same assumptions on the initial distributions of close binaries over mass ratios of components and their separations and have rather similar initial IMF for the primary components. The variations of the birthrates and numbers of objects in different classes (within factor ~ 3) arise mainly from the differences in the treatment of mass loss and transfer, in initial-final mass relations and other more minor details of population synthesis codes. Since the assumptions in the different studies are generally in agreement with results obtained from the modelling of stellar evolution, Tab. 2 illustrates the limits of the accuracy of predictions by the state of the art population synthesis studies for binary stars.

Observational data which may help to constrain the models are rather scarce and uncertain, e.g., the estimates of local space density of white dwarfs $\rho_{wd,\odot}$ differ by a factor 5 (Knox et al., 1999; Festin, 1998; Oswalt et al., 1995; Ruiz and Takamiya, 1995). Our basic model A, as well as model C, complies with these observational limits. Model A gives the annual birthrate of planetary nebulae in somewhat better agreement with the observations (Pottasch, 1996) than model C.

Model A, as well as the rest of the models in Tab. 2, agrees reasonably with the occurrence of SNe Ia in galaxies similar to the Milky Way (E. Cappellaro, this volume). The difference in SNe Ia rates between the models has to be attributed mainly to the different treatment of mass loss and to the different initial-final mass relations for the components of binaries.

6. MODELS *VS.* OBSERVATIONS

1. Orbital periods and masses of close double white dwarfs. The parameters which are determined for all known CDWD are the orbital period and the mass of the brighter component of the pair. We plot in Fig. 4 the $P_{\rm orb} - m$ distributions of the occurrence rate for the currently born CDWD and for the simulated magnitude limited samples ($V_{\rm lim} = 15$) for the models with different cooling prescriptions.

In general, the white dwarf observed as the brighter member of the pair, is the one that was formed last, but occasionally, it may be the one that was formed first, due to the effect of differential cooling, see Fig. 3. The top right panel of Fig. 4 shows that if DSBH cooling curves are taken at face value, the observed sample contains predominantly low mass ($M \lesssim 0.3\,{\rm M}_\odot$) white dwarfs, in contradiction to observations. However, as we mentioned above, low-mass WD may experience thermal flashes, which may reduce the mass of their hydrogen envelopes and extinguish hydrogen burning. This may be especially true for WD in close pairs ($a \sim R_\odot$), where one easily expects the formation of a common envelope during a flash. Since estimates for the amount of mass which may be lost in a flash is not yet available , we make an *ad hoc* extreme assumption that white dwarfs with masses below $0.3\,{\rm M}_\odot$ cool like the most massive helium ($0.46\,{\rm M}_\odot$) white dwarfs (hereafter we call this *modified DSBH cooling*). As can be seen from the bottom right panel of Fig. 4, this assumption brings the model in a much better agreement with observations. All model distributions which follow, are given for the modified DSBH cooling.

For comparison, we plot in the left bottom panel of Fig. 4 the "observed" distribution assuming (like in the studies of other authors) that

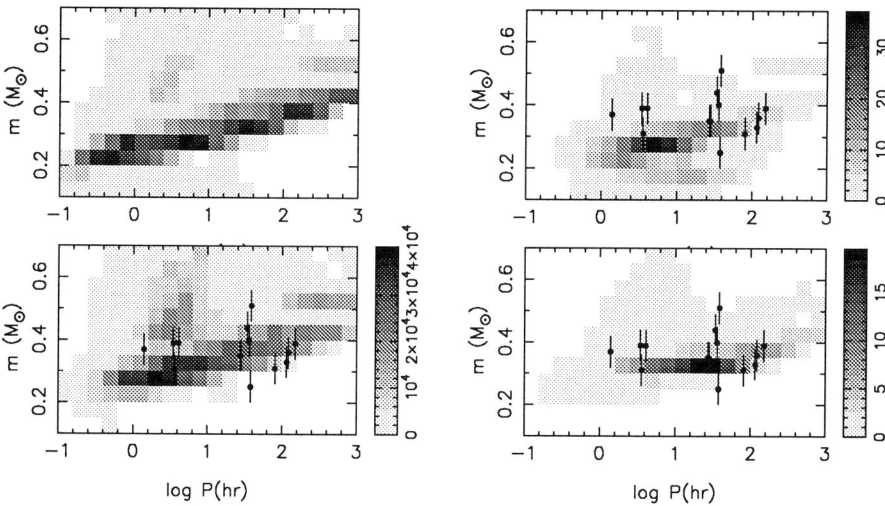

Figure 4 The model population of CDWD as a function of the orbital period and mass of the brighter dwarf of the pair. Top left: distribution of the currently born CDWD in model A. Top right: "observed sample" ($V_{\rm lim} = 15$), with cooling according to DSBH and Blöcker (1995). Bottom right: the same sample for the modified DSBH cooling. Bottom left: the total Galactic population of CDWD younger than 100 Myr. Dots with error bars: observed CDWD.

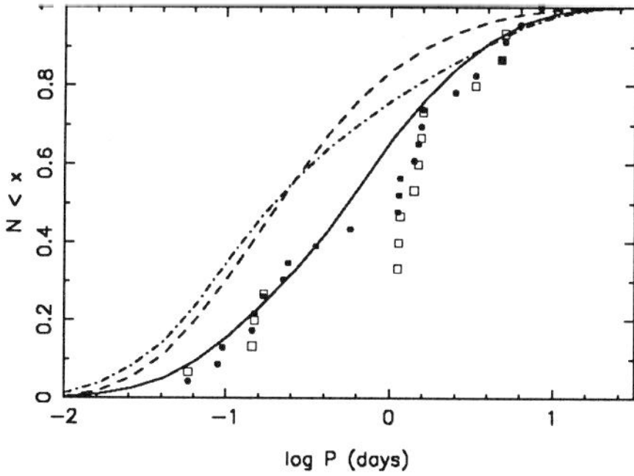

Figure 5 Cumulative distribution over the periods. The solid line is for the model with reduced cooling time for low-mass WD. The dashed line is for DSBH cooling without modifications and the dash-dotted line is for a model with constant "observability" time of 100 Myr. Open squares give distribution of observed CDWD, filled circles give the distribution including the sdB+wd binaries.

all WD are visible for 10^8 yr, unless a pair merges earlier due to GWR. Since in this case cooling curves are not applied, i.e. no magnitudes are computed for the white dwarfs, we can not construct a magnitude limited sample and this panel gives the total number of "potentially visible" CDWD in the Galaxy. The better agreement with observations compared to the modified DSBH case is only apparent, as can be easily seen from the cumulative distributions (Fig. 5).

Figure 5 shows a deficiency of observed systems between ~ 5 hr and ~ 1 day. No selection effects are known that prevent detection of white dwarf binaries with such periods. This "gap" may be partially filled if we plot also subdwarf B stars with suspected white dwarf companions, thus assuming that the current sdB star is a white dwarf in the making. In addition to systems listed in Tab. 1, we include the binaries for which only P_{orb} is determined. However, the number of detected systems is still too small to decide whether the "gap" is real and whether revisions of the stellar evolution models or CDWD formation scenarios are required.

The merger of white dwarfs is one of the models for SNe Ia (see, e.g., Livio 1999 for a review). We estimate that one merger candidate with $M_1 + M_2 \geq 1.4\,\mathrm{M}_\odot$ may be found in a WD sample complete to $V \approx 15$ which contains $\sim$ 200 CDWD among a total of $\sim$ 1000 WD. In view of the data in Tab. 2, this estimate is probably uncertain within a factor of at least $2 \div 3$.

2. Period - mass ratio distribution. Our treatment of the first phase of mass transfer between a giant and a main-sequence star, which doesn't result in a significant spiral-in, leads to a concentration of the mass ratios of the model systems around $q \sim 1$. In practice, q in the observed systems can only be determined if the ratio of the luminosities of the components is $\lesssim 5$ (Moran et al., 2000). The $P_{orb} - q$ distribution for the theoretical magnitude limited sample which obeys this criterion, is shown in Fig. 6. In model A$'$ (with a common envelope in the first mass transfer) CDWD have predominantly $q \sim (1.75 - 2)$, which is not observed and there are hardly any systems with $q \sim 1$.

3. Mass spectrum of white dwarfs. The left panel of Fig. 7 shows the spectrum of white dwarf masses in a sample limited by $V = 15$ and based on model A. It includes white dwarfs in close pairs which are brighter than their companions and genuine single objects, white dwarfs which are components of the initially wide pairs, merger products, white dwarfs which became single due to disruption of binaries by SNe explosions. In the same Figure we plot the cumulative distribution for DA white dwarfs masses, as estimated by Bergeron et al. (1992) and Bragaglia et al. (1995)[1].

---

[1]These masses may have to be increased by about $0.05\,\mathcal{M}_\odot$, if one uses models of white dwarfs with thick hydrogen envelopes for the estimates (Napiwotzki et al., 1999).

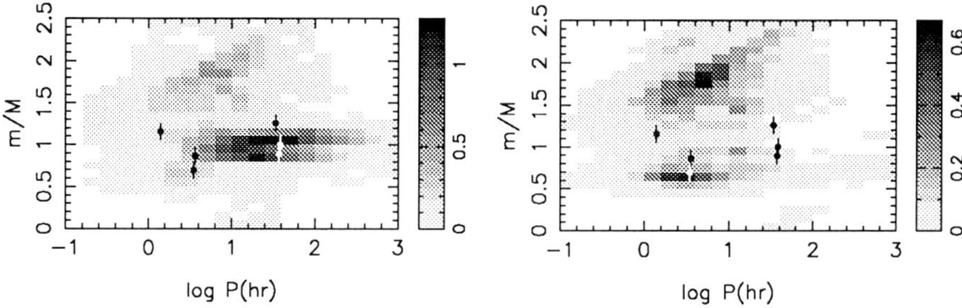

Figure 6 Model sample of CDWD as a function of the orbital period and mass ratio of components for $V_{lim} = 15$ and the ratio of the luminosities of the components ≤ 5. Left panel is for model A with the first phase of mass transfer treated by Eq. (25.3), right panel is the same for the model A′ with standard Eq. (25.1) with $\alpha\lambda = 2$ in both mass transfer phases. Dots with error bars are observed systems.

The expected fraction of CDWD among all "observed" WD in our preferred model A is $\sim$ 23%, slightly higher than the upper limit of the range suggested by Maxted and Marsh (1999) for DA white dwarfs: 1.7 to 19% with 95% confidence. Since lowering the initial fraction of binaries below 50% would contradict the observations, the high percentage of CDWD in the model sample may mean that, even after our modification, DSBH results overestimate cooling times of the lowest mass white dwarfs. The right panel of Fig. 7 shows results of a simple numerical experiment in which we assign to all helium white dwarfs the cooling curve of a $0.414\,\mathcal{M}_\odot$ dwarf from DSBH and a cooling curve of a $0.605\,\mathcal{M}_\odot$ dwarf (Blöcker, 1995) to all CO white dwarfs. This gives 17% for the fraction of CDWD. Even so, the contribution of the lowest mass WD to the total model sample still seems to be overestimated. However, there may exist yet unknown selection effects which prevent their detection.

Yet another problem is the deficiency of $(0.45-0.50)\,\mathrm{M}_\odot$ white dwarfs in the model sample. In this part of the theoretical mass distribution only hybrid white dwarfs, descending via case B mass transfer from the stars in a narrow range of $\sim(2.5-5)\,\mathrm{M}_\odot$ do occur (Iben and Tutukov, 1985). Their formation rate is relatively low. On the other hand, cooling of these objects which have almost pure helium envelopes with mass $\sim(0.01-0.1)\,\mathrm{M}_\odot$ was not yet studied and it's possible that the simple interpolation between cooling curves for CO and He white dwarfs is not appropriate and gives misleading results.

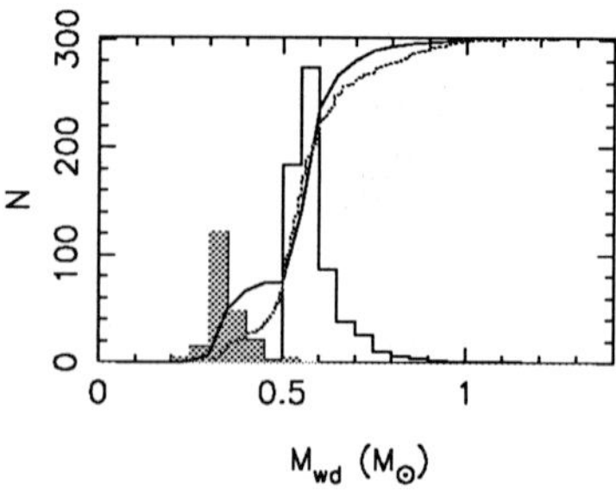

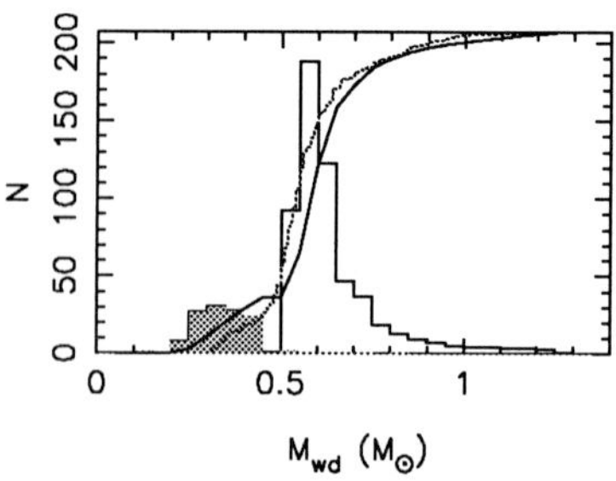

Figure 7 Left panel - mass spectrum of all white dwarfs in model A with modified DSBH cooling. Right panel - the same for a model in which all helium white dwarfs cool like a $\sim 0.414\,\mathcal{M}_\odot$ dwarf and all CO white dwarfs cool like a $0.605\,\mathcal{M}_\odot$ dwarf. In both panels the theoretical cumulative distribution is shown as a solid black line and the cumulative distribution of observed systems as a grey line.

7. CONCLUSIONS

1. The standard Algol-type evolution and spiral-in in common envelopes cannot explain the wide orbits of the progenitors of close double WD after the first stage of mass loss. In low-mass binaries with similar masses of components, the loss of the envelope of the initial primary probably doesn't cause a strong spiral-in. In the absence of a physical picture of this process, we suggest to describe the change of separation of the components by an angular momentum loss law [Eq. (25.3)]. An attempt to model the parameters of three observed CDWD with two helium components results in $1.4 \lesssim \gamma \lesssim 1.8$.

2. Modelling of the observed sample of close binary WD shows that low-mass WD have to cool faster than is suggested by the recently published cooling curves for helium WD with thick hydrogen envelopes (Driebe et al., 1998; Sarna et al., 2000). We suggest that an additional mass loss by WDs in winds or thermal flashes may strongly diminish the masses of envelopes and reduce the cooling times.

3. A reasonable agreement of the population synthesis results with observations in the estimates of the local space density of WD, and the $P_{orb}-m$, $P_{orb}-q$, and $N(P_{orb})$ distributions for CDWD may be achieved if the parameter γ in Eq. (25.3) is ~ 1.7 and it's assumed that WD with $M \leq 0.3\,M_\odot$ cool like the most massive helium WD. Further reduction of cooling times for the helium WD may even improve agreement.

4. The model with an initial binary fraction of 50% (2/3 of stars are in binaries) fits the observations better than the model in which initially 100% of stars are in binaries.

5. We estimate that the detection of at least one WD pair with $M_1 + M_2 \geq M_{Ch}$ may require a survey of $\sim$ 1000 white dwarfs. But

the uncertainties in the input data for the population synthesis studies and badly known selection effects, most probably make this estimate accurate only within a factor of $\sim 2 \div 3$.

Acknowledgments

LRY acknowledges the support of the LOC and Astronomical Institute "Anton Pannekoek" which enabled him to attend the conference. This work was supported by NWO Spinoza grant 08-0 to E.P.J. van den Heuvel, RFBR grant 99-02-16037, "Astronomy and Space Research" Program (project 1.4.4.1), and NASA through Hubble Fellowship grant HF-01112.01-98A awarded (to SPZ) by the Space Telescope Science Institute, which is operated by the Association of Universities for Research in Astronomy, Inc., for NASA under contract NAS 5-26555.

References

Abt, H. A. (1983). *ARA&A*, 21:343.
Althaus, L., Serenelli, A. M., and Benvenuto, O. G. (2000). astro-ph/0010169
Benvenuto, O. G. and Althaus, L. G. (1999). *MNRAS*, 303:30.
Bergeron, P., Saffer, R. A., and Liebert, J. (1992). *ApJ*, 394:228.
Blöcker, T. (1995). *A&A*, 299:755.
Bragaglia, A., Renzini, A., and Bergeron, P. (1995). *ApJ*, 443:735.
Dewi, J. D. M. and Tauris, T. M. (2000). *A&A*, 360:1043.
Driebe, T., Schönberner, D., Blöcker, T., and Herwig, F. (1998). *A&A*, 339:123.
Duquennoy, A. and Mayor, M. (1991). *A&A*, 248:485.
Festin, L. (1998). *A&A*, 336:883.
Han, Z. (1998). *MNRAS*, 296:1019.(HAN).
Hansen, B. M. S. (1999). *ApJ*, 520:680.
Iben, I. Jr. and Tutukov, A. V. (1984). *ApJS*, 54:335.
Iben, I. Jr. and Tutukov, A. V. (1985). *ApJS*, 58:661.
Iben, I. Jr. and Tutukov, A. V. (1986). *ApJ*, 311:742.
Iben, I. Jr., Tutukov, A. V., and Yungelson, L. R. (1996). *ApJ*, 456:750.
Iben, I. Jr., Tutukov, A. V., and Yungelson, L. R. (1997). *ApJ*, 475:291. (ITY).
Knox, R., Hawkins, M. R. S., and Hambly, N. C. (1999). *MNRAS*, 306:736.
de Kool, M., van den Heuvel, E. P. J., and Pylyser, E. (1987). *A&A*, 183:47.
Livio, M. (1999). In Truran, J. A. and Niemeyer, J., editors, *Type Ia Supernovae: Theory and Cosmology*, Cambridge. CUP. astro-ph/9903264.
Marsh, T. R. (2000). *NewAR*, 44:119.
Maxted, P. F. L. and Marsh, T. R. (1999). *MNRAS*, 307:122.

Maxted, P. F. L., Marsh, T. R., and Moran, C. K. J. (2001). *MNRAS*, in press. astro-ph/0007129v2.

Maxted, P. F. L., Marsh, T. R., Moran, C. K. J., and Han, Z. (2000a). *MNRAS*, 314:334.

Maxted, P. F. L., Marsh, T. R., and North, R. C. (2000b). *MNRAS*, 317:L41.

Miller, G. E. and Scalo, J. M. (1979). *ApJS*, 41:513.

Moran, C. K. J., Maxted, P. F. L., and Marsh, T. R. (2000). *MNRAS*, submitted.

Napiwotzki, R., Green, P. J., and Saffer, R. A. (1999). *ApJ*, 517:399.

Nelemans, G., Verbunt, F., Yungelson, L. R., and Portegies Zwart, S. F. (2000a). *A&A*, 360:1011.

Nelemans, G., Yungelson, L. R., Portegies Zwart, S. F., and Verbunt, F. (2000b). *A&A*, in press.

Nelemans, G., Portegies Zwart, S. F., Verbunt, F., and Yungelson, L. R. (2001). *A&A*, submitted.

Nomoto, K. and Sugimoto, D. (1977). *PASJ*, 29:765.

Orosz, J. A. and Wade, R. A. (1999). *MNRAS*, 310:773.

Oswalt, T., Smith, J., Wood, M. A., and Hintzen, P. (1995). *Nat*, 382:692.

Portegies Zwart, S. and Verbunt, F. (1996). *A&A*, 309:179.

Pottasch, S. R. (1996). *A&A*, 307:561.

Rana, N. C. (1991). *ARA&A*, 29:129.

Refsdal, S. and Weigert, A. (1970). *A&A*, 6:426.

Ruiz, M. T. and Takamiya, M. (1995). *AJ*, 109:2817.

Sackett, P. D. (1997). *ApJ*, 483:103.

Sarna, M., Ergma, E., and Gerškevitš-Antipova, E. (2000). *MNRAS*, 316:84.

Sweigart, A. V., Greggio, L., and Renzini, A. (1990). *ApJ*, 364:527.

Tutukov, A. and Yungelson, L. (1993). *SvA*, 36:266.

Tutukov, A. and Yungelson, L. (1994). *MNRAS*, 268:871.

van den Hoek, L. B. and de Jong, T. (1997). *A&A*, 318:231.

Webbink, R. F. (1975). *MNRAS*, 171:555.

Webbink, R. F. (1984). *ApJ*, 277:355.

BINARY POPULATION SYNTHESIS: LOW- AND INTERMEDIATE-MASS X-RAY BINARIE

Ph. Podsiadlowski
Oxford University
OX1 3RH, U. K.
podsi@astro.ox.ac.uk

S. Rappaport, E. Pfahl
Department of Physics,
Massachusetts Institute of Technology,
Cambridge, MA 02139
sar@mit.edu, pfahl@mit.edu

Keywords: X-ray binaries, millisecond pulsars, binary population synthesis, Cygnus X-2

Abstract As has only recently been recognized, X-ray binaries with intermediate-mass secondaries are much more important than previously believed. To assess the relative importance of low- and intermediate-mass X-ray binaries (LMXBs and IMXBs), we have initiated a systematic study of these systems consisting of two parts: an exploration of the evolution of LMXBs and IMXBs for a wide range of initial masses and orbital periods using detailed binary stellar evolution calculations, and an integration of these results into a Monte-Carlo binary population synthesis code. Here we present some of the main results of our binary calculations and some preliminary results of the population synthesis study for a "standard" reference model. While the inclusion of IMXBs improves the agreement with the observed properties of "LMXBs", several significant discrepancies remain, which suggests that additional physical processes need to be included in the model.

1. INTRODUCTION

Until recently it was generally believed that X-ray binaries fall into two distinct classes, low-mass X-ray binaries (LMXBs) and high-mass X-ray

D. Vanbeveren (ed.), The Influence of Binaries on Stellar Population Studies, 355–369.

binaries (HMXBs), with Her X-1 being the only exception that does not really belong to either class (for a general review see, e.g., Bhattacharya & van den Heuvel 1991). However, in the systems classified as "LMXBs" the nature of the secondary is usually unknown. Whenever it has been possible in recent years to determine the nature of the secondaries in so-called "LMXBs", these did not actually comply with the standard paradigm for LMXBs. First, van Kerkwijk et al. (1992) showed that the X-ray binary Cyg X-3, until then generally considered a "proto-type LMXB", actually contained a Wolf-Rayet star and hence, if anything, should be classified as a HMXB. More recently, it was recognized that the second-brightest X-ray source in the same constellation, Cyg X-2, also does not belong to the group of LMXBs, but is the descendant of an intermediate-mass X-ray binary (IMXB; see Sect. 2.1). This latter discovery immediately raises the question whether many other, perhaps even most systems classified as "LMXBs" are in actual fact "IMXBs".

In order to determine the relative importance of LMXBs and IMXBs we have initiated a systematic binary population synthesis study (Pfahl, Rappaport & Podsiadlowski 2001; Podsiadlowski, Rappaport & Pfahl 2001). Our approach is different from most previous studies in so far as we use realistic binary evolution calculations instead of simplified analytical prescriptions to model individual binary sequences.

In Section 2 of this contribution we first discuss the developments that have led to the re-assessment of the importance of IMXBs; in Section 3 we describe our comprehensive set of binary calculations highlighting some of the main results of these calculations, and in Section 4 we show how we are implementing these sequences into a binary population synthesis code and illustrate this with some preliminary results. For a discussion of the related problem, the formation of X-ray binaries and millisecond pulsars in globular clusters, we refer to the contribution by Rasio (Rasio 2001).

2. THE IMPORTANCE OF INTERMEDIATE-MASS X-RAY BINARIES (IMXBS)

Until recently IMXBs had received relatively little attention (see, however, Pylyser & Savonije 1988, 1989 and in the context of black-hole transients Kolb 1998; Regös et al. 1998). One of the main reasons for this neglect is the fact that only one such system, Her X-1, had unambiguously been identified in the past. This changed recently with two new developments. First, Davies & Hansen (1998) studying dynamical interactions in globular clusters found that IMXBs are much easier to

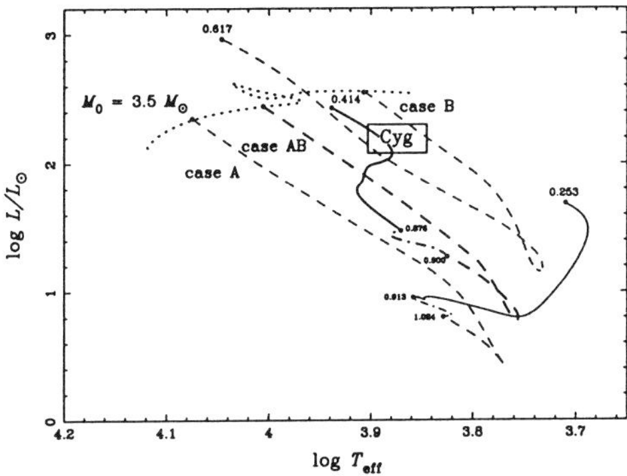

Figure 1 Evolutionary tracks of the secondaries in three binary calculations in the Hertzsprung-Russell diagram. The secondary has an initial mass of 3.5 $M_\odot$ and the primary, assumed to be a neutron star, an initial mass of 1.4 $M_\odot$ in all calculations. The dotted curve shows the track of a 3.5 $M_\odot$ star without mass loss. The mass-loss tracks, labelled case A, AB and B, start at different evolutionary phases of the secondary ('case A': the middle of the main sequence; 'case AB': the end of the main sequence; 'case B': just after the main-sequence). The dashed portions in each track indicate the rapid initial mass-transfer phase, the dot-dashed and dashed portions the slow phases where mass-transfer is driven by hydrogen core burning and hydrogen shell burning, respectively (only in case A and case AB). The beginning and end points of the various phases are marked by solid bullets, the small figures next to them give the mass of the secondary at these points. The boxed region labelled 'Cyg' indicates the observationally determined parameter region for the secondary in Cyg X-2. The tracks after mass transfer has ceased are not shown (from Podsiadlowski & Rappaport 2000).

form dynamically than LMXBs and speculated that these IMXBs, which do not exist in globular clusters at the present epoch, might be the progenitors of the observed millisecond pulsars rather than the presently observed LMXBs.

The second development, which spurred our own investigation, was a re-assessment of the evolutionary status of the X-ray binary Cyg X-2.

2.1. THE CASE OF CYGNUS X-2

The spectroscopic observations of Cyg X-2 by Casares, Charles & Kuulkers (1998) combined with the modelling of the ellipsoidal light curve (Orosz & Kuulkers 1999) showed that the secondary in Cyg X-2 was indeed a low-mass star with $M_2 = 0.6 \pm 0.13\,M_\odot$. However, the same observations also showed that the spectral type of the companion is, to within two subclasses, A9III and that *the spectral type does not vary with orbital phase.* Since this almost certainly constitutes the intrinsic spectral type of the secondary, it means that the secondary is far too hot and almost a factor of 10 too luminous to be consistent with a low-mass subgiant with an orbital period of 9.84 days (see Podsiadlowski & Rappaport 2000).

A resolution of this paradox was found independently by King & Ritter (1999) and Podsiadlowski & Rappaport (2000) (also see Kolb et al. 2000; Tauris, van den Heuvel, & Savonije 2000) who showed that the characteristics of Cyg X-2 can be best understood if the system was the descendant of an IMXB where the secondary initially had a mass of $\simeq 3.5\,M_\odot$ and lost most of its mass in very non-conservative case AB or case B mass transfer. This is illustrated in Figures 1 and 2 (from Podsiadlowski & Rappaport 2000).

Thus, Cyg X-2 provides observational proof that IMXBs can eject most of the mass that is being transferred from the secondary (perhaps in a way similar to what happens in SS433) and subsequently resemble classical LMXBs. This raises the immediate question of what fraction of so-called "LMXBs" are in reality IMXBs or their descendants. It is also worth noting that IMXBs are much easier to form than LMXBs since they do not require the same amount of fine-tuning as LMXBs (see Bhattacharya & van den Heuvel 1991). Furthermore, if there is a large fraction of IMXBs this will not only affect the period and the luminosity distribution of "LMXBs", but also has important implications for the so-called "LMXB" – ms pulsar birthrate problem (e.g., Kulkarni & Narayan 1988), all issues that can only be properly addressed with the help of binary population synthesis.

3. BINARY SEQUENCES FOR LMXBS AND IMXBS

In order to systematically investigate LMXBs and IMXBs, we performed a series of ~100 binary evolution calculations, covering the mass range of 0.8 to $7\,M_\odot$ and all evolutionary phases from early case A to late case B mass transfer.

3.1. MODEL ASSUMPTIONS

All calculations were carried out with an up-to-date, standard Henyey-type stellar evolution code (Kippenhahn, Weigert, & Hofmeister 1967), which uses OPAL opacities (Rogers & Iglesias 1992) complemented with those from Alexander & Ferguson (1994). We use solar metallicity ($Z = 0.02$), a mixing-length parameter $\alpha = 2$ and assume 0.25 pressure scale heights of convective overshooting from the core (Schröder, Pols, & Eggleton 1997). The mass-transfer rate, $\dot{M}$ is calculated explicitly using the formalism of Ritter (1988). We further assume that

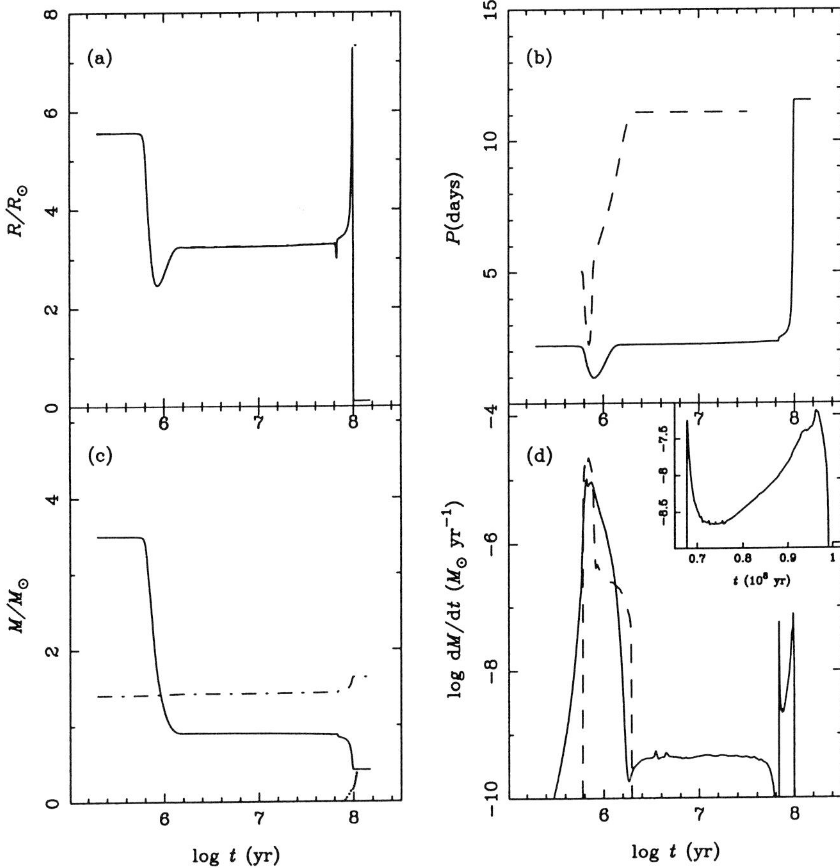

Figure 2 Key binary parameters for the case AB binary calculation as a function of time (with arbitrary offset). Panel (a): radius (solid curve) and Roche-lobe radius (dot-dashed curve) of the secondary; panel (b): the orbital period (solid curve); panel (c): the mass of the secondary (solid curve) and the primary (dot-dashed curve); panel (d): the mass-loss rate from the secondary (solid curve); the inset shows a blow-up of the second slow mass-transfer phase (hydrogen shell burning). The dashed curves in panel (b) and (d) show the orbital period and mass-transfer rate for the case B calculation for comparison (from Podsiadlowski & Rappaport 2000).

half the mass transferred is accreted by the neutron star up to the Eddington limit, taken to be $2 \times 10^{-8}\, M_\odot\, \mathrm{yr}^{-1}$. Any matter leaving the system carries with it the specific angular momentum of the accretor. Furthermore, we include angular-momentum losses due to gravitational radiation and magnetic braking (for stars with convective envelopes), for the latter using the formalism of Verbunt & Zwaan (1981).

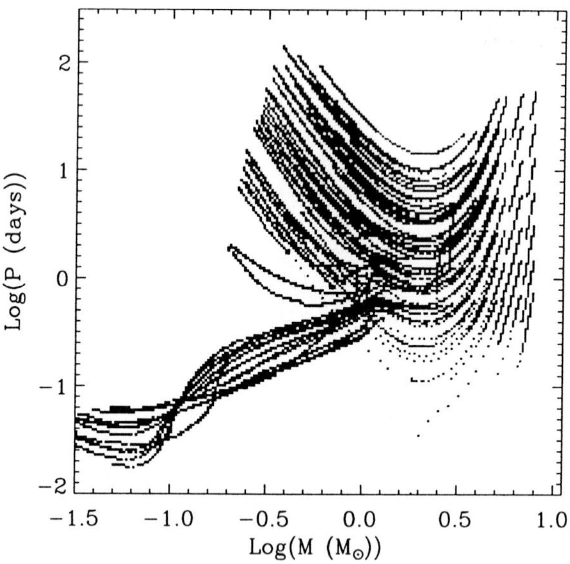

Figure 3 Binary evolution tracks in the $\log M_2 - \log P_{\rm orb}$ plane. The degree of shading measures the amount of time spent in a particular part of the diagram.

3.2. SELECTED RESULTS

Figure 3 presents the evolutionary tracks of our binary sequences in the secondary-mass orbital-period ($\log M_2 - \log P_{\rm orb}$) plane. Broadly the sequences can be divided into three classes: (1) and (2) systems evolving to long periods and short periods, respectively, separated by the well-known bifurcation period of just under 1 day, and (3) more massive systems experiencing dynamical mass transfer and spiral-in (more easily discerned in Fig. 4). What this figure does not show, however, is the actual variety in these sequences. Some 70 of the ~100 sequences are qualitatively different with respect to the importance and the order of different mass-transfer driving mechanisms, the occurrence of detached phases, the final end products, etc. In this context it is worth noting that there are very few sequences that resemble the classical cataclysmic-variable (CV) evolution where mass transfer is solely driven by gravitational radiation and magnetic braking. This may be important in explaining some of the fundamental differences observed between CVs and LMXBs/IMXBs.

Delayed dynamical instability. The majority of systems in which the initial secondary mass is $4\,M_\odot$ and all systems more massive than

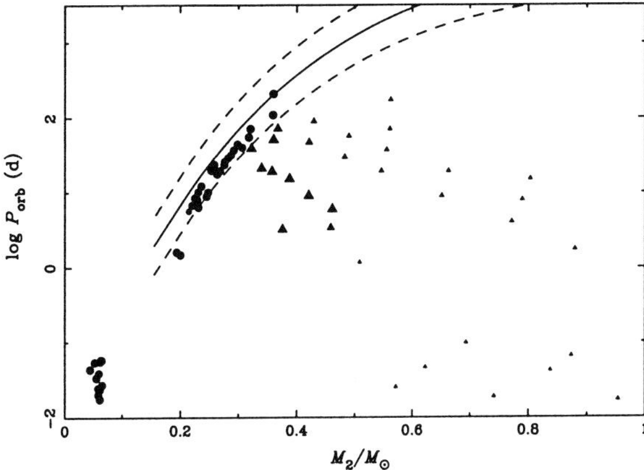

Figure 4 The end points of the binary sequences in the $M_2 - \log P_{\rm orb}$ plane. Circles and triangles indicate that the secondaries are He white dwarfs and HeCO white dwarfs, respectively. The size of the symbols indicates how much mass the neutron stars have accreted (systems with large symbols may be reasonably expected to contain millisecond pulsars). The low-mass, ultracompact systems ($M_2 < 0.1\,M_\odot$, $P_{\rm orb} < 0.1\,$d) are plotted when the systems pass through the orbital-period minimum. The more massive, ultracompact systems have experienced dynamical mass transfer and a spiral-in phase and are plotted at the orbital period at which a common envelope can be ejected energetically (although in reality most of these systems are likely to merge completely). The solid and dashed curves give the range of the white-dwarf mass – orbital-period relation for wide binary radio pulsars (from Rappaport et al. 1995).

$4.5\,M_\odot$ experience dynamical mass transfer, resulting in the spiral-in of the neutron star inside the secondary. The final outcome of this channel is very uncertain at present, but it may be a spun-up neutron star or a black hole, single or in a very compact binary. In all systems, where the secondary is still on the main sequence when mass transfer starts, this dynamical instability is delayed (see Hjellming & Webbink 1987) since the secondaries initially have radiative envelopes which are stable against dynamical mass loss. Dynamical instability occurs once the radiative part of the envelope with a rising entropy profile has been lost and the core with a relatively flat entropy profile starts to determine the reaction of the star to mass loss. This delay may last for up to 10^6 yr; during this time the system should still be detectable as a X-ray binary, with a very high mass-transfer rate and quite possibly some unusual properties (such as SS433?) in the last $10^4 - 10^5$ yr before the onset of the dynamical instability.

Evolutionary end products. In Figure 4 we show the final position of the majority of the sequences calculated in the $M_2 - \log P_{\rm orb}$

plane, where the size of the symbols indicates how much mass the neutron stars have accreted (systems with large symbols may be reasonably expected to contain millisecond pulsars). Circles indicate that the secondary ends its evolution as a He white dwarf. Note that all systems with $P_{\rm orb} > 0.1\,$d lie very close to the relation between white-dwarf mass and orbital period for wide binary radio pulsars calculated by Rappaport at al. 1995 (solid and dashed curves). In systems with triangle symbols, the secondaries burn helium in the core in a hot OB subdwarf phase (after mass transfer has been completed) and become HeCO white dwarfs (i.e., non-standard white dwarfs with a CO core and a large He envelope). Note that these HeCO white dwarfs can have a mass as low as $0.3\,M_\odot$, the minimum mass for helium ignition in non-degenerate cores (see, e.g., Kippenhahn & Weigert 1990). Most of these systems lie well below the white-dwarf – orbital period relation without having experienced a common-envelope phase (also see Podsiadlowski & Rappaport 2000; Tauris et al. 2000). It is not immediately clear whether the fact that many of these relatively low-mass white dwarfs have CO cores has detectable, observational consequences.

Finally, it is worth noting that most of the white dwarfs with masses $< 0.6\,M_\odot$ experience several dramatic hydrogen shell flashes (typically 2 to 4) before settling on the white-dwarf cooling sequence (these are reminiscent of the final helium shell flashes for relatively massive post-AGB stars; Iben et al. 1983). During these flashes, the luminosity typically rises by a factor of 1000 and the radius increases by a factor of 10 or more on timescales of a few decades. Indeed, during these flashes the secondaries tend to fill their Roche lobes again, leading to several short mass-transfer phases with very high mass-transfer rates driven entirely by the dynamics of these flashes.

3.3. THE FORMATION OF ULTRA-COMPACT BINARIES

As Figures 3 and 4 show, systems with initial orbital periods somewhat less than $\sim 1\,$d lead to ultra-compact binaries with minimum orbital periods in the range of $20 - 90\,$min. The shortest period is not much longer than the 11 min period in the X-ray binary 4U 1820-30 in the globular cluster NGC 6624 (Stella et al. 1987). Unlike the two popular models for the formation of this system, this evolutionary channel involves neither a direct collision (Verbunt 1987) nor a common-envelope phase (Bailyn & Grindlay 1987; Rasio, Pfahl, & Rappaport 2000) and therefore constitutes an attractive alternative scenario for 4U 1820-30 (this model was

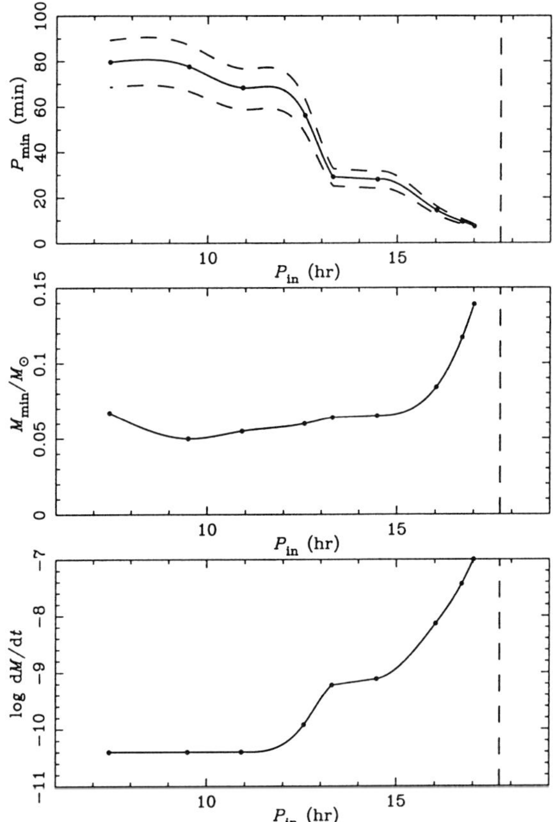

Figure 5 The formation of ultra-compact LMXBs. *Top Panel:* The minimum orbital period versus initial orbital period. *Middle and Bottom Panels:* The mass of the secondary and the mass-transfer rate, respectively, at the minimum period. All calculations start with a $1.4\,M_{\odot}$ neutron star and a $1\,M_{\odot}$ secondary. The vertical dashed line indicates the initial orbital period above which systems become wider rather than more compact (i.e., the bifurcation period). The dots indicate the results of the calculated sequences. The dashed curves in the top panel indicate the range of minimum periods if mass transfer is assumed to be either fully conservative (upper curve) or fully non-conservative (lower curve).

first suggested by Tutukov et al. 1987 and was examined in some detail by Fedorova & Ergma 1989).

To determine the shortest orbital period that can be attained through this channel we performed a separate series of binary calculations for a $1\,M_{\odot}$ secondary (as would be appropriate for the initial mass of the secondary in 4U 1820-30). We found, confirming the earlier results of Fedorova & Ergma (1987), that, if the secondaries start mass transfer near the end of core hydrogen burning (or, in fact, just beyond), the secondaries transform themselves into degenerate helium stars and that orbital periods as short as 5 min can be attained without the spiral-in of the neutron star inside a common envelope.

The results of this exploration are summarized in Figure 5. As the top panel shows, there is a fairly large range of initial orbital periods (13 – 18 hr) which leads to ultra-compact LMXBs with a minimum orbital period less than 30 min. Indeed, it is remarkable that one of our binary sequences, with an initial orbital period around 17 hr, appears to be a suitable sequence to explain all five LMXBs in globular clusters whose orbital periods are presently known, from the system with the longest period (AC211/X2127+119 in M15; $P_{\rm orb} = 17.1$ hr; Ilovaisky et al. 1993) to the 11-min binary. Furthermore, systems with an initial period in the range of 13 – 18 hr are quite naturally produced as a result of the tidal capture of a neutron star by a main-sequence star (Fabian, Pringle, & Rees 1975) and are not the generally expected outcome of a 3- or 4-body exchange interaction (see, e.g., Rasio et al. 2000; Rasio 2001). This suggests to us that it may be premature to rule out tidal capture as a formation scenario for LMXBs in globular clusters as has often been done in recent years.

4. BINARY POPULATION SYNTHESIS

For our population synthesis study, we determine, starting from primordial binaries, the mass of the secondary and the orbital period at the beginning of the X-ray binary phase. We then choose the binary sequence from our two-dimensional grid that is closest to this point in the $(M_2, P_{\rm orb})$ plane to represent the X-ray binary phase. As our systematic exploration of LMXB/IMXB evolution has shown, there is such a large variety of possible binary sequences that one cannot hope that all of them can be well represented by simple (semi-)analytic prescriptions — as is often done in such studies. We have therefore chosen this improved approach, wherein we directly integrate our grid of realistic binary sequences into the Monte Carlo code. Apart from this, most of our model assumptions are fairly standard.

4.1. MODEL ASSUMPTIONS

For our standard reference model, we pick an initial orbital period distribution that is logarithmically flat, adopt a Miller-Scalo initial mass function for the primary (Miller & Scalo 1979) and a flat distribution for the mass-ratio distribution. For the supernova-kick distribution we choose the Maxwellian distribution with $\sigma = 190\,{\rm km\,s^{-1}}$ from Hansen & Phinney (1997). We plan to investigate the effects of varying these

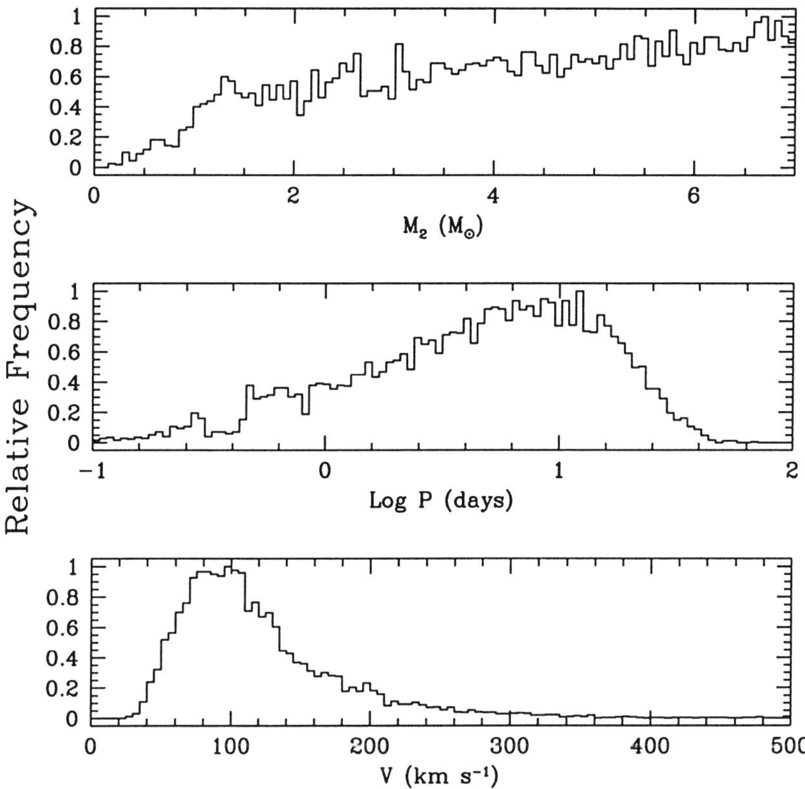

Figure 6 The distributions of secondary mass (top panel), orbital period (middle panel) and system velocity (bottom panel) at the beginning of the mass-transfer phase onto the neutron star for our standard reference model.

assumptions as part of our study in the future. To determine the orbital parameters for post-common-envelope systems, we use binding energies for the common envelope obtained from realistic stellar calculations with or without the inclusion of the ionization energy (Han et al. 1995; Dewi & Tauris 2000).

4.2. PRELIMINARY RESULTS

In Figures 6 and 7 we present some of the early results of our population synthesis simulations for our standard set of assumptions. Figure 6 displays the distributions of the secondary mass, orbital period and system space velocity at the beginning of the X-ray binary phase. The mass distribution is dominated by intermediate-mass systems; the more mas-

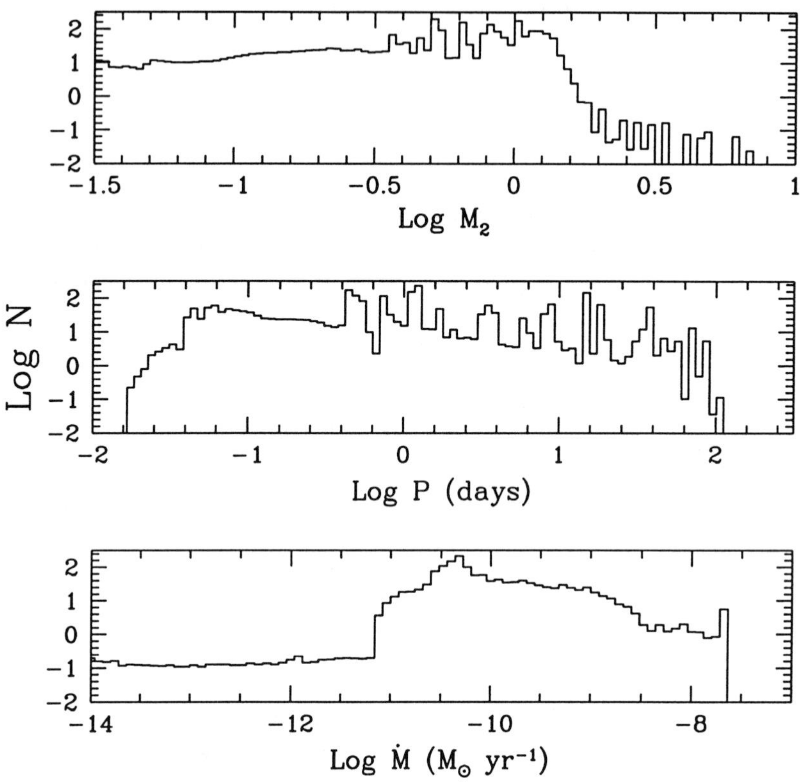

Figure 7 The distributions of secondary mass (top panel), orbital period (middle panel), and neutron-star mass-accretion rate at the current epoch for our standard reference model. The jagged structure of the histograms is a consequence of the discrete nature of the grid of binary sequences.

sive systems (above $\sim 4\,M_\odot$) will experience (delayed) dynamical mass transfer and will not contribute significantly to the population of X-ray active systems. At the other end of the mass spectrum, we find very few low-mass, CV-like systems.

Figure 7 in contrast shows the distributions of the secondary mass, orbital period and neutron-star mass-accretion rate during the X-ray binary phase at the current epoch. Now the mass distribution is dominated by relatively low-mass systems, and there are hardly any systems above $\sim 2\,M_\odot$. The reason is simply that the intermediate-mass systems lose most of their mass rapidly in the initial super-Eddington phase and spend most of their lives at relatively low masses (see Figs 2 and 3). The orbital-period distribution shows no period gap and extends to very low periods. Finally, the luminosity distribution displays a fairly strong peak around $5 \times 10^{-11}\, M_\odot\, \mathrm{yr}^{-1}$ and has a sharp cut-off at $\sim 10^{-11}\, M_\odot\, \mathrm{yr}^{-1}$.

While these distributions show many of the characteristics of the observed distributions of "LMXBs", in fact more so than a model that only includes CV-like systems, there are still some fairly obvious discrepancies. First, there are too many short-period systems to be consistent with the observed period distribution (e.g., Ritter & Kolb 1998), although we note that, in this preliminary calculation, we did not consider whether systems are transients or not. Second, while the distribution of mass-accretion rate (and hence X-ray luminosity) has a sharp cut-off at $\sim 10^{-11}\, M_\odot\, \mathrm{yr}^{-1}$ — as is desirable, the peak in the distribution is probably too low by about an order of magnitude.

5. OUTLOOK

The binary-population synthesis results presented in this contribution represent only a first step in our investigation. By defining a standard reference model, we can quantitatively identify the problems with this model (e.g., concerning the period and luminosity distributions of LMXBs/IMXBs, the properties of binary millisecond pulsars, etc.). Having identified the problems, we can then examine possible solutions. Some of the modifications we plan to consider are cyclical mass transfer due to external irradiation, interrupted mass transfer and the role of the pulsar turn-on. To first order, we intend to implement these using a perturbation-style approach where we use the available grid of binary sequences to define the secular evolution.

Once we have developed a better model, we can then extend the present calculations with the improved model on a much finer grid. With the improvements in computing power and implementing a 'minimalistic' version of our binary stellar-evolution code, we estimate that even with present-day workstations we should be able to calculate one whole binary sequence in about 2 minutes. An attractive alternative is to use the 'Eggleton' supercluster at the Lawrence Livermore Laboratory whose potential was impressively demonstrated at this meeting (Eggleton 2001). The ultimate goal, of course, is to obtain a realistic modelling of the X-ray binary population, a goal that now appears achievable within the next few years.

References

Alexander, D. R., & Ferguson, J. W. 1994, ApJ, 437, 879
Bailyn, C. D., & Grindlay, J. E. 1987, ApJ, 316, L25

Bhattacharya, D., & van den Heuvel, E. P. J. 1991, Phys. Rep., 203, 1
Casares, J., Charles, P., & Kuulkers, E. 1998, ApJ, 493, L39
Davies, M. B., & Hansen, B. M. S. 1998, MNRAS, 301, 15
Dewi, J. D. M., & Tauris, T. M. 2000, A&A, 360, 1043
Eggleton, P. P. 2001, these proceedings
Fabian, A. C., Pringle, J. E., Rees, M. J. 1975, MNRAS, 172, 15
Fedorova, A. V., & Ergma, E. V. 1989, Ap&SS, 151, 125
Han, Z., Podsiadlowski, Ph., & Eggleton, P. P. 1995, MNRAS, 172, 15
Hansen, B., & Phinney, E. 1997, MNRAS, 291, 569
Hjellming, M. S., & Webbink, R. F. 1987, ApJ, 318, 794
Iben, I., Jr., Kaler, J. B., Truran, J. W., & Renzini, A. 1983, ApJ, 264, 605
King, A. R., & Ritter, H. 1999, MNRAS, 309, 253
Kippenhahn, R., & Weigert, A. 1990, Stellar Structure and Evolution (Berlin: Springer)
Kippenhahn, R., Weigert, A., & Hofmeister, E. 1967, in Methods in Computational Physics, Vol. 7, ed. B. Alder, S. Fernbach, & M. Rothenberg (New York: Academic), 129
Kolb, U. 1998, MNRAS, 297, 419
Kolb, U., Davies, M. B., King, A., & Ritter, H. 2000, MNRAS, 317, 438
Kulkarni, S. R., & Narayan, R. 1988, ApJ, 335, 755
Ilovaisky, S. A. et al. 1993, A&A, 270, 139
Miller, G. E., & Scalo, J. M. 1979, ApJS, 41, 513
Orosz, J. A., & Kuulkers, E. 1999, MNRAS, 305, 132
Pfahl, E. D., Rappaport, S., & Podsiadlowski, Ph. 2001, in preparation
Podsiadlowski, Ph., & Rappaport, S. 2000, ApJ, 529, 946
Podsiadlowski, Ph., Rappaport, S., & Pfahl, E. D 2001, in preparation
Pylyser, E. H. P., & Savonije, G., J. 1988, A&A, 191, 57
Pylyser, E. H. P., & Savonije, G., J. 1989, A&A, 208, 52
Rappaport, S., Podsiadlowski, Ph., Joss, P. C., DiStefano, R., & Han, Z. 1995, MNRAS, 273, 731
Rasio, F. A. 2001, these proceedings
Rasio, F. A., Pfahl, E., & Rappaport, S. 2000, ApJ, 532, L147
Regös, E., Tout, C. A., & Wickramasinghe, D. 1998, ApJ, 509, 362
Ritter, H. 1988, A&A, 202, 93
Ritter, H., & Kolb, U. 1998, A&AS, 129, 83
Rogers, F. J., Iglesias, C. A. 1992, ApJS, 79, 507
Schröder, K.-P., Pols, O. R., Eggleton, P. P. 1997, MNRAS, 285, 696
Stella, L., White, N., & Priedhorsky, W. 1987, ApJ, 315, L49
Tauris, T. M., van den Heuvel, E. P. J., & Savonije, G. J. 2000, ApJ, 530, L93

Tutukov A., Fedorova A., Ergma E., & Yungelson L. 1987, Sov. Astron. Lett., 13, 328
van Kerkwijk, M. H. et al. 1992, NAT, 355, 703
Verbunt, F. 1987, ApJ, 312, 23
Verbunt, F., & Zwaan, C. 1981, A&A, 100, L7

SIMULATING OPEN STAR CLUSTERS

Simon F. Portegies Zwart
Massachusetts Institute of Technology, Cambridge MA 02139, USA, **Hubble Fellow**
spz@space.mit.edu

Stephen L. W. McMillan
Department of Physics,Drexel University, Philadelphia, PA 19104, USA

Piet Hut
Institute for Advanced Study, Princeton, NJ 08540, USA

Junichiro Makino
Department of Astronomy, University of Tokyo, 7-3-1 Hongo, Bunkyo-ku,Tokyo 113-0033, Japan

Keywords: binaries: close — blue stragglers — stars: evolution — stars: mass loss — star clusters: general — star clusters: Pleiades, Hyades, Praesepe

Abstract We study the evolution of relatively young Galactic open clusters, such as the Pleiades, Praesepe and the Hyades. The calculations include a realistic mass function, primordial binaries and the external potential of the parent Galaxy. The equations of motions of all stars are computed using the GRAPE-4 while taking the evolution of single stars and binaries into account consistently.

Our model clusters are generally significantly flattened in the Galactic tidal field, and dissolve before deep core collapse occurs. At late times, the cluster core is quite rich in giants and white dwarfs. There is no evidence for preferential evaporation of old white dwarfs, on the contrary the formed white dwarfs are likely to remain in the cluster. Stars tend to escape from the cluster through the first and second Lagrange points, in the direction of and away from the Galactic center.

Mass segregation manifests itself in our models well within an initial relaxation time. As expected, giants and white dwarfs are much more strongly affected by mass segregation than main-sequence stars.

D. Vanbeveren (ed.), The Influence of Binaries on Stellar Population Studies, 371–386.

The combined effect of stellar mass loss and evaporation of stars from the cluster potential drives its dissolution on a much shorter time scale than if these effects are neglected. The often-used argument that a star cluster is barely older than its relaxation time and therefore cannot be dynamically evolved is clearly in error for the majority of star clusters.

An observation of a blue straggler in an eccentric orbit around an unevolved star or a blue straggler of more than twice the turn off mass might indicate past dynamical activity. We find two distinct populations of blue stragglers: those formed above the main-sequence turn off, and those which appear as blue stragglers as the cluster's turnoff drops below the mass of the rejuvenated star.

1. INTRODUCTION

Star clusters are remarkable laboratories for the study of fundamental astrophysical processes spanning the range of stellar evolution, from birth to death. Open clusters are relatively small (typically $\lesssim 10^4$ stars), young (generally $\lesssim 1$ Gyr old) systems found primarily in the Galactic disk. Roughly 1100 are known to lie within a few kiloparsecs of the Sun (Lynga 1987a). As a class, they afford us the opportunity to witness star formation and explore the complex evolution of young stellar systems. Since stars are born in star clusters and become part of the general Galactic population only when the parent cluster dissolves in the Galactic tidal field, it is of considerable interest to understand how this process occurs, on what time scale it operates, and what are the detailed observable consequences of cluster evolution.

At least as a first approximation, many star clusters are quite well characterized dynamically as pure N-body systems, in that there are few collisions or close encounters between stars and the clusters themselves are relatively isolated in space. The same argument can be made for the evolution of the stars and binaries without interactions, which may also quite well explain the properties of open clusters. This simple picture rapidly becomes inadequate when both effects are combined—cluster evolution is actuality an intricate mix of dynamics, stellar evolution, and external (tidal) influences, and the subtle interplay between stellar dynamics and stellar physics makes for a formidable modeling problem. Of the many physical processes influencing cluster evolution, probably the most important are the effects of stellar evolution, the tidal field of the Galaxy, and the evolution and dynamics of binary stars. We present here a series of models in which all these effects are taken self-consistently into account.

The processes just mentioned are tightly coupled, complicating the evolution and making it difficult to isolate the importance of each individual effect. We refer to Meylan & Heggie (1997) for a recent review. Generally, speaking, we can say mass loss from stellar evolution is of greatest importance during the first few tens of millions of years of cluster evolution, and may well result in the disruption of the entire cluster (cf. Chernoff & Weinberg 1990, Fukushige & Heggie 1995, Takahashi & Portegies Zwart 2000). If the cluster survives this early phase, stellar evolutionary time scales soon become longer than the time scales for dynamical evolution, and two-body relaxation and tidal effects become dominant. Ultimately, these effects cause the cluster to dissociate and the stars to become part of the general "field" population of the disk.

There is strong observational evidence that star clusters contain substantial binary populations, quite possibly as rich as those found in the field (see Rubenstein 1997). The properties and numbers of observed cluster binaries cannot be explained by internal formation processes, such as three body dynamics or two body tidal capture (Aarseth & Lecar 1975). The majority must be primordial, i.e. formed together with the single stars at the cluster's birth. These observations are important to cluster dynamics, as a cluster's evolution depends strongly on its binary population and even a small initial binary fraction can play a pivotal role in governing cluster dynamics (Goodman & Hut 1989, McMillan et al. 1990, 1991a and 1991b, Hut et al. 1992, McMillan & Hut 1994). Binaries are also crucial to cluster stellar evolution. The possibility of mass transfer between binary components permits wholly new stellar evolutionary states to arise; in addition, the presence of binaries will enhance the rate of stellar collisions and close encounters, through the temporary capture of single stars and other binaries in three-body resonance encounters (Verbunt & Hut 1987; Portegies Zwart et al. 1998).

Until quite recently, dynamical models have tended to exclude binaries, for the good practical reasons that (1) binaries slow down the calculations dramatically and tend to induce numerical errors, and (2) their internal evolution is much more complicated than the evolution of single stars. However, it has also long been known that cluster models lacking adequate treatments of binary systems are of at best limited validity. In this paper we begin to address these limitations by performing calculations of open clusters containing substantial numbers of primordial binaries, with the goal of studying the mutual influence of stellar evolution and stellar dynamics in these systems—the "ecology" of star clusters (Heggie 1992).

We concentrate on open star clusters near the Sun (between ~ 6 and 12 kpc from the Galactic center), which are less than 1 billion years old, and which initially contain a few thousand ($\sim$ 2000–3000) stars, roughly half of them in binaries. Well known clusters fitting this general

Table 1 Observed and derived parameters for several open star clusters with which our simulations may be compared. Subsequent columns give (3) the distance to the cluster (in pc), (4) the cluster age (in Myr), (5) the half mass relaxation time (in Myr), (6) the total mass (in $\mathcal{M}_\odot$), (7) the tidal radius (in pc), (8) estimate for the half mass radius (in pc), and (9) the core radius (in pc). In cases where the parameters (relaxation time, mass, etc.) are not accessible in the literature, we calculate it; these entries are printed *in italics.* In most cases these numbers can be calculated (see Portegies Zwart et al. 2000). Dashes and question marks indicate that we cannot derive these numbers from the literature. The final column contains information on the cluster stellar content. The last column ($N_{\rm bss:gs:wd}$) gives the number of observed blue stragglers, giants and white dwarfs, separated by colons.[2]

name	ref.	d [pc]	t	$t_{\rm rlx}$ [Myr]	M [$\mathcal{M}_\odot$]	$r_{\rm tide}$	$r_{\rm hm}$ [pc]	$r_{\rm core}$	$N_{\rm bss:gs:wd}$
NGC 2516	a	373	110	*220*	*1000*	*13*	2.9	—	6:4:4
Pleiades	b	135	115	150	~1500	16	2–4	1.4	0:3:3
Praesepe	c	174	650	*370*	1160	12	*3.5*	2.8	5:5:11
Hyades	f	46	625	*390*	750	10.3	*3.7*	2.6	1:4:10

References to the literature (second column) are: (a) Abt & Levy (1972); Dachs, J & Kabus (1989); Hawley et al. (1999). (b) Pinfield et al. (1998); Raboud & Mermilliod, (1998); Bouvier et al (1998); (c) Andrievsky (1998); Jones & Stauffer (1991); Mermilliod & Mayor (1999); Mermilliod et al. (1990); Hodgkin et al. (1999). (d) Perryman et al. (1998 and references therein) Reid & Hawley (1999); Data on numbers of white dwarfs was taken from Anthony-Twarog (1984) for Praesepe, from Koester & Reimers (1981).

description are the Pleiades, Praesepe and the Hyades. Such systems are small enough that multiple simulations can be performed in order to improve statistical coverage of their properties, yet they are large enough and old enough that both stellar evolution and stellar dynamics have had time to play significant roles in determining their present structure and appearance. More details about the simulations reported in this paper are presented by Portegies Zwart et. al. 2000 (Paper IVa).

2. INITIAL CONDITIONS

In order to begin a simulation, a number of critical cluster parameters must be chosen: the mass, virial radius, and tidal radius (or its equivalent), the initial mass function, and the initial distributions of binary spatial density and orbital properties. Table 1 presents an overview of observed parameters for some star clusters with similar overall masses, stellar membership and half-mass radii. The ages of these clusters vary widely, from about 110 Myr (Pleiades) to 625 Myr (Hyades). Their core and half mass radii suggest that they may be described approximately by King models (King 1966) with dimensionless depths in the range $W_0 \sim 4 - 6$, where larger W_0 corresponds to higher central concentration.

We define our mass scale by arbitrarily adopting a "Hyades-like" model, in which the mass of the system at an age of 625 Myr is $\sim 1000\mathcal{M}_\odot$. We then estimate the initial mass of a cluster by applying a number of corrections. We adopt the initial mass function for the solar neighborhood described by Scalo (1986).

We adopt an initial cluster mass of $M_0 \sim 1600\,\mathcal{M}_\odot$. This is close to the $1800\,\mathcal{M}_\odot$ for the initial mass of Hyades derived by Weidemann (1992). We assume a Scalo (1986) initial mass function, with minimum and maximum masses of $0.1\,\mathcal{M}_\odot$ and $100\,\mathcal{M}_\odot$, respectively, and mean mass $\langle m\rangle \simeq 0.6\mathcal{M}_\odot$. Consistent with the above estimates, our simulations are performed with 1024 single stars and 1024 binaries, for a total of 3096 (3k) stars and a binary fraction of 50%. This binary fraction is comparable to the currently observed binary fractions in Pleades (43.8%), Praesepe (33%) and Hyades (40%).

Stars and binaries within our model are initialized as follows. A total of 2k single stars are selected from the initial mass function and placed in an equilibrium configuration in the selected density distribution (see below). We then randomly select half the stars and add a second companion star to them. The masses of the companions are randomly selected between $0.1\,\mathcal{M}_\odot$ and the primary mass. For low-mass primaries, the mass ratio distribution peaks at unity, whereas the distribution is flat for more massive primaries. Once stellar masses are chosen, other binary parameters are determined. Binary eccentricities are selected from a thermal distribution between 0 and 1. Orbital separations a are selected with equal probability in $\log a$ with the lower limit set by the separation at which the primary fills its Roche lobe or at $1\,\mathrm{R}_\odot$, whichever is smaller. The upper limit for the initial semi-major axis is taken at $10^6\,\mathrm{R}_\odot$ (about 0.02 pc, Duquennoy & Mayor 1991). When a binary appears to be in contact at pericenter, new orbital parameters are selected.

We select initial density profiles from the anisotropic density distributions described by Heggie & Ramamani (1995) with $W_0 = 4$ and $W_0 = 6$. The Heggie-Ramamani models are derived from King (1966) models, but take into account the velocity anisotropy and non-spherical shape of the critical zero-velocity (Jacobi) surface of the cluster in the Galactic tidal field.

All models are started with a virial radius of $r_{\rm vir} = 2.5$ pc. For our adopted parameters, the initial cluster dynamical time scale is then

Table 2 Initial conditions and parameters for the selected models. The columns give the model name, initial mass ($\mathcal{M}_\odot$), number of stars, dimensionless central depth of the potential well (W_0), the distance from the Galactic center (kpc), the initial relaxation and half mass crossing times (both in Myr), Jacobi radius (r_J) and the half mass and core radius (all in parsec).

name	M	N	W_0	R_{Gal}	t_{rlx}	t_{hm}	r_{Jacob}	r_{hm}	r_{core}
W4	1600	3k	4	6.3	109	4.07	14.5	2.14	0.83
W6	1600	3k	6	12.1	102	4.15	21.6	2.00	0.59

$t_{hm} \equiv (GM/r_{vir}{}^3)^{-1/2} \sim 1.5$ Myr. Each cluster is assumed to precisely fill its Jacobi surface at birth. (Expressed less precisely, we could say that the limiting radius of the initial King model is equal to the Roche radius of the cluster in the Galactic tidal field.) Given the Oort constants in the solar neighborhood, we find that the models with $W_0 = 6$ are somewhat farther (~12.1 kpc) from the Galactic center than is the Sun, while a model with $W_0 = 4$ is slightly closer (~6.3 kpc).

For a total cluster mass of 1600 $\mathcal{M}_\odot$, the Lagrange points of our two standard clusters lie, respectively, at 14.5 pc ($W_0 = 4$) and 21.6 pc ($W_0 = 6$) from the cluster center.

Table 2 reviews the adopted parameters and initial conditions of our models. In order to improve statistics, we performed five calculations for each set of initial conditions.

3. DISCUSSION OF THE RESULTS

The aim of our simulations is to study the evolution of open star clusters such as the Pleiades, Praesepe and the Hyades. These clusters differ significantly in age, but have comparable physical characteristics, stellar membership, total magnitude and internal velocity dispersions. Our N-body calculations incorporate, in a fully self-consistent fashion, mass loss from single stars, binary evolution, dynamical encounters among single stars and binaries and the effect of the Galactic tidal field. The calculations are performed on the GRAPE-4 (Makino et al 1997).

We have compared the luminosity functions, isochrones and projected luminosity profiles of our models with observations. For some clusters, it is hard to find a good match between model and observed luminosity functions. Mass segregation and observational limitations significantly reduce our ability to find a match, and restrict our understanding of the differences we see. However, we find that for models for which we compared the luminosity functions with Praesepe and the Hyades, the observations show evidence for significantly more mass segregation than is seen in the models. We conclude that these clusters may have been born somewhat mass segregated. Alternatively these clusters may have started out somewhat more massive than assumed here, but with a shallower density profile. The selective evaporation of lower mass stars then results in a "dynamically old" appearance (see also Takahashi & Portegies Zwart 2000).

The global luminosity function of Praesepe and its degree of mass segregation suggest an age greater than 800 Myr, which is at the high end of the observed range. Our age estimates for the Pleiades and Hyades, based on the structure and dynamical state of these clusters, are consistent with ages derived from isochrone fitting.

3.1. MASS SEGREGATION

The first effects of mass segregation in our models are discernible in the cluster core after only a few million years, a small fraction of an initial half-mass relaxation time. After about a relaxation time, mass segregation becomes measurable in the cluster outskirts. The luminosity function of the inner parts of the cluster provides a useful tool for studying mass segregation, although one has to select regions which are well separated in radius in order to make the effect visible. As expected, giants and binary stars are most strongly affected by mass segregation and it is easiest to identify the effect by comparing the radial distribution of giants with that of the lower mass main-sequence stars.

One important effect of mass segregation is that older clusters become rich in white dwarfs and giants relative to the Galactic field. These stars may be single or members of binary systems. The main reason for the overabundance of giants and white dwarfs in clusters is the the depletion of low mass main-sequence stars by evaporation. This flattening of the mass function due to mass segregation has also been studied by Takahashi & Portegies Zwart (2000) for more massive clusters. They also find that clusters which are close to disruption are rich in compact objects and giant stars.

The W4 models have shallower initial potentials, lie closer to the Galactic center, and are more strongly affected by mass segregation because of their more rapid evaporation. We suggest that the best place to look for evidence of mass segregation is in star clusters with larger half-light radii, which have shallower potentials. Since mass segregation manifests itself more clearly in dynamically evolved systems, it is also better to look at older clusters. A relatively old cluster (age $\gtrsim 500$ Myr) with a relatively small mass ($M_{\rm tot} \lesssim 500\,\mathcal{M}_\odot$) would be ideal.

3.2. AGE ESTIMATES

The "dynamical ages"[3] of our model clusters are often inconsistent with ages determined by isochrone fitting. The instantaneous relaxation time is a poor estimator of a cluster's dynamical age. The cores of star clusters lose their memory of the initial relaxation time within a few million years, well within an initial half mass relaxation time, but the time scale for global cluster amnesia to set in is far larger than the initial relaxation time. The half-mass relaxation time tends to increase by a factor of two or three, reaching a maximum near the cluster's half-life and decreasing thereafter. The increase is caused by the internal heating of the cluster; the decrease mainly by loss of stars.

3.3. ESCAPING STARS

Stars tend to escape in the direction of the L_1 and L_2 Lagrange points of the Galaxy–cluster system. The velocities of the stars at these points are highly anisotropic and, as expected, pointing mostly radially away from the cluster center. The velocities of the escaping stars are comparable to the cluster velocity dispersion; very few stars are ejected with high velocities following a strong dynamical encounter.

Neutron stars are ejected isotropically and with much higher velocities, owing to the asymmetric kicks they receive during the supernovae that create them. Binaries containing the progenitors of neutron stars are generally disrupted by the first supernova explosion; in all the calculations presented in this paper, only one binary survived the first supernova. However, the existence of that single binary does suggest that it may be possible to form X-ray binaries in open clusters. Such binaries are only expected to exist in star clusters which are younger than ~ 45 Myr (the turnoff age of a $7\,\mathcal{M}_\odot$ star), because mass loss and the velocity kick imparted to the binary causes it to escape.

3.4. TIDAL FLATTENING

The tidal field of the Galaxy flattens the cluster significantly in the z direction, and to a lesser extent along the y axis. A cluster which is spherical at birth develops this flattening in its outer regions within

[3] We define the dynamical age of a cluster as its lifetime expressed in units of the initial relaxation time.

a few crossing times; the inner parts remain fairly spherical. All our models show this flattening, but its observability from Earth depends on the orientation of the cluster in the Galactic plane.

Ellipticities reported for the Pleiades ($\epsilon \sim 0.3$; van Leeuwen et al. 1986) and Hyades ($\epsilon \sim 0.5$ in the outer regions; Oort 1979) are consistent with our model calculations.

3.5. CORE COLLAPSE

Our models experience rather shallow core collapse during their early evolution. The clusters then expand more or less homologously, preserving their initial density profiles. The expansion is driven by stellar mass loss and, to a lesser extent, by binary heating; dynamical models without stellar mass loss but which include a tidal field and primordial binaries do show core collapse (McMillan & Hut 1994). Once the cluster has lost a considerable fraction of its stars the system shrinks again. The remnant with a few remaining stars may become quite compact before it dissolves, but we find no evidence for late core collapse before complete disruption.

3.6. STELLAR POPULATIONS

Table 4 presents, for several cluster ages, the numbers of single stars and binaries by generic stellar types, normalized to the number expected from a non dynamically evolving population (see below). The numbers are averaged over the calculations performed for each set of initial conditions, chosen with a different random seed from the same probability distribution. Table 3 gives the same data for a population of evolving binaries without dynamics, calculated using SeBa (Portegies & Verbunt 1996).

Overall, the evolutionary differences in the populations of single stars and binaries between models W6 and W4 are quite small. Clearly, as already noted, the W4 clusters evaporate more rapidly, resulting in a generally more rapid decrease in the numbers of both stars and binaries. The W4 models also show a larger numbers of white-dwarf binaries compared to other stellar types. Table 3 presents the stellar and binary properties of an evolving population of isolated stars and binaries. The differences between these binaries and the dynamically evolving population is considerable. We comparing the results of the population synthesis without dynamics with our dynamical models W4 and W6. For practical reasons, however, we only give the number of single main sequence stars and the number of main-sequence binaries for the model W4 (top two rows in Tab. 4). The other numbers in Tab. 4 are renor-

Table 3 Stellar types from population synthesis studies of 1.5×10^5 single stars and 1.5×10^5 binaries. The numbers of single stars and binaries are renormalized to 1024 because these are the number of stars and primordial binaries in each of our dynamical calculations. We added the extra class of binaries which includes a neutron star or a black hole, such binaries are omitted in the dynamical models due their small number.

time [Myr]:	0	100	200	400	600	800
ms	1024	1015.49	1008.12	997.62	988.64	982.41
gs	0	3.14	4.81	6.93	7.30	5.91
wd	0	3.87	10.06	21.17	30.99	39.25
ns/bh	0	8.02	8.02	8.02	8.02	8.02
(ms, ms)	1024	985.14	976.72	965.18	956.62	949.82
(ms, gs)	0	1.81	3.01	4.45	4.82	4.41
(ms, wd)	0	1.89	4.35	8.48	12.25	15.61
(ms, ns/bh)	0	0.10	0.08	0.05	0.05	0.04
(gs, gs)	0	0.13	0.25	0.24	0.27	0.21
(gs, wd)	0	0.27	0.78	1.33	1.59	1.63
(gs, ns/bh)	0	0.01	0.01	0.01	0.00	0.00
(wd, wd)	0	0.52	1.98	4.55	6.78	8.88
(wd, ns/bh)	0	0.27	0.25	0.23	0.22	0.21
(ns/bh, ns/bh)	0	0.11	0.10	0.10	0.09	0.09

malized to the number of single main sequence stars and main sequence binaries in the non-dynamically evolving population. For a comparison with model W6 we refer to paper IVa.

Table 4 gives the fractions of various types of stars and binaries in the dynamical calculations, relative to the corresponding numbers from the population synthesis studies. The latter are normalized to the same numbers of single main-sequence stars and main-sequence binaries as in the initial dynamical calculations. Neutron stars and black holes are omitted from the comparison because the differences are directly obvious: Neutron stars escape from open star clusters, but they are retained in the non-dynamical models. Black holes are omitted because of their small numbers.

Although the numbers of giants and white dwarfs are also small, a trend is clearly visible: Single white dwarfs and dwarfs in binaries are overrepresented at later stages in the dynamical calculations, especially for the W4 models. The basic reason for this overabundance of white dwarfs and giants is their larger mass, which makes them more likely

Table 4 Relative numbers of stars and binaries in the dynamical models with $W_0 = 4$, as fractions of the numbers found in the non-dynamical population synthesis studies (from Tab. 3). The models with $W_0 = 6$ are presented in Portegies Zwart et al. (2000). The first two rows give the numbers of single main-sequence stars and binaries consisting of two main sequence stars in the various models. The other rows are normalized. The normalization is such that the dynamical and non-dynamical calculations contain equal numbers of single main-sequence stars and main-sequence binaries. Numbers greater than 1 indicate excesses of those stellar type in the dynamical calculation; numbers less than 1 represent depletion.

	Normalized data for the W4 models					
time [Myr]:	0	100	200	400	600	800
ms	1024	998.8	938.8	771.2	604.5	164.0
(ms, ms)	1024	502.8	600.6	453.0	333.0	187.5
ms	1	1	1	1	1	1
(ms, ms)	1	1	1	1	1	1
gs	0	1.0	1.0	1.5	2.3	6.3
wd	0	1.2	1.0	1.3	1.9	4.6
(ms, gs)	0	0.6	1.2	0.5	1.4	3.3
(ms, wd)	0	1.7	0.8	0.6	1.2	1.8
(gs, gs)	0	0.0	0.0	1.6	0.0	2.2
(gs, wd)	0	2.6	0.1	0.6	1.7	0.0
(wd, wd)	0	0.9	0.5	1.3	1.6	2.7

to be retained by the cluster. White dwarfs have a very complicated evolution within the stellar system, their progenitors being among the most massive objects while on the main sequence and the giant branch, but the white dwarfs having masses comparable to the mean once their envelopes are lost. White dwarfs are therefore preferentially formed in the cores of star clusters. Once the white dwarf is formed, it is hard to extract it from the core. These effects are more pronounced in smaller clusters ($t_{\rm rlx} \lesssim 1\,{\rm Gyr}$). The W4 models retain more white dwarfs than the W6 models because the latter relax on a longer time scale and evolve dynamically less rapidly than the former.

3.7. GIANTS AND WHITE DWARFS IN OPEN CLUSTERS

A cluster's single-star population is not noticeably affected by cluster dynamics until the age of the system exceeds ~ 2 initial half-mass relaxation times. However, the binary populations are measurably influenced by dynamics even at early times. At later times ($t \geq 400\,\mathrm{Myr}$), our model clusters tend to become rich in giants and white dwarfs. Small-number statistics on the giants limit the degree to which we can quantify this statement, but the white-dwarf populations in our models increased by factors of 1.3 to 4.6 (for model W6 and W4, respectively) relative to what one would expect for a non-dynamically evolving population of single stars and binaries. Most of the excess white dwarfs are members of binary systems.

Table 1 presents an overview of the numbers of white dwarfs observed in the clusters studied in this paper. Explaining the number of white dwarfs in the Hyades is a long-standing problem, starting with discussions in the mid-1970s by Tinsley (1974) and van den Heuvel 1975), and continuing into the 1990s (Eggen 1993; Weidemann 1993). All these papers conclude that the observed number of white dwarfs is too small, by about a factor of three, for the inferred cluster mass and age. The three leading explanations were: (1) the upper mass limit for the production of white dwarfs may be as low as $\sim 4\,\mathcal{M}_\odot$ (Tinsley 1974), white dwarfs are born with a velocity kick as neutron stars do (Weidemann et al. 1992) and (3) mass segregation selectively ejects white dwarfs (van den Heuvel 1975). By studying the white dwarfs in NGC 2516, Koester & Reimers (1996) firmly conclude stars up to $8\,\mathcal{M}_\odot$ can form white dwarfs, removing the first solution to this conundrum. However, our calculations are inconsistent with the idea that white dwarfs are preferentially ejected from star clusters. On the contrary, the white dwarfs in our simulations are more easily retained than main-sequence stars of the same mass, causing the older clusters to become relatively white-dwarf *rich* for their mass.

We can study this problem further by comparing the ratio of the number of white dwarfs to the number of giants: $f_{\frac{\rm wd}{\rm gs}} \equiv N_{\rm wd}/N_{\rm gs}$. For the first five open clusters in Tab. 1 this fraction ranges from $f_{\frac{\rm wd}{\rm gs}} = 1$ for NGC 2516 and the Pleiades to $f_{\frac{\rm wd}{\rm gs}} = 2.5$ for the Hyades. For an evolving population of single stars without dynamics (see Tab. 3), we find that the ratio ranges from 1.1 at 100 Myr (the age of NGC 2516) to 4.2 at 600 Myr (comparable to the age of Hyades).

Binary evolution complicates the comparison due to an obvious selection effect—the giants can probably all be seen, but white dwarfs are easily hidden near a main-sequence or giant companion. We obtain estimates for $f_{\frac{wd}{gs}}$ in a field population with 50% primordial binaries from Tab. 3. We calculate an upper limit to $f_{\frac{wd}{gs}}$ by accounting for all white dwarfs [counting (wd, wd) binaries as two objects and including (ms, wd) and (gs, wd) binaries]. A lower limit is obtained by counting (wd, wd) binaries as one object and excluding white-dwarf binaries containing a main-sequence or giant companion. For the field (no dynamics) population, we find $f_{\frac{wd}{gs}} = 0.80$–1.3 at 100 Myr, and $f_{\frac{wd}{gs}} = 2.7$–4.1 at 600 Myr. Combining models W4 and W6 (see Tab. 4) we obtain $f_{\frac{wd}{gs}} = 1.2$–1.7 at 100 Myr and $f_{\frac{wd}{gs}} = 3.0$–4.0 at 600 Myr.

The observed value of $f_{\frac{wd}{gs}} \sim 1$ at around 100 Myr seems somewhat low, but probably not inconsistent with our models. However, the value of $f_{\frac{wd}{gs}} \sim 2.5$ of the Hyades cluster (at about 600 Myr) is smaller than our models predict, suggesting that a considerable fraction of the white dwarfs are hidden in binaries.

The value of the fraction $f_{\frac{wd}{gs}}$ does not seem to pose a serious problem to understand any of the clusters discussed here. But according to the evolution of the field stars and binaries (Tab. 3), at an age of 600 Myr we expect a total of about 15 giants and 59 white dwarfs. The comparable models W4 and W6 contain, respectively, 13 and 12 giants, and 48 and 57 white dwarfs at that age. Model W6 has lost about 40% of its mass and model W4 has lost about 60%, yet the numbers of giants and white dwarfs have decreased by only 4% – 20%. Averaging over time, we find a decrease in the number of giants and white dwarfs of about 2% per 100 Myr. Apparently, the dynamical evolution of these clusters has little effect on the number of giants and white dwarfs. (In fact, this was assumed by von Hippel [1998] in estimating the mass fractions of white dwarfs in open clusters.) Number counts of giants and/or white dwarfs may therefore provide a reasonable estimate of a cluster's initial mass.

3.8. COMPARISON WITH OTHER WORK

Our models evaporate on time scales generally consistent with dissolution times reported in previous calculations. The models of Terlevich (1987) and de la Fuente Marcos (1997) compare well with the evaporation times of our W6 models. Terlevich's models evolve somewhat more slowly due to the lack of massive stars and the rather cool initial conditions, which drive the cluster to core collapse and therefore extend its lifetime somewhat. The models of McMillan & Hut (1994) dissolve more slowly than ours. This discrepancy can be completely explained by the absence of stellar evolution mass loss and binary evolution in their models, along with the small size of those models relative to the clusters' tidal radii in the Galactic potential.

The only real discrepancy is in the work of Kroupa (1995), whose models dissolve somewhat more rapidly than ours. Possibly the small numbers of stars and the large binary fractions drive a more rapid evaporation than one might naively expect. The evaporation rate of star clusters is known to depend on the total number of stars (Heggie et al. 1997; Portegies Zwart et al. 1998). The presence of primordial binaries has little effect on the cluster lifetime. It is, however, not clear how this trend propagates in the scaling of cluster lifetimes with respect to the number of stars.

Acknowledgments

We thank Anthony Brown, Douglas Heggie, Ken Janes, Gijs Nelemans, Koji Takahashi and Lev Yungelson for numerous discussions. This work was supported by NASA through Hubble Fellowship grant HF-01112.01-98A awarded (to SPZ) by the Space Telescope Science Institute, which is operated by the Association of Universities for Research in Astronomy, Inc., for NASA under contract NAS 5-26555, and by ATP grant NAG5-6964 (to SLWM). SPZ is grateful to Drexel University, Tokyo University and the University of Amsterdam (under Spinoza grant 0-08 to Edward P.J. van den Heuvel) for their hospitality. Special thanks goes to the University of Tokyo for providing time on their GRAPE-4 system.

References

Aarseth, S. J., Lecar, M. 1975, ARAA, 13, 1
Abt, H. A., Levy, S. G. 1972, ApJ, 172, 355
Andrievsky, S. . M. 1998, A&A, 334, 139
Anthony-Twarog, B. J. 1984, AJ, 89, 267
Bouvier, J., Rigaut, F., Nadeau, D. 1997, A&A, 323, 139
Chernoff, D. F., Weinberg, M. D. 1990, ApJ, 351, 121
Cox, A. N. 1954, ApJ, 119, 188
Dachs, J., Kabus, H. 1989, A&As, 78, 25
de la Fuente Marcos, R., 1997, A&A 322, 764
Duquennoy, A., Mayor, M. 1991, A&A, 248, 485
Eggen, O. J. 1993, AJ, 106, 642
Fukushige, T., Heggie, D. C. 1995, MNRAS, 276, 206
Goodman, J., Hut, P. 1989, Nat, 339, 40
Hawley, S. L., Tourtellot, J. G., Reid, I. N. 1999, AJ, 117, 1341

Heggie, D. C. 1992, Nat, 359, 772
Heggie, D. C., Giersz, M., Spurzem, R., Takahashi, K. 1997, in Highlights of Astronomy Vol. 11. Kluwer, Dordrecht (ed. J. Andersen)
Heggie, D. C., Ramamani, N. 1995, MNRAS, 272, 317
Hodgkin, S. T., Pinfield, D. J., Jameson, R. F., Steele, I. A., Cossburn, M. R., Hambly, N. C. 1999, MNRAS, 310, 87
Hut, P., McMillan, S., Goodman, J., Mateo, M., Phinney, S., Pryor, C., Richer, H., Verbunt, F., Weinberg, M. 1992, PASP, 104, 981
Jones, B. F., Stauffer, J. R. 1991, AJ, 102, 1080
King, I. R. 1966, AJ, 71, 64
Koester, D., Reimers, D. 1981, A&A, 99, L8
Koester, D., Reimers, D. 1996, A&A, 313, 810
Kroupa, P. 1995, MNRAS, 277, 1507
Lynga, G. 1987, in Catalog of open cluster data, Computer based catalogue available through the CDS, Strasbourg, FRance and through NASA Data Center, GReenbelt, Maryland, USA. 5th edition, Vol. 4, p. 121
Makino, J., Taiji, M., Ebisuzaki, T., Sugimoto, D. 1997, ApJ, 480, 432
McMillan, S., Hut, P. 1994, ApJ, 427, 793
McMillan, S., Hut, P., Makino, J. 1990, ApJ, 362, 522
McMillan, S., Hut, P., Makino, J. 1991a, ApJ, 372, 111
McMillan, S. L. W., Hut, P., Makino, J. 1991b, Star Cluster Evolution with Primordial Binaries, 421, ASP Conf. Ser. 13: The Formation and Evolution of Star Clusters
Mermilliod, J. C., Mayor, M. 1999, in 10 pages, 3 tables, 7 eps figures. Accepted for A&A;. LaTeX Preprint no. IAUL preprint no 85., 11405
Mermilliod, J. C., Weis, E. W., Duquennoy, A., Mayor, M. 1990, A&A, 235, 114
Meylan, G., Heggie, D. C. 1997, A&Ar, 8, 1
Perryman, M. A. C., Brown, A. G. A., Lebreton, Y., Gomez, A., Turon, C., De Strobel, G. C., Mermilliod, J. C., Robichon, N., Kovalevsky, J., Crifo, F. 1998, A&A, 331, 81
Pinfield, D. J., Jameson, R. F., Hodgkin, S. T. 1998, MNRAS, 299, 955
Portegies Zwart, S. F., Hut, P., Makino, J., McMillan, S. L. W. 1998, A&A, 337, 363
Portegies Zwart, S. F., Verbunt, F. 1996, A&A, 309, 179
Portegies Zwart, S. F., Yungelson, L. R. 1998, A&A, 332, 173
Portegies Zwart, S. F., McMillan, S. L. W., Hut, P., Makino, J. 2000, MNRAS *in press* (astro-ph/0005248)
Raboud, D., Mermilliod, J. C. 1998, A&A, 329, 101
Reid, I. N., Hawley, S. L. 1999, AJ, 117, 343
Rubenstein, E. P. 1997, PASP, 109, 933

Scalo, J. M. 1986, Fund. of Cosm. Phys., 11, 1
Takahashi, K., Portegies Zwart, S. F. 1998, ApJ, in press
Terlevich, E. 1987, MNRAS, 224, 193
Tinsley, B. M. 1974, PASP, 86, 554
Van den Heuvel, E. P. J. 1975, ApJl, 196, L121
Verbunt, F., Hut, P. 1987, in IAU Symp. 125: The Origin and Evolution of Neutron Stars, Vol. 125, 187
von Hippel, T. 1998, AJ, 115, 1536
Weidemann, V. 1993, A&A, 275, 158
Weidemann, V., Jordan, S., Iben, I., J., Casertano, S. 1992, AJ, 104, 1876

BINARIES AND GLOBULAR CLUSTER DYNAMICS

Frederic A. Rasio, John M. Fregeau, & Kriten J. Joshi
Dept of Physics, MIT, Cambridge, MA 02139, USA
rasio@ensor.mit.edu

Keywords: Globular clusters, dynamical evolution, binaries

Abstract We summarize the results of recent theoretical work on the dynamical evolution of globular clusters containing primordial binaries. Even a very small initial binary fraction (e.g., 10%) can play a key role in supporting a cluster against gravothermal collapse for many relaxation times. Inelastic encounters between binaries and single stars or other binaries provide a very significant energy source for the cluster. These dynamical interactions also lead to the production of large numbers of exotic systems such as ultracompact X-ray binaries, recycled radio pulsars, double degenerate systems, and blue stragglers. Our work is based on a new parallel supercomputer code implementing Hénon's Monte Carlo method for simulating the dynamical evolution of dense stellar systems in the Fokker-Planck approximation. This new code allows us to calculate very accurately the evolution of a cluster containing a realistic number of stars ($N \sim 10^5 - 10^6$) in typically a few hours to a few days of computing time. The discrete, star-by-star representation of the cluster in the simulation makes it possible to treat naturally a number of important processes, including single and binary star evolution, all dynamical interactions of single stars and binaries, and tidal interactions with the Galaxy.

1. INTRODUCTION

The dynamical evolution of dense star clusters is a problem of fundamental importance in theoretical astrophysics, but many aspects of the problem have remained unresolved in spite of years of numerical work and improved observational data (see Meylan & Heggie 1997 for a recent review). The realization over the last 10 years that primordial binaries are present in globular clusters in dynamically significant numbers has completely changed our theoretical perspective on these systems (see, e.g., Gao et al. 1991; Hut et al. 1992; Sigurdsson & Phinney 1995). Most im-

D. Vanbeveren (ed.), The Influence of Binaries on Stellar Population Studies, 387–401.

portantly, dynamical interactions between hard primordial binaries and other single stars or binaries are now thought to be the primary mechanism for supporting a globular cluster against gravothermal contraction and avoiding core collapse. In addition, exchange interactions between primordial binaries and compact objects can explain very naturally the formation of large numbers of X-ray binaries and recycled pulsars in globular cluster cores (see, e.g., Camilo et al. 2000). Dynamical interactions involving primordial binaries can also result in dramatically increased collision rates in globular clusters. This is because the interactions are often *resonant*, with all the stars involved remaining together in a small volume for a long time ($\sim 10^2 - 10^3$ orbital times). For example, in the case of an interaction between two typical hard binaries with semi-major axes $\sim 1\,$AU containing $\sim 1\,M_\odot$ main-sequence stars, the effective cross section for a direct collision between any two of the four stars involved is essentially equal to the entire geometric cross section of the binaries (Bacon, Sigurdsson & Davies 1996; Cheung, Portegies Zwart, & Rasio 2001). This implies a collision rate ~ 100 times larger than for single stars. Direct observational evidence for stellar collisions and mergers in globular clusters comes from the detection of large numbers of blue stragglers concentrated in the dense cluster cores (see, e.g., Bailyn 1995).

2. MONTE CARLO SIMULATIONS OF CLUSTER DYNAMICS

The first Monte Carlo methods for calculating the dynamical evolution of star clusters in the Fokker-Planck approximation were developed more than 30 years ago. They were first used to study the development of the gravothermal instability (Hénon 1971a,b; Spitzer & Hart 1971a,b). More recent implementations have established the Monte Carlo method as an important alternative to direct N-body integrations (see Spitzer 1987 for an overview). The main motivation for our recent work at MIT was our realization a few years ago that the latest generation of parallel supercomputers now make it possible to perform Monte Carlo simulations for a number of objects equal to the actual number of stars in a globular cluster (in contrast, earlier work was limited to using a small number of representative "superstars," and was often plagued by high levels of numerical noise). Therefore, the Monte Carlo method allows us to do right now what remains an elusive goal for N-body simulations (see, e.g., Aarseth 1999): perform realistic, star-by-star computer simulations of globular cluster evolution. Using the correct number of stars in a dynamical simulation ensures that the relative rates of different dynamical processes (which all scale differently with the number of stars) are correct. This is particularly crucial if many different dynamical processes are to be incorporated, as must be done in realistic simulations.

Our implementation of the Monte Carlo method is described in detail in the papers by Joshi, Rasio, & Portegies Zwart (2000), Joshi, Nave, & Rasio (2001), and Joshi, Portegies Zwart, & Rasio (2001). We adopt the usual assumptions of spherical symmetry (with a 2D phase space distribution function $f(E, J)$, i.e., we do *not* assume isotropy) and standard two-body relaxation in the weak scattering limit (Fokker-Planck approximation). In its simplest version, our code computes the dynamical evolution of a self-gravitating spherical cluster of N point masses whose orbits in the cluster are specified by an energy E and angular momentum J, with perturbations ΔE and ΔJ evaluated on a timestep that is a fraction of the local two-body relaxation time. The cluster is assumed to remain always very close to dynamical equilibrium (i.e., the relaxation time must remain much longer than the dynamical time). Our main improvements over Hénon's original method are the parallelization of the basic algorithm and the development of a more sophisticated method for determining the timesteps and for computing the two-body relaxation from representative encounters between neighboring stars. Our new method allows the timesteps to be made much smaller in order to resolve the dynamics in the cluster core more accurately.

We have performed a large number of test calculations and comparisons with direct N-body integrations, as well as direct integrations of the Fokker-Planck equation in phase space, to establish the accuracy of our basic treatment of two-body relaxation (Joshi et al. 2000). Fig. 1 shows the results from a typical comparison between Monte Carlo and N-body simulations. Fig. 2 shows the results obtained with our code for Heggie's Collaborative Experiment run (Heggie et al. 1998).

3. SUMMARY OF RECENT RESULTS

Our recent work has focused on the addition of more realistic stellar and binary processes to the basic Monte Carlo code, as well as a simple but accurate implementation of a static tidal boundary in the Galactic field (Joshi et al. 2001a). As a first application, we have studied the dependence on initial conditions of globular cluster lifetimes in the Galactic environment. As in previous Fokker-Planck studies (Chernoff & Weinberg 1990; Takahashi & Portegies Zwart 1998), we include the effects of a power-law initial mass function (IMF), mass loss through a tidal boundary, and single star evolution, and we consider initial King models with varying central concentrations. We find that the disruption and core-collapse times of our models are significantly longer than

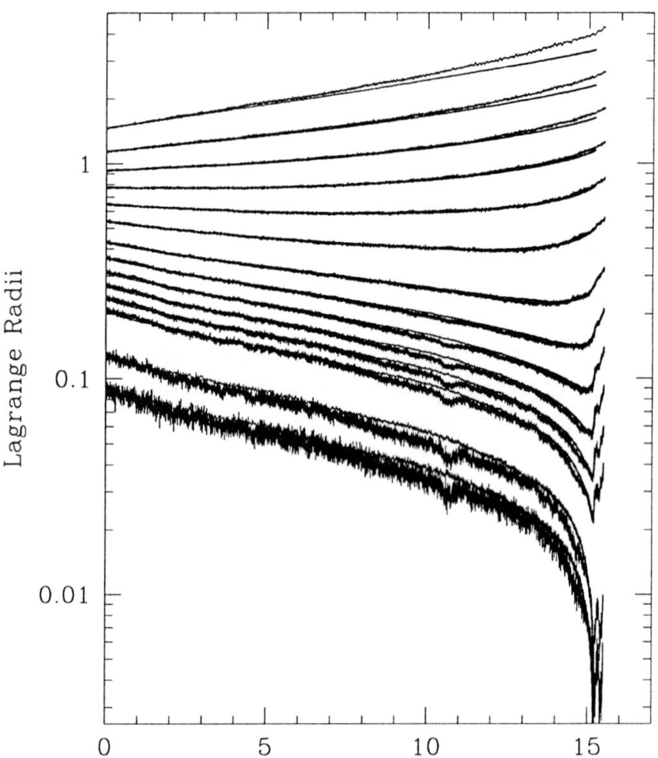

Figure 1 Evolution of the Lagrange radii for an isolated, single-component Plummer model (from bottom to top: radii containing 0.35%, 1%, 3.5%, 5%, 7%, 10%, 14%, 20%, 30%, 40%, 50%, 60%, 70%, and 80% percent of the total mass are shown as a function of time, given in units of the initial half-mass relaxation time). The results from a direct N-body integration with $N = 16,384$ (noisier lines) and from a Monte Carlo integration with $N = 10^5$ stars (smoother lines) are compared. The Monte Carlo simulation was completed in less than a day on a Cray/SGI Origin2000 parallel supercomputer, while the N-body integration ran for over a month on a dedicated GRAPE-4 computer. The agreement between the N-body and Monte Carlo results is excellent over the entire range of Lagrange radii and time. The small discrepancy in the outer Lagrange radii is caused mainly by a different treatment of escaping stars in the two models. In the Monte Carlo model, escaping stars are removed from the simulation and therefore not included in the determination of the Lagrange radii, whereas in the N-body model escaping stars are not removed. Note also that the Monte Carlo simulation is terminated at core collapse, while the N-body simulation continues beyond core collapse.

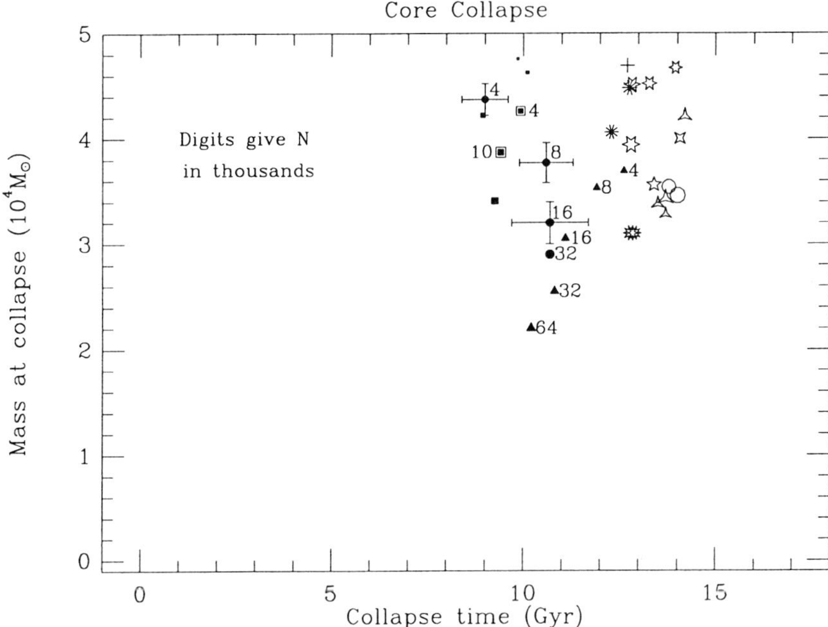

Figure 2 Core collapse time t_{cc} and cluster mass at core collapse M_{cc} as computed by many different numerical codes in Heggie's Collaborative Experiment, and with our Monte Carlo code. The solid round dots, triangles and squares are from various sets of direct N-body simulations. Open symbols are from Fokker-Planck simulations and the stars are from anisotropic gas models. Our data point is indicated by the plus symbol, corresponding to a core collapse time $t_{cc} = 12.86\,\mathrm{Gyr}$ and a mass at core collapse $M_{cc} = 4.73 \times 10^4\,M_\odot$. The initial condition is a King model with a dimensionless central potential $W_0 = 3$, a tidal radius $r_t = 30\,\mathrm{pc}$, and containing a mass $M = 6 \times 10^4\,M_\odot$. The cluster contains single stars only with a power-law IMF of slope -2.35 (Salpeter mass function) between $0.1\,M_\odot$ and $1.5\,M_\odot$. All simulations are done assuming no stellar evolution. Heating by "3-body binaries" is not included in our Monte Carlo simulation (this only affects the evolution beyond core collapse). See Heggie et al. 1998 for more details.

those obtained with previous 1D (isotropic) Fokker-Planck calculations, but agree well with more recent results from direct N-body simulations and 2D Fokker-Planck integrations. In agreement with previous studies, our results show that the direct mass loss due to stellar evolution causes most clusters with a low initial central concentration to disrupt quickly in the Galactic tidal field. The disruption is particularly rapid for clusters with a relatively flat IMF. Only clusters born with high central concentrations or with very steep IMFs are likely to survive to the present and undergo core collapse.

In another recent study, we have used our Monte Carlo code to examine the development of the Spitzer "mass stratification instability" in simple two-component clusters (Watters, Joshi, & Rasio 2000). We have performed a large number of dynamical simulations for star clusters containing two stellar populations with individual masses m_1 and $m_2 > m_1$, and total masses M_1 and $M_2 < M_1$. We use both King and Plummer model initial conditions and we perform simulations for a wide range of individual and total mass ratios, m_2/m_1 and M_2/M_1, in order to determine the precise location of the stability boundary in this 2D parameter space. As predicted originally by Spitzer (1969) using simple analytic arguments, we find that unstable systems never reach energy equipartition, and are driven to rapid core collapse by the heavier component. These results have important implications for the dynamical evolution of any population of primordial black holes or neutron stars in globular clusters. In particular, primordial black holes with $m_2/m_1 \sim 10$ are expected to undergo very rapid core collapse independent of the background cluster, and to be ejected from the cluster through dynamical interactions between single and binary black holes (see Portegies Zwart & McMillan 2000 and references therein). We have also used Monte Carlo simulations of simple two-component systems to study the evaporation (or retention) of *low-mass* objects in globular clusters, motivated by the surprising recent observations of planets and brown dwarfs in several clusters (Fregeau et al. 2001).

Much of our current work concerns the treatment of dynamical interactions with primordial binaries. We are in the process of completing a first study of globular cluster evolution with primordial binaries (Joshi et al. 2001b), based on the same set of approximate cross sections and recipes for dynamical interactions used in the Fokker-Planck simulations of Gao et al. (1991). Typical results are illustrated in Figs. 3–6. The heating of the cluster core generated by a small population of primordial binaries can support the cluster against core collapse for very long times, although the details of the evolution depend sensitively both on the number of stars N and on the initial binary fraction f_b. Clusters

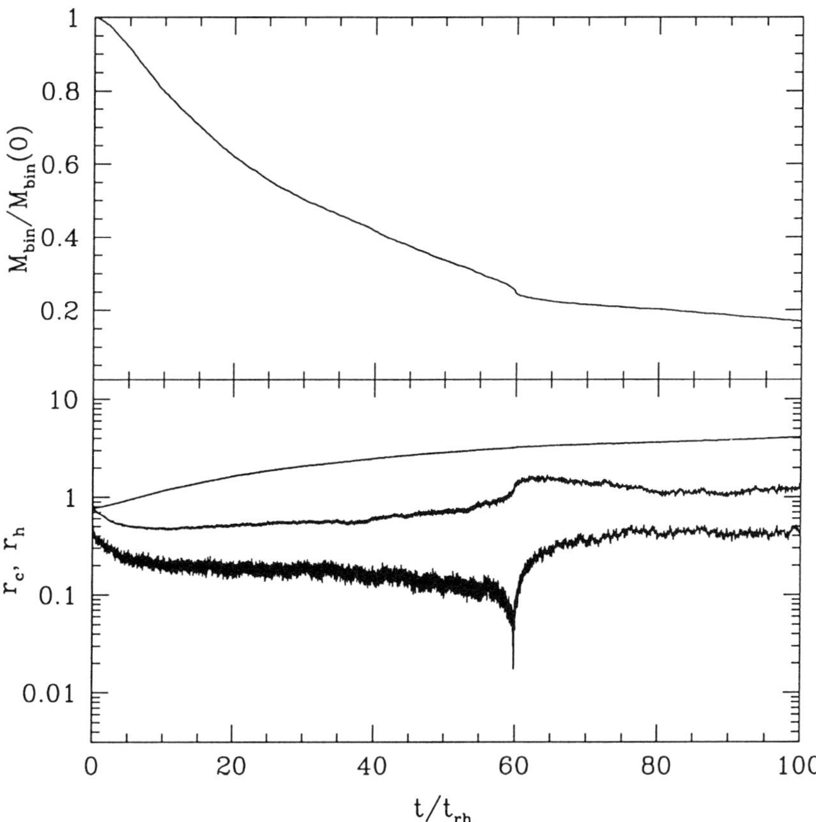

Figure 3 Results of a Monte Carlo simulation for the evolution of an isolated Plummer model containing $N = 3 \times 10^5$ equal-mass stars, with 10% of the stars in primordial binaries. The binaries are initially distributed uniformly throughout the cluster, and with a uniform distribution in the logarithm of the binding energy (roughly between contact and the hard-soft boundary, i.e., no soft binaries are included). The simulation includes a treatment of energy production and binary destruction through binary-single and binary-binary interactions. Stellar evolution and tidal interactions with the Galaxy are not included. Time is given in units of the initial half-mass relaxation time $t_{\rm rh}$. The upper panel shows the evolution of the total mass (or number) of binaries. The lower panel shows, from top to bottom initially, the half-mass radius of the entire cluster, the half-mass radius of the binaries, and the cluster core radius. These quantities are in units of the virial radius of the cluster. Note the long, quasi-equilibrium phase of "binary burning" lasting until $t \simeq 60\, t_{\rm rh}$, followed by a brief episode of core contraction and re-expansion to an even longer quasi-equilibrium phase with an even larger core. By $t \sim 100\, t_{\rm rh}$, only about 15% of the initial population of binaries remains in the cluster, but this is enough to support the cluster against core collapse for another $\sim 100\, t_{\rm rh}$. For most globular clusters, $t_{\rm rh} \sim 10^9$ yr, and this is well beyond a Hubble time. The evolution shown here should be contrasted to that of an identical cluster, but containing single stars only (Fig. 1), where core collapse is reached at $t \simeq 15\, t_{\rm rh}$.

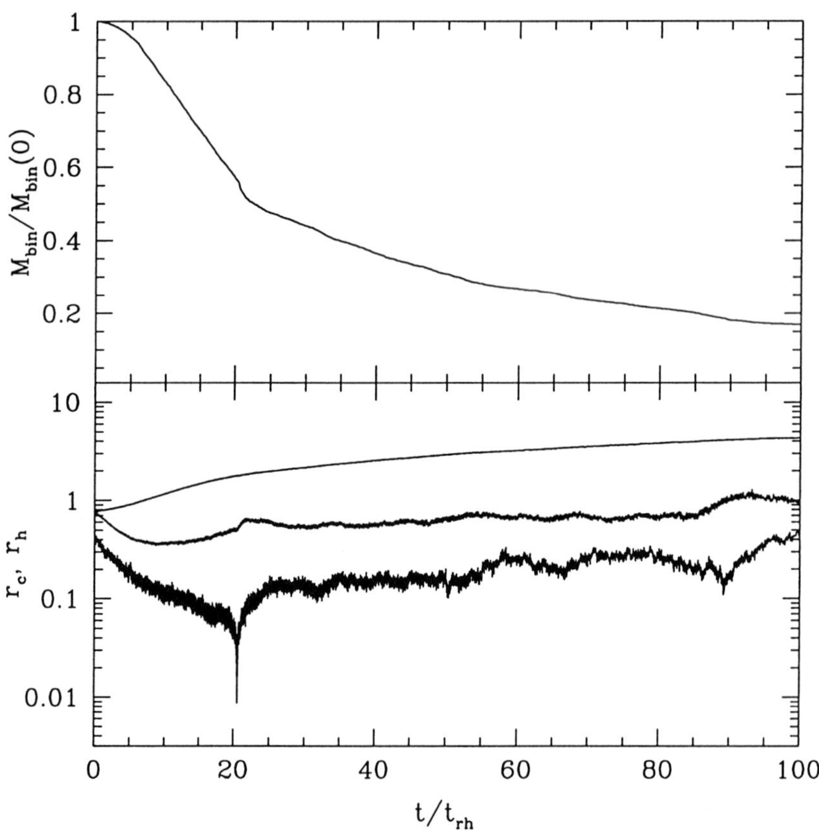

Figure 4 Same as Fig. 3 but for $N = 10^5$ equal-mass stars with 10% binaries. Note the somewhat faster and deeper initial contraction, but still followed by re-expansion into a long-lived, quasi-equilibrium phase of binary burning.

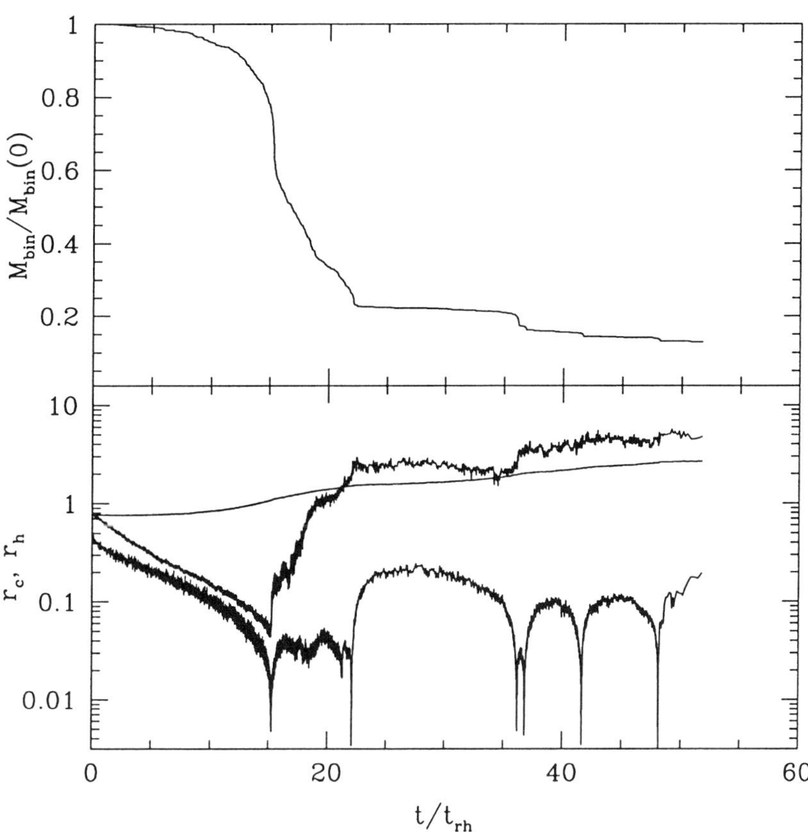

Figure 5 Same as Fig. 3 but for $N = 10^5$ equal-mass stars with only 1% binaries. Here the initial core contraction is followed by more rapid gravothermal oscillations, but the time-averaged core size remains similar to what is seen in Figs. 3 and 4.

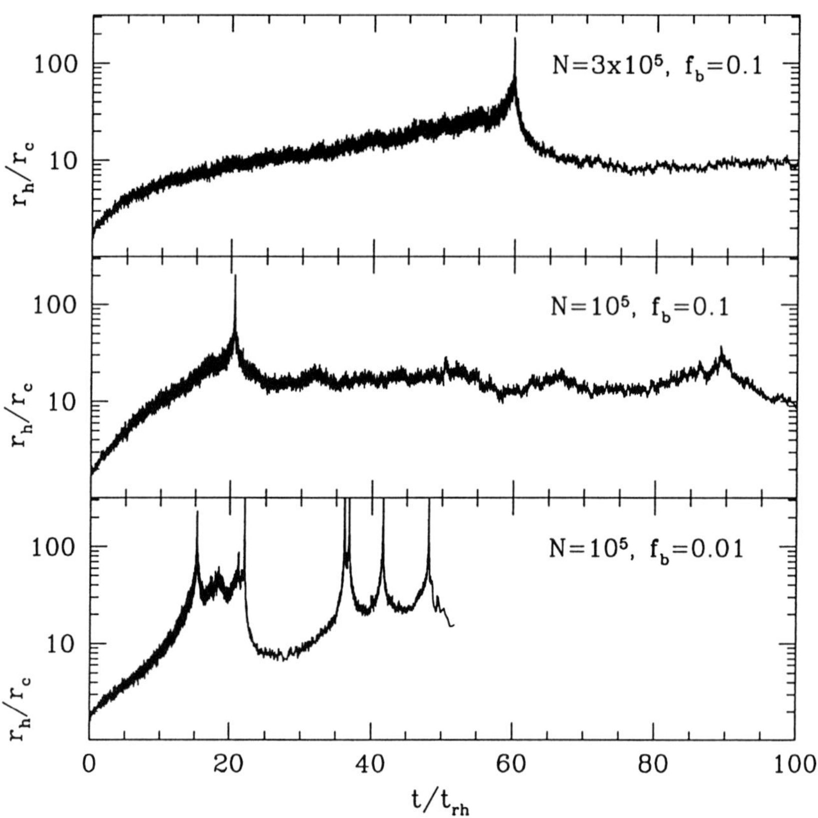

Figure 6 Evolution of the half-mass radius to core radius ratio for the three cases illustrated in Figs. 3–5. For comparison, most observed globular clusters with resolved cores have $r_{\rm h}/r_{\rm c} \simeq 2 - 10$.

with smaller N tend to evolve faster towards deeper core collapse (compare Figs. 3 and 4, and note that the main dependence on N through $t_{\rm rh} \propto N/\log N$ has been scaled out), and so do clusters with smaller initial binary fractions (Fig. 5). After the initial core collapse, gravothermal oscillations powered by primordial binaries can continue for very long times $\sim 50 - 100\,t_{\rm rh}$ even with an initial binary fraction as small as 1%. With an initial binary fraction of 10%, we observe a single, very moderate phase of core collapse during which the core shrinks by a factor $\sim 5 - 10$ in radius and then re-expands rapidly on a timescale of a few central relaxation times (this occurs at $t/t_{\rm rh} \simeq 60$ in Fig. 3 and at $t/t_{\rm rh} \simeq 20$ in Fig. 4). The cluster then enters a second quasi-equilibrium phase of primordial binary burning with the core radius *increasing slowly* until well beyond $\sim 100\,t_{\rm rh}$.

Since $t_{\rm rh} \sim 10^9\,$yr for most clusters, if the type of evolution illustrated in Fig. 3 applied to all globular clusters, there would be no "core-collapsed" clusters in the Galaxy (10–15% of all Galactic globular clusters are classified observationally as "core collapsed"). However, the timescale on which real clusters will exhaust their primordial binary supply and undergo (deeper) core collapse depends on a number of factors not considered here, including the orbit of the cluster in the Galaxy (the simulations are for an isolated cluster, but mass loss and tidal shocking can accelerate the evolution dramatically), and the stellar IMF (the clusters shown here contain all equal-mass stars and binaries; a more realistic mass spectrum will also accelerate the evolution). In addition, some clusters may be formed with much fewer primordial binaries than even the 1% considered in Fig. 5. However, the simple picture that emerges from these simulations may well, to first approximation, describe the dynamical state of most Galactic globular clusters observed today. Indeed, for a cluster in the stable "binary burning" phase, the ratio of half-mass radius to core radius $r_{\rm h}/r_{\rm c} \sim 2 - 10$ (Fig. 6), which is precisely the range of values observed for the $\sim 80\%$ of globular clusters that have a well-resolved core and are well-fitted by King models (see, e.g., Djorgovski 1993). Some of these clusters may have gone in the past through a brief episode of "moderate core collapse" (as shown around $t \simeq 60\,t_{\rm rh}$ in Fig. 3). Yet, they should not be called "core-collapsed" (nor would they be classified as such by observers). Unfortunately some theorists will even call "core-collapsed" clusters that have just reached the *initial phase* of binary burning ($t \simeq 10 - 50\,t_{\rm rh}$ in Fig. 3). Since the core has just barely contracted by a factor $\sim 2-3$ by the time it reaches this phase, it seems hardly justified to speak of a "collapsed" state.

The addition of binary stellar evolution processes to our simulations will allow us to study in detail the dynamical formation mechanisms for many exotic objects such as X-ray binaries and millisecond radio pulsars, which have been detected in large numbers in globular clusters. For example, exchange interactions between neutron stars and primordial binaries can lead to common-envelope systems and the formation of short-period neutron-star / white-dwarf binaries that can become visible both as ultracompact X-ray binaries and binary millisecond pulsars with low-mass companions (see, e.g., Camilo et al. 2000, on observations of 20 such millisecond radio pulsars in 47 Tuc). Rasio, Pfahl, & Rappaport (2000) present a preliminary study of this formation scenario, based on simplified dynamical Monte Carlo simulations. In a dense cluster such as 47 Tuc, most neutron stars acquire binary companions through exchange interactions with primordial binaries. The resulting systems have semimajor axes in the range $\sim 0.1 - 1\,\mathrm{AU}$ and neutron star companion masses $\sim 1 - 3\,M_\odot$. For many of these systems it is found that, when the companion evolves off the main sequence and fills its Roche lobe, the subsequent mass transfer is dynamically unstable. This leads to a common envelope phase and the formation of short-period neutron-star / white-dwarf binaries. For a significant fraction of these binaries, the decay of the orbit due to gravitational radiation will be followed by a period of stable mass transfer driven by a combination of gravitational radiation and tidal heating of the companion. The properties of the resulting short-period binaries match well those of observed binary pulsars in 47 Tuc (Fig. 7). A similar dynamical scenario involving massive CO white dwarfs in place of neutron stars could explain the recent detections of double degenerate binaries containing He white dwarfs concentrated in the core of a dense globular cluster (Hansen et al. 2001; see Edmonds et al. 1999 and Taylor et al. 2001 on observations of He white dwarfs in the core of NGC 6397).

We are also currently working on incorporating into our Monte Carlo simulations a more realistic treatment of tidal interactions, and, in particular, tidal shocking through the Galactic disk (based on Gnedin, Lee, & Ostriker 1999). Tidal shocks can accelerate significantly both core collapse and the evaporation of globular clusters, reducing their lifetimes in the Galaxy (Gnedin & Ostriker 1997). Future work may include a fully dynamical treatment of all strong binary-single and binary-binary interactions (exploiting the parallelism of the code to perform separate numerical 3- or 4-body integrations for all dynamical interactions) as well as a fully dynamical treatment of tidal shocking (performing short, N-body integrations for each passage of the cluster through the Galactic disk or near the bulge).

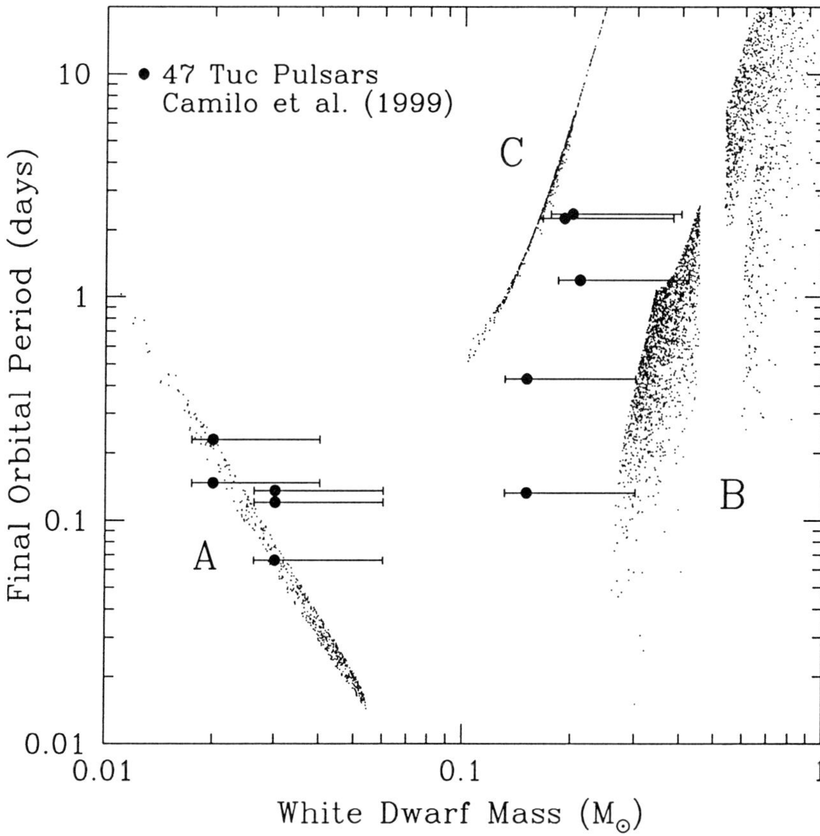

Figure 7 Results of our initial Monte Carlo study of binary millisecond pulsar formation in a dense globular cluster such as 47 Tuc. Each small dot represents a binary system in our simulation, while the filled circles are the 10 binary pulsars in 47 Tuc with well measured orbits (the error bars extend from the minimum companion mass to the 90% probability level for random inclinations). There are 3 principal groups of simulated binaries. Systems in the diagonal band on the left (A) are neutron star – white dwarf binaries that decayed via gravitational radiation to very short orbital periods (~ mins), then evolved with mass transfer back up to longer periods under the influence of both gravitational radiation and tidal heating. The large group labeled B on the right contains neutron star – white dwarf binaries which had insufficient time to decay to Roche-lobe contact via the emission of gravitational radiation. The neutron stars in this group are not likely to be recycled since they may not have accreted much mass during the common envelope phase. Finally, the systems lying in the thin diagonal band toward longer periods (C) are those in which the mass transfer from the giant or subgiant to the neutron star in the initial binary would be stable. These have not been evolved through the mass transfer phase; the mass plotted is simply that of the He core of the donor star when mass transfer commences. There are many more systems in this category that have longer periods but lie off the graph. See Rasio, Pfahl, & Rappaport 2000 for more details.

Acknowledgments

This work was supported in part by NSF Grant AST-9618116 and NASA ATP Grant NAG5-8460. Our computations were performed on the Cray/SGI Origin2000 supercomputer at Boston University under NCSA Grant AST970022N.

References

Aarseth, S.J. 1999, PASP, 111, 1333

Bacon, R., Sigurdsson, S. & Davies, M.B. 1996, MNRAS, 281, 830

Bailyn, C.D. 1995, ARA&A, 33, 133

Camilo, F., Lorimer, D.R., Freire, P., Lyne, A.G., & Manchester, R.N. 2000, ApJ, 535, 975

Chernoff, D.F. & Weinberg, M.D. 1990, ApJ, 351, 121

Cheung, P., Portegies Zwart, S., & Rasio, F.A. 2001, in preparation

Djorgovski, S. 1993, in Structure and Dynamics of Globular Clusters, eds. S. Djorgovski & G. Meylan (ASP Conf. Series Vol. 50), 373

Edmonds, P.D., Grindlay, J.E., Cool, A., Cohn, H., Lugger, P., & Bailyn, C. 1999, ApJ, 516, 250

Fregeau, J.M., Joshi, K.J., Portegies Zwart, S., & Rasio, F.A. 2001, in preparation

Gao, B., Goodman, J., Cohn, H., & Murphy, B. 1991, ApJ, 370, 567

Gnedin, O.Y., & Ostriker, J.P. 1997, ApJ, 474, 223

Gnedin, O.Y., Lee, H.M., & Ostriker, J.P. 1999, ApJ, 522, 935

Hansen, B.M.S., Kalogera, V., Pfahl, E., & Rasio, F.A. 2001, ApJ, submitted

Heggie, D.C., Giersz, M., Spurzem, R., Takahashi, K. 1998, HiA, 11, 591; for a more complete description of the experiment, go to www.maths.ed.ac.uk/people/douglas/experiment.html

Hénon, M. 1971a, Ap. Space Sci., 13, 284

Hénon, M. 1971b, Ap. Space Sci., 14, 151

Hut, P., McMillan, S., Goodman, J., Mateo, M., Phinney, E.S., Pryor, C., Richer, H.B., Verbunt, F., & Weinberg, M. 1992, PASP, 104, 981

Joshi, K.J., Rasio, F.A., & Portegies Zwart, S. 2000, ApJ, 540, 969

Joshi, K.J., Nave, C.P., & Rasio, F.A. 2001a, ApJ, in press [astro-ph/9912155]

Joshi, K.J., Portegies Zwart, S., & Rasio, F.A. 2001b, in preparation

Meylan, G., & Heggie, D.C. 1997, Astron. Astrophys. Rev., 8, 1

Portegies Zwart, S.F., & McMillan, S.L.W. 2000, ApJ, 528, L17

Rasio, F.A., Pfahl, E.D., & Rappaport, S.A. 2000, ApJ, 532, L47

Sigurdsson, S., & Phinney, E.S. 1995, ApJS, 99, 609

Spitzer, L., Jr. 1969, ApJ, 158, L139

Spitzer, L., Jr. 1987, Dynamical Evolution of Globular Clusters (Princeton: Princeton University Press)
Spitzer, L., Jr. & Hart, M.H. 1971a, ApJ, 164, 399
Spitzer, L., Jr. & Hart, M.H. 1971b, ApJ, 166, 483
Takahashi, K., & Portegies Zwart, S.F. 1998, ApJ, 503, L49
Taylor, J.M., Grindlay, J.E., Edmonds, P.D., & Cool, A.M. 2001, ApJ, submitted
Watters, W.A., Joshi, K.J., & Rasio, F.A. 2000, ApJ, 539, 331

THE EFFECT OF TIDES AND STELLAR DYNAMICS ON BINARY AND CLUSTER POPULATIONS

Onno R. Pols
Dept. of Mathematics, P.O. Box 28M, Monash University, Victoria, 3800, Australia
o.pols@sci.monash.edu.au

Jarrod R. Hurley
Dept. of Astrophysics, American Museum of Natural History, Central Park West at 79th Street, New York, NY 10024, USA
jhurley@amnh.org

Christopher A. Tout and Sverre J. Aarseth
Institute of Astronomy, Madingley Road, Cambridge CB3 0HA, UK
cat@ast.cam.ac.uk; sverre@ast.cam.ac.uk

Keywords: binary evolution, population synthesis, blue stragglers, globular clusters

Abstract We present a rapid binary star evolution (BSE) algorithm that enables modelling of even the most complex binary systems. In addition to all aspects of single star evolution, features such as mass transfer and accretion, angular-momentum loss mechanisms, common-envelope evolution, collisions and supernova kicks are included. In particular, circularization and synchronization of the orbit by tidal interactions are calculated in detail. We investigate the effect that tidal friction has on the outcome of binary evolution and show that modelling of tidal evolution in binary systems is necessary in order to draw accurate conclusions from population synthesis work.

The BSE algorithm is included into a state-of-the-art N-body code which allows direct modelling of stellar and binary evolution together with stellar dynamics. This code is ideal for investigating the evolution of open and globular star clusters and their stellar populations. We use the N-body code to model the blue straggler population of the old open cluster M67. Calculations with our binary population synthesis code show that binary evolution alone cannot explain the observed numbers or properties of the blue stragglers. On the other hand, our N-body

D. Vanbeveren (ed.), The Influence of Binaries on Stellar Population Studies, 403–418.

model of M67 generates the required number of blue stragglers and provides formation paths for all the various types found in M67. This demonstrates the effectiveness of the cluster environment in modifying the nature of the stars it contains and highlights the importance of combining dynamics with stellar evolution.

1. INTRODUCTION

The effects of close binary evolution are observed in many systems, such as cataclysmic variables, X-ray binaries and Algols, and in the presence of stars such as blue stragglers which cannot be explained by single star evolution. While many of the processes involved are not understood in detail we do have a qualitative picture of how binaries evolve and may construct a model that follows them through the various phases of evolution. In order to conduct statistical studies of complete binary populations, i.e. population synthesis, such a model must be able to produce any type of binary that is observed in enough detail but at the same time be computationally efficient. By comparing results from the model with observed populations we can enhance our understanding of both binary evolution and the initial distributions.

Many models for binary evolution have been presented (e.g. Whyte & Eggleton 1985; Pols & Marinus 1994; Portegies Zwart & Verbunt 1996; and many contributions in this volume). Our present model supersedes the work of Tout et al. (1997) primarily by including eccentric orbits and stellar spins, which are subject to tidal circularization and synchronization. The model incorporates the detailed single star evolution (SSE) formulae of Hurley, Pols & Tout (2000, hereinafter PapI) which include a wide range of metallicities and allow for a large variety of stellar types. In the first part of this paper we briefly describe the model and concentrate on the inclusion of tidal interaction and how this affects binary evolution and the study of binary populations.

The rich environment of a star cluster provides important tests for stellar evolution theory and the formation of exotic stars and binaries. However, in the case of dense or dynamically old clusters, the appearance of the colour-magnitude diagram (CMD) can be significantly altered by dynamical encounters between the cluster stars. Therefore it is necessary to combine population synthesis with a description of the cluster dynamics. Our approach in this area is to include a consistent treatment of stellar and binary evolution in a state-of-the-art N-body code so as to allow the generation and interaction of the full range of stellar populations within a cluster environment. The variable metallicity included in our SSE algorithm enables us to produce realistic cluster models for comparison with observed cluster populations of any age. We are now embarking on a major project to investigate the dynamical evolution of star clusters and their populations. In the second part of this paper we describe the results of the first stage of this project: the construction of an N-body model for M67 to investigate the incidence and distribution of blue stragglers.

2. TIDAL INTERACTION

Observations of close binary stars reveal that stellar rotation tends to synchronize with the orbital motion, and that the orbital eccentricity tends to decrease. These orbital changes occur even in detached binaries so a mechanism such as tidal friction must be responsible. The presence of a companion introduces a tidal, or differential, force which acts to elongate the star along the line between the centres of mass, producing tidal bulges. If the rotational period of the star is longer than the orbital period then frictional forces on the surface of the star will drag the bulge axis behind the line of centres, and vice versa. The resulting torque transfers angular momentum between the stellar spin and the orbit. Energy is dissipated in the tides which diminishes the total energy. Thus the orbital parameters change and either asymptotically approach an equilibrium state or lead to an accelerated spiralling in of the two stars (Hut 1980). The equilibrium state is characterized by co-rotation and a circular orbit, corresponding to a minimum energy for a given total angular momentum.

The main difficulty in a treatment of tidal evolution arises in identifying the physical processes which are responsible for the tidal torque, that is, the dissipation mechanisms which cause the tides to deviate from an instantaneous equilibrium shape and thus to be misaligned with the line of centres.

The Equilibrium Tide with Convective Damping. The equilibrium tide is described by assuming that the star is always very close to hydrostatic equilibrium. The most efficient form of dissipation which may operate on the equilibrium tide is turbulent viscosity in the convective regions of a star. All other forms of dissipation produce timescales that normally exceed stellar nuclear lifetimes. This mechanism provides a satisfactory interpretation of the behaviour of stars with relatively deep convective envelopes. Even though no completely satisfactory description of stellar convection is available, estimates based on the mixing-length theory seem to adequately represent the tidal timescales for such stars (Zahn 1989).

In our model we use the tidal circularization and synchronization timescales in the form given by Rasio et al. (1996),

$$\frac{1}{\tau_{\rm circ}} \propto \frac{1}{\tau_{\rm conv}}\, q\,(1+q)\left(\frac{R}{a}\right)^8, \qquad \frac{1}{\tau_{\rm sync}} \propto \frac{1}{\tau_{\rm conv}}\, q^2\left(\frac{R}{a}\right)^6. \tag{29.1}$$

where $\tau_{\rm conv}$ is the convective eddy turnover timescale, R is the stellar radius, q is the ratio of stellar mass to companion mass and a is the orbital semi-major axis. The full expression contains a correction factor for the fact that if $\tau_{\rm conv} > P_{\rm orb}$ then the largest convective cells can no longer contribute to viscosity.

The Dynamical Tide with Radiative Damping. For stars with radiative envelopes, another mechanism is required to explain the observed synchronization of close binaries. A star can experience a range of oscillations that are driven by the tidal field, the dynamical tide, and which may be damped by radiative cooling. This provides another mechanism for tidal dissipation, although it is much less efficient than the equilibrium tide. Zahn (1977) derived circularization and synchronization timescales of the form

$$\frac{1}{\tau_{\rm circ}} \propto q\,(1+q)^{11/6}\left(\frac{R}{a}\right)^{21/2}, \qquad \frac{1}{\tau_{\rm sync}} \propto q^2\,(1+q)^{5/6}\left(\frac{R}{a}\right)^{17/2} \tag{29.2}$$

for radiative damping of the dynamical tide. These timescales are generally orders of magnitude larger than those characterizing the equilibrium tide in convective-envelope stars.

The most remarkable feature of the above equations is the very strong dependence of tidal interaction on the ratio of stellar radius to separation. The rapid decrease of the timescales as a star expands is usually enough to ensure that co-rotation and circularization are achieved by the time Roche-lobe overflow (RLOF) starts, even for binary stars with radiative envelopes.

3. BINARY EVOLUTION MODEL

Our rapid binary-star evolution (BSE) algorithm is described in detail in a forthcoming paper (Hurley, Tout & Pols 2001a, hereinafter PapII) and here we only highlight its main characteristics. The inclusion of tidal synchronization and circularization necessitates the introduction of the stellar spins as independent variables. Hence the state of a binary system is characterized by six parameters: the orbital period P (or orbital frequency $\Omega_{\rm orb}$), eccentricity e, component masses M_1 and M_2 and component spin frequencies Ω_1 and Ω_2.

Single star evolution. The SSE algorithm (PapI) provides the stellar luminosity L, radius R, core mass M_c, core radius R_c, and moment of inertia I, as a function of mass, metallicity and age, for each of the component stars as they evolve. The algorithm covers all the evolution phases from the zero-age main-sequence (ZAMS), up to and including the remnant stages, and is valid for masses in the range 0.1 to 100 $M_\odot$ and metallicities from $Z = 10^{-4}$ to $Z = 0.03$. The formulae are based on the detailed stellar evolution models for same range of metallicity computed by Pols et al. (1998), which include a moderate amount of convective overshooting consistent with observations.

A prescription for mass loss from stellar winds is included in the SSE algorithm, parameterized in terms of the overall stellar quantities. Angular momentum losses associated with stellar winds and magnetic braking are also taken into account, so that the algorithm accurately keeps track of the evolution of the spin frequency.

Wind accretion. The secondary star in a binary can accrete some of the material lost by the primary in a wind as it orbits through it. We estimate the mean accretion rate according to the Bondi & Hoyle (1944) mechanism, multiplied by a factor of order unity to mimic deviations from the simplifying assumptions in the Bondi-Hoyle prescription. The wind accretion rate is very sensitive to the wind velocity, a quantity that is difficult to determine accurately from observations, especially for cool stars.

When the companion accretes wind material a fraction of the angular momentum lost from the the intrinsic spin of the primary goes into the spin angular momentum of the companion. We assume that the specific angular momentum in the wind is transferred with perfect efficiency to the spin of the accreting star. We also keep track of the changes to the orbital angular momentum and eccentricity resulting from stellar wind mass loss and wind accretion. Finally we allow for the possibility that mass loss may be tidally enhanced by the presence of a moderately close companion (Tout & Eggleton 1988).

Tidal evolution. We implement tidal evolution by calculating the changes to e and $\Omega_{\rm spin}$ from the equations given by Hut (1981), based on the weak tidal friction model. The expressions for $\tau_{\rm circ}$ and $\tau_{\rm sync}$ discussed in Sect. 2 are used to calibrate the timescale appearing in Hut's equations. The change in semi-major axis is taken care of by the change in orbital angular momentum, which exactly compensates for the change in spin angular momentum.

Other binary interactions. The BSE algorithm also includes prescriptions for angular momentum loss owing to gravitational wave radiation, supernova explosions and associated velocity kicks, Roche-lobe overflow, common-envelope (CE) evolution, coalescence of contact binaries and stellar collisions. The treatment of these processes is described in PapII. We stress that the details of many of these processes are not well understood, and consequently the algorithm contains a fairly large number of free parameters. Probably the most important of these is the CE efficiency parameter $\alpha_{\rm CE}$, because common-envelope evolution is crucial for the production of many types of binary.

4. THE EFFECT OF TIDES ON BINARY EVOLUTION

To illustrate the BSE algorithm and demonstrate the effect of including tidal interaction, we present an example scenario for Algol evolution. Consider a system with $Z = 0.02$ in which the initial masses are 2.9 $M_\odot$ and 0.9 $M_\odot$ in an 8 d orbit with $e = 0.7$. Figure 1 shows the evolution of the binary stars in the Hertzsprung-Russell diagram. On the MS tidal friction arising from convective damping of the tide raised on the 0.9 $M_\odot$ secondary acts to circularize the system so that when the 2.9 $M_\odot$ primary reaches the end of its MS lifetime (after 413 Myr), the eccentricity has fallen to 0.28 and P to 3 d. The radius of the primary then increases rapidly as it evolves across the Hertzsprung gap (HG) until $R_1 = 6.2\ R_\odot$ when it fills its Roche-lobe. By then $P = 2.7$ d and the orbit has circularized (position a in Figure 1). Mass transfer proceeds on a thermal timescale, with the secondary accreting 80% of the transferred mass, until the primary reaches the end of the HG with $M_1 = 0.53\ M_\odot$, $M_2 = 2.74\ M_\odot$ and $P = 17$ d. Mass transfer continues as the primary begins to ascend the giant branch (GB), but now on a nuclear timescale which allows the secondary to accrete all the transferred mass, until the primary's envelope mass becomes so small that it shrinks inside its Roche-lobe (position b). Figure 2 shows the intrinsic spin of the stars and the orbital frequency as a function of the eccentricity, up to this point. We see clearly that co-rotation is achieved as the orbit circularizes. When RLOF ends $M_1 = 0.42\ M_\odot$, $M_2 = 2.85\ M_\odot$ and $P = 30$ d. While the primary was transferring mass on the HG the mass-ratio of the system inverted, $q_1 < 1$, so that the more evolved component is now the least massive and the binary has become an Algol system.

Shortly after the mass transfer phase has ended the GB primary loses all of its envelope leaving a 0.42 $M_\odot$ naked helium star with a 2.85 $M_\odot$ MS companion which would be a blue straggler in a star cluster (posi-

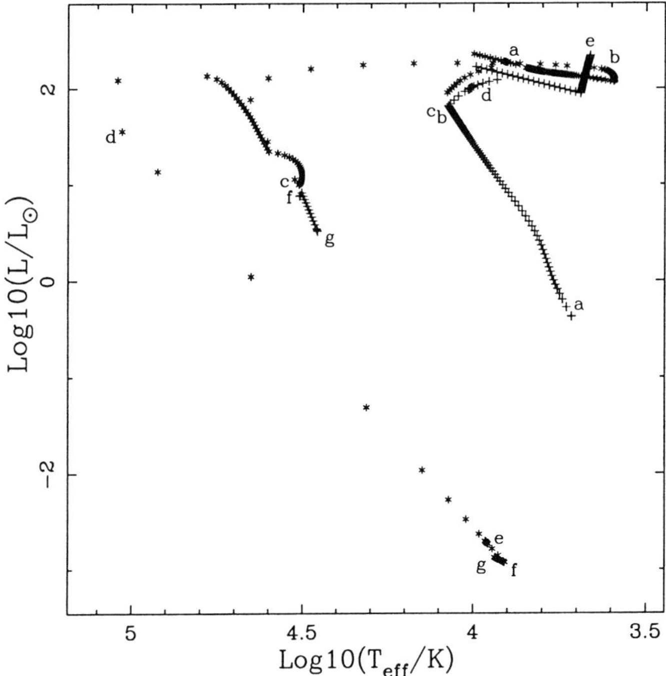

Figure 1 Hertzsprung-Russell diagram showing the evolution of a 2.9 $M_\odot$ star ($*$ points) and a 0.9 $M_\odot$ star (+ points) in a binary system with an initial orbit of $P = 8$ d and $e = 0.70$. The letters represent various times in the evolution of the binary system (see text for details). Each point represents an iteration of the BSE algorithm except in regions of high density, i.e. rapid evolution.

tion c). The He star evolves and becomes a 0.417 $M_\odot$ carbon-oxygen white dwarf (COWD) after losing its helium envelope (position d). The MS star, now the primary, evolves to the GB and fills its Roche-lobe (position e) when the period has reduced to 18 d as a result of the transfer of angular momentum from the orbit to the spin of the star. As a result of the extreme mass ratio, the mass transfer is dynamical, a CE forms and leaves a binary containing the COWD and the 0.416 $M_\odot$ He core of the primary in an orbit of 0.037 d (where we have assumed $\alpha_{\rm CE} = 3$). The He star evolves to fill its Roche-lobe (position f) which has shrunk by gravitational radiation, and transfers 0.1 $M_\odot$ of He-rich material to the COWD before the system reaches a contact state at $P = 1$ min (position g). The two stars merge to a single shell-burning He star.

In a second example we take the same initial configuration but start with an already circular orbit and $P = 2.9$ d, so that the orbital angular momentum is the same as in example 1 above. Because tidal interaction conserves total angular momentum, by the time RLOF occurs and the

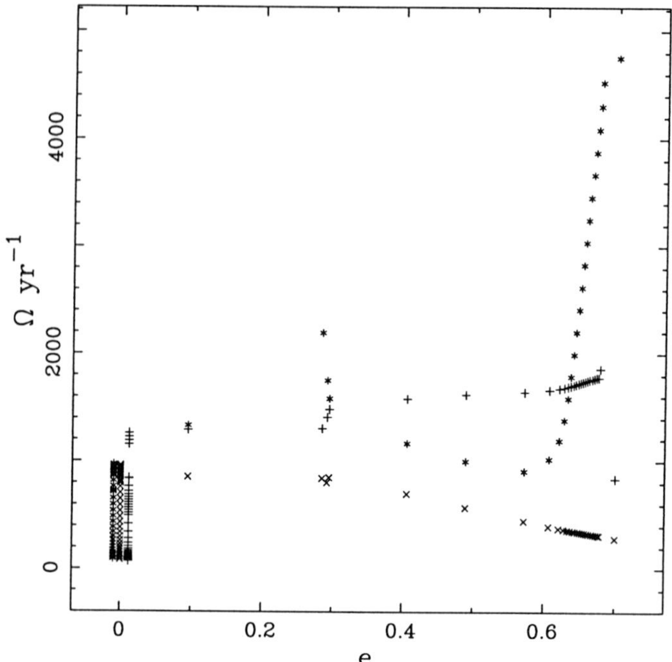

Figure 2 The evolution of the spins of the primary (∗ points) and secondary (+ points) stars, and the orbital spin (× points), as a function of the eccentricity of the orbit for the example of Algol evolution, shown until the end of the first phase of mass transfer. The points at zero eccentricity have been slightly displaced so that the co-rotation of the stars with the orbit can be clearly seen.

orbit in example 1 has circularized owing to tidal interaction, the orbit of this system becomes identical to that of example 1. As a result, the subsequent evolutionary paths in these two examples are also identical. This illustrates an important consequence of tidal interaction: *the outcome of the evolution of an interacting binary does not depend on the initial eccentricity, but only on the initial angular momentum.* This is a result of the very strong dependence of the tidal circularization timescale on R/a, so that almost all systems circularize before RLOF.

In a third example we switch off tidal interaction altogether, so that the orbital and spin evolutions are decoupled. This is the usual assumption in other binary population synthesis models, and only makes sense in initially circular orbits. We therefore start this example with the same initial configuration as in example 2. RLOF starts in a slightly wider orbit because orbital angular momentum is conserved up to this point (P remains 2.9 d) but mass transfer proceeds at a slightly higher rate so that only 75% of the transferred mass is accreted by the companion and slightly more angular momentum is lost. At the end of RLOF $M_1 = 0.42\ M_\odot$, $M_2 = 2.75\ M_\odot$ and $P = 29$ d. Subsequent evolution also leads to a CE when the original secondary has reached the GB but the period remains at 29 d prior to the CE event. As a result the WD+He-star binary emerges from the common envelope with a larger period, 0.061 d. In order to produce a post-CE binary with a similar orbit as in example 1, a smaller value of $\alpha_{CE} = 2.1$ would be required. This example illustrates a second important consequence of tidal interaction: *tidal synchronization tends to bring binary components closer together and cause them to follow an evolutionary path similar to that of a closer binary if tides were ignored.* The reason is that as a star expands, its spin frequency decreases and tidal friction will transfer angular momentum from the orbit to the stellar rotation. However, we find that as a result of the sometimes very convoluted evolutionary paths of interacting binaries, this general principle does not always hold.

Binary population synthesis. The primary differences in our population synthesis results (described in PapII) when tidal evolution within binary systems is ignored are as follows: (1) a 35% decrease in the birth rate of cataclysmic variables; (2) a 10% increase in the birth rate of double-degenerates; (3) an increase of 190% in the incidence of exploding helium WDs; and (4) 55% more Algols currently in the Galaxy.

While these differences may seem substantial they are generally not enough to confirm one way or another whether tidal evolution helps to explain the observed binary populations. This is mainly due to the many uncertain free parameters in binary evolution, as well as to the lack of comprehensive binary searches and the selection effects involved that make discrimination on the basis of observational evidence a risky business. However, tides are present in binary systems whether we like it or not. What we would like to emphasise is that the outcome of binary evolution is sensitive to the physical processes of tidal circularization and synchronization. Therefore, any attempt to constrain uncertain parameters in binary evolution by the method of population synthesis must utilise a binary evolution algorithm that incorporates a working model of tidal evolution, such as that presented here.

5. BLUES STRAGGLERS IN M 67

Blue stragglers are cluster MS stars that seem to have stayed on the main sequence for a time exceeding that expected from standard single-star evolution theory for their mass: they lie above and blueward of the turn-off in a cluster CMD. Ahumada & Lapasset (1995) conducted an extensive survey of blue straggler (BS) candidates in Galactic open

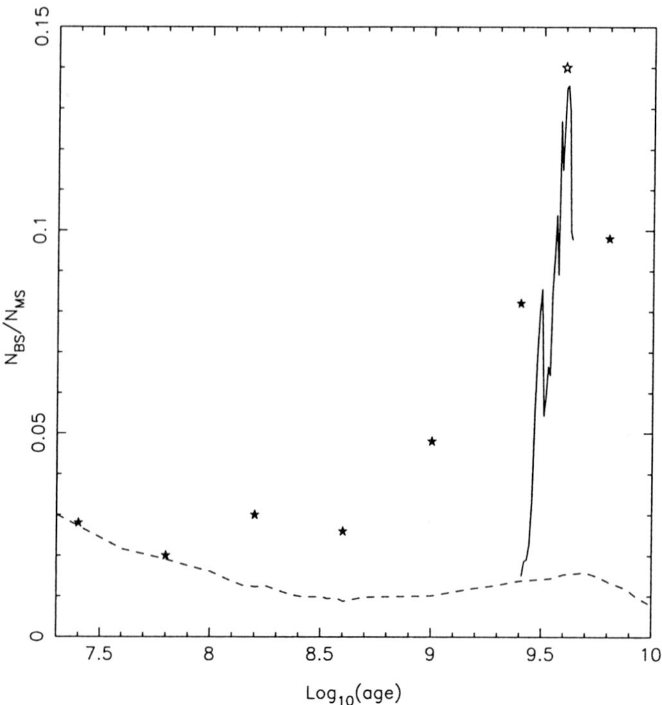

Figure 3 The number of blue stragglers relative to the number of MS stars in the two magnitudes below the turn-off as a function of the population age. The stars represent the open cluster data of Ahumada & Lapasset (1995) with the M67 point an open symbol. The dotted line represents our population synthesis with a 50% binary population. The solid line represents our M67 *N*-body simulation (note that the log-scale does not clearly indicate the length of the simulation).

clusters. Their results are shown in Figure 3. The point for the old open cluster M67 (NGC 2682) stands out above the mean for its age, containing 28 proposed blue stragglers with high probability of cluster membership.

Of a total of ten well-observed blue stragglers Milone & Latham (1992) find that six are members of spectroscopic binaries. One of these is a short-period binary with a period of 4.183 d and an eccentricity of 0.205. The others are long-period binaries with periods from 846 to 4913 d; three have eccentric orbits and two have orbits consistent with being circular (Latham & Milone 1996). Thus, if this sample is representative of the overall M67 blue straggler population, then for every ten blue stragglers we expect four to be single stars and six to be found in binaries with about two of these circular and at least four eccentric. Of particular interest among the blue stragglers in M67 is the super-BS F81 which is a single star and has a mass of $\simeq 3\ M_\odot$ (Leonard 1996), more than twice the cluster turn-off mass, $M_{\rm TO} \simeq 1.3\ M_\odot$. At least three BS in the sample are fast rotators and one is an X-ray source, possibly with a hot subdwarf companion (Van den Berg et al. 1999).

This raises the question of how the blue stragglers formed with the most obvious scenarios involving binary evolution. Case A mass transfer involves a main-sequence star filling its Roche-lobe and transferring mass to its companion, followed by coalescence of the two stars as the orbit shrinks owing to angular momentum loss. The result is a more massive main-sequence star that is rejuvenated relative to other stars of the same mass and thus evolves to become a blue straggler. This is an efficient method of producing single blue stragglers provided that a large population of close binaries exists in the cluster. Case B mass transfer involves a main-sequence star accreting material from a more evolved companion and thus could be a likely explanation for blue stragglers in short-period spectroscopic binaries. Blue stragglers in long-period binaries could be produced by Case C mass transfer when the primary is an asymptotic giant branch (AGB) star that has lost much of its mass. Wind accretion in binaries that initially have fairly large periods could also be responsible for such systems.

In all these cases, except perhaps wind accretion, the binary orbits should be circularized by tides before and during mass transfer. Binary interaction can only produce BSs with at most twice the turn-off mass. A number of population synthesis runs employing the BSE algorithm confirm that binary interaction alone cannot explain the blue straggler population of M67 (see Hurley et al. 2001b for details, hereinafter PapIII). Even if all binaries that produce a BS are retained by the cluster, population synthesis fails to reproduce the number of blue stragglers found in M67 by a large margin (dashed line in Fig. 3). Moreover, only about 25% of the BSs are in binaries, all of these have circular orbits, and practically none are found in binaries with periods greater than a year. So dynamical encounters during cluster evolution are required to explain BSs found in wide binaries and those found with eccentric orbits.

Physical stellar collisions during binary-binary and binary-single interactions can produce blue stragglers in eccentric orbits as well as allowing the possibility of an existing blue straggler being exchanged into an eccentric binary. The probability of such encounters depends on the binary fraction in the core as well as on the density of the cluster. It is also possible that perturbations from passing stars may induce an eccentricity in a previously circular orbit. Another effect of dynamical interactions is the hardening of primordial binaries (i.e., a decrease of their orbital separations) which increases the incidence of Case A and B mass transfer events. As discussed by Leonard (1996) it is unlikely that any one formation mechanism dominates and in the case of the diverse blue straggler population of M67 it seems probable that all the above scenarios play a role.

6. N-BODY MODELS FOR M67

The BSE algorithm was implemented into the N-body code `NBODY4` described by Aarseth (1999). This code has been adapted to run on the HARP-3 special-purpose computer (Makino, Kokubo & Taiji 1993). For details regarding the treatment of close stellar encounters, stellar collisions and hierarchical multiple systems we refer to PapIII.

In a cluster environment from time to time a gravitational encounter gives enough energy to a star that it can escape from the system. Such a dynamical evaporation of the cluster is enhanced by the presence of an external tidal field as well as the inclusion of a full spectrum of stellar masses. In order to quantify this effect we performed a number of N-body runs with 10 000 stars including 1 000 binaries in a Galactic tidal field. We find that the timescale of evaporation is proportional to the instantaneous half-mass relaxation timescale $t_{\rm rh}$ of the cluster. This result depends very little on the initial concentration of the cluster or on the inclusion of binaries. Owing to mass loss from massive stars and core contraction, the cluster quickly expands to fill its tidal radius which depends only on its mass. After about 6 times $t_{\rm rh}$ the cluster mass has decreased by a factor of two, and the evolution becomes remarkably self-similar with the half-mass radius at about $\frac{1}{4}$ of the tidal radius.

We use these properties to estimate that, to evolve to its presently observed mass of about 1000 $M_\odot$ in stars with masses above 0.5 $M_\odot$ and the present half-mass radius of 2.5 pc (Fan et al. 1996), M67 initially had about 40 000 stars and a half-mass radius of about 1 pc. This is unfortunate because a simulation with $N_0 > 20\,000$ takes a prohibitively long time with currently available equipment. On the other hand the main interest of this work is in the BS population near the current M67 age of 4 200 Myr so it should be possible to extract meaningful results using a semi-direct method.

Hence we start the simulation after the self-similar phase has begun, using an initial model that reflects the preceding stellar, binary and cluster evolution. For a first attempt using this method we begin a simulation with $N = 15\,000$ stars at 2 500 Myr using a 50% binary fraction. This model is substantially smaller in population number than the $N \simeq 40\,000$ required to start from the birth of the cluster, so it can be evolved directly to the age of M67 in a reasonable time. This is prefer-

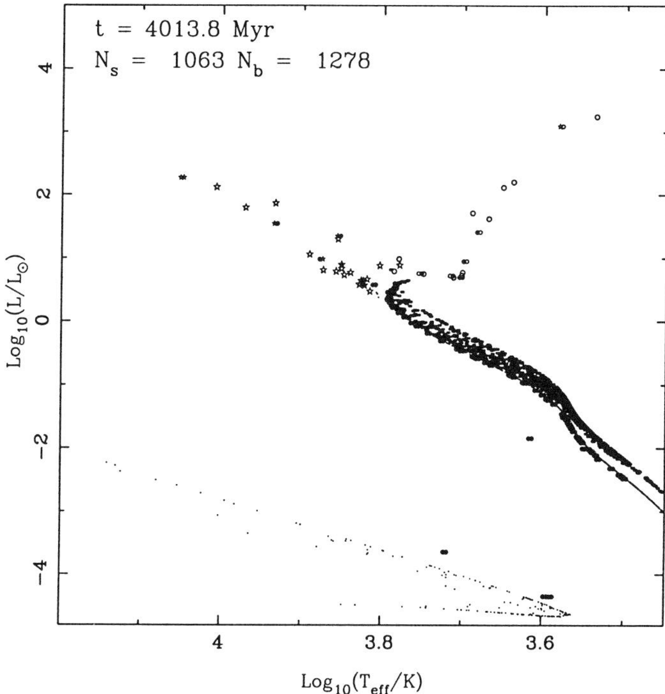

Figure 4 Hertzsprung-Russell diagram for the M67 N-body simulation at an age of 4.0138 Gyr when 1 063 single stars and 1 278 binaries remain. Present in the cluster at this time are 27 blue stragglers with six of these in binaries. There are two super-BSs and one of these forms part of a BS-BS binary (see text). There are nine cataclysmic variables and three double-degenerate systems. There are 19 giants, six of which are in binaries. Main-sequence stars (dots), blue stragglers (open stars), sub-giants, giants and naked helium stars (open circles) and white dwarfs (dots) are distinguished. Binary stars are denoted by overlapping symbols appropriate to the stellar type of the components, with main-sequence binary components depicted with filled circles and white dwarf binary components as $\oplus$ symbols.

able to scaling up the results of a smaller simulation starting from initial conditions. We estimate the dynamically modified mass function at 2 500 Myr from the $N = 10\,000$ runs, and we populate the starting model according to this mass function. The stars are distributed according to a multi-mass King (1966) model with concentration parameter $W_0 = 7$.

We evolve the model to an age of 4 310 Myr using NBODY4. The solid line in Fig. 3 shows the evolution of the number of BSs present in the simulation. This is in good agreement with the observed number of blue stragglers in M67. Between 3.5 and 4 Gyr there are on average 25 BSs in the model with about 7 of these in binaries. If it had been possible to start a full simulation from $T = 0$ then hopefully the relative number of

BSs would increase gradually with time to reach the M67 point, rather than the rapid increase produced by the semi-direct simulation (although the log-scale in Figure 3 distorts the actual range of the simulation).

Among all the BS binaries formed in the simulation there appear to be two fairly distinct populations, one consisting of close circular orbits and the other with wider eccentric orbits. The BSs in circular orbits formed by stable Case B mass-transfer, and these still have their primordial companion which evolves to become a WD. No instance of a BS formed from Case C mass transfer, or as a result of wind accretion, occurs in this simulation. For the relatively wide binaries, $P > 100\,\mathrm{d}$, the distribution appears uniform for $e > 0.2$. These binaries form in hierarchical systems or as a result of an exchange interaction. A star is more likely to be exchanged into a wide orbit than into an existing hard binary. Furthermore, since exchange interactions tend to take place in the dense central regions of the cluster it is not surprising to find a paucity of wide circular binaries.

Of the single BSs formed in the model, about half formed by coalescence of a Case A binary, in a few cases only after the orbit was strongly modified by a number of close encounters. The other half formed as a result of collisions within binaries, either after a star was exchanged into a highly eccentric orbit or after perturbations to the existing binary increased the eccentricity and caused the orbit to become chaotic. The latter process can also occur in a hierarchical triple which occasionally led to the formation of a close eccentric BS binary.

Figure 4 shows the Hertzsprung-Russell diagram of the cluster model at a time of 4 014 Myr. At this time there are two super-BSs in the cluster. One of these has a mass 2.3 times the then turn-off mass of 1.32 $M_\odot$. The other super-BS has a mass of $2.7 M_{\mathrm{TO}}$ and is in a binary with another BS. This is an interesting case. At the beginning of the simulation it is a single star with mass $M_1 = 1.33\ M_\odot$. At 3 190 Myr it enters a triple system in which it exchanges into the original binary. The resulting binary has orbital parameters $P = 9\,120\,\mathrm{d}$ and $e = 0.52$ and survives to 3 740 Myr when it is involved in a binary-binary encounter. This leaves the proto-BS in an orbit of $P = 153\,\mathrm{d}$ and $e = 0.99$ with a companion of mass $M_2 = 1.0\ M_\odot$. Owing to the large eccentricity the binary components collide and merge to form a BS of mass $M_1 = 2.33\ M_\odot$. At 3 870 Myr this BS is involved in a three-body interaction and forms another binary, this time with a star of mass $M_2 = 2.28\ M_\odot$ which is itself a BS. The BS-BS binary has $P = 1\,350\,\mathrm{d}$ and $e = 0.67$. At 3 990 Myr this binary forms a stable four-body hierarchy with yet another binary, the original BS collides with one of the stars in the other binary, forms the super-BS with $M_1 = 3.45\ M_\odot$, and remains bound to the other BS. The fourth body escapes and leaves the binary that is observed at 4014 Myr. Later at 4080 Myr the super-BS and BS collide when the eccentricity of the orbit has grown to 0.93 as a result of external perturbations. This makes a BS with $M_1 = 5.7\ M_\odot$, i.e. $4.4 M_{\mathrm{TO}}$. The new super-BS captures a companion at 4112 Myr, with which it then collides to form a super-BS with $M_1 = 7.7\ M_\odot \simeq 6 M_{\mathrm{TO}}$. Soon after it strongly interacts with a hard binary and gets ejected from the cluster.

7. DISCUSSION AND CONCLUSION

An obvious problem with our M67 model is the lack of blue stragglers in wide circular binaries because two such systems are observed in the real cluster. Wide binaries can form via exchange in the cluster but the results show this is unlikely to produce circular orbits. Perhaps this indicates that wind accretion is more efficient than we have assumed. With a higher efficiency, or a lower wind velocity, the required number of BSs in wide circular binaries might be produced by wind accretion.

Another discrepancy is that, although the total number of BSs in M67 can be reproduced, the frequency of model BSs found in binaries is too low. The BS half-mass radius in the model is smaller than the half-mass radius of the MS stars, in qualitative agreement with the observations, but is itself a factor of 2 less than is observed for M67. This suggests that the population within the half-mass radius is too centrally condensed. A decrease in the central concentration implies that wide binaries formed from exchange interactions have longer lifetimes.

Ideally it would be desirable to perform more simulations and investigate the effects of varying parameters such as the central concentration in the starting model. It would also be helpful if we could begin a complete simulation from initial conditions but, considering that a single semi-direct simulation took a month to perform, the capability of the current hardware is already pushed to its limit. Future improvements in computing efficiency, such as the availability of GRAPE-6, will enable a leap forward in the field of star cluster modelling. The ability to perform direct simulations of clusters like M67 comfortably will decrease substantially the uncertainties involved, and allow observed and model parameters to be matched iteratively.

We have shown that binary evolution alone cannot account for the numbers of observed blue stragglers in open clusters, or the binary properties of these blue stragglers. The influence of the cluster environment can increase the number of blue stragglers produced, leading to good agreement with the observations. Our N-body model of M67 demonstrates that blue stragglers are most likely generated by a variety of processes and in particular we find formation paths for all the BSs observed in M67.

References

Aarseth S.J., 1999, PASP, 111, 1333

Ahumada J., Lapasset E., 1995, A&AS, 109, 375

Bondi H., Hoyle F., 1944, MNRAS, 104, 273

Fan X., Burstein D., Chen J.-S., et al., 1996, AJ, 112, 628

Hurley J.R., Pols O.R., Tout C.A., 2000, MNRAS, 315, 543 (PapI)

Hurley J.R., Tout C.A., Pols O.R., 2001a, MNRAS, submitted (PapII)

Hurley J.R., Tout C.A., Aarseth S.J., Pols O.R., 2001b, MNRAS, in press (PapIII)

Hut P., 1980, A&A, 92, 167

Hut P., 1981, A&A, 99, 126

King I.R., 1966, AJ, 71, 64

Latham D.W., Milone A.A.E., 1996, in Milone E.F., Mermilliod J.-C., eds., The Origins, Evolution, and Destinies of Binary Stars in Clusters. ASP Conf. Series 90, p. 385

Leonard P.J.T., 1996, ApJ, 470, 521

Makino J., Kokubo E., Taiji M., 1993, PASJ, 45, 349

Milone A.A.E., Latham D.W., 1992, in Kondo Y., Sistero R.F., Polidan R.S., eds., Evolutionary Processes in Interacting Binary Stars. IAU Symposium 151, Kluwer, Dordrecht, p. 473

Pols O.R., Marinus M., 1994, A&A, 288, 475

Pols O.R., Schröder K.P., Hurley J.R., Tout C.A., Eggleton P.P., 1998, MNRAS, 298, 525

Portegies Zwart S.F., Verbunt F., 1996, A&A, 309, 179

Rasio F.A., Tout C.A., Lubow S.H., Livio M., 1996, ApJ, 470, 1187

Tout C.A., Aarseth S.J., Pols O.R., Eggleton P.P., 1997, MNRAS, 291, 732

Tout C.A., Eggleton P.P., 1988, MNRAS, 231, 823

van den Berg M., Verbunt F., Mathieu R.D., 1999, A&A, 347, 866

Whyte C.A., Eggleton P.P., 1985, MNRAS, 214, 357

Zahn J.-P., 1977, A&A, 57, 383

Zahn J.-P., 1989, A&A, 220, 112

MASSIVE STAR POPULATION SYNTHESIS OF STARBURSTS

Joris Van Bever and Dany Vanbeveren
Astrophysical Institute, Vrije Universiteit Brussel, pleinlaan 2, 1050 Brussels, Belgium
jvbever@vub.ac.be; dvbevere@vub.ac.be

Keywords: Population synthesis, starburst, close binaries, X-ray binaries, supernova Remnants

Abstract Using the Brussels population synthesis code which includes a realistic fraction of close binaries, we study the effect of massive stars on spectral properties of starburst regions, for the metallicities Z=0.02 and 0.001. We conclude that binary interaction may cause a so called 'rejuvenation' of a starburst, producing a stellar population seemingly younger than in reality. The consequences for the age determination of these objects through $W_{H\beta}$ are discussed. We then discuss the properties of the red and blue 'Wolf-Rayet bump' in the spectra of some extra-galactic starbursts and we argue that a comparison between theory and observation seems to require the presence of a non-negligible amount of interacting binaries. Finally, we show the results of theoretical modelling of the hard X-ray luminosity produced by a starburst (due to high mass X-ray binaries (HMXBs) and young supernova remnants (YSNRs)) as a function of its age.

1. INTRODUCTION

The effect of interacting binaries on massive star population (number) synthesis [further abbreviated as P(N)S] of starbursts has been discussed by Van Bever and Vanbeveren (1998). Our P(N)S code and the massive star evolutionary computations that we use in the code have been described by Vanbeveren (present proceedings). Recall that P(N)S of massive stars depends critically on

- the stellar wind mass loss rate formalisms during the various phases of massive star evolution,

D. Vanbeveren (ed.), The Influence of Binaries on Stellar Population Studies, 419–434.

- the treatment of mass transfer and accretion of matter in interacting binaries (requiring a large library of binary evolutionary calculations),
- the adopted degree of asymmetry of the supernova (SN) explosion and the effect on the binary disruption probability when one of the components explodes.

The present paper deals with massive star population synthesis of starbursts and we focus on three topics:

- the time evolution of the population of O and Wolf-Rayet (WR) type stars, showing the fundamentally different behaviour between single stars and close binaries, leading to the so-called 'rejuvenation' of a starburst (section 2).
- the emission features of the massive star population in starbursts (section 3). More specifically, we concentrate on the collection of WR emission lines, grouped into the so-called blue and red WR bump at 4650 and 5808 Å respectively, which have been observed in the spectra of 'Wolf-Rayet galaxies'.
- hard X-rays (E $\geq$ 2keV) emitted by a starburst due to a population of high mass X-ray binaries and young supernova remnants (YSNRs) and the evolution of the luminosity as a function of age.

The implementation of binaries in P(N)S of starbursts implies the implementation of a number of binary parameters. It is not the scope of the present paper to discuss the effects of all these parameters. However, to illustrate typical binary effects we will compare predictions of a 'single stars only' model with a *standard binary starburst model* where binaries are included.

The standard binary starburst (SBS) model. As usual, we classify binaries as Case A, Case B_r, Case B_c and Case C depending on the onset of RLOF. Case A binaries are treated as Case B_r. Case B_c and Case C evolve through a common envelope phase and binaries with a compact companion experience spiral-in. Most of the binaries with initial mass ratio q $= M_2/M_1 \leq 0.2$ merge. From that point on, all of them are treated as single stars with mass = sum of the masses of both components.

We have an extended library of massive single star and massive close binary (Case B_r) evolutionary calculations for different metallicities (0.001 $\leq$ Z $\leq$ 0.02) where the effects of stellar wind mass loss are included in detail as discussed by Vanbeveren (present proceedings). In the SBS model we use evolutionary computations where *the mass loss rates of red supergiants are Z-dependent* ($\propto Z^{0.5}$) *but those of Wolf-Rayet (WR) stars are independent of Z.*

Primaries of close binaries and single stars have the same initial mass function $\Psi(M) \propto M^{-2.35}$ ($M_{min} = 1\,M_\odot$ and $M_{max} = 100\,M_\odot$), the binary mass ratio distribution is flat and the binary orbital period distribution is flat in log P (periods between 1 day and 10 years).

The effect of the SN explosion is studied in full 3-D where we adopt a χ^2-distribution for the kick-velocity v_{kick} (describing the asymmetry of the SN explosion) with average = 450 km/s.

The binary frequency at birth, defined as the ratio at birth of the number of binaries N_{bin} and the sum of the number of binaries and the number of single stars N_{single}:

$$f = \frac{N_{bin}}{N_{bin} + N_{single}} \tag{30.1}$$

Observations tell us that about 30% of the OB type stars in the solar neighbourhood are primary of a close binary with a period less than 100 days and mass ratio $\geq$0.2 (Garmany et al., 1980). Reproducing this number with our PNS code requires a large initial overall binary frequency. In the SBS model, we use the representative value = 80%.

And finally, a Monte-Carlo method simulates the instantaneous formation of 300000 massive objects (either a single star or a binary where at least the primary is massive), accounting for the distributions discussed above.

2. REJUVENATION OF STARBURSTS

The massive star population in starbursts consists entirely of OB stars during the first 3 Myrs. Then WR stars appear. In starbursts with single stars only, the O star features disappear after about 6 Myr. The WR stage ends when the lowest mass WR-progenitor (about 20 $M_\odot$ for Z=0.02 and 40 $M_\odot$ for Z=0.001) explodes. This is illustrated in Fig 1 which shows the time-evolution of the number of O and WR stars for different metallicities.

The situation for the SBS model is significantly different. After about 4 to 5 Myr the first binaries (with primary mass $\leq 40\,M_\odot$) start to interact through RLOF. A star that accretes matter during its core hydrogen burning phase becomes bluer and more luminous (a Blue Straggler): it rejuvenates. This is the reason why the O star population in the starburst increases again : *the starburst is rejuvenated due to binary interac-*

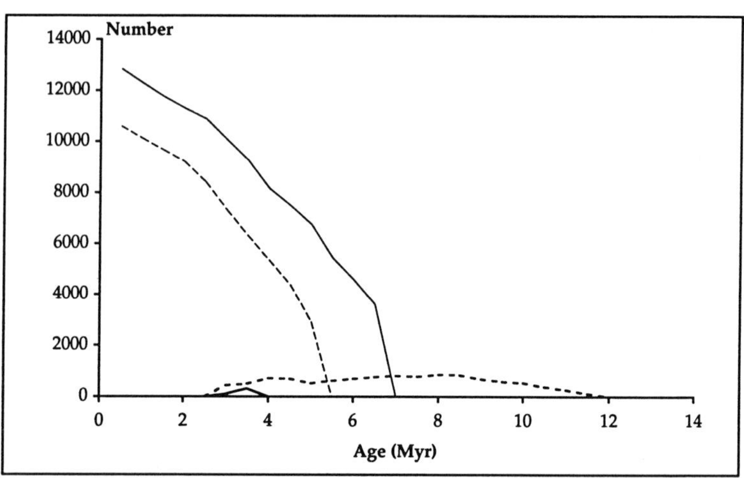

Figure 1 The evolution with time of the number of O (thin lines) and WR stars (thick lines) in a single stars only-model for Z=0.02 (dashed lines) and for Z=0.001 (full lines). The starburst is instantaneous and we used the Salpeter IMF.

tion. As a consequence, the O stage and the WR stage of the starburst lasts much longer compared to starbursts with single stars only. These effects are illustrated in Fig 1 as well.

RLOF alters the time-evolution of the O type and WR stellar population and therefore also all the emission properties associated with them. One of the bright emission lines in spectra of HII regions is the Hβ line. Its equivalent width $W_{H\beta}$ is a good age indicator when starbursts consist of single stars only. The number of O type stars decreases monotonically as a function of time and so does $W_{H\beta}$. However, when binaries are included, one can expect $W_{H\beta}$ to increase along with the massive star population when binaries start to interact. Fig 2 shows that $W_{H\beta}$ ceases to be monotonically decreasing in the standard binary model, as expected. As a consequence, binary evolution complicates the age determination through this quantity. For instance, when a starburst has a $W_{H\beta}$ of about 100 Å (150 Å at Z=0.001) one concludes that the age of the burst is about 4 to 5 Myr when we use the model with single stars only, but it could be twice as old when the starburst has a significant population of binaries.

To illustrate the importance of this effect, we show in Fig 3 a histogram of the $W_{H\beta}$ of a large sample of extragalactic HII regions, compiled by Terlevich (1991). When we consider those cases where $W_{H\beta}$ is smaller than ~100 Å, it is clear from the figure that a large fraction of the HII regions may be older than originally thought. One there-

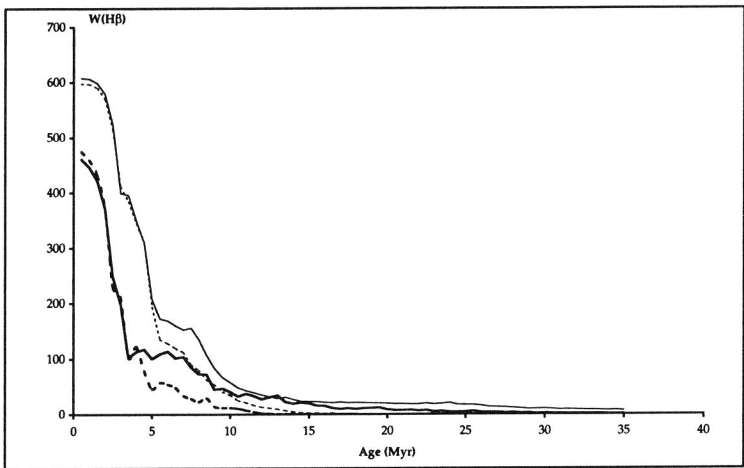

Figure 2 The evolution with time of $W_{H\beta}$ at Z=0.02 (thick lines) and for Z=0.001 (thin lines), for the standard binary model (full lines) compared to a single star model (dashed lines) with the same IMF.

fore needs to be very careful when deducing ages of starburst regions by comparing data with predictions of starbursts models where only single stars are considered.

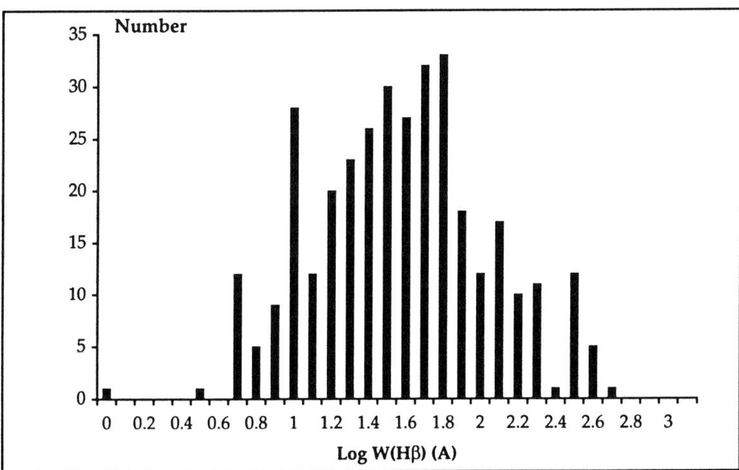

Figure 3 Histogram of the measured equivalent widths of the Hβ emission line of 200 extragalactic HII regions, compiled by Terlevich (1991).

3. WOLF-RAYET EMISSION LINES

The presence of WR stars in extragalactic systems is betrayed by a strong complex of emission lines around 4650 Å, the so-called *blue WR bump.* It consists mainly of NIII/NV λ4640, CIII/CIV λ4650, HeII λ4686, as well as a nebular contribution to the latter. Because of the variety in elements, all WR subtypes may contribute to the bump.

In recent years a second but weaker *red bump* has been detected in many starburst regions. It consists of a carbon multiplet; CIVλ5808-5812, prominent in WC type stars. The strenghts of these bumps have been synthesised in the past using single star models by a.o. Krueger et al. (1992, 1993), Schaerer (1996), Schaerer & Vacca (1998). Our code also allows the comparison with the standard binary model.

All lines considered are stellar in origin, except the contribution of the surrounding gas to HeII λ4686. To compute the nebular line strength and continuum we use the atomic data of Osterbrock (1989) and make the following assumptions concerning the physical state of the gas:

- helium to hydrogen number density ratio $N_{He}/N_H = 0.1$
- electron number density ne $= 100cm^{-3}$,
- radiation bounded nebula,
- Case B recombination,
- electron temperature $T_e = 10000$K.

The emitted luminosities in the stellar WR lines is taken from the compilation of Schaerer & Vacca (1998). We use the (theoretical) stellar continua from Schmutz (1992) for WR stars and from Lejeune (1997) otherwise.

3.1. NEBULAR HEII λ4686

Fig 4 (resp. Fig 5) shows the equivalent width W_{HeII} of the nebular HeII λ4686 line vs. $W_{H\beta}$ (resp. the intensity ratio $I_{HeII}/I_{H\beta}$ of the nebular HeII λ4686 line/$H\beta$ line vs. $W_{H\beta}$) at Z=0.001, for the SBS model compared to the single star starburst model. The observations (black dots) are from Guseva et al. (2000). Apparently, the theoretical predictions at large $W_{H\beta}$ (meaning young starbursts) are much too high for both models. This is due to a large population of hot WC stars in the early stages, producing a large flux of He^+ ionizing photons.

Two solutions for this discrepancy. The theoretical models of Schmutz et al. (1992) were calculated for pure He atmospheres. We apply them for all WR subtypes. One may wonder whether these models are good approximations for WC stars, which have comparable He, C and O abundances. It may be that the increased opacity due to the presence of C and O decreases the ionizing luminosity, and this would obviously reduce the discrepancy between our models and the observations. For this reason, we also computed a SBS model without including the He^+ ionizing power of the WC type stars. As can be seen, the difference between model and observations at large $W_{H\beta}$ is eliminated.

A second possibility is that our models predict too many WC type stars in low Z regions. The WR population in these regions may be dominated by WR binaries (Foellmi and Moffat, present proceedings) and this means that the theoretically predicted relative populations of the various WR subtypes depend mainly on the mass loss rate formalism during the WR stage (Vanbeveren, present proceedings). Our standard model assumes that the WR mass loss rates are Z-independent. However, if they would have the same dependence on metallicity as the mass loss rates of OB type stars, the number of WC (binary)stars in starbursts with Z=0.001 would be substantially reduced, idem dito the HeII λ4686 discrepancy.

Interestingly, single star models are unable to produce nebular HeII λ4686 at $W_{H\beta}$ smaller than about 150 Å, whereas many systems are observed in this region. The SBS model does predict nebular emission but not enough to correspond to the observations. Nevertheless, it is tempting to conclude that at least some starbursts require a non negligible fraction of binaries to explain the nebular HeII emission.

The situation at Z=0.02 is unclear. Although a significant nebular contribution is predicted at high $W_{H\beta}$ none has been observed by Guseva et al. (2000). However, given that the only four systems observed with about solar metallicity have low $W_{H\beta}$, this need not be problematic.

3.2. RED AND BLUE WOLF-RAYET BUMPS

Fig 6 resp. Fig 7 show the equivalent width of the total blue bump (at 4650 Å) resp. red bump (at 5808 Å) relative to the equivalent width of Hβ, for the SBS model compared to the single star only starburst model, at Z=0.001. The total blue bump seems to behave in a way similar to the nebular HeII 4686 line. For high values of $W_{H\beta}$, i.e. young bursts, the models predict to much emission. We also plot the result of the SBS model that ignores the WC star contribution, and notice that the correspondence to the observations is better, but not yet satisfactory. At

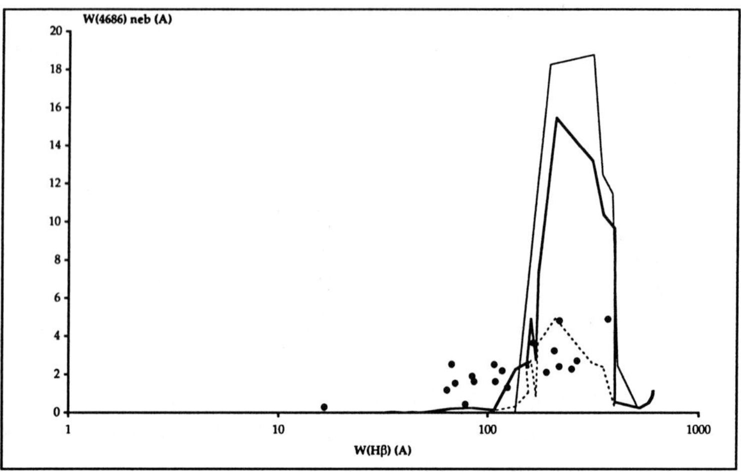

Figure 4 W(neb. HeII λ4686) vs. $W_{H\beta}$ at Z=0.001. The thick full line corresponds to the SBS model, the narrow full line to the single star only starburst model with the same IMF and the dashed line the SBS model but without emission of WC stars. The dots are the observations from Guseva et al. (2000).

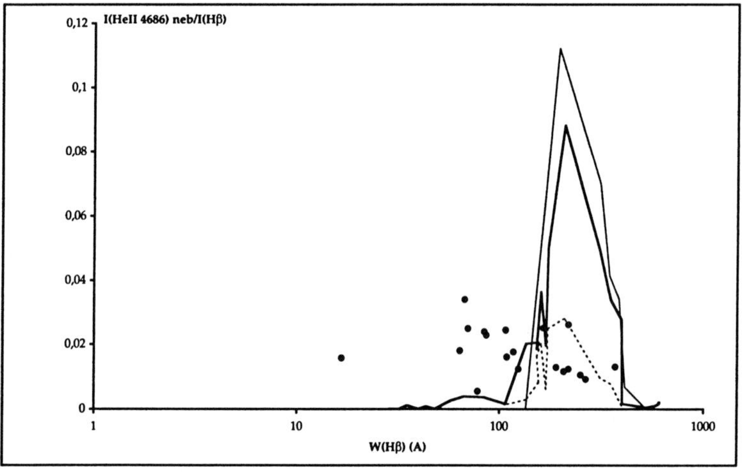

Figure 5 Same as in Fig 4 but here I(neb. HeII λ4686)/I(Hβ) vs. $W_{H\beta}$ is shown.

lower $W_{H\beta}$ values, the predictions are a good match to the observations. Note also, that the single star only starburst model does not predict any blue (or red) bump emission below $W_{H\beta}$ = 150 Å, which indicates that at least some observed bursts require the presence of a sufficient number of binaries. Similar conclusions can be drawn concerning the red bump results. Fig 8 resp. Fig 9 are similar to Fig 6 resp. Fig 7, but for Z=0.02. The number statistics here is very low, with only four bursts observed. The predictions of the SBS model however, provide an almost perfect fit for the blue bump as well as for the red bump, whereas the single star only starburst model predicts slightly too much emission. So here again, one can state that there seems to be a need for a non negligible binary fraction to explain the Wolf-Rayet emission features, at least for these four observed bursts. But even though the SBS model is succesful in some cases and should be preferred over the single star only model, there are still situations for which the discrepancy with observations is too large (especially for young burst at low Z). Therefore, efforts to improve our knowledge about stellar evolution (e.g. mass loss rates) as well as stellar atmospheres (e.g. emission properties of WC stars), should continue.

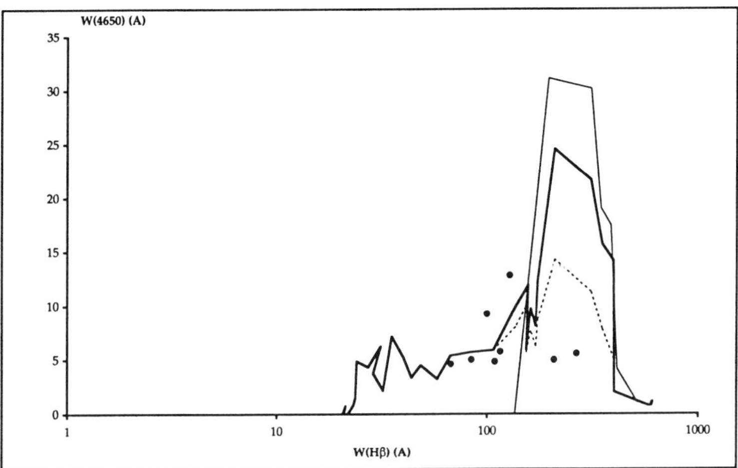

Figure 6 Equivalent width of the total blue bump (at 4650 Å) versus the equivalent width of Hβ, for the SBS model (thick line) compared to the single star only model (thin line), at Z=0.001. The dashed line denotes the standard binary model without the emission of WC stars. The dots are the observations from Guseva et al. (2000).

4. STARBURSTS AND HARD X-RAYS

A population of massive binaries evolves into a population of high mass X-ray binaries (HMXB). HMXBs emit hard X-rays ($\geq$2keV). X-ray observations of starbursts may therefore be very important to decide upon the importance of binaries in starbursts.

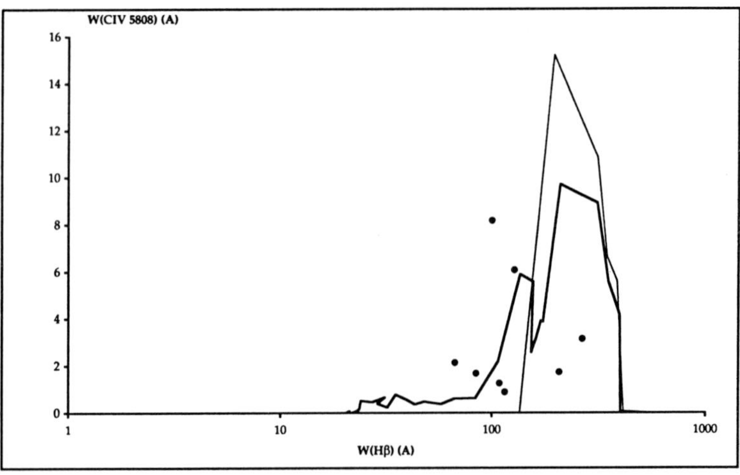

Figure 7 Same as in Fig 6 but for the red bump.

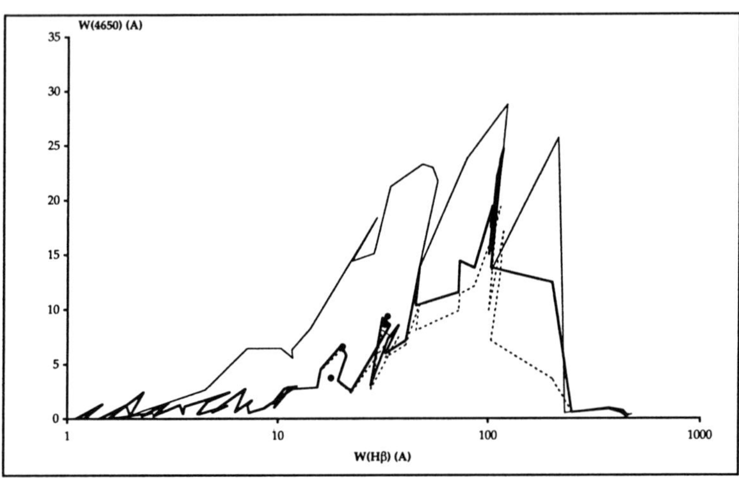

Figure 8 Same as in Fig 6 but for Z=0.02.

X-ray observations of starbursts have been reviewed by I. Stevens in the present proceedings. Question: can we reproduce these observations by PNS, and, do we need binaries to explain them?

X-rays with E≥2keV are produced by HMXBs, low mass X-ray binaries (LMXB), young supernova remnants (YSNR) and active galactic nuclei (AGN). With a PNS model we are able to predict the HMXB, LMXB and YSNR population. Since we are interested in young star-

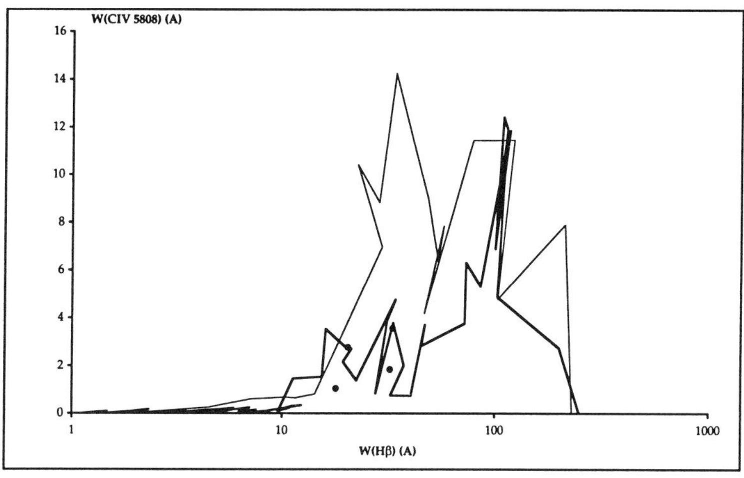

Figure 9 Same as in Fig 8 but for the red bump.

bursts, LMXBs will not be considered. Notice however that the integrated X-ray observations of starbursts may be affected by the underlying older populations and the latter may obviously harbour LMXBs.

The results that will be discussed in the next sections are a summary of the results of Van Bever and Vanbeveren (2000).

4.1. THE HMXB POPULATION

Our PNS code simulates the population of OB+compact companion (OB+cc) binaries. We like to remind the reader that the stellar wind mass loss rate formalisms that are used in our evolutionary calculations imply that black holes (BHs) form out of

- primaries in Case B_r binaries with initial mass larger than 40 $M_\odot$,
- single stars and primaries of Case B_c and Case C systems with initial mass larger than ~25 $M_\odot$,
- mass gainers who became single after the first SN explosion and who's mass after mass transfer became larger than ~25 $M_\odot$ (see also Vanbeveren, present proceedings).

Although BH formation may be preceded by a SN explosion (Israelian et al., 1999; Nelemans et al., 1999), we do not account for the effect in the PNS code.

X-rays are formed when matter lost by the optical star is trapped by the cc. An essential condition is that an accretion disc can be formed

(Shapiro and Lightman, 1976) and/or a very strong magnetic field is present (Langer and Rappaport, 1982). A formalism was presented by Iben et al. (1995). Once the conditions are fulfilled, we use the model of Davidson and Ostriker (1973) when the matter is trapped from the stellar wind of the optical component or the model of Savonije (1979, 1983) when the cc accretes matter during its spiral-in phase in the envelope of the optical star.

4.2. THE YSNR POPULATION

The formation of X-rays from young single pulsars (young supernova remnants=YSNRs) has been outlined in classical papers like Goldreich and Julian (1969), Goldreich et al. (1971), Rees and Gun (1974). A rotating magnet emits magnetic dipole radiation and slows down. Part of this radiation is in the form of hard X-rays.

To estimate the X-radiation produced by YSNRs we therefore need to know the distribution of the magnetic field B and the distribution of initial spin period P_o of neutron stars. The B-distribution appears to be Gaussian in Log B with average = 12.5 and standard deviation = 0.3 (Bhattacharya et al., 1992). This is used in our PNS code. However, the initial period of neutron stars is unknown. We adopt a model where the initial spin period is the same for all pulsars and study the effect by considering different values of P_o between 0.001sec and 0.1sec. Since the total rotational energy of a pulsar $\propto P^2$, it can be expected that the X-ray results may depend on P_o.

4.3. THE HARD X-RAY LUMINOSITY OF A STARBURST

Fig 10 illustrates the time variation of the X-ray luminosity of a SBS model. The initial rotation pariod of the pulsars P_o = 0.01 s. We separately consider the HMXBs and the YSNRs. We also show the variation for a starburst with single stars only. In Fig 11 we compare the time variation of the X-ray luminosity of the YSNRs for the SBS when it is assumed that the neutron stars are born with a rotation period P_o = 0.01 sec and P_o = 0.001 sec. The luminosities are given divided by the total mass of massive stars in the burst (29.6 million $M_\odot$ in the standard model and 21.1 $M_\odot$ million in the starburst with single stars only).

We list a few important conclusions.

- In our model single stars with initial mass $\geq$ 20-25 $M_\odot$ finally collapse to form a BH. This implies that our starburst with single

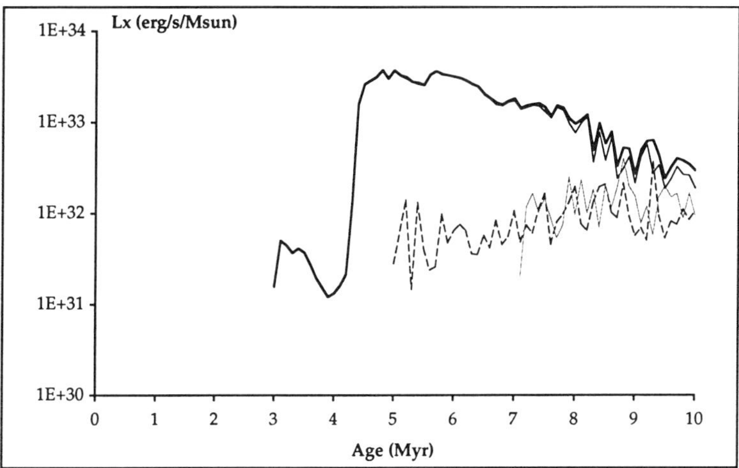

Figure 10 The evolution with time of the X-ray luminosity of a standard starburst model. The full thick line represents the total luminosity. The full thin line (resp. dashed line) is the contribution of the HMXBs (resp. YSNRs). The dotted line is the total luminosity of the 'single star only' model.

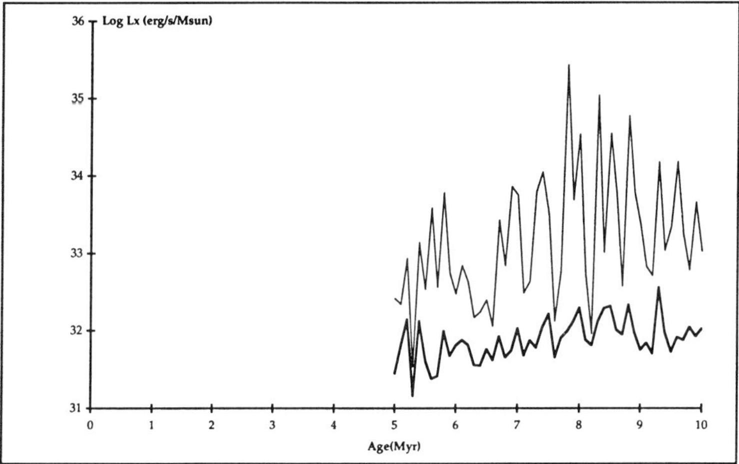

Figure 11 X-ray luminosity of YSNRs in case of P_o=0.01s (thick line) and 0.001s (thin line).

stars only emits hard X-rays (due to YSNRs) after 7-8 million years. We are therefore tempted to conclude that when a starburst which is younger than 7-8 million years emits hard X-rays, binaries were present initially in the burst.

- As illustrated by Fig 10, a starburst that emits hard X-rays with energies $\geq$ 2 keV is at least 3-4 million years old.
- WR+cc binaries hardly contribute to the overall X-ray luminosity of a starburst.
- YSNRs contribute significantly to the total X-ray luminosity of a starburst only when most of the neutron stars are born with a rotation period $\leq$ 0.01 sec.
- During the first 8-9 million years, the hard X-ray luminosity of starbursts (with binaries) comes from YSNRs and from HMXBs with a black hole component. This means that the X-ray spectrum of the starburst will be charaterized by a kT$\sim$2keV component due to YSNRs and a somewhat harder component due to black hole HMXBs. HMXBs with a neutron star dominate the HMXB population after 8-9 million years. The X-ray spectrum may be significantly harder then.
- The number of YSNRs and HMXBs in starburst is always quite small. This explains the stochastic character of the figures showing the X-ray luminosity. This means that even when interacting binaries play an essential role in the evolution of a starburst, it does not always contain HMXBs.

5. CONCLUSIONS

With a PNS code adapted to follow the evolution of young starbursts with binaries, we have investigated the time evolution of

- The emission line Hβ
- The nebular He II λ4686 line
- The stellar emission lines He II λ4686, NIII/NV λ4640 and CIII/CIV λ4650 (the blue WR bump)
- The stellar emission lines CIV λ5808-5812 (the red WR bump)
- The hard X-ray emission (energies $\geq$ 2 keV)

We conclude

- The He II λ4686, the red and blue WR bumps observed in starbursts with low Z is better reproduced by theoretical models where a significant percentage of the massive stars are components of interacting close binaries.

- The age determination of starbursts based on Hβ is not unique when a significant percentage of massive interacting close binaries is present in the burst.
- The WR population in starbursts with low Z consists maily of WR+OB binaries. As a consequence, the distribution of WR subtypes depends maily on the stellar wind mass loss formalism during the WR phase.

  The observed He II λ4686, the red and blue WR bumps in starbursts with low Z is best reproduced by theoretical starbursts models where the number of WC type stars is small. Since the number of WC stars depends on the stellar wind rates, this can be considered as indirect evidence that WR rates are Z-dependent as predicted by the radiatively driven wind theory.
- The observed hard X-ray properties of starbursts are strong evidence that interacting binaries were present initially in the burst. However, at present, it is difficult to propose a unique percentage. To illustrate, most of the properties of the starburst He2-10 can be explained with a starburst which initially contained 51000 (resp. 300000) massive stars and a binary frequency of 80% (resp. 10%) which is 11 Myr (resp. 7 Myr) old.
- The hard X-ray spectrum emitted by YSNRs depends strongly on the initial spin period of neutron stars. If most of the neutron stars are born with a spin period ~0.001 sec, then the hard X-radiation of starbursts will be dominated by YSNRs. Detailed X-ray observations of starbursts may therefore give important hints concerning these initial spin periods.

Final remark. N-body close encounters in massive starbursts may be very frequent. One of the most spectacular events is of course the merging of two stars. This merging may obviously have the same effect as binary mass transfer, i.e. the rejuvenation of the burst. One interesting case is the compact star cluster R136 in 30 Dor (LMC) which contains a large number of very young O3 type stars. N-body simulations (Portegies Zwart et al., 1999) show that mergers are a possible explanation. Future starburst simulations will have to account for this effect.

References

Bhattacharya, D., Wijers, R., Hartman W. J., Verbunt, F.: 1992, A&A, 254, 198.

Davidson, K., Ostriker, J. P.: 1973, ApJ, 179, 585.
Garmany, C., Conti P. S., Massey P.: 1980, ApJ 242, 1063.
Goldreich, P., Julian, W. H.: 1969, ApJ 157, 869.
Goldreich, P., Pacini, F., Rees, M. J.: 1971, Comm. Astrophys. Space Phys., 3, 185.
Guseva, N. G., Izotov, Y. I., Thuan, T. X.: 2000, ApJ, 531, 776.
Iben, I., Tutukov, A. V., Yungelson, L. R.: 1995, ApJS 100, 217.
Israelian, G., Rebolo, R., Basri, G., Casares, J., Martin, E. L.: 1999, Nat., 401, 142.
Krueger, H., Fritze-v. Alvensleben, U., Fricke, K. J., Loose, H.-H.:1992, A&A 259, L73.
Krueger, H., Fritze-v. Alvensleben, U., Fricke, K. J., Loose, H.-H.:1993, Astroph. and Space Science 205, no. 1, 57.
Langer, S. H., Rappaport, S.: 1982, ApJ 257, 733.
Lejeune, T., Cuisinier, F., Buser R.: 1997, A&AS, 125, 229.
Nelemans, G., Tauris, T. M., van den Heuvel, E. P. J.: A%A, 352, L87.
Osterbrock, D.E.: 1989, in 'Astrophysics of Gaseous Nebulae and Active Galactic Nuclei', University Science Books, California.
Portegies Zwart, S. F., Makino, J., McMillan, S. L., Hut, P: A&A 348, 117.
Rees, M. J., Gunn, J. E.: 1974, MNRAS, 167, 1.
Savonije, G. J.: 1979, A&A, 71, 352.
Savonije, G. J.: 1983, in 'Accretion Driven Stellar X-Ray Sources', Cambridge University Press, Lewin W. G. H., van den Heuvel E. P. J. (eds).
Schaerer, D.:1996, ApJL 467, L17.
Schaerer, D., Vacca, W.:1998, ApJ 497, 618.
Schmutz, W., Leitherer, C., Gruenwald, R.: 1992, PASP, 104, 1164.
Shapiro, S. L., Lightman, A. P.: 1976, ApJ, 204, 555.
Terlevich, R., Melnick, J., Masegosa, J., Moles, J., Copetti, M. V. F.: 1991, A&AS 91, 285.
Van Bever, J., Vanbeveren D.: 1998, A&A 334, 21.
Van Bever, J., Vanbeveren D.: 2000, A&A 358, 462.

RADIOACTIVITIES IN POPULATION STUDIES: ^{26}Al AND ^{60}Fe FROM OB ASSOCIATIONS

S. Plüschke, R. Diehl, K. Kretschmer
MPI f. extraterrestrische Physik, PO-Box 1312, D-85741 Garching, Germany

D. H. Hartmann
Department of Physics & Astronomy, Clemson University, Clemson, SC 29634, USA

U. Oberlack
Astrophysics Laboratories, Columbia University, New York, NY 10027, USA

Keywords: OB Associations, Nucleosynthesis, Mass loss, ISM

Abstract The observation of the interstellar 1.809 MeV decay-line of radioactive ^{26}Al by the imaging gamma-ray telescope COMPTEL have let to the conclusion, that massive stars and their subsequent core-collapse supernovae are the dominant sources of the interstellar ^{26}Al abundance. Massive stars are known to affect the surrounding interstellar medium by their energetic stellar winds and by the emission of ionising radiation. We present a population synthesis model allowing the correlated investigation of the gamma-ray emission characteristics with integrated matter, kinetic energy and extreme ultra-violet radiation emission of associations of massive stars. We study the time evolution of the various observables. In addition, we discuss systematic as well as statistical uncertainties affecting the model. Beside uncertainties in the input stellar physics such as stellar rotation, mass loss rates or internal mixing modifications due to a unknown binary component may lead to significant uncertainties.

1. INTRODUCTION

During recent years spatially resolved observations of decaying radioactive isotopes spread in the interstellar medium of the Milky Way by their characteristic γ-ray emission became feasible. The Compton telescope COMPTEL (Schönfelder et al., 1993) aboard NASA's Compton Gamma-Ray Observatory surveyed the sky in the MeV regime during it's 9 year mission time. One of the key results of this mission was the generation of the first all sky image in the 1.809 MeV γ-ray line

D. Vanbeveren (ed.), The Influence of Binaries on Stellar Population Studies, 435–445.

from radioactive ^{26}Al (Oberlack etal, 1996, Plüschke et al., 2000). The COMPTEL observations confirm a integral ^{26}Al content of the Milky Way of 2 to 3 $M_{\odot}$. In principle ^{26}Al can be produced by various nucleosynthesis sites. Since it's first detection by HEAO-C (Mahoney et al., 1984) non-explosive sites such as AGB stars (Mowlavi & Meynet, 2000) and Wolf-Rayet stars (Meynet et al., 1997) as well as explosive environments (novae (Jos et al, 1997) and core-collapse supernovae (Woosley & Weaver, 1995; Woosley et al, 1995) are discussed as possible source candidates for interstellar ^{26}Al.

The reconstructed γ-ray intensity distribution correlates best with tracers of massive stars such as thermal free-free emission from the ionised medium (Knödlseder et al., 1999). Due to its long lifetime freshly released ^{26}Al travels quite some distances until its final decay. Therefore the 1.809 MeV line emission is rather diffusive and it was not possible to observe isolated candidate sources. γ^2 Velorum, which is the nearest Wolf-Rayet star, is the only possible source for which significant upper limits on the 1.8 MeV flux could be extracted so far (Oberlack et al., 2000). Because of the close correlation between the galactic distribution of massive stars and the observed 1.8 MeV intensity pattern, aggregations of massive stars such as young open clusters and OB associations seem to be suitable laboratories to study these candidate sources. In particular, the by combined analysis of the measured 1.8 MeV intensity together with observables in other wavelength bands one may gain insight in the physical processes involved. Massive stars not only eject freshly synthesised material into the surrounding ISM but also impart a huge amount of kinetic energy by stellar winds and their subsequent core-collapse supernovae. Furthermore they affect the state of the interstellar medium by emitting a large fraction of their electromagnetic radiation in the photoionizing extreme ultra-violet regime. Therefore correlations between the γ-ray line emission of ^{26}Al and observables being related to one or the other phenomenon of massive stars are expected.

We make use of a detailed population synthesis model in combination with an 1-dimensional thin shell expansion model of evolving superbubbles to study the time evolution of the different emission parameters and their resultant observables. In the following sections we present the model and discuss the predictions and their uncertainties in detail. Finally, we apply the model to OB associations in the Cygnus region and discuss the results for the expected ^{26}Al content of this area.

2. MODELLING THE OB STAR EMISSION

2.1. RELEASE OF RADIOACTIVE ISOTOPES

^{26}Al is synthesised via proton capture onto ^{25}Mg and therefore is a secondary product of the CNO-cycle in massive stars. The efficient production of ^{26}Al requires temperatures of at least 3.5×10^7 K, which are reached by stars initially more massive than 25 $M_\odot$. The production rate increases as the core temperature approaches values of $\sim 10^8$ K and depends strongly on the initial metallicity of the stellar material (Meynet et al., 1997). During later burning stages ^{26}Al is produced by proton captures of secondary protons onto freshly synthesised ^{25}Mg. The main destruction channels at temperatures below 220 keV are the β^+-decay and subsequently proceeding reactions with ^{26}Al as input whereas for the late burning phases the direct decay of the isomeric form of ^{26}Al which is 220 keV above the ground-state destroys the remaining 26Al very efficiently. The internal mixing processes bring freshly synthesised ^{26}Al into areas of the massive star which will be expelled by the strong stellar winds during the Of- and WR-phases. (Meynet et al., 1997) computed a stellar evolution grid with detailed nucleosynthesis of ^{26}Al. Their yields could be sufficiently well described by a power-law fit, which is used in our population synthesis model to describe the ^{26}Al release of Wolf-Rayet stars. The uncertainties in the description of the mixing processes, stellar rotation and mass loss give raise to uncertainties in the expected yields of factors up to 3.

Beside Wolf-Rayet stars core-collapse supernovae are demonstrated to be efficient sources of interstellar ^{26}Al (Woosley & Weaver, 1995; Woosley et al., 1995; Woosley & Heger, 1999). In these events ^{26}Al is produced by hydrostatic and explosive Ne-burning as well as the ν-process. Whereas for the mass range below 30 $M_\odot$ detailed explosive nucleosynthesis models exist, which cover a wide range of input physics including different mixing schemes and stellar rotation, for type Ib/c supernovae this aspect is rather unexplored. The typical type II yields are fit reasonably well by a power-law. In contrast, the convergence of the core-masses for the most massive stars lead to the assumption of a non-mass-dependent 26Al yield, which is supported by the findings of Woosley et al. (Woosley et al., 1995). As in the case of the WR contribution the theoretical uncertainties give raise to an inaccuracy of a factor 2 to 3. In addition, core-collapse supernovae release a large amount of radioactive ^{60}Fe, which might be observed by the γ-ray lines of its daughter nucleus ^{60}Co at 1.173 and 1.332 MeV. Due to its solely production by supernovae and the longer lifetime, which reduces the intensity of the expected γ-rays, the Iron lines are still undetected so far. Nevertheless, the ejection of ^{60}Fe is included in our population synthesis model.

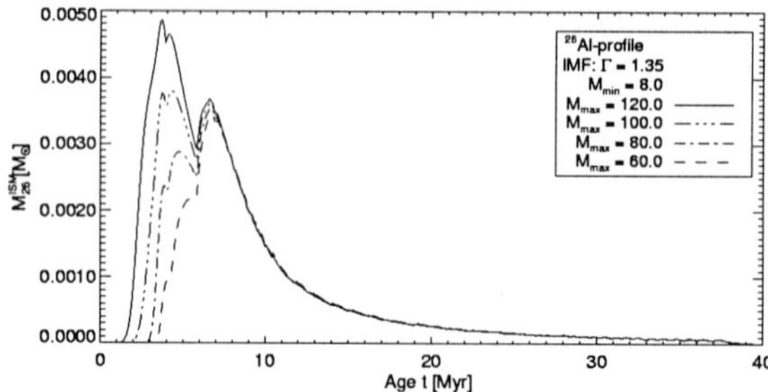

Figure 1 Time evolution of the interstellar mass of ^{26}Al. The curves assume an OB association of 500 stars initially more massive than $8\,M_\odot$ with a parameterised upper mass limit distribute as a Salpeter-IMF (see legend).

Figure 1 shows the time evolution of interstellar content of ^{26}Al for an OB association assuming an instantaneous starburst with Salpeter-IMF. The resulting time profile is convolution of the mass spectrum and the mass- and time-dependent ejection rates of ^{26}Al taking the radioactive decay into account. The profile shows the typical two-peaked structure, where the first peak is due to the WR contribution and the second originates from the SN activity. Usually OB associations have limited population statistics, leaving the observer with a poor sampling of the initial mass spectrum. A Monte Carlo version of our population synthesis code allows the estimation of the uncertainties introduced by this effect. Our finding is, that for populations being initially richer than 100 stars more massive than $8\,M_\odot$, the resulting time profiles are dominated by theoretical uncertainties instead of statistical effects.

An additional source of uncertainty is a possible contribution to the interstellar ^{26}Al content by peculiar massive close binaries. Langer et al. (1998) have shown that some of these systems may give raise to a large overproduction of ^{26}Al in the supernova explosion of secondary. Their contribution may be enhanced by factors up to 1000. Up to now, it is not possible to really incorporate these effects in a population synthesis model due to a lack of a detailed investigation of the appropriate parameter space. Therefore we have to leave this as open question.

2.2. IMPARTING KINETIC ENERGY TO THE ISM

A well known phenomenon of massive stars is their very strong stellar wind imparting a fairly large amount of kinetic energy to the ISM. In addition, their subsequent supernova explosions release on a very short time-scale even larger amounts of energy to the ISM. In OB associations these effects can be expected to overlap and serve as an energy source for blowing large-scale gas structures, known as superbubbles, into the surrounding ISM. Our population synthesis model predicts the flux of kinetic energy as well as matter due to an underlying massive star population. The stellar wind part is modelled using the semi-empirical mass loss rates from the Geneva stellar evolution models (Meynet et al., 1994) in combination with a semi-analytical wind velocity formula from Prinja and colleagues (Horwarth & Prinja, 1989; Prinja et al., 1990). The energy released by a single supernova is assumed to be 10^{51} erg per event. By using a similar scheme as for the nucleosynthesis part our population synthesis model predicts the integral flux of energy and matter from an association.

The integral wind power as well as the flux of matter can only be observed by their impact on the surrounding ISM, therefore we used the results from our population synthesis model as input for an 1-dimensional, numerical model of an expanding supershell. The model is based on the thin shell approximation and incorporates radiative cooling as well as a parameterised description of evaporation from the inner shell or cloudlets which passed the shock and made it into the hot bubble medium. Our description of a possible evaporative poisoning as it was labelled by Shull & Saken (Shull & Saken, 1995) aims on a study of the relevance of this process and uses two parameters, which are the transmission efficiency ϵ_{poros} and the evaporation time-scale τ_{evap}. Depending on the ambient density and pressure evaporation becomes critical for quite low transmission efficiencies of the order of few thousands relative to the swept up mass sitting in the thin shell. This is especially true if the evaporation time-scale is short. If the bubble medium reaches a critical density at sufficiently low temperatures of some 10^6 K the interior energy of the bubble is converted into radiation and the bubble medium is cold down to some 10^4 K rather immediately. If the evaporation time-scale is long compared with the dynamic time-scales then the transmission of some fractions of the ambient medium reduces the mass of the shell at a given time and the bubble expansion stays faster than in cases where no material is transmitted through the shell. In this cases the bubbles become larger. Figure 2 shows the time evolution of the ratio of the observable kinetic energy of an expanding supershell relative to the kinetic energy from source. In this case intermediate values have been chosen for the evaporation parameters.

During the early phase the ratio approaches the canonical value of 20-25% rather quickly, whereas after approximately 15 Myr the ratio drops significantly and approaches values of 5% or less in the late phase. This behaviour depends strongly on the chosen parameter set for the environment as well as evaporation process. If evaporation is efficiently suppressed due to magnetic fields or some other processes the ratio stays near 20% over the whole evolution. But, the critical parameter range for

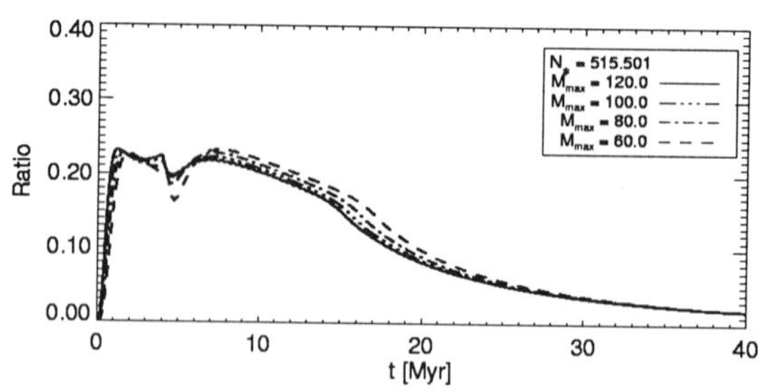

Figure 2 Ratio of observable to injected kinetic energy of an expanding superbubble around an OB association. The surrounding density was chosen to 100 cm-3 whereas the evaporation parameters were set to $\tau_{\mathrm{poros}} = 0.005$ and $\tau_{\mathrm{evap}} = 0.5$ Myr.

efficient cooling and subsequently stalling the expansion are limited by very modest values, which might be typical for interstellar conditions.

The combination of these gas dynamic effects with the emission characteristics for the release of freshly produced radioactive isotopes such as ^{26}Al and ^{60}Fe allows to predict the spatial intensity distribution of the expected γ-ray emission. This is of special importance in the physical understanding why the observed 1.809 MeV intensity distribution correlates so well with strongly blurred tracers. Beside the uncertainties due to the evaporation process and therefore the efficiency of radiative cooling in the early stages, the results of the population synthesis are strongly affected by the uncertainties in the stellar wind models. The Geneva stellar evolution which have been used for consistency reasons apply an enhancement factor of 2 to the mass loss rates during the main-sequence and WR-phase. Therefore the expected wind power is roughly a factor of 2 higher than for typical observed stars. In contrast, an underestimation of the wind velocities of only 30% will already restore the wind power in our model. We therefore estimate our results to be reliable within a factor of 2 or so.

2.3. IONISING RADIATION

As mentioned in the introduction the observed 1.809 MeV intensity pattern correlates very nicely with the galactic free-free emission observed with COBE DMR at 53 GHz (Knödlseder et al., 1999; Bennett et al., 1994). The interpretation of this close correlation points to massive, hot stars being the dominant sources of interstellar ^{26}Al. The free-free emission originates from the ionised portions of the interstellar medium which is found in compact HII regions as well as in from of the diffuse ionised medium. It is strongly believed that in both cases the ionising extreme ultra-violet radiation from young, massive stars is the dominant source of keeping the medium ionised (Dove Shull, 2000). In the framework of our population synthesis model we therefore incorporated the emission of Lyman continuum photons by means of a simple fit-function in dependence of the initial mass of the emitting star.

$$Q_{0/1} = \left[\exp\left(a_1 + \frac{a_2}{M_i} \right) \right] \cdot 10^{49}\,\mathrm{s}^{-1} \tag{31.1}$$

This function is fitted to the Lyman continuum photon fluxes from detailed stellar atmosphere calculations (Vacca et al., 1996). For main-sequence stars the error is less then 5% by using this fit instead of the appropriate model values for the given initial mass. Nevertheless, the error of the resulting time-profile of the Lyman continuum emission of an OB association is considerably larger due to disregarding the effects of stellar evolution. However, a comparison with an alternative EUV emission model based on direct interpolation of the appropriate fluxes from stellar atmosphere models reveals an overestimation of the cumulative Lyc flux in our association model of less than 30% which is of the order as theoretical uncertainties of the underlying stellar models.

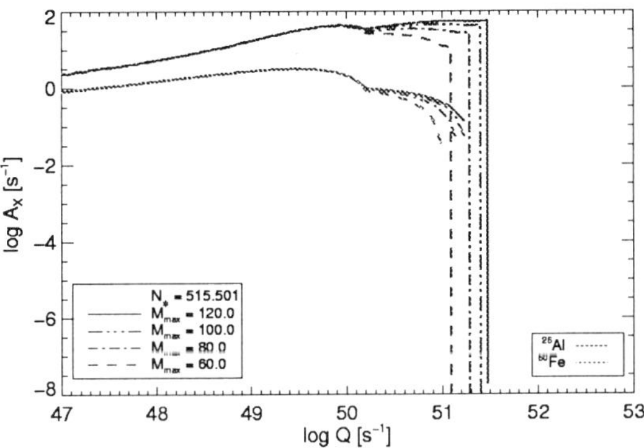

Figure 3 Evolution of the correlation of the Lyman continuum flux vs. the decay rate of interstellar ^{26}Al and ^{60}Fe, respectively

Figure 3 shows the time-evolution of the integral Lyc flux versus the decay rate of interstellar ^{26}Al and ^{60}Fe, respectively, for the same association model discussed earlier. The plotted trajectories start with low decay rates at high EUV luminosities and evolve to considerable decay rates at vanishing Lyc fluxes. In principle, this behaviour could be exploit to construct a very sensitive age indicator for the underlying population. The ratio of the observed 1.809 MeV flux from an astronomical source population and the respective Lyc flux, which in principle could be extracted from the observed free-free intensity by using a proper ionisation model, should be independent of the distance to the population. After an additional normalisation of this ratio relative to the Lyc flux of an O7V star one gets the O7V equivalent yield for an radioactive isotope for a given population (Knödlseder, 1997). Indeed, this quantity can be used as sensitive age indicator as shown in figure 4 for the case of ^{26}Al. Especially in the regime between 3 and 10 Myr the O7V equivalent yield shows a strong increase over 3 orders of magnitude. This is due to the strong decrease of the number of O stars because of supernovae and the

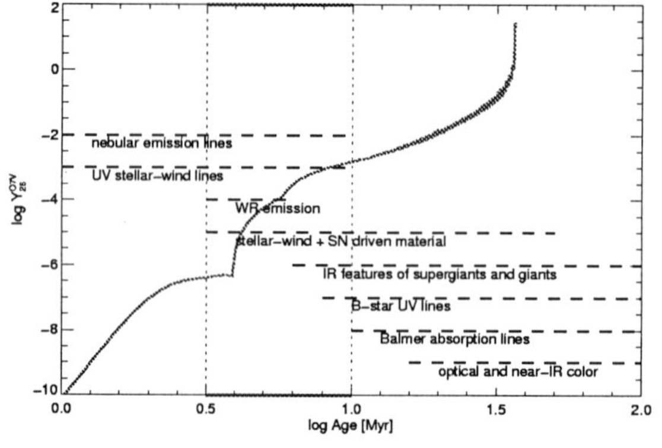

Figure 4 Time evolution of the O7V equivalent yield for ^{26}Al in comparison to the age ranges for other age indicators in astronomical populations.

two emission peaks in the time profile of the interstellar ^{26}Al content (cf. Fig. 1).
Additionally, the released Lyc flux can be used as input to an ionisation model to calculate a prediction map of expected free-free intensity of an area under investigation. This could in turn be compared to the observed one and therefore may give an additional constrain on the real population.

3. STATISTICS

As already discussed in section 2.1 the application of our population synthesis model, which assumes a continuous mass function, to a real astronomical population is expected to be significantly disturbed by population statistics. Whereas for galaxies the mass function is sampled sufficiently dense to justify a quasi-analytic treatment this might not be the case for OB associations or even worth open clusters. We therefore studied the statistical errors as function of the richness by means of Monte Carlo version of the model. Figure 5 shows the spread of the resulting interstellar ^{26}Al masses at time of the emission maximum in dependence of the richness.

For populations richer than 100 stars initially more massive than 8 $M_\odot$ the statistical error drops below the theoretical uncertainties. In addition, the Monte Carlo code allows the determination of probability density functions for direct application of the association model to observed populations by means of a Bayesian analysis.

4. 1.8 MEV FROM CYGNUS

The Cygnus region is the most significant isolated structure in the COMPTEL 1.8 MeV maps beyond the inner galaxy. Figure 6 shows the Cygnus

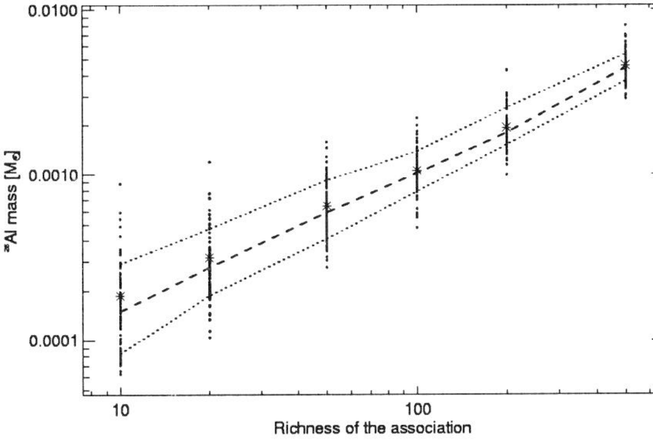

Figure 5 Statistical error of the interstellar ^{26}Al mass as function of the richness of the association

OB associations superimposed as circles on the latest COMPTEL Maximum Entropy image of the Cygnus region (Plüschke et al, 2000).

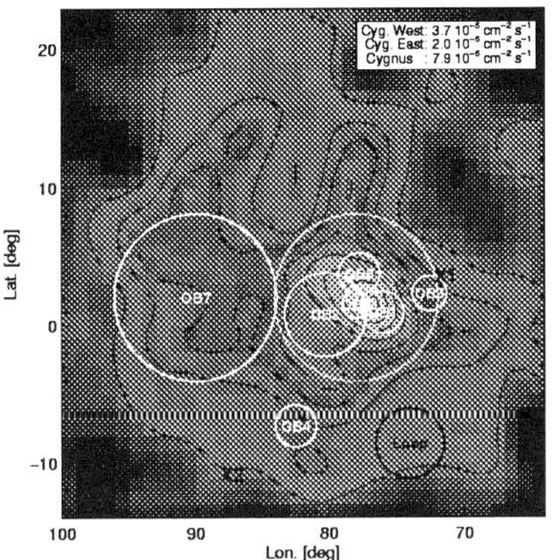

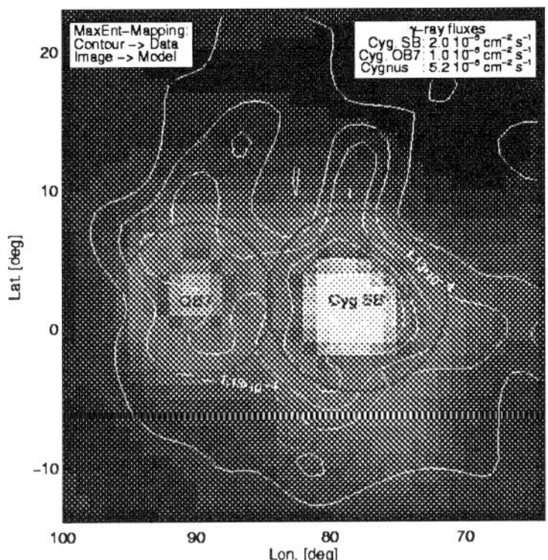

Figure 6 1.809 MeV emission from the Cygnus Region as observed with COMPTEL (Plüschke et al., 2000)

Figure 7 1.809 MeV intensity model based on population synthesis of the Cygnus OB associations

As a first test we applied our model to the given OB associations encircled in Figure 6 and generated an 1.809 MeV intensity model, which for simplicity neglects the gas dynamic effects discussed in section 2.2. The population synthesis is based on a database built from recent literature on population studies of the Cygnus region. Figure 7 shows the resulting intensity model. At first glance this model already reproduces the observed emission peak and the extension to higher longitudes. But, significant discrepancies are still remaining. First, the overall flux is underestimated by the model. After a variation of the slope of the initial mass spectrum turned out to be insignificant in resolving this problem, the most plausible explanation is an underestimation of populations due to strong visual extinction towards Cygnus. Second, the modelled intensity distribution is significantly narrower than the observed pattern, which could be understood if one reminds the neglect of gas dynamic effects due to bubble formation.

5. SUMMARY

We have presented an extension of population synthesis studies to the domain of γ-ray line astronomy. The discussed theoretical uncertainties are far from being insignificant. In addition, it is questionable if the observation and modelling of OB associations as laboratories for testing specific aspects of the theoretical understanding of the chemical evolution leading to interstellar enrichment of radioactive isotopes are sufficient due to possible tremendous statistical uncertainties. For stellar populations of massive stars being richer than 100 objects our model shows that theoretical uncertainties begin to dominate. So, one can conclude that indeed areas with rich associations may be used for testing theoretical aspects. At least, the observations can be used to check the consistency by extracting constrains from combined analyses at different wavelengths.

It was shown explicitly that the combined analysis of free-free emission and the observed 1.809 MeV intensity may serve as a sensitive age indicator for young stellar populations. Furthermore, the rather simplistic application of the model to the 1.809 MeV emission in the Cygnus region already showed the potential in extracting constrains from the γ-ray line observations. However, due to the neglect of additional possible source candidates such as peculiar massive close binary systems these constrains are not very stringent up to now.

References

Schönfelder, V., et al.: 1993, ApJS 86, 657
Oberlack, U., et al.: 1996, A&AS, 120, 311
Plüschke, S., et al.: 2000, accepted for publ. in 'Proc. of the 4th INTEGRAL Workshop', Alicante/Spain, publ. by ESA
Mahoney, W.A., et al.: 1984, ApJ, 286, 578
Mowlavi, N. & Meynet, G.: 2000, A&A, 361, 959
Meynet, G., et al.: 1997, A&A, 320, 460
José, J., et al.:1997, ApJ, 479, L55
Woosley, S. & Weaver, T.: 1995, ApJS, 101, 181
Woosley, S., et al.: 1995, ApJ, 448, 315
Woosley, S. & Heger, A.: 1999, "Proc. of Astronomy with Radioactivities II", edt. by R. Diehl & D. Hartmann, MPE Rep. 274, 133

Knödlseder, J., et al.: 1999, A&A, 344, 68
Oberlack, U., et al.: 2000, A&A, 353, 715
Langer, N., et al.: 1998, "Proc. of the 9th Workshop on Nuclear Astrophysics", edt. by W. Hillebrandt & E.Müller, MPA Garching, p. 18+
Meynet, G., et al.: 1994, A&AS, 103, 97
Horwarth, I. & Prinja, R.: 1989, ApJS, 69, 527
Prinja, R., et al.: 1990, ApJ, 361, 607
Shull, J.M. & Saken, J.: 1995, ApJ, 444, 663
Bennett, C.L., et al.: 1994, ApJL, 464, L1
Dove, & Shull, J.M.: 2000, ApJ, 531, 846
Vacca, W., et al.: 1996, ApJ 460, 914
Knödlseder, J.: 1997, PhD Thesis Univ. Toulouse
Plüschke, S., et al.: 2000, accepted for publ. in "Proc. of the 4th INTEGRAL Workshop", Alicante/Spain, publ. by ESA

BINARY POPULATION SYNTHESIS: METHODS, NORMALIZATION, AND SURPRISES

Vassiliki Kalogera
Harvard-Smithsonian Center for Astrophysics, Cambridge, MA 02138, USA
vkalogera@cfa.harvard.edu

Krzysztof Belczynski
Harvard-Smithsonian Center for Astrophysics, Cambridge, MA 02138, USA; Nicolaus Copernicus Astronomical Center, 00-716 Warszawa, Poland
kbelczynski@cfa.harvard.edu

Keywords: Binary populations, supernova rates, close double neutron stars

Abstract In this paper we present a brief overview of population synthesis methods with a discussion of their main advantages and disadvantages. In the second part, we present some recent results from synthesis models of close binary compact objects with emphasis on the predicted rates, their uncertainties, and the model input parameters the rates are most sensitive to. We also report on a new evolutionary path leading to the formation of close double neutron stars (NS), with the unique characteristic that none of the two NS ever had the chance to be recycled by accretion. Their formation rates turn out to be comparable to or maybe even higher than those of recycled NS–NS binaries (like the ones observed), but their detection probability as binary pulsars is much smaller because of their short lifetimes. We discuss the implications of such a population for gravitational-wave detection of NS–NS inspiral events, and possibly for gamma-ray bursts and their host galaxies.

1. INTRODUCTION

Binary systems with compact objects (neutron stars or black holes) have provided us with some of the most surprising stellar configurations discovered in galaxies: from high– and low–mass X-ray binaries, persistent or transient and mini–quasars to binary millisecond radio pulsars and double neutron stars. For decades a lot of effort has been devoted to

D. Vanbeveren (ed.), The Influence of Binaries on Stellar Population Studies, 447–461.

understanding the evolutionary history of these systems, identifying the dominant formation processes and factors that determine their properties. This multi–faceted research effort has followed two main directions. One includes studies of specific observed systems with the goal to understand their characteristics and origin, and provide us with clues about the general population properties. Although it is possible to tailor an evolutionary model to reproduce the properties of an isolated system, this exercise provides little perspective on whether the putative initial conditions and subsequent tailoring are plausible. A different and more general approach is to model the evolution of an entire ensemble of primordial binaries under a common set of assumptions. Such *population synthesis* models (e.g., Dewey & Cordes 1987; Politano 1988; Lipunov & Postnov 1988; de Kool 1992; Kolb 1993; and many more) can be very useful in analyzing the statistical properties of the population under study and allows comparisons to observed samples of objects.

In the first part of the paper, we discuss in some detail two different methods used in binary population synthesis calculations and the reasons for choosing one or the other (§ 2). In the second part, we focus on some recent results we have obtained from studies of the formation of X-ray binaries and double compact objects, without attempting at any level to review this area of research (§ 3, 4). In our discussion of the results we focus on issues related to the absolute normalization of models and predicted formation rates, ways of constraining them and their sensitivity to model input parameters. In § 4, we report on a new class of double neutron star systems with implications for gravitational–wave detection and gamma-ray bursts.

2. POPULATION SYNTHESIS METHODS

The basic goal in population synthesis calculations is to follow the evolution of an ensemble of primordial binaries through all possible evolutionary phases until the formation of the systems of interest. Typical examples of such phases include: wind mass loss, stable or unstable and conservative or non–conservative mass and angular momentum transfer phases, formation of compact objects with associated mass loss and supernova kicks, circularization, angular momentum loss through gravitational radiation and/or magnetic braking. The details of some of these phases is not well understood, but in general the choice of prescriptions and assumptions about their treatment in the models is guided by detailed hydrodynamic or stellar evolution studies. Most often for the modeling of stellar evolution and the physical processes involved, analytic approximations adequate for statistical estimates are employed.

In what follows we discuss in some more detail two very different methods that have been used so far in population synthesis calculations.

2.1. EVOLUTION OF DISTRIBUTION FUNCTION IN PHASE SPACE

A class of semi-analytical methods have been employed for population synthesis calculations in various flavors over the years (e.g., Kolb 1993; Politano 1996; Kalogera & Webbink 1998; Kalogera 1998). These methods are based on the idea of evolving a distribution function through phase space using Jacobian transformations. In the specific case of binaries, the "phase space" is the space of binary parameters, i.e., masses, orbital separations, and eccentricities. A distribution function describing the population of binaries at the beginning of a given evolutionary phase, $F(M_1, M_2, A, e)$, can be transformed into another function $F'(M_1', M_2', A', e')$ describing the binaries at the end of this phase. The transformation can be performed using Jacobians (involves partial derivatives of final binary parameters with respect to initial parameters) and provided that the functional relationships between final and initial quantities are known:

$$F\left(M_1', M_2', A', e'\right) \;=\; F'\left(M_1, M_2, A, e\right)\left[\mathcal{J}\left(\frac{\mathcal{M}_1', \mathcal{M}_2', \mathcal{A}', \mathrm{e}'}{\mathcal{M}_1, \mathcal{M}_2, \mathcal{A}, \mathrm{e}}\right)\right]^{-1} . \tag{32.1}$$

In the vast majority of cases in binary evolution, these relationships are not just known but they are such that the partial derivatives can be calculated analytically! It is often the case that the evolution of a distribution function can be calculated analytically through a whole sequence of evolutionary phases (e.g., wind mass loss, common envelope, stable mass transfer, supernova explosions). Some of the mathematical details of such cases have been derived and discussed, for example, by Kalogera (1996) and Kalogera & Webbink (1998). In its general form this computational method is *semi-analytical* because at times it is required that integrations are performed over physical parameters that are not of interest. Typically such integrations can be performed only numerically.

These methods of phase–space evolution have great advantages in providing us with high–accuracy results, high resolution in distributions of binary parameters at low computational costs. The numerical implementation is done at the final stage of evolution, after the binary population has been evolved through a given evolutionary sequence analytically. The grid over which the distribution function is calculated is set up on the actual set of parameters and evolutionary stage of interest and the density of this grid can be chosen to be high enough so that numerical noise is not an issue. This is to be contrasted with the concept of setting up a "grid" on the initial parameters of the population and anticipating that this initial grid will be dense enough to describe well the final properties of the population. In addition to high numerical accuracy and low computational cost, the semi-analytical population synthesis method offers physical insight to how the final properties of the population depend on the properties of their progenitors and on the physical parameters that enter the description of the various evolutionary phases. This invaluable physical insight originates from the analytical derivation of the distribution functions in each evolutionary stage and it allows us to identify the dominant physical quantities that dictate the final results. Often it also allows us to derive analytical, strict limits of physical properties (e.g., Kalogera 1996; Kalogera & Lorimer 2000) or functional dependencies between parameters.

There is however one major disadvantage to the phase-space evolution method: it can be used only if the exact evolutionary sequence is known *a priori*. In other words, the calculation of the evolution (in practice of the Jacobian transformations) has to be "tailored" to a specific path and, for populations of systems that are formed through a multitude of evolutionary paths, the analysis has to be repeated for each one of them. Another consequence is that population studies performed solely with this method cannot lead to the discovery of new unexplored channels of evolution.

2.2. MONTE CARLO TECHNIQUES

Monte Carlo methods have a straightforward application in population synthesis calculations and has been used most often in the study of binary systems (e.g., Dewey & Cordes 1987; Lipunov & Postnov 1988; Portegies–Zwart & Verbunt 1996; Han, Podsiadlowski, & Eggleton 1995; and many others). The basic design relies on the idea of generating a large number of unevolved binary and single stars with physical properties generated according to pre-determined distributions. The evolution of each "Monte Carlo object" is followed through a wide variety of evolutionary stages until the formation of systems of interest (i.e., binary compact objects). The evolution modeling relies on assumptions and prescriptions for relevant physical processes, in a way very similar to phase–space evolution calculations.

Probably the most important disadvantage of Monte Carlo techniques in population synthesis studies is that of statistical accuracy of the results. Even with current computational resources, statistical accuracy can still be a problem for studies of rare populations with relatively low formation rates, such as binary compact objects and low–mass X–ray binaries. Reducing statistical noise in predicted rates of double neutron stars, for example, to less than even 10% requires a total number of primordial binaries in excess of $\sim 10^6$ and poses very serious computational–cost requirements (typically 100 hours on fast workstation). Consequently, achieving a satisfactory statistical accuracy for multi-dimensional distribution of rare populations over physical parameters and not just their rate needs careful examination of the models and top–level computational resources. With respect to gaining physical insight to the properties of the population of interest and what determines them, Monte Carlo techniques are less suitable than phase–space evolution methods. For example, a strict limit on physical characteristics cannot be *rigorously* identified in Monte Carlo simulations. Typically only statistical statements can be made and there is always the question whether a certain area of the parameter space is physically excluded or just underpopulated.

On the other hand, among the most important advantages of Monte Carlo techniques used in population studies are: (i) they are rather simple in their conceptual design and implementation, (ii) numerical calculations can be parallelized in a straightforward manner, (iii) they allow the study of a large number of evolutionary sequences and stellar populations simultaneously, and (iv) they can lead to the discovery of *new*, previously unappreciated or ignored formation channels for objects of interest (see § 4).

3. POPULATION SYNTHESIS OF DOUBLE COMPACT OBJECTS: PREDICTED RATES

In what follows we discuss preliminary results from a Monte Carlo population study of double compact objects, neutron stars (NS) or black holes (BH), forming through a multitude of evolutionary channels (Belczynski, Kalogera, & Bulik 2001). In particular we focus on the absolute normalization of the models and the predicted formation rates of close NS–NS, BH–NS, and BH–BH binaries. Interest in these formation rates arises from the importance of such close systems for gravitational–wave astronomy and possibly for gamma–ray bursts. The late inspiral of close binary compact objects are primary sources for the ground–based laser interferometers currently under construction (such as LIGO, VIRGO, GEO600), and predictions of detection rates depend sensitively on the formation rates of these objects. On the other hand, the question of the origin of gamma–ray bursts (GRB) and their possible connection to compact object mergers can to be addressed to some extent through a comparison of GRB and binary compact object formation rates.

To address the above questions, though, knowledge of theoretical formation rates are essential. Population synthesis models, however, often lack a good calibration and predicted merger rates available in the literature tend to cover a large range (many orders of magnitude; for a recent review see Kalogera 2001). Therefore, it is important to use a number of ways to better constrain the absolute normalization of synthesis models.

3.1. OBSERVATIONAL CONSTRAINTS

For the calibration of the population synthesis models for double compact object formation presented here, we have used a number of independent constraints on stellar populations derived from observations: (i) rate of Type II supernovae thought to be core collapse events of hydrogen–rich massive stars, (ii) rate of Type Ib/c thought to be core collapse events of hydrogen–poor stars, (iii) star formation history for our Galaxy, (iv) empirical merger rates of NS–NS binaries obtained based on the observed NS–NS sample. Additional constraints can also be obtained from (i) observed frequency of helium stars in isolation and in binaries, (v) frequency and lifetime of compact object binaries with helium stars, and (vi) formation rates of X–ray binaries.

This list of constraints includes populations and events that are directly related to the formation of double compact objects but also some others that are not, but their formation can be followed (as side products) in population synthesis calculations. We use the recent determination of supernova rates of Cappellaro, Evans, & Turatto (1999). The derived rates for Type II and Ib/c supernovae for a galaxy like the Milky Way lead to a rate ratio (II/Ibc) of 6 ± 4. Estimates of the Milky Way star formation rate lie in the range $1 - 3\,\mathrm{M}_{\odot}\,\mathrm{yr}^{-1}$ (Blitz 1997; Lacey & Fall 1985). Empirical estimates of the NS–NS merger rate when taking all uncertainties into account see to lie in the range $10^{-6} - 10^{-4}\,\mathrm{yr}^{-1}$ (Kalogera et al. 2000).

3.2. MODEL ASSUMPTIONS

Here, we give a brief description of our population synthesis code. More details about the treatment of various evolutionary processes will be presented in Belczynski, Kalogera, & Bulik (2001).

To describe the evolution of single or non–interacting binary stars (hydrogen– and helium–rich) from the zero age main sequence (ZAMS) to carbon–oxygen (CO) core formation, we employ the analytical formulae of Hurley, Pols, & Tout (2000), whose results are in good agreement with earlier stellar models (e.g., Schaller et al. 1992). To calculate masses of compact objects formed at core–collapse events, we have adopted a prescription based on the relation between CO core masses and final FeNi core masses (Woosley 1986). Our progenitor–remnant mass relation is in agreement with the results of Fryer & Kalogera (2001) based on hydrodynamic calculations of core collapse of massive stars.

Concerning the evolution of interacting binaries, we model the changes of mass and orbital parameters (separation and eccentricity) taking into account mass and angular momentum transfer between the stars or loss from the system during Roche–lobe overflow, tidal circularization, rejuvenation of stars due to mass accretion, wind mass loss from massive and/or evolved stars, dynamically unstable mass transfer episodes leading to common–envelope (CE) evolution and spiral–in of the stars. We extend the usual treatment of CE evolution based on energy considerations (Webbink 1984) to include cases where both stars have reached the giant branch and have convective envelopes (hydrogen or low–mass helium stars). As suggested by Brown (1995) for hydrogen–rich stars, we expect the two cores to spiral–in until a merger occurs or the combined stellar envelopes are ejected. We also account for the *possibility* that compact objects accrete mass during CE phases (following Brown 1995). At NS formation, we model the effects of asymmetric supernovae (SN) on binaries, i.e., mass loss and natal kicks, for both circular and eccentric orbits (e.g., Kalogera 1996; Portegies–Zwart & Verbunt 1996). We assume that kicks are isotropic with a given magnitude distribution.

In the synthesis calculations, we evolve a population of primordial binaries and single stars through a large number of evolutionary stages, until *coalescing* double compact objects are formed (merger times $<$ 10 Gyr). The total number of binaries (typically a few million) in each simulation is determined by the requirement that the statistical (Poisson) fractional errors ($\propto 1/\sqrt{N}$) of the final populations are lower than 10%.

In our standard model, the properties of primordial binaries follow certain assumed distributions: for primary masses ($5 - 100\,\mathrm{M}_\odot$), $\propto M_1^{-2.7} dM_1$; for mass ratios ($0 < q < 1$), $\propto dq$; for orbital separations (from a minimum, so both ZAMS stars fit within their Roche lobes, up to $10^5\,\mathrm{R}_\odot$), $\propto dA/A$; for eccentricities, $\propto 2e$. Each of the models is also characterized by a set of assumptions, which, for our standard model, are:

- *Kick velocities.* We use a weighted sum of two Maxwellian distributions with $\sigma = 175\,\mathrm{km\,s^{-1}}$ (80%) and $\sigma = 700\,\mathrm{km\,s^{-1}}$ (20%)

(Cordes & Chernoff 1997); Black–hole kicks are weighted based on the amount of fallback mass during their formation.

- *Maximum NS mass.* We adopt a conservative value of $M_{\rm max} = 3\,{\rm M}_\odot$ (e.g., Kalogera & Baym 1996). It affects the relative fractions of NS and BH and the outcome of NS hyper–critical accretion in CE phases.

- *Common envelope efficiency.* We assume $\alpha_{\rm CE} \times \lambda = 1.0$, where α is the efficiency with which orbital energy is used to unbind the stellar envelope, and λ is a measure of the central concentration of the giant.

- *Non–conservative mass transfer.* In cases of dynamically stable mass transfer between non–degenerate stars, we allow for mass and angular momentum loss from the binary (see Podsiadlowski, Joss, & Hsu 1992), assuming that half of the mass lost from the donor is also lost from the system ($1 - f_{\rm a} = 0.5$) with specific angular momentum equal to $\beta 2\pi {\rm A}^2/{\rm P}$ ($\beta = 1$).

- *Star formation history.* We assume that star formation has been continuous in the disk of our Galaxy for the last 10 Gyr (e.g., Gilmore 2001).

An extensive parameter study is essential in assessing the robustness of population synthesis results. Apart from our standard case, we examine the results for 25 additional models, where we vary all of the above parameters within reasonable ranges. The complete set of models and the assumptions that are different from our standard choices are:

- A: standard model

- B1–7: zero kicks, single Maxwellian with $\sigma = 50, 100, 200, 300, 400\,{\rm km\,s^{-1}}$, "Paczynski" kicks with $\sigma = 600\,{\rm km\,s^{-1}}$

- C: no hyper–critical accretion onto NS and BH in CEs

- D: maximum NS mass: $M_{\rm max} = 2, 1.5\,{\rm M}_\odot$

- E1–6: $\alpha_{\rm CE} \times \lambda = 0.1, 0.25, 0.5, 0.75, 2, 3$

- F1–4: mass fraction accreted: ${\rm f_a} = 0.1, 0.25, 0.75, 1$

- G1–2: specific angular mom. $\beta = 0.5, 2$

- H: primary mass: $\propto M_1^{-2.35}$

- I1–2: mass ratio: $\propto q^{-2.7}$, $\propto q^{3.0}$

3.3. MOST IMPORTANT MODEL PARAMETERS AND THEIR EFFECTS

In assessing the sensitivity of our results to the model input parameters varied as described in § 3.2, we examine the behavior of (i) the calculated ratio of Type II to Type Ib/c rates, and (ii) the formation rates of three types of coalescing double compact objects: NS–NS, BH–NS, and BH–BH.

We find that the SN rate ratio is by far most sensitive to the assumed binary fraction for the stellar population. In agreement with earlier results (De Donder & Vanbeveren 1998), the ratio increases with decreasing binary fraction, primarily because the majority of Type Ib/c supernovae occur in binary systems, where loss of the hydrogen envelope is facilitated. This behavior tentatively points to a low binary fraction for our Galaxy (about 25%), or conversely to the galaxies used to determine the SN rates having a binary fraction lower than that in the Milky Way. We should note however, that the predicted SN rate ratios are consistent with the observational estimates within the one-sigma error bar.

Variations of the mass ratio distribution appear to be the most important factor for the predicted coalescence rates. A distribution that strongly favors high mass ratios leads to an increase of the NS–NS rate by a factor of about 30 relative to the rate in the case that low mass ratios are strongly favored. The BH–NS and BH–BH rates are less sensitive and show variations by factors of about 3 and 6, respectively.

The common–envelope efficiency is well known to be an important model parameter in population synthesis. Rates of NS binaries decrease by about an order of magnitude, for a similar decrease of the product $\alpha_{\rm CE}\lambda$, which leads to an increased rate of merger prior to the formation of the binaries of interest. However, the effect on the coalescence rate of BH–BH binaries is non-monotonic. The reason is that these binaries tend to form with wide orbits. Decreasing the CE efficiency initially causes the rate to *increase*, as more tight binaries are allowed to form. With further decrease of the efficiency though, the early merger rate dominates and the formation rate of BH–BH binaries drops.

Last, the effect of kicks (imparted to NS but also to BH with smaller magnitudes) affects the double compact object formation in many different ways: supernova survival, formation of tight binaries, mergers. The formation typically peak for $\sigma = 100 - 200\,{\rm km\,s^{-1}}$ and decrease with increasing average kick magnitudes. The rate of close BH–NS systems is found to vary the most in our parameter study by a factor of about 3.

4. NEW CLASS OF CLOSE NS–NS: WITHOUT NS RECYCLING

Empirical rate estimates of NS–NS mergers, are derived under an important underlying assumption that all NS–NS binaries have contained a rather long–lived radio pulsar at some point in their lifetime (see Kalogera et al. 2000). However, in our very recent Monte Carlo population synthesis calculations (Belczynski & Kalogera 2000), we discovered a *new* evolutionary path leading to NS–NS formation that does not involve recycling of any of the two neutron stars. The implication of these findings is that there may be a significant population of old NS–NS binaries that could not be detected as binary pulsars. We use the calculated formation rate of non–recycled NS–NS relative to that of recycled NS–NS systems to derive upward revisions for the rates estimates based on the current observed sample.

4.1. MODEL CALCULATIONS AND RESULTS

We use our population synthesis models to investigate all possible formation channels of NS–NS binaries realized in the simulations. We find that a significant fraction of *coalescing* NS–NS systems are formed through a new, previously not identified evolutionary path.

In Figure 1 we describe in detail the formation of a typical NS–NS binary through this new channel. The evolution begins with two phases of Roche–lobe overflow. The first, from the primary to the secondary, involves non–conservative but dynamically stable mass transfer (stage II) and ends when the hydrogen envelope is consumed. The second, from the initial secondary to the helium core of the initial primary, involves dynamically unstable mass transfer, i.e., CE evolution (stage IV). The post–CE binary consists of two bare helium stars of relatively low masses. As they evolve through core and shell helium burning, the two stars acquire "giant–like" structures, with developed CO cores and convective envelopes (e.g., Habets 1987). Their radial expansion eventually brings them into contact and the system evolves through a double CE phase (stage VI; similar to Brown (1995) for hydrogen–rich stars). During this double CE phase, the combined helium envelopes are ejected at the expense of orbital energy. The tight, post–CE system consists of two CO cores, which eventually end their lives as Type Ic supernovae. The survival probability after the two supernovae is quite high, given the tight orbit before the explosions. The end product in this example is a close NS–NS with a merger time of $\simeq 5\,\mathrm{Myr}$ (typical merger times are found in the range $10^4 - 10^8\,\mathrm{yr}$).

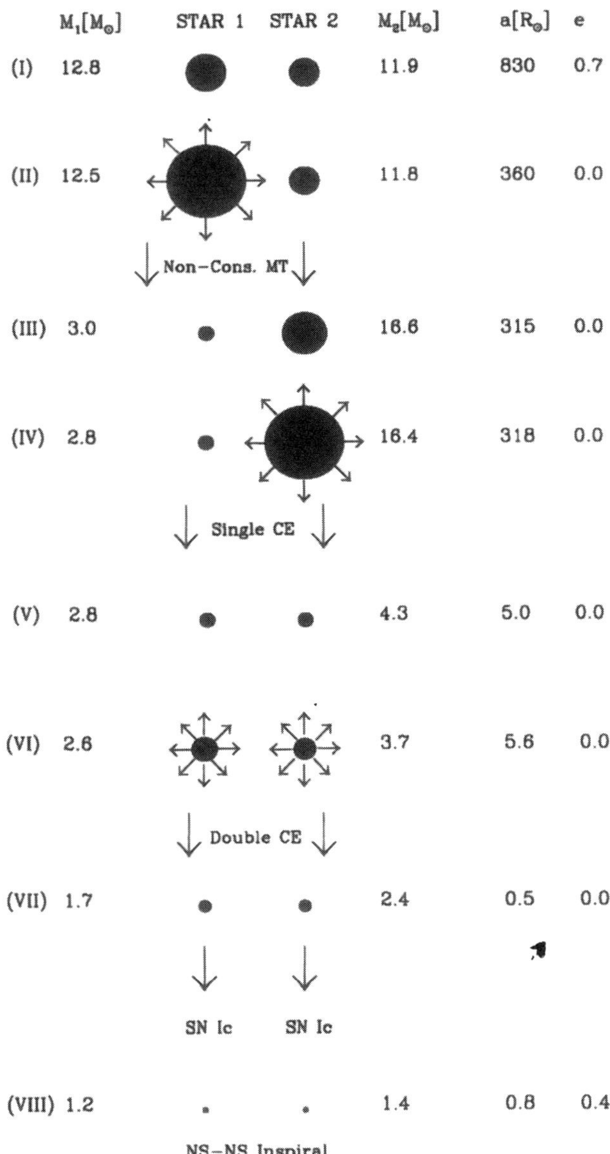

Figure 1 Stages of the new non–recycled NS–NS formation path: (I) Zero Age Main Sequence, (II) star 1 fills its Roche lobe and non–conservative mass transfer begins, (III) at the end of stage II the helium core of Star 1 is exposed, (IV) star 2 fills its Roche lobe on the giant branch leading to dynamically unstable mass transfer and CE evolution, (V) at the end of stage IV the helium core of Star 2 is exposed, (VI) both helium stars fill their Roche lobes on the giant branch, leading to a double CE phase, (VII) at the end of stage VI, the binary consists of two bare CO cores in a tight orbit (VIII) after two subsequent supernovae and 20 Myr since ZAMS a close NS–NS binary forms with a merger time of about 5 Myr. (From Belczynski & Kalogera 2000.)

The unique qualitative characteristic of this NS–NS formation path is that both NS have avoided recycling. The NS progenitors have lost both their hydrogen and helium envelopes prior to the two supernovae, so no accretion from winds or Roche–lobe overflow is possible after NS formation. Consequently, these systems are detectable as radio pulsars only for a time ($\sim 10^6$ yr) much shorter than recycled NS–NS pulsar lifetimes ($\sim 10^8 - 10^{10}$ yr in the observed sample). Such short lifetimes are of course consistent with the number of NS–NS binaries detected so far and the absence of any *non–recycled* pulsars among them.

Given the uncertain absolute normalization of population synthesis models, we focus primarily on the formation rate of non–recycled NS–NS binaries *relative* to that of recycled pulsars, formed through other, qualitatively different evolutionary paths. Based on this comparison, for each of our models, we derive a correction factor for empirical estimates of the Galactic NS–NS coalescence rate. Since these estimates are derived based on the observed sample, they can account only for NS–NS systems with recycled pulsars, and they must be increased to include any non–recycled systems formed. For the majority of the examined models, we find this correction factor to lie in the range $1.5 - 3$ although for some models it can be as high as 10 or even higher (for more details see Belczynski & Kalogera 2000).

As already mentioned, the realization of the newly identified path through a double helium CE phase depends on the final stages of a helium star evolution. It has long been known that low mass helium stars, after core helium exhaustion, expand significantly and develop a "giant-like" structure with a clearly defined core and a convective envelope (Habets 1987; Avila-Reese 1993; Woosley, Langer, & Weaver 1995; Hurley et al. 2000). We further examined in detail models of evolved helium stars (Woosley 1997, private communication) and found that helium stars below $4.0\,M_\odot$ have deep convective envelopes and that slightly more massive helium stars ($\sim$ 4–4.5 $M_\odot$) still form convective envelopes although shallower. Evolved stars with convective envelopes, overfilling Roche lobes in binary systems, transfer mass on a dynamical time scale, and as a consequence CE evolution ensues. The development of CE phases was proposed first in the context of cataclysmic variable formation (Paczynski 1976) and is now supported by detailed hydrodynamic calculations in a variety of binary configurations (e.g. Rasio & Livio 1996; Taam & Sandquist 2000 and references therein). At present no hydro calculations exist for the case of two evolved stars. Based on our basic understanding of CE development, it seems reasonable to expect that, if two stars with convective envelopes are involved in a mass transfer episode, a double core spiral-in can occur leading to double CE ejection (Brown 1995). Based on these earlier calculations, we adopt a maximum helium–star mass for CE evolution (double or single) of $4.5\,M_{\odot}$. The formation rates of both types of NS–NS binaries are somewhat sensitive to this value, because they depend on whether helium stars evolve through CE phases (single and double, for recycled and non-recycled systems, respectively). Reducing the maximum mass to $4\,M_{\odot}$ actually increases the rate correction factor, although by less than 20%, as it reduces the recycled NS–NS rate by a factor larger than the non–recycled rate.

We note that the identification of the formation path for non–recycled NS–NS binaries stems entirely from accounting for the evolution of helium stars and for the possibility of double CE phases, both of which have typically been ignored in previous calculations (with the exception of Fryer et al. 1999, although formation paths for non–recycled NS–NS were not discussed).

4.2. DISCUSSION

We have identified a new possible evolutionary path leading to the formation of close NS–NS binaries, with the unique characteristic that both NS have avoided recycling by accretion. The realization of this path is related to the evolution of helium–rich stars, and particularly to the radial expansion and development of convective envelopes (on the giant branch) of low–mass ($\lesssim 4.5\,M_{\odot}$) helium stars. We find that a significant fraction of coalescing NS–NS form through this new channel, for a very wide range of model parameters. In some cases, the non–recycled NS–NS systems strongly dominate the total close NS–NS population.

Since both NS are non–recycled, their pulsar lifetimes are too short (by $\sim 10^3$), and hence their detection probability is negligible relative to recycled NS–NS. However, intermediate progenitors of these systems may provide evidence in support of the evolutionary sequence. Examples are close binaries with two low–mass helium stars or a helium star with an O,B companion. Although there are selection effects against their detection too (e.g., short lifetimes, high–mass helium stars are brighter, broad spectral lines due to winds, high luminosity contrast between binary members, etc.), it has been estimated that so far only about 1% of binary helium stars have been detected (Vrancken et al. 1991). Therefore, it is reasonable to expect that the observed sample will increase in future years as observational techniques improve. It is worth noting that theoretical calculations indicate that explosions of low–mass helium stars formed in binaries reproduce the light curves of type Ib supernovae (Shigeyama et al. 1990).

The results of our population modeling show that, if one accounts for the formation of non–recycled NS–NS binaries, the total number of coalescing NS-NS systems could be higher by factors of at least 50%, and up to 10 or even higher. Such an increase has important implications for prospects of gravitational wave detection by ground–based interferometers. Using the results of Kalogera et al. (2000) on the empirical NS–NS coalescence rate, we find that their *most optimistic* prediction for the LIGO I detection rate could be raised to at least 1 event per 2–3 years, and their *most pessimistic* LIGO II detection rate could be raised to 3–4 events per year or even higher.

Our results also have important implications for gamma–ray bursts (GRBs), if they are associated with NS–NS coalescence. We find that the typical merger times of the non–recycled NS–NS are considerably shorter than those of recycled binaries, whereas their center–of–mass velocities are higher. The balance of these two competing effects could alter the current consensus for the location of GRB progenitors relative to their host galaxies (e.g., Belczynski, Bulik, & Rudak 2000; Bloom, Kulkarni, & Djorgovski 2000; Fryer et al. 1999).

Acknowledgments

Support is acknowledged by the Smithsonian Institution through a CfA Predoctoral Fellowship to KB and a Clay Fellowship to VK, and by a Polish Nat. Res. Comm. (KBN) grant 2P03D02219 to KB.

References

Avila-Rees, V. A. 1993, Rev. Mex. Astron. Astrofis., 25, 79

Belczynski, K., Bulik, T., & Rudak, B. 2000, A&A, submitted [astro-ph/0011177]

Belczynski, K., Kalogera, V. 2000, ApJ Letters, submitted [astro-ph/0012172]

Belczynski, K., Kalogera, V. & Bulik T., 2001 in preparation

Blitz, L. 1997, in "CO: Twenty–Five Years of Millimeter-Wave Spectroscopy", eds. W.B. Latter, et al. (Kluwer Academic Publishers), 11

Bloom, J.S., Kulkarni, S.R., & Djorgovski, S.G. 2000, ApJ, submitted [astro-ph/0010176]

Brown, G.E. 1995, ApJ, 440, 270

Cappellaro, E., Evans, R., & Turatto, M. 1999, ApJ, 351, 459

Cordes, J., & Chernoff, D.F. 1997, ApJ, 482, 971

de Donder, E., & Vanbeveren, D. 1998, A&A, 333, 557

de Kool, M. 1992, A&A, 261, 188

Dewey, R. J., & Cordes, J. M. 1987, ApJ, 321, 780

Fryer, C.L., & Kalogera, V. 2001, ApJ, in press [astro-ph/9911312]

Fryer, C. L., Woosley, S. E., & Hartmann, D. H. 1999, ApJ, 526, 152
Gilmore, G. 2001, to appear in "Galaxy Disks and Disk Galaxies", eds. J.G. Funes & E.M. Corsini (San Francisco: ASP)
Habets, G. M. H. J. 1987, A&A Supplement Series, 69, 183
Han, Z., Podsiadlowski, P., & Eggleton, P.P. 1995, MNRAS, 272, 800
Hurley, J.R., Pols, O.R., & Tout, C.A. 2000, MNRAS, 315, 543
Kalogera, V. 1996, ApJ, 471, 352
Kalogera, V. 1998, ApJ, 493, 368
Kalogera, V. 2001, to appear in *Workshop on Astrophysical Sources for Ground-Based Gravitational Wave Detectors*, ed. J. Centrella
Kalogera, V., & Baym, G.A. 1996, ApJ, 470, L61
Kalogera, V., & Lorimer, D.R. 2000, ApJ, 530, 890
Kalogera, V., Narayan, R., Spergel, D.N., & Taylor, J.H. 2000, ApJ, submitted [astro-ph/0012038]
Kalogera, V., & Webbink, R.F. 1998, ApJ, 493, 351
Kolb, U. 1993, A&A, 271, 149
Lacey, C.G., & Fall, S.M. 1985, ApJ, 290, 154
Lipunov, V. M., & Postnov, K. A. 1988, Astrophysics & Space Science, 145, 1
Paczynski, B. 1976, in "The Structure and Evolution of Close Binary Systems", eds. P. Eggleton, S. Mitton, & J. Whelan, Dordrecht, Reidel, p. 75
Podsiadlowski, P., Joss, P.C., & Hsu, J.J.L. 1992, ApJ, 391, 246
Politano, M. J. 1996, ApJ, 465, 338
Portegies–Zwart, S. F., & Verbunt, F. 1996, A&A, 309, 179
Portegies–Zwart, S. F., & Yungel'son, L. R. 1998, A&A, 332, 173
Rasio, F. A., & Livio, M. 1996, ApJ, 471, 366
Schaller, G., Schaerer, D., Meynet, G., & Maeder, A. 1992, A&A Supplement Series, 269, 331
Shigeyama, T., Nomoto, K., Tsujimoto, T., & Hashimoto, M. 1990, ApJ, 361, L23
Tamm, R. E., & Sandquist, E. L. 2000, ARA&A, 38, 113
Vrancken, J., DeGreve, J.P., Yungel'son, L., & Tutukov, A. 1991, A&A, 249, 411
Webbink, R.F. 1984, ApJ, 277, 355
Woosley, S.E. 1986, in "Nucleosynthesis and Chemical Evolution", 16th Saas-Fee Course, eds. B. Hauck et al., Geneva Obs., p. 1
Woosley, S. E., Langer, N., & Weaver, T. A. 1995, ApJ, 448, 315

POPULATION SYNTHESIS AND GAMMA-RAY BURST PROGENITORS

Chris L. Fryer
Los Alamos National Laboratories
Theoretical Astrophysics (T-6) Los Alamos, NM 87544
clf@t6-serv.lanl.gov

Keywords: Binary Population Synthesis, Gamma-Ray Bursts, Compact Binaries

Abstract Population synthesis studies of binaries are always limited by a myriad of uncertainties from the poorly understood effects of binary mass transfer and common envelope evolution to the many uncertainties that still remain in stellar evolution. But the importance of these uncertainties depends both upon the objects being studied and the questions asked about these objects. Here I review the most critical uncertainties in the population synthesis of gamma-ray burst progenitors. With a better understanding of these uncertainties, binary population synthesis can become a powerful tool in understanding, and constraining, gamma-ray burst models. In turn, as gamma-ray bursts become more important as cosmological probes, binary population synthesis of gamma-ray burst progenitors becomes an important tool in cosmology!

1. GAMMA-RAY BURST PROGENITORS

One of the fastest moving fields in astronomy today is the study of gamma-ray bursts (GRBs). In the past few years, observations have proven that we are seeing these powerful outbursts at far distances (beyond a redshift of 4) and it is not surprising that more and more scientists are seeing GRBs as a possible probe of our early universe, both to study the star formation history (Fryer, Woosley, & Hartmann 1999) and as a tool to determine cosmological parameters (Lamb & Reichart 2000). But what can observations really allow us to learn about (and with) GRBs? In this article, I will review our current understanding of GRBs and show that, by understanding the progenitor (which requires accurate binary population synthesis), we can not only learn about the engines of GRBs, but also begin to use GRBs as cosmological probes.

D. Vanbeveren (ed.), The Influence of Binaries on Stellar Population Studies, 463–478.

First, let's review what we have learned the past few years about GRBs. The rapid variability of the bursts suggest a mechanism involving a compact object (e.g. a neutron star or black hole), and over 100 models have been suggested using compact objects to power bursts. However, without an estimate of the distance of these objects, astronomers did not know where these bursts were coming from, and hence, had no idea of the energy requirements for these bursts. Redshift measurements (particularly the absorption lines in the GRB afterglow) have placed GRBs (at least the long-duration bursts) beyond the Milky Way. This means that GRBs must be extremely energetic (if isotropic, the predicted energies range from 10^{48} to 10^{54} ergs).

The energies alone prove challenging for theorists, but these outbursts have the additional constraint that most of the energy is emitted in photons with energies above 50 keV. What has become the standard emission mechanism for GRBs, the fireball model (Meszaros & Rees 1992), assumes that an explosion near the black hole produces a relativistic jet and it is the material directed at us (highly doppler shifted) which produces the large number of observed gamma rays. The fireball helps to explain a variety of GRB observations (time variability when the photons are optically thin, high energy of photons, etc.) and is now considered to be *the* emission mechanism for GRBs. Unfortunately, it places an additional constraint on the GRB engine; the engine must only eject a small amount of material (less than $\sim 10^{-5}\,M_\odot$) in the $\sim 10^{51}$ erg outburst.

These observations have caused one class of GRB engines to rise above the others in the past few years: black hole accretion disk (BHAD) GRBs. This class invokes, as a power source, the rapid accretion ($>$ $0.01\,M_\odot$) from a disk onto a black hole. How does one power a fireball outburst from this rapid accretion? As this material accretes, it emits a sizable portion of its binding energy in the form of neutrinos, and these neutrinos can annihilate (preferentially along the disk axis due to the angle dependence of neutrino/anti-neutrino annihilation) and produce a rapidly expanding jet of material. Alternatively, some magnetic field mechanism (akin to the still unknown mechanism which drives active galactic nuclei) could drive a relativistic jet along the disk axis. In either mechanism, the BHAD GRB engine will produce a relativistic jet along the disk axis, and it is this jet which becomes the fireball and ultimately produces the gamma-ray outburst (for a review, see Popham, Woosley, & Fryer 1998).

The merger of double neutron star (DNS) systems is probably the best studied representative of the BHAD GRB class, but a variety of progenitor systems lead to similar conditions: mergers of other compact systems such as binaries consisting of a neutron star and a black hole

(BH/NS) or of a white dwarf and black hole (WD/BH), the inspiral of a compact object into its companion's helium core during common envelope evolution (helium-mergers), and the accretion of material into a black hole in a "failed supernova" (collapsar). It may be that all of these systems produce GRB outbursts, in which case we would like to distinguish the progenitor either through the nature of their gamma-ray bursts, their afterglows, or their host galaxies. Hydrodynamical simulations of the formation of these BHAD systems do allow us to distinguish between different bursts. The accretion disks produced in DNS and BH/NS mergers are more compact than the rest, and it is believed that they will produce the short duration bursts ($< 1\,\mathrm{s}$) whereas the WD/BH mergers, helium-mergers, and collapsars are believed to produce the long-duration GRBs (e.g Popham et al. 1998).

But understanding the progenitor evolution of the GRB engine is likely to provide us much more insight into the differences between each engine. For instance, the afterglows of GRBs provide information about the medium into which the fireball is moving. Livio & Waxman (2000) and Chevalier & Li (2000) found that the afterglows of some long-duration GRBs are consistent with a fireball moving through a wind-swept medium, whereas others are consistent with a medium of constant density. We know that collapsar GRBs are likely to be surrounded by a wind-swept medium, whereas helium mergers are more often surrounded by the constant density of the interstellar medium. Some DNS and BH/NS mergers should occur in the intergalactic medium, and this too will affect their afterglows (and maybe introduce some selection effects). From binary population synthesis studies of GRB progenitors, we also know that some short duration bursts will occur far from their host galaxies, whereas nearly all long-duration will occur within their host. Böttcher & Fryer (2001) have also suggested that, by understanding the nature of the GRB progenitor, one can distinguish between different BHAD GRBs by observations of the iron line emission.

In this paper, I will describe why binary population synthesis studies are not accurate enough to estimate rates of BHAD GRBs. However, binary population synthesis studies provide us with the best handle of GRB progenitors which allows us to determine when and where these bursts occur. This knowledge not only allows us to test our current understanding of GRBs, but also allows us to use GRBs as tracers of star formation, and perhaps, as cosmological probes.

2. FORMATION SCENARIOS AND THEIR UNCERTAINTIES

For each of the different BHAD GRB models, their exist several formation scenarios, and no doubt the number will continue to grow. However, different scenarios are dominated by different uncertainties. In this section, I review the different GRB models and their various formation scenarios. For each formation scenario, I discuss the dominant uncertainties in determining the formation rate of GRBs. Figure 1 summarizes all of the different formation paths of these scenarios (for more details, see Fryer et al. 1999).

2.1. DOUBLE NEUTRON STAR MERGERS

Double Neutron Star mergers are probably the best studied BHAD GRB model (see Fryer et al. 1999 for an extensive list of papers on this particular model) and are believed to produce short-duration GRBs. Despite this intense study, the formation of DNS mergers may not be any better understood than the other BHAD GRB scenarios. The "standard" formation scenario of DNS systems may not produce a significant number of DNS mergers. In addition, the observed DNS binaries may only represent a fraction of the total number of DNS systems.

The standard scenario (Sc. I in Fig. 1) for forming close orbit DNS systems begins with two massive stars. The more massive star (primary) evolves off the main sequence, overfills its Roche-lobe, and transfers mass to its companion (secondary). The primary then evolves to the end of its life, forming a neutron star in a supernova explosion. If the binary remains bound after the supernova explosion, a binary consisting of a neutron star and massive companion is formed. This system passes through an X-ray binary phase which then evolves through a common envelope as the secondary star expands. During this common envelope phase, the neutron star spirals into the massive secondary and the orbital energy released ejects the secondary's hydrogen mantle, forming a neutron star/helium star binary. Mass accretion onto the neutron star primary is alleged to add angular momentum to the neutron star, "recycling" it as a pulsar. After the explosion of the helium star as a supernova, a close DNS binary remains which, in time, merges via gravitational wave emission. This formation scenario ignores what we know about hypercritical accretion (see §3), and it may be that this formation process is more likely to produce BH/NS binaries, not DNS binaries (Sc. V).

Due to concerns about accretion during the inspiral phase, Brown (1995) suggested an alternative scenario in which the initial binary sys-

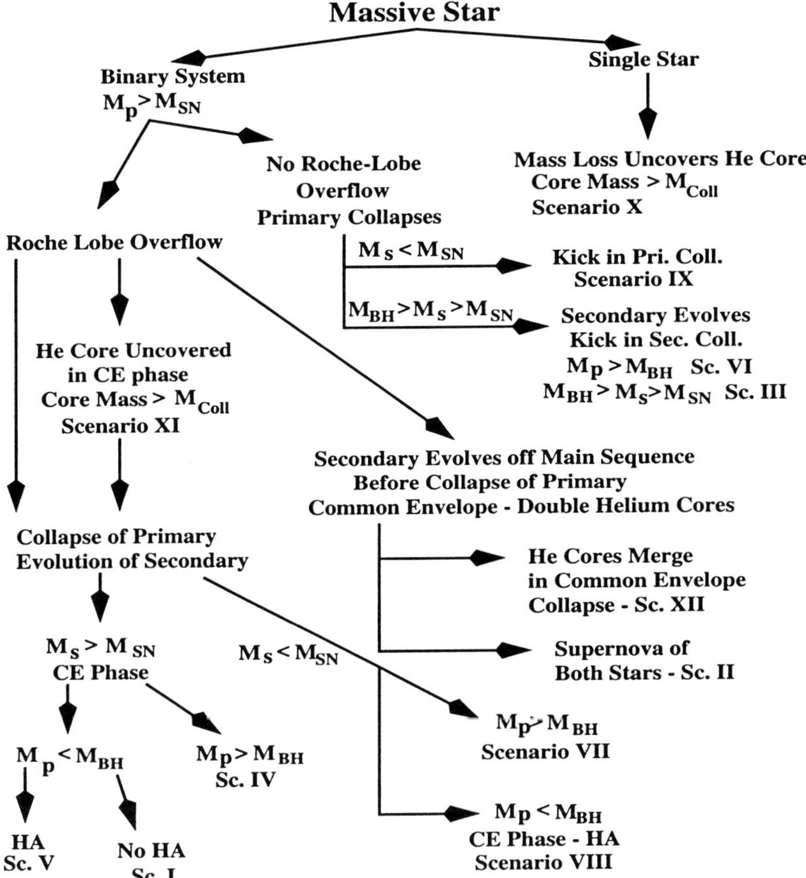

Figure 1 Summary of the BHAD GRB formation scenarios discussed in this paper. Many of the scenarios are quite similar. For instance, scenarios III, VI, and IX are differentiated simply by the initial mass of the companion stars. Scenarios I and IV differ only by the mass of their primary star. Scenarios I and V are identical except that, for the case of scenario I, we assume that photons limit the accretion onto the neutron star and, for scenario V, hypercritical accretion occurs. The helium merger models are not listed here but occur when a compact object (either black hole or neutron star) merges with its helium companion during a common envelope phase. We use the following abbreviations: $M_{\rm p}$, $M_{\rm s}$ are the primary and secondary masses respectively, $M_{\rm SN} \approx 10{\rm M}_\odot$are the critical masses above which a massive stars collapse to form supernovae, $M_{\rm BH} \approx 25{\rm M}_\odot$is the critical mass where massive stars form black holes, $M_{\rm Coll} \approx 40{\rm M}_\odot$is the mass of helium cores above which no explosion occurs and the entire star collapses to form a black hole, CE indicates a common envelope phase, and HA stands for hypercritical accretion.

tem is comprised of two massive stars of nearly equal mass (Sc. II). The secondary (whose mass is assumed to be within 5% of the primary) evolves off the main sequence before the explosion of the primary as a supernova. The two stars then enter a common envelope phase with two helium cores orbiting within one combined hydrogen envelope. After the hydrogen envelope is ejected, first one, then the other, helium stars explode. To generate the observed systems, after the first supernova, the newly formed neutron star must accrete material, possibly from the helium star wind of its companion, to become the observed "recycled" radio pulsar. Since the progenitors of these systems are so close in mass, they may not have time to recycle the pulsar and these "silent" DNS binaries (Fryer et al. 1999) would not show up in any rates derived from observations (but are important as GRB progenitors).

Vicky Kalogera (at this meeting) presented a derivative of this formation scenario in which the helium envelope is also ejected leaving to C/O cores which produce DNS binaries. Such a scenario will produce mostly silent DNS binaries and will depend upon the same uncertainties as Sc. II.

A third scenario avoids any common envelope phase, but requires that the neutron star formed in the explosion of the secondary receive a kick that places it into an orbit that will allow the two neutron stars to merge within a Hubble time. However, in general, to avoid a common envelope phase, the pre-supernova orbital separation must be extremely wide (greater than 1 AU). The range of kick magnitudes and directions which will produce such orbits is so small that, as we shall show in this paper, this formation scenario is probably not important.

2.2. BLACK HOLE + NEUTRON STAR MERGERS

BH/NS mergers are probably the second best studied GRB progenitor. Their GRB outbursts are similar to those of DNS mergers (short duration, etc.) and it will be difficult to distinguish the two from GRB observations. Unfortunately, their formation scenarios are also closely linked, and whether or not a binary star system becomes a DNS or a BH/NS system may require a better understanding of binary star evolution.

The "standard" formation scenario (Sc. IV) for BH/NS systems begins with two massive stars, the primary having a mass large enough that it collapses to form a black hole and not a neutron star ($> M_{\rm BH}$). As with the standard DNS formation scenario, the primary evolves off the main sequence, overfills its Roche-lobe, and transfers mass to its companion. The primary continues to evolve to the end of its life, forming a black hole and, possibly, a supernova explosion. Withour large black hole kicks, the system remains bound, creating a binary consisting of a black hole and a massive star. This system passes through an X-ray binary phase (e.g., Cyg X-1, LMC X-1) and then evolves through a common envelope as the secondary star expands. During this common envelope phase the black hole spirals into the massive secondary and ejects the secondary's hydrogen envelope. After the secondary's supernova explosion, a BH/NS binary forms which then merges via gravitational wave emission.

BH/NS binaries can also form from neutron star binaries in which the neutron star of the primary gains too much matter during common the envelope phase and collapses into a black hole (Sc. V). This scenario occurs where Sc. I (for DNS systems) fails due to hypercritical accretion. It will not produce black hole X-ray binaries and so its formation rate is not limited by the observed black hole X-ray binary fraction and it may dominate (by several orders of magnitude) the total BH/NS merger rate.

Finally, just as with the DNS systems, a BH/NS binary can form under a third mechanism avoiding any common envelope evolution for appropriate kick magnitudes and directions (Sc. VI). Like the DNS kick formation scenario, the kick scenario is probably not important for the total BH/NS formation rate.

2.3. WHITE DWARF + BLACK HOLE MERGERS

WD/BH mergers seem quite similar to the mergers of NS/BH binaries, but the bursts they produce are very different. Because the white dwarf is disrupted much further from the black hole than a neutron star, the disk it forms has much more angular momentum and accretes relatively slowly (0.01-0.05 $\mathrm{M}_\odot\,\mathrm{s}^{-1}$ instead of 0.1-1.0 $\mathrm{M}_\odot\,\mathrm{s}^{-1}$). With magnetic fields, it may produce a long duration burst. WD/NS mergers may also produce GRB outbursts and their rates may well be much higher than BH/WD mergers (Bulik, Belczynski & Zbijewski 1999)

Our first scenario (Sc. VII) begins with a binary consisting of a primary with mass greater than M_{BH} and a low-mass ($< 10\mathrm{M}_\odot$) companion and mimics Sc. IV of NS/BH mergers very closely. After the primary collapses to form a black hole, one has a binary consisting of a black hole and a main sequence star. If the mass of the main sequence companion is less than $\sim 2-3M_\odot$, magnetic braking will drive the binary together, forming a low-mass X-ray binary (LMXB). More massive companions are more likely to form X-ray binaries after they have evolved off the main sequence (e.g. LMC X-3, 2023+338 Nova Cyg, J0422+32 Nova Per). As the star expands off the main-sequence, a common envelope phase can occur which decreases the orbital separation. This occurs preferentially for stars in the mass range of $5 - 10\,M_{\odot}$and these stars eventually evolve into high mass white dwarfs, a fraction of which will merge with the black hole primary in a Hubble time.

Alternatively, these systems may form from neutron star binaries with low-mass ($< 10M_{\odot}$) companions (mimicking Sc. V). If the neutron star enters a common envelope phase when the low-mass companion expands into a giant, it can accrete too much material and collapse to a black hole.

BH/WD binaries may also form without the first common envelope phase (where the low-mass companion spirals into the massive primary) by using a "well-placed" kick (Sc. IX). Unlike the other kick scenarios (Sc. III, VI), this scenario may make a sizable fraction of the BH/WD binaries.

2.4. COLLAPSARS

Collapsars have gained much of the spotlight over the past few years as one of the most successful GRB models. Collapsars predicted that long-duration GRBs should be coupled to supernova explosions (observed in GRB 1998bw) and that GRBs should preferentially form in star forming regions. What is not well known is that a) there are 3 types of collapsar and b) most of the formation scenarios for these three types of collapsar require binaries.

First we'll go through the different collapsar types, and then address their formation scenarios. Type I collapsars were designed to produce the most energetic bursts (MacFadyen & Woosley 1999). They invoked a rotating progenitor which would collapse directly to a black hole (without a supernova explosion). Most GRB fireballs have considerable energy (10^{51} ergs) but eject little matter ($\sim 10^{-5}\,M_{\odot}$). To minimize the amount of matter ejected, Type I collapsars also require that the hydrogen envelope of the massive star has been ejected before collapse. So, the progenitor of a Type I collapsar is a rotating, helium star which is massive enough to collapse directly into a black hole.

But most black holes form in supernova explosions where the fallback is high enough to drive the remnant to a black hole (Fryer 1999). Rotating stars which produce black holes via fallback may also produce GRBs, but the accretion rate through the disk will be much lower and will likely yield weaker GRBs. These models are Type II collapsars (MacFadyen, Woosley & Heger 2001).

In both of these models[1], it is assumed that the hydrogen envelope is lost before collapse, and binaries may well be the most efficient means to eject this envelope. Hence, binary population synthesis can be very important in determining collapsar rates. The formation scenario for both types of collapsar are similar, so we will discuss both together. Note, however, that Fryer et al. (1999) only calculated the rates of Type I collapsars.

In Sc. X, the hydrogen envelope is assumed to be removed through stellar winds. The difficulty with this mechanism is that winds remove the angular momentum of the star, and it is unlikely that stars with such strong winds will be rotating rapidly enough. Also, the winds must lie on a very sensitive boundary: strong enough to remove the entire hydrogen, but not so strong that the core is drastically changed and the star collapses to form a neutron star instead of a black hole. This is particular difficult with Type I collapsars which must collapse directly to a black hole.

Alternatively, a binary companion may remove the primary's hydrogen envelop in a mass transfer phase (Sc. XI). If the hydrogen envelope is removed early, the star may retain much of its angular momentum and rotation is less of a problem in this scenario.

Concerns about the angular momentum distribution led Fryer et al. (1999) to propose a third formation scenario (Sc. XII) akin to Sc. II for DNS mergers. If the secondary evolves off the main sequence before its primary collapses, the two helium cores could go through a joint common envelope phase. If they merge during this phase, the will produce a rapidly rotating helium star. Although the rate for this scenario is low, it has the best chance of actually producing a massive, rotating core which will produce either a Type I or Type II collapsar.

2.5. HE-MERGERS

He-merger GRBs produce outbursts which are very similar to those produced by collapsars. They produce long-duration bursts, preferentially in star forming regions, and depending upon the mass of the helium star, can be the most energetic of all the GRB models. A He-merger GRB is produced when a compact remnant (either neutron star or black hole) spirals into the center of its companion in a common envelope

[1]A third collapsar model also exists, but we will not discuss it here. see Fryer, Woosley, & Heger (2001) for details.

phase (either as the companion evolves into a giant or supergiant) [2]. Unfortunately, the name He-merger often causes people to confuse this progenitor with Sc. XII of collapsars.

The advantage of this formation scenario is that it naturally provides an explanation for the rotation required to form a disk (the rotation comes from the inspiral process). Also, the inspiral process almost always ejects the hydrogen envelope, but it reaches the core of the star so quickly that Wolf-Rayet winds do not eject the helium envelope as well.

3. MAJOR UNCERTAINTIES

With all of the assumptions and uncertainties, binary population synthesis codes begin to look like black boxes with a number of knobs and dials (Fig. 2). But for specific binary systems (and specific questions about these binary systems) many of these dials and knobs do not affect the results dramatically. In this section, I will discuss the uncertainties which have the largest affects upon the formation of BHAD GRBs. These uncertainties include neutron star and black hole kicks, accretion onto neutron stars in common envelope phases, and black hole formation (affected by winds and stellar radii).

3.1. KICKS

The list of evidence for neutron star kicks (velocities imparted to the neutron star at birth) continues to grow (see Fryer et al. 1999 for a review). The distribution of kicks imparted onto neutron stars is generally determined from the observed distribution of pulsar velocities. Unfortunately, extracting the actual 3-dimensional kick from the observed proper motions of nearby pulsars suffers from a range of difficulties, from uncertainties in the effects of binaries and the Galactic potentional to low number statistics and observational systematics and biases. The most recent attempts at determining the *pulsar* velocity distribution lead to the following results: Maxwellian distribution with mean velocity of $\sim 450, 250 - 300\ \mathrm{km\,s^{-1}}$ (Lyne & Lorimer 1994; Hansen & Phinney 1997 respectively), and a bimodal distribution with a mean velocity of $\sim 450\,\mathrm{km\,s^{-1}}$ (Cordes & Chernoff 1998). Fryer, Burrows & Benz (1998) went one step further and found that the neutron star *kick* distribution is also bimodal with a mean velocity of $\sim 450\,\mathrm{km\,s^{-1}}$. When kicks are large, the net effect of binaries on the mean pulsar velocity is minimal. However, binaries do spread the velocity distribution, so the neutron

[2] Alternatively, the compact remnant could be driven into its companion by a kick, but the rate of this scenario is negligible

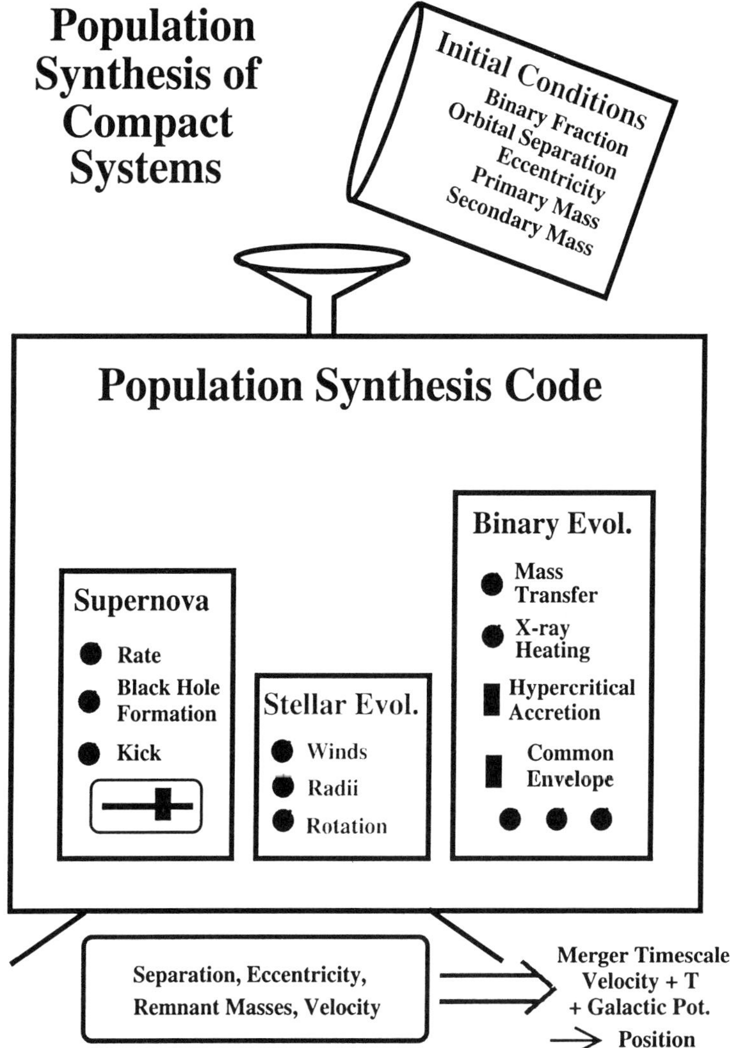

Figure 2 Binary population synthesis machine with some of its free parameters.

star kick distribution may not have the same form as the observed pulsar velocity distribution.

As the number of pulsars with measured proper motions increases, the statistics on pulsar velocities will improve, hopefully settling the question of bimodal kick distributions. With sufficient funding, the number of reliable proper motion measurements could more than double in the next 5-10 years (Cordes - pvt. communication). But even if we can accurately measure the pulsar velocity distribution, there remain a number of unanswered questions about kicks. Do pulsars represent the neutron star population as a whole? Do anomolous X-ray pulsars receive different kicks than neutron stars? What about neutron stars born with low magnetic fields? Unfortunately, no proposed mechanism has been shown to produce large kicks, and theory can not help provide insight into the neutron star kick mechanism.

There is even less data on the kicks imparted to black holes. Given that the majority of black hole systems appear to have low space velocities, Fryer (2000) found that a BH kick of at least $\sim 50\,\mathrm{km\,s^{-1}}$ is required to explain the velocities of the complete set of BH X-ray transients. But with only binary systems to go by (and only 10 of them at that), it is difficult to estimate biases or systematics in the observationally determined kick distribution. For instance, there could be a population of black holes which receive strong kicks. None of these black holes would remain in binaries, and hence we would not ever detect them. Black hole kick mechanisms tend to be just extrapolations of neutron star kick mechanisms, and theory provides *no* handle on black hole kick distributions.

3.2. ACCRETION IN COMMON ENVELOPE

It is unclear to me why population synthesis modellers have for so long assumed that the accretion onto the neutron star during the inspiral phase is limited to the Eddington rate. For accretion to be Eddington limited, all of the potential energy released as material accretes onto the neutron star must be converted into photons which then prevent further accretion. Unfortunately, for any reasonable accretion scenarios in common envelope evolution, the temperatures and densities at the surface of the neutron star get so high that neutrino emission dominates cooling and the accretion rate is set to nearly the Bondi-Hoyle rate. For a neutron star accreting material from a $20\,\mathrm{M_\odot}$giant star (if it is within 10^{12} cm of the the center of the giant star), it takes over 1 day for photons emitted at the surface of the neutron star to diffuse 1 km from the surface. Neutrinos remove most of the energy (including the radiation energy) of this material in less than a minute (see Fryer, Benz, & Herant 1996 for details). How can photons exert any pressure?!?

3.3. BLACK HOLE FORMATION

One of the most troubling uncertainties in GRB population synthesis comes in determining black hole formation. The problem is that if one assumes that a) mass-loss rates from stellar winds are correct, and b) stellar radii are correct, then it is extremely difficult to form black hole binaries (and impossible to form V404 Cyg). This occurs because if mass-loss rates from winds are correct, then no stars which lose their hydrogen envelope prior to helium ignition can form black holes. Stars more massive than $\sim 25M_{\odot}$don't expand after helium ignition, and hence if these stars don't undergo mass-transfer before helium ignition, they never will. But if they don't go through a mass-transfer phase which tightens their orbit (e.g. common envelope evolution), the resultant black hole binary will never merge! Either the errors in stellar winds or stellar radii must be greater than what is believed by stellar modellers. However, it is not clear how large these uncertainties are (indeed, population synthesis provides probably the best constraints at the moment). Until then, it is will be impossible to place any firm constraints on those BHAD GRB models which require the collapse of a star into a black hole (e.g. BH/NS mergers, WD/BH mergers, and collapsars).

3.4. STELLAR RADII

For some reason, binary population synthesis theorists have assumed that the radii of giant stars are well defined. One would think that supernova 1987A would have taught us otherwise since its progenitor was a compact blue supergiant, not the expanded red supergiant that theorists predicted. Indeed, it is not even clear what we mean by the "radius" of a massive star. We certainly don't mean the photosphere of the star. What we want is the radius that, using our nice equations for mass transfer, will allow us to accurately calculate the orbital separation at which Roche Lobe overflow occurs. Unfortunately, stellar evolution codes do not calculate this radius and none of our population synthesis codes fully account for these uncertainties.

Even worse, many codes do (did) not check to see if, during the common envelope phase, the neutron star merged with its companion's helium core (that is, they don't even check the helium star radius). If so, it should not form a DNS binary. However, because many codes did not check for this possibility, these codes produced numerous close DNS systems (even without kicks). Hopefully, most codes have corrected for this error, but I have seen some papers out as late as 1999 which still have this bug! The fact that this coding error prevailed in nearly all population synthesis codes for so many decades makes one wonder what other global "bugs" we all have!

4. SOME OPTIMISM

Although I do not believe we will be able to tie down the formation rates of any of the BHAD GRB models in the near future, binary population synthesis can provide important clues into our understanding of GRB progenitors. One example is the distribution (in and beyond the host galaxy) of GRBs. Regardless of the many uncertainties, DNS and BH/NS binaries receive much higher systematic velocities than the other GRB models and bursts (probably short-duration bursts) from these models will occur farther from their host galaxy. Nearly all of the long-duration GRB models will occur within their host galaxy and many will occur in their star formation sites. The current observations of long-duration bursts (we currently only have data of long-duration bursts) seem to match thes predictions extremely well (Bloom, Kulkarni, & Djorgovski 2000). Observations of short-duration bursts (which HETE-II will provide) may allow us to rule out (or justify) DNS and BH/NS binaries as the source of short-duration GRBs.

If we can convince ourselves that we truly can distinguish GRB engines, we can use observations of GRBs to trace the star formation history of the universe. Collapsars and helium mergers should trace the type II supernova rate (and hence star formation rate) of the universe very closely, and we could presumably use the redshift distribution of GRBs to study the star formation history. One must approach this problem with some caution, because we still do not understand the progenitors of these GRBs well enough. For instance, the collapsar rate is sensitive to to mass-loss from winds which, in turn, may depend on metallicity (and hence redshift). The delay in binary mergers is even more sensitive to a range of binary uncertainties, and a better understanding of their evolution will also be important if we wish to probe the early universe with GRBs. But GRBs will be used as cosmological probes, and because of this, GRBs may be able to revive activity in binary stellar evolution so that many of the current uncertainties in binary population synthesis may be solved.

This trip was partially funded by an international travel grant from the AAS.

References

Bloom, J.S., Kulkarni, S.R., Djorgovski, S.G., 2000, submitted to ApJ, astro-ph/0010176

Brown, G.E. 1995, ApJ, 440, 270

Bulik, T., Belczynski, K., & Zbijewski, W. 1999, MNRAS, 309, 629

Chevalier, R.A., & Li, Z.-Y. 2000, ApJ, 536, 195

Cordes, J.M., & Chernoff, D.F. 1998, ApJ, 505, 315

Fryer, C.L., Benz, W., & Herant, M. 1998, ApJ, 496, 333

Fryer, C.L., Burrows, A., & Benz, W. 1998, ApJ, 496, 333

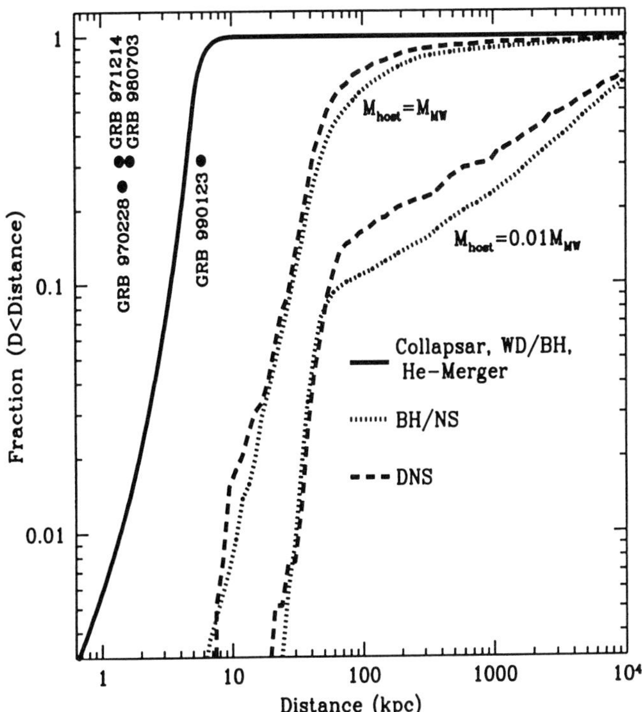

Figure 3 Merger times for DNS and BH/NS binaries using a range of population synthesis parameters. Higher stellar radii or winds produce wider binaries and hence, longer merger times. The double peaked merger distribution for the standard model (solid line) is a signature of the double peaked kick distribution from Fryer, Burrows, & Benz (1998). >90% of the binaries with merger times less than 30 Myr received kicks of 300km s^{-1} or greater.

Fryer, C.L., Woosley, S.E., & Hartmann, D.H. 1999, ApJ, 526, 152

Fryer, C.L. 2000, "Highly Energetic Physical Processes and Mechanisms for Emission from Astrophysical Plasmas", Proceedings of IAU Symposium 195, held at Montana State University - Bozeman, 6-10 July 1999. Published by Astronomical Society of the Pacific, San Francisco, p. 339.

Hansen, B.M.S., & Phinney, E.S. 1997, MNRAS, 291, 569

Kulkarni, S. R., *et al.* 1998, Nature 393, 35

Lamb, D.Q., & Reichart, D.E. 2000, ApJ, 536, L1

Livio, M., & Waxman, E. 2000, ApJ, 538, 187

Lyne, A.G., Lorimer, D.R., 1994, Nature, 369, 127L

Meszaros, P., & Rees, M.J. 1992, MNRAS, 257, 29

Popham, R., Woosley, S. E., & Fryer, C. L., 1999, accepted by ApJ, astro-ph/9807028

GENERAL DISCUSSION ON POPULATION SYNTHESIS MODELLING

led by E.P.J. van den Heuvel
Sterrenkundig Instituut 'Anton Pannekoek', Universiteit van Amsterdam, Kruislaan 403, 1098 SJ Amsterdam.
edvdh@astro.uva.nl

1. INTRODUCTION

Van den Heuvel: I am sure you will agree with me that from the talks we heard today we are all very impressed by the progress made in our understanding of the role of binaries in the evolution of stellar populations. The last talk, by Onno Pols as well as the talks this morning by Simon Portegies Zwart and by Fred Rasio clearly showed that in studying the evolution of dense star clusters real understanding can only be achieved if the dynamics is included in combination with the evolution of the individual stars and binaries. The talk of Pols showed that even in an open cluster such as M67 the dynamical effects are not small perturbations, and that the presence of a variety of very peculiar objects in that cluster, such the Super Blue Stragglers, could never be understood if dynamics was not included. This mornings talk by Portegies Zwart on the Hyades showed the same. We have seen that, in order to include the evolution of binaries in Population Synthesis, many input distributions are needed and many input-assumptions have to be made. A systematic list of these was given in several of the talks, particularly those of Kalogera, Fryer and Van Bever. Fryer gave on one transparency a very nice overview of all ingredients that go into these simulations and he was so kind to lend it to me to show to you again: one sees that the number of input assumptions that have to be made is impressive.(Van den Heuvel shows the transparancy). There are, of course, a number of input assumptions on which we all more or less agree, but there are also assumptions that have great uncertainties and which we should discuss. Furthermore, there are still many open questions on which we would like to find answers. In order to somewhat structure the discussion, I first of all would like to discuss with you the input assumptions for Population

D. Vanbeveren (ed.), The Influence of Binaries on Stellar Population Studies, 479–487.

Synthesis and after that I would like to reserve the second half of our discussion to the open questions.

2. INPUT ASSUMPTIONS WHERE WE CAN AGREE ON

Van den Heuvel: I have made a list of all input assumptions that I recorded today in the different talks and I would like to go through them with you and see on which of them there is general agreement and on which not. (Van den Heuvel shows list on a transparency). On the first four assumptions we can probably all more or less agree. These concern:

- (1) the IMF to be used,
- (2) the distribution of semi-major axes a ,
- (3) the distribution of the mass ratios q and
- (4) the binary frequency.

As to (1) we in general use the single-star IMF for the distribution of masses of the primary stars of binaries. A word of caution is in place here, however, as we did hear from Mermillod that the observations show that the IMF of primaries of binaries differs somewhat from that of single stars.

(2) As to the a-distribution: everybody uses F(log a) = Const., which is known already for over 30 years. It was implicit in the results in van Albadas (1968) Ph.D.thesis and I used it in my first calculations of the evolution of the HRDs of open clusters with a large percentage of binaries when I was a postdoc at Lick in 1968/69 (cf. van den Heuvel 1994).

(3) As to the q-distribution we heard, as van Rensbergen showed so nicely this week, that several distributions are possible: flat or $(1+q)^{-\alpha}$. (The results seem not to depend strongly on them.) The latter one, with $\alpha = 2$ goes back to Kuiper (1935). I used it in my Ph.D. thesis (van den Heuvel 1968) and in the above-mentioned 1968/69 simulations. As Kuiper noticed, you get this distribution if you randomly chop a piece of something (anything, e.g. a piece of cheese) into two parts.

(4) As to the binary frequency, we all agree that it has be somewhere between 0.5 and 1.0.

3. INPUT INGREDIENTS WHICH REQUIRE DISCUSSION

Van den Heuvel: The next 5 input ingredients require more discussion, which I now would like to start. We go through them one by one:

- (5) Evolutionary tracks.

We heard about the importance of the stellar radii and that different codes may give somewhat different radii for stars of the same mass and composition, which will influence the results of binary evolution. But by far the most important point is that of the stellar winds of massive stars. Here there is clearly no full agreement. I would like to invite Norbert Langer to comment on this.

Langer: Particularly regarding the Wolf-Rayet winds there is considerable uncertainty about clumpiness, which has a strong influence on the derived mass-loss rates: this gives an uncertainty in the mass-loss rates which may be as much as a factor of three. On the other hand, the WR mass-loss rate also depends on the luminosity of the star (larger for larger luminosity) so, if you decrease the mass-loss rate the star stays massive longer, and thus more luminous longer, and stays at higher mass-loss rate longer, so lowering the mass-loss rate has also counterproductive effects. As a result the outcome from varying the mass-loss rate seems always to be that the masses of WR stars converge to low final values. But one has to do detailed calculations with varying mass-loss formulae and perhaps also final mass values as high as 5 to 10 solar masses are still possible.

Vanbeveren: I prefer lower mass-loss rates for WR stars - the values that you use in your recent paper with Wellstein in my opinion are rather high.

Van den Heuvel: I think there is some convergence of opinion here. Norbert clearly leaves room for lower mass-los rates. What about the dependence of stellar wind mass-loss rates of normal OB-stars on metallicity? It seems clear that the winds of these stars in the SMC weaker. Can we agree on this?

Vanbeveren: The fact that, as has been reported here, all ten WR stars in the SMC are in binaries clearly indicates that in the SMC the only way to make a WR star is by mass exchange in a binary, and not directly from single OB-stars via red supergiants by high stellar wind mass loss. We know the latter to be a viable channel for producing (single) WR stars in our Galaxy with its much higher metallicity. This seems a clear indication that the winds from OB stars and red supergiants in the SMC are much weaker.

Van den Heuvel: If the mass-loss rate from OB stars is so much lower in the SMC, does this then imply that more stars become black holes? It superficially seems that in the LMC (which also has much lower metallicity than our Galaxy) there are relatively more black-hole X-ray binaries than in our Galaxy. As black-hole formation seems to be related with Gamma Ray Bursts (GRBs), as seems indicated by the correlation of several GRBs with peculiar supernovae, does this then imply that we expect relatively more GRBs in low-Z galaxies? Fryer seemed to imply that in his talk this morning. The blue compact "starburst" galaxies where we tend to see most of the high-redshift GRBs indeed are low-metallicity objects. (This even holds for the blue compact galaxies at low redshift).

- (6) Evolutionary models of binaries.

Van den Heuvel: Here there are the big uncertainties on mass and angular-momentum losses from the systems. I would, however, like to jump over this rather quickly as we know so little about it.

Vanbeveren: Here there is the famous parameter beta (the fraction of the exchanged matter lost from the system; the remaining part is transferred to the companion). We do not know the value of that parameter. Does anyone have an idea about its value? We often just take it to be 0.5. Here in Brussels we often have used that parameter.

Eggleton: I know that comparisons of the parameters of observed Algol systems with the models calculated by De Greve with beta = 0.5 did not fit at all. Clearly beta must for these systems be much smaller, perhaps 0.1 or 0.2.

Van den Heuvel: In fact, one has to do a complete calculation of the effects of the accretion on the radius of the companion. One will have to do the hydrodynamics of the accretion through a disk, but I presume that for not too low q a quite large fraction of the matter may land on the companion. But it will also depend on the orbital separation and on the radius of the accretor, as was shown so beautifully by the observations shown by Geraldine Peters. I think it is hard at present to make a more specific statement about the value of beta. Concerning the alpha-parameter of Common Envelope (CE) evolution there seems to be a quite nice unanimity at this conference and, by the way, also about the CE formalism to be used: everybody seems to be converging on Webbinks (1984) formalism. From the presentation of Yungelson we heard that he suggests alpha to be between 1 and 4, Pols suggests that there is good agreement between his M67 modelling and the observations with alpha = 3. This all seems to indicate that there are other energy sources in addition to the drop in orbital potential energy at work in driving off the common envelope.

- (7) The lower mass limit for black hole formation.

Van den Heuvel: Here we saw from Fryers presentation that in single stars this limit may be as low as 20 to 30 solar masses, but in binaries it may be much higher due to the mass loss. Norbert Langer even suggested that we may never be able to make a black hole in systems that evolve according to the cases A and B. But I must say that I see great difficulties in making a system like Cyg X-1 through case C.

Vanbeveren: I have the same difficulty.

Langer: I agree, Cyg X-1 may have been a case B.

- (8) Kick velocity distributions.

Van den Heuvel: we heard that to understand the large numbers of pulsars found in globular clusters we need a population of low-velocity neutron stars. This may, however, be due to these neutron stars having gone though a High Mass X-ray Binary phase early in the evolution of the clusters. This will despite high natal kicks, have kept these neutron stars at low system velocity, which after complete spiral-in in the HMXB resulted in a low velocity single neutron star.

Fryer: You should keep in mind that the Lyne-Lorimer distribution and its variants need not be the initial kick velocity distribution at birth. They are the observed pulsar space velocity distributions which must also contain the effects of binary evolution/binary disruption.

Van den Heuvel: I agree, this is a good point. Still, it is quite remarkable that Hartmann and Verbunt, using similar observational data, get a distribution that with its considerable low-velocity part, is so different from those of Lyne and Lorimer and of Cordes and Chernoff and others. On the other hand, the H-V distribution is quite similar to that of Paczynski and of Hansen and Phinney. It shows how large the uncertainties are in converting proper motion distributions into space velocity distributions.

Rasio: The low-velocity part of the distribution may also be intrinsic, i.e.: not due to evolution through a High-Mass X-ray Binary/ Thorne-Zytkow phase. There may well be a genuine low-velocity part of the distribution at birth of some 20 to 30 per cent. You then will automatically retain thousands of neutron stars in a cluster like 47 Tuc.

- (9) Rotation.

Van den Heuvel: With regard to kicks I would also like to draw attention to rotation. The beautiful Chandra pictures of the Crab and Vela supernova remnants, with their pulsar-driven plerions, show that in both cases there are jets, which are perpendicular to a ring-like structure around the pulsar. As the ring-like structures are expected to be in the plane of the pulsars rotational equator, the jets are aligned with the rotation axis of the neutron star. As in both cases the jets are exactly aligned with the pulsars proper motion vector, it seems that both pulsars received their kicks along the rotation axis. This strongly suggests that the kicks are causally related to the stars rotation.

4. OPEN QUESTIONS

Van den Heuvel (showing a transparancy): I would now like to come to the list of open questions that I prepared. As we implicitely already discussed part of them, I would like to go through them rather quickly. I skip the first two (about the globular cluster neutron stars and the ways to determine the kick velocity distribution) as we already quite extensively discussed them and would like to go to question 3: Where are the globular cluster black holes?

Portegies Zwart: My numerical simulations show that they have kicked each other out of the clusters by dynamical interactions. They were soon much more massive than the other cluster stars, therefore they evolved dynamically independently, sank to the cluster core, where they formed binaries with each other, which kicked each other subsequently out of the cluster, swapping partners. So they are all gone.

Van den Heuvel: Is none left?

Portegies Zwart: Perhaps a lonely one in the core.

Van den Heuvel: Does that mean that there are many black hole binaries (double black holes) in the galactic halo? When these merge they must give quite some gravitational radiation signal for LIGO or LISA, I presume.

Portegies Zwart: Indeed, we calculate an observable signal of one per one or two years out to the detectability limit of LIGO. There are lots of galaxies with such binaries out to that distance.

Van den Heuvel: Question 4: Does supercritical accretion onto a neutron star occur in Common Envelope evolution (neutron star in envelope of massive companion) and does this lead to a black hole? As mentioned by Rasio, a good way to retain neutron stars in a globular cluster might be to have them to be descendants of High Mass X-ray Binaries which went through a CE phase followed by a Thorne-Zytkow phase (massive star with a neutron star in its center). For this to be a viable way, the neutron star should survive the CE and Thorne-Zytkow phases and not be converted into a black hole by supercritical accretion during these phases, as has been suggested to occur by Chevalier(1993) and by Brown (1995).

Fryer: You cannot avoid supercritical accretion and the formation of a black hole during these phases. If the neutron star is in the dense helium core of the companion there is not enough energy to get the matter out. It is the same problem as with fall-back of matter on a young neutron star formed in the center of a massive star. This also leads to a black hole.

Van den Heuvel: All these calculations were done for a non-rotating non-magnetized neutron star. I could imagine that a strong magnetic field together with rapid rotation might drastically change the outcome of this process and might prevent supercritical accretion to happen.

Fryer: You would need fields like in Magnetars (larger than 10^{15} Gauss) and probably also rapid rotation. That combination could very well be rare.

Van den Heuvel: Question 5 is related to this: Do Thorne-Zytkow stars exist and what happens to them?

Fryer: That is the same problem: supercritical accretion will turn them into black holes.

Van den Heuvel: I presume Philip Podsiadlowski might wish to say something about this?

Podsiadlowski: You know what in my opinion happens: a black hole forms and the envelope with its large angular momentum (the orbital angular momentum of the binary progenitor) will collapse to a disk around this black hole. Out of this disk one or more low-mass companions will condense, and later on the system will become a Black Hole Low Mass X-ray Binary.

Van den Heuvel: Next I would like to turn to the problem of supernovae. First the question of the progenitors of Type Ia Supernovae. We are all convinced that they are related to binary systems. We have two models: the merging of Double Degenerates (two CO white dwarfs with total mass above the Chandrasekhar limit) and the Super Soft binary X-ray Sources as progenitors. As was argued at this meeting by Yungelson and Marsh, respectively, both provide rates that are of the same order as the observed Type Ia supernova rate.

Secondly, I would like to recall the relation between supernovae and Gamma Ray Bursts, which Fryer so nicely covered in his talk. Here we are just at the beginning of a new field which is opening up and which may give us for the first time information on the birth events of stellar-mass black holes. These birth events may be very much more common than the GRBs themselves, since there is strong evidence that the gamma rays and other emissions are very strongly beamed, so the probability of the Earth being in the beam may be very low. Nevertheless, this information on black hole formation is very important as input for the Population Synthesis programmes.

Time is up, and I would like to round off and summarize this great Population Synthesis day. We heard excellent talks and many exciting new results. Particularly it was great to see the first results form the combined stellar dynamics and stellar evolution codes. GRAPES, used by both the Hut-Makino-MacMillan-Portegies Zwart collaboration and by the Aarseth-Tout-Pols-Han-Eggelton collaboration have brought us much progress and exciting new insights.

You will all agree, I think, with what Chris Fryer said so nicely this morning: binaries are of great importance for fundamental physics and cosmology:

- Binaries consisting of two compact stars already have shown us that gravitational radiation exists and that Einsteins form of General Relativity is correct. And they are the most promising sources for the first actual detection of gravitational waves with experiments such as LIGA, VIRGO and LISA.

- Without binaries we would hardly see any X-ray sources in galaxies, and we would not know that stellar black holes exist. At the IAU General Assembly in Manchester last week, Fabbiano showed beautiful Chandra pictures of nearby galaxies, each showing 80 to hundred or more point X-ray sources, clearly X-ray binaries!

- Double degenerates and Supersoft Binary X-ray sources are thought to be the progenitors of Type Ia supernovae, which nowadays are the prime "standard candles" with which the rate of expansion of the Universe is being measured. Without binaries we would not know (and cosmologists would not know) that we now need a cosmological constant to explain why the Universe is so flat. We would only have some fuzzy ripples in the cosmic microwave background radiation which would be insufficient to draw definite conclusions from.

- Gamma Ray Bursts, as Fryer mentioned, come from the high redshift Universe and at least half of them (the short ones with hard spectra) are thought to be related to binaries (mergers of double neutron stars and/or of neutron star plus black hole binaries).

I must say that I am very pleased to see the enormous progress reported in this session. Simon Portegies Zwart mentioned this morning that already the GRAPE 4 computers are addictive. I presume that the GRAPE 6 will be even more addictive. They apparently are at least as addictive as the fermented juice of real grapes! We are therefore just seeing the beginning of a new era in our understanding the full evolution of stellar systems that contain a realistic population of interacting binaries. We can therefore expect many new and exciting results in the near future!

I would like to thank all the speakers of today for their marvellous contributions and herewith close this session.

References

Chevalier, R.A., 1993, ApJ, 411, L33

Brown, G.E., 1995, ApJ, 440, 270

Van Albada, T.S., 1968, Bull.Astron.Inst.Neth, 20, 47

Van den Heuvel, E.P.J., 1968, Bull.Astron.Inst.Neth., 19, 326

Van den Heuvel, E.P.J., 1994, in "Interacting Binaries" (S.N. Shore, M. Livio and E.P.J. van den Heuvel), Springer, Heidelberg, p. 263-474

VI

CHEMICAL EVOLUTION OF GALAXIES

ELEMENT ABUNDANCES AND GALACTIC CHEMICAL EVOLUTION

Sean G. Ryan
Dept of Physics and Astronomy, The Open University, Walton Hall, Milton Keynes MK7 6AA, United Kingdom.
s.g.ryan@open.ac.uk

Abstract Recent developments in the study of element abundances and their impact on our understanding of the evolution of stars and the chemical evolution of the Galaxy are described. In particular, problems that may benefit, or benefit from, work on binary stellar populations are highlighted. These include:
• different mixing processes in globular cluster and field stars, evidenced by Li and nuclei in the NeNa- and MgAl-cycles.
• the range of carbon overabundances in metal-poor stars, accompanied by a range of s- and r-process element abundances.
• the timescale for the evolution of SN Ia progenitors is poorly known.
• the production of nitrogen is very poorly understood.
• the nucleosynthesis of many iron-peak elements are poorly understood.

1. INTRODUCTION

The ideal model of Galactic evolution would match numerous observational characteristics. More correctly we should regard them as characteristics *derived* from observational data, since uncertainties affect the handling and interpretation even of perfect data.

The ideal model would reproduce for *each* stellar population:
• the abundance distribution (metallicity distribution function);
• the kinematics;
• the age-metallicity relations — which need not be single valued;
• isotopic abundances in stars, meteorites, and the interstellar medium;
• the current gas and stellar fractions; and
• the present day mass function of stars.

No doubt other characteristics can be identified which a model should match. Our Galaxy should also be viewed in the context of other galaxies including high-redshift systems. Ryan (2000, Figure 1) illustrates the parallels between abundances seen in:

D. Vanbeveren (ed.), The Influence of Binaries on Stellar Population Studies, 491–506.

• the old, metal-poor Galactic halo and the Lyman-α-forest clouds of the intergalactic medium;
• the moderately metal-poor Galactic thick disk and damped-Lyman-α systems; and
• the old, metal-rich bulge stars and the high redshift, metal-rich Lyman-break galaxies (based on the observation of Steidel et al. 1999).

This paper describes a subset of the abundance data from measurements of Galactic stars which are believed to preserve the material from which they formed, as well as a few stars whose abundances have been substantially altered but which nevertheless provide valuable insights into previous epochs of Galactic evolution. The review is complete neither in terms of elements discussed nor works cited, but some effort has been made to report on elements where substantial problems are revealed or recently have been solved, and where binary interactions are or might be important to understanding the observations.

The first section of this paper will discuss the overall metallicity of the Galactic halo as summarised by iron. Subsequent sections will examine element abundances in greater detail. To simplify the universe, astronomers have reduced the number of elements from $\simeq$100 to three: hydrogen, helium, and metals. A less-crude simplification is to consider groups of element according to the nucleosynthetic processes by which they originate. These classes are identifiable as distinct groupings of elements in a plot of abundance versus atomic number. One possible categorisation is as follows (see Figure 1):
• primordial elements: H, He, and some ^{7}Li;
• light elements: Li, Be, and B, formed in interstellar gas by spallation, and in the case of Li by stellar fusion and synthesis;
• life elements: C, N, O, formed during H- and He-burning;
• intermediate-mass species: Na-Ca, and possibly Sc and Ti, formed during hydrostatic and explosive burning of C and O;
• iron-peak elements: Sc-Ni, and possibly Co and Zn, formed in supernovae by explosive Si-burning;
• heavy s-process elements: Zn-Pb, formed by slow neutron-capture in AGB stars and during He- and C-burning in $M > 15\ M_\odot$ stars;
• heavy r-process elements: Sr-U, formed by rapid neutron-capture in SN and possibly more exotic sites (e.g. colliding neutron stars); and
• heavy p-process species.

Thermal broadening of stellar spectral lines exceeds the isotope shifts of most elements, so only in rare cases will atomic (e.g. Li) or molecular spectral features (e.g. CH, MgH) yield stellar isotopic ratios. Consequently, the isotopic analysis of the pre-solar nebula possible with me-

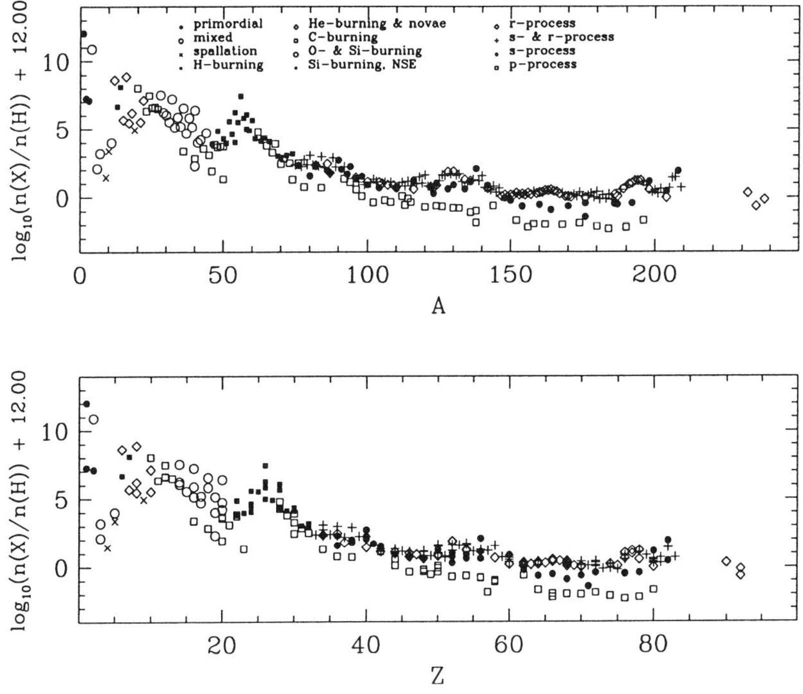

Figure 1 Isotopic abundances arranged by atomic-mass number A and atomic number Z. The legend indicates the main nucleosynthesis pathway for each species. (Based on data from Lang 1980, Table 38).

teoritic samples, and of some stars via interstellar grains, cannot be extended to earlier epochs of the Galaxy.

2. THE HALO METALLICITY DISTRIBUTION

The Simple Closed-Box Model of the chemical evolution of the Galaxy (e.g. Hartwick 1976) predicts star counts to fall towards lower metallicity by a factor of ten per decade of metallicity. Despite its crude assumptions, including complete mixing and instantaneous recycling, the model fits the majority of halo stars remarkably (e.g. Ryan & Norris 1991). However, the accumulation of larger samples of halo stars numbering some 1300 with [Fe/H] < -2.0 shows only two stars with [Fe/H] < -4, whereas the Simple Model leads to the expectation of 11 (Norris 1999, Figure 2). Finding no more than two stars in a regime where 11 are expected has a Poisson probability of only 0.12%, suggesting that the deficit has high statistical significant. We are left then with the possibility that a stellar generation leaving no sub-0.8 $M_\odot$ main-sequence stars after $\simeq$13 Gyr enriched the Galactic halo to [Fe/H] = -4.0 prior to the formation of the first long-lived 0.8 $M_\odot$ stars. This could in principal be achieved if the lower mass limit for the formation of zero-metallicity stars exceeds 0.8 $M_\odot$, as some theoretical calculations suggest (e.g. Nakamura & Umemura 1999).

Note also that the pre-main-sequence contraction time for 0.8 $M_\odot$ stars is $\sim 10^8$ yr, a factor of ten longer than the main-sequence lifetime of massive SN II progenitors ($\sim 10^7$ yr for 25 $\mathrm{M}_\odot$), so another possibility is that the very first SN exploded and enriched the earliest star-forming regions even before the 0.8 $M_\odot$ stars of that same generation could form. In this race-against-the-clock, it is necessary that enough high-mass SN II progenitors existed to complete such a pollution process. It is easy to show that that was probably the case. It is possible to estimate the mass M_{Max} of the most massive star in a generation of mass M_* by integrating the IMF from a low limit of around 0.08 $\mathrm{M}_\odot$ up to M_{Max}, such that only M_{Max} remains of the initial M_*. As an illustrative example, for a Scalo IMF $\propto M^{-2.7}$, we find $M_{\mathrm{Max}} \simeq 0.35 M_*^{0.6}$. Thus, even if only 1% of the mass of a 10^6 $\mathrm{M}_\odot$ cluster is caught up in the first episode of star formation, stars up to 90 $\mathrm{M}_\odot$ should be formed. (A steeper IMF for more massive stars would reduce the mass; an IMF $\propto M^{-3.2}$ implies $M_{\mathrm{Max}} \simeq 0.25 M_*^{0.45}$, or $M_{\mathrm{Max}} = 16$ $\mathrm{M}_\odot$ for 1% of a 10^6 $\mathrm{M}_\odot$ cluster.)

3. DETAILED ABUNDANCES OF SELECTED ELEMENTS

Recent studies have provided data on a wide range of Galactic stars. An incomplete sample of such works includes:

- very metal-poor stars: McWilliam et al. (1995), Ryan, Norris & Beers (1996);
- halo-disk transition and thick disk stars: Nissen & Schuster (1997), Bonifacio, Centurion & Molaro (1999), Jehin et al. (1998, 1999), Gratton et al. (2000), Prochaska et al. (2000);
- bulge stars: McWilliam & Rich (1994);
- disk stars: Edvardsson et al. (1993), Chen et al. (2000);
- metal-rich stars: Feltzing & Gustafsson (1998), Feltzing & Gonzales (2000).

Galactic chemical evolution models covering a wide range of elements include Timmes, Woosley & Weaver (1995), Chiappini et al. (1999), and Goswami & Prantzos (2000).

3.1. LITHIUM IN DENSE STELLAR ENVIRONMENTS AND THE GALACTIC DISK

Two problems related to Li will be discussed. Li has several production mechanisms (see above) and also a wide range of proposed destruction mechanisms, the latter coming about because it burns in stellar interiors where the temperature exceeds $T \simeq 2.5 \times 10^6$ K. Mixing can cause Li-depleted material to appear at the surface of stars. In many studies, the conclusions depend on whether Li depletion (and in some stars production) has occurred *and* is recognised.

Different behaviour is found in the halo field-star and globular cluster samples. In the halo field, only a very small star-to-star spread is found in Li abundance, and that is consistent with an underlying dependence on [Fe/H] (Ryan, Norris & Beers 1999).[1] In M92, on the other hand, Boesgaard et al. (1998) find a substantially larger spread (e.g. Ryan et al. 2001c, Figure 1). This suggests that environmental effects in the dense globular cluster have generated a range of Li-evolution histories in this object. Whether this happens during the early pre-main-sequence phase when the radii of the protostellar disks are still very large, or later, is unclear, but *studies of the interactions of newly-forming binary stars with similar proximity might be instructive.*

A second problem with Li (see Figure 2) is its rate of evolution in the Galactic disk. Enrichment during the halo and thick-disk formation epochs did not exceed $\simeq 0.4$ dex (e.g. Ryan et al. 2001c, Figure 7), but then increased by 0.9 dex when the Galaxy was enriched from [Fe/H] = -0.5 to solar. Models of this phase of Galactic chemical evolution do not provide completely satisfactory fits to the observations, despite inclusion of the yields of novae and AGB stars (e.g. Romano et al. 1999; Travaglio, Randich & Galli 2000; Ryan et al. 2001c). Cool-bottom processing also appears unlikely as a suitable explanation due to the short-lived nature of Li produced this way (Charbonnel & Balachandran 2000). *A currently unrecognised source of Li in the disk appears to be required.*

3.2. UBIQUITOUS CARBON OVERABUNDANCES IN THE HALO

A surprise finding in the Beers, Preston & Shectman (1992) survey was the discovery that $\simeq 10\%$ of stars with $[Fe/H] < -2$ have significant carbon excesses, up to $[C/Fe] = +2.0$ (Rossi, Beers & Sneden 1999).

[1]The issue of whether this [Fe/H] dependence is real or artificial is contested by some, but that debate does not detract from the finding that the inferred intrinsic spread is $\sigma_{\rm int} < 0.02$ dex.

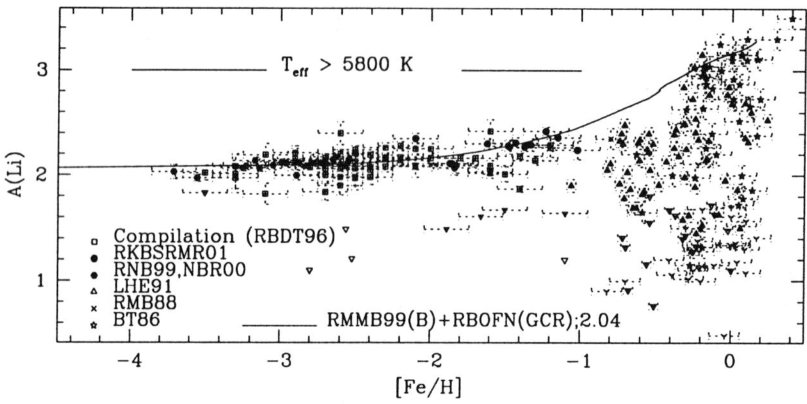

Figure 2 Li evolution with metallicity in halo main-sequence-turnoff stars and a wider range of disk stars. The solid curve is a hybrid model incorporating primordial nucleosynthesis, spallation, supernovae, AGB stars and novae. See text for details.

Detailed studies of these and other C-rich, metal-poor stars have shown a variety of neutron-capture-element anomalies. At least four are s-process rich (Norris, Ryan & Beers 1997a; Hill et al. 2000), at least one other has a strong r-process signature (Cowan et al. 1995; Sneden et al. 1996), and at least one has no neutron-capture excesses at all (Norris, Ryan & Beers 1997b). This suggests a variety of origins, including SN nucleosynthesis for the (apparently rare) r-process-rich object, and mass transfer from evolved companions for the others, where s-process enhancements suggest an AGB donor, and objects without s-process enhancements may have accreted material from an RGB donor. Fujimoto, Ikeda & Iben (2000) suggest another mechanism for C-enrichment. They compute that low-metallicity ($[Fe/H] < -2.5$), low-mass stars may undergo extensive mixing at the He flash, and drag envelope H down to He-burning levels where not only primary ^{12}C can be made (by the triple-α reaction) but also ^{13}C via $^{12}C(p,\gamma)^{13}C$, allowing activation of the $^{13}C(\alpha,n)^{16}O$ neutron source. Thus the stars would be capable of synthesising C and s-process elements, which could reach the surface at second dredge-up, and be observed in situ or after transfer to a longer-living companion by mass loss. *Mass transfer between binaries may play a crucial role in the ubiquitous carbon overabundances observed in very metal-poor stars.*

LP 625-44 is presented as a case study to illustrate the analysis of these systems. It was discovered by Norris et al. (1997a) as a C-rich member of a proper-motion-selected halo sample. Its strong CH bands indicated $[C/Fe] = +2.0$. A high resolution spectrum showed a strong overabundance of neutron-capture elements, with [Ba/Fe] = +2.6, but [Eu/Ba] $\simeq -1.0$. These and other neutron-capture element abundances reveailed that the star has a strong s-process overabundance, not an r-process overabundance as is seen in CS 22892-052. Norris et al. invoked an explanation similar to that for CH-subgiants generally, that a star undergoing or having completed the thermally-pulsing AGB phase transferred material to a lower-mass companion and subsequently faded beyond the luminosity of the accretor. That companion, now the more luminous star, thus bears the signature of AGB star carbon and s-process nucleosynthesis, despite it having the luminosity of a subgiant and hence too low a mass to synthesise these elements itself. Norris et al. were troubled by the lack of convincing radial-velocity variations in the object over several years, but two more recent epochs of data (Aoki et al. 2000) have confirmed radial velocity variations with a period T $\gtrsim$ 12 years and velocity semi-amplitude $\gtrsim$ 6 km s^{-1}. Assuming M_1 = 0.8 M$_{\odot}$ we infer a semi-major axis $a > 5$ AU. These orbital parameters are consistent with those of CH-subgiants in which mass-transfer is believed to occur via wind-accretion rather than Roche-lobe overflow (Han et al. 1995), supporting the suggestion that the star now observed was enriched by a previous AGB companion. If we assume a limiting eccentricity for the system $e < 0.2$, which is consistent with other CH-subgiants (McClure 1997), we derive a lower limit on the mass of the evolved companion $M_2 > 0.35$ M$_{\odot}$, which is again consistent with it being the white dwarf remnant of an AGB star.

The improved spectra have also permitted the measurement of additional s-process elements including lead (Pb). The identification of the spectral line can be confirmed by comparing the spectrum of LP 625-44 with that of the ^{12}C- and ^{13}C-rich star CS22957-027, which does not have neutron-capture element overabundances. One surprise with the lead detection was its *low* abundance. We derived [Pb/Ba] = 0, whereas one might expect a higher value in such a metal-poor star, since [Ba/Sr], and [hs/ls] more generally, increases towards lower metallicity due to the increase in the number of neutrons per seed nucleus in metal-poor (seed-poor) stars (e.g. Gallino et al. 1998). However, s-process yields are extremely sensitive to the poorly-modelled ^{13}C pocket (Busso, Gallino & Wasserburg 1999), and a range of pocket normalisations is required by the observed [hs/ls] ratios in disk stars. Lead is much more sensitive to the normalisations than Ba, and a range of normalisations producing a 1 dex spread in Ba at the metallicity of LP 625-44 produces a 2 dex range in Pb (Ryan et al. 2001a). Consequently, the solar value of [Pb/Ba] in this object is compatible with the models, albeit with a different normalisation to that of Busso et al's "ST" model.

This and other stars, such as LP 706-7 which has the same [Fe/H] and [C/Fe] and in which we also detect lead, enables us to provide missing constraints on the parametrisation of the ^{13}C pocket in metal-poor AGB-star models.[2]

3.3. OXYGEN

Oxygen is the most abundant metal and of particular importance in stars, because of its role in energy generation via the ON-cycle and because of its role as an electron-donor for the opacity of stellar material. It is formed in $M > 10$ $M_\odot$ stars and it marks the last step in helium burning, where the chain $^{12}C(\alpha,\gamma)^{16}O$ terminates because of the slow rate of the $^{16}O(\alpha,\gamma)^{20}Ne$ reaction at stellar temperatures and densities. It is produced well outside the mass-cut of a SN II progenitor. In contrast, Fe is produced in the core, and its yield is very sensitive to the mass-cut, whose position is uncertain and must be adjusted to reproduce the Fe yield of observed SN IIs, especially SN1987A, and to reproduce the element ratios seen in other iron-peak elements which form close to Fe but with slightly different radial profiles (Nakamura et al. 1999). Typical iron yields of SN IIs are of order 0.1 $M_\odot$, with oxygen yields an order of magnitude larger (e.g. Nomoto et al. 1997). On the other hand, the numbers are essentially reversed in SN Ia, which produce far more Fe and little oxygen, with iron yields of $\simeq 0.7$ $M_\odot$ (Iwamoto et al. 1999). The result of the different evolutionary timescales of massive SN II progenitors compared to lower mass SN Ia progenitors in binary systems is that the O/Fe ratio decreases once the timescale for the evolution of SN Ia is reached. The result is that oxygen is over-abundant (relative to iron) in the galactic halo and falls to a solar ratio later in the evolution of the Galaxy. The rate of the decline depends on the relative yields of O and Fe in the various SN progenitors and their evolutionary timescales (e.g. Tinsley 1980). *Although values around 1 Gyr are often given for this timescale, is is not well constrained, and would benefit from theoretical calculations from the binary characteristics of SN Ia progenitors.* The recent opening of a second pathway for producing single-degenerate SN Ia progenitors, via a WD + RGB($\sim$1 $M_\odot$) binary (Hachisu et al. 1999; cf. WD + MS($\sim$2—3.5 $M_\odot$) binary), may lead to much longer progenitor lifetimes due to the lower mass and slower evolution of the mass donor (see also De Donder 2001). Kobayashi et al. (1998) similarly show that

[2]Mowlavi (2000) notes that the Geneva group uses a different parametrisation of the ^{13}C source in their AGB models, via "diffusive" (extra) mixing of H below the H-burning shell. The measurements of Pb in LP 625-44 and LP 706-7 will permit their models to be tested and constrained in a similar fashion to those of the Torino group.

the mass window giving rise to SN Ia progenitors depends on the metallicity of the mass donor; this metallicity-dependent mass-window will also lead to a metallicity-dependent mean timescale.

Models have tended to show that [O/Fe] should remain below +0.5 even down to [Fe/H] = −3, not dissimilar to the appearance of other [α-element/Fe] values (e.g. Timmes, Woosley & Weaver 1995). However, the observational picture has differed considerably. Four observational diagnostics are used to measure oxygen abundances: UV OH bands, the weak red "forbidden" line of atomic oxygen [O I], the O I triplet of atomic oxygen at 7771-7775 Å, and IR OH bands. The agreement between diagnostics is poor (e.g. Ryan 2000, Figure 4). On the basis of radiation transfer, the [O I] line was often advocated as preferable a priori, but the weakness of this diagnostic in very metal-poor stars or dwarfs caused the UV OH bands to be exploited as well. The latter showed strong [O/Fe] evolution, with [O/Fe] = 1.0–1.2 at [Fe/H] = −3 (Israelian, García López & Rebolo 1998; Boesgaard et al. 1999), whereas the [O I] line in similar metallicity giants gave [O/Fe] = +0.35 (Barbuy 1988; Fulbright & Kraft 1999).

Diverse results were obtained for the O I triplet. Progress in reconciling these disparate observational results may have been made with 3D hydrodynamical calculations by Asplund et al. (2001; also 1999). The outer layers of their metal-poor 3D models are much cooler than the 1D models of the same $T_{\rm eff}$, and since OH forms in the outer layers, the OH analysis is particularly sensitive to this effect. The use of 3D models would consequently reduce the overabundance inferred from the UV OH bands to values more in keeping with the [O I] lines. These models are however still in the preliminary stages — they are available only for a few stars, since they must be constructed on a star by star basis, and do not yet include all likely effects such as NLTE — so a definitive statement cannot be made yet. At the same time, the commissioning of the new UVES spectrograph on the ESO VLT has made possible additional observations of the challenging [O I] line, which should increase our ability to utilise that diagnostic in coming years.

3.4. CNO, NE-NA, AND MG-AL CYCLES

The convective envelopes of low-mass giants at the base of the RGB should not reach down to ON-cycled material according to classical stellar models. However, globular-cluster giants often have a range of O abundances, deficient relative to field-stars (e.g. Pilachowski, Wallerstein & Leep 1980; Kraft et al. 1993). Anti-correlated with the O deficiencies are Na and Al overabundances (Peterson 1980; Kraft et al.

1993). In ω Cen, for example, Norris & Da Costa (1995) find C to be deficient, N overabundant, and O deficient, but all giving C+N+O constant, which is to say that the nuclei have been cycled without net creation or destruction, and Na and Al are overabundant.

These abundances show the signature of proton capture reactions in the CNO cycle — the CN cycle converts ^{12}C to ^{14}N, while the ON cycle converts ^{16}O to ^{14}N — and the less-well-known NeNa and MgAl cycles. The last two are unimportant energetically, but they are important for the transmutation of Ne, Na, Mg, and Al nuclei.

The NeNa and MgAl cycles occur along with the CNO cycle, converting the minor isotope ^{22}Ne to ^{23}Na, and the minor isotope ^{26}Mg to ^{27}Al. As the reactants are minor isotopes, the five-fold increase in Na and two-fold increase in Al that can result from these proton capture cycles have almost no impact on Ne or Mg *element* abundances. (In any case, Ne is not observable in the spectra of giants.)

The CNO, NeNa, and MgAl cycles occur deep in the stellar interiors, and it is only when the proton-burnt zone is mixed to the surface that a star will reveal it. The signature may also appear if a star forms from gas in which these elements were deposited.

Despite the clear nucleosynthetic signature, several mysteries surround the observations:

• Why do some globular cluster giants and not others show the proton-capture signature?

• Why do some globular cluster turnoff stars, whose core temperatures have been too low to initiate the NeNa cycle, also have Na-Al-N enhancements (e.g. Cottrell & Da Costa 1981; Briley et al. 1996)?

These observations require that at least some of the stars formed from material in which the proton-capture products were present. However, a pre-stellar origin for the anomalies does not explain everything.

• Why do only globular cluster stars, and not field stars, exhibit these signatures? If proton capture products were routinely present in the gas from which stars form, then they would be evident in the field population as well, but they are not.

It appears that the CNO, NeNa, and MgAl cycles do happen, but where *and* when *are poorly known. As for Li, there are interactions between stellar evolution and the stellar environment that are not yet understood.*

4. THE PRIMARY PROBLEM WITH NITROGEN

In unevolved stars, nitrogen is best observed via the UV NH band at 3360 Å. The short wavelength of this feature, implying a lack of stellar flux and traditionally a poor response of detectors, has meant that few stars have in practice been observed. Data from Laird (1985), Carbon et al. (1987), and most recently Israelian, García López & Rebolo (2000) show that [N/Fe] $\sim$ 0.0 down to [Fe/H] = −3, indicating primary production throughout Galactic history. The problem with this is that massive star models alone currently under-produce the solar system N-abundance by a factor of three, suggesting that low-to-intermediate mass stars must contribute the bulk. However, such stars are ineffective in Galactic chemical evolution models until [Fe/H] $\sim$ −1.5 (e.g. Timmes et al. 1995, Figure 14; Goswami & Prantzos 2000), in contradiction to the observations. *Clearly, massive metal-poor stars do synthesise nitrogen effectively, i.e. in the Pop I ratio, but how?* Maeder (2001; Venn 1999) discusses the high N-overabundances in the (slightly metal-poor) SMC, and suggests that rapid rotation (but probably not binarity) in massive stars could be responsible. This mechanism may yet provide a means of routinely producing solar effective yields of N even in the most metal-poor massive stars.

5. THE α-ELEMENTS AND THE IRON PEAK

It was established long ago that the α-elements Mg, Si, Ca, and even Ti show overabundances relative to iron by $\sim$ 0.3 dex in metal poor halo stars. That result has been extended down to the cutoff of the metallicity distribution ([Fe/H] = −4) by McWilliam et al. (1995) and Ryan et al. (1996). The statistically-robust fits of Ryan et al. show that the trends appear to be flat to within 0.2 dex over the range $-4 <$ [Fe/H] < -1. (The weak trends within this 0.2 dex range could be due to metallicity-dependent systematic errors due to the different spectral lines studied in metal-rich and metal-poor halo stars.)

Furthermore, the robust fits to the observations show that the spread of the [X/Fe] data, as measured by robust quartiles, does *not* increase appreciably towards more metal-deficient stars. This quantitative finding is at odds with a common perception that below [Fe/H] = −3, relative abundances vary wildly. Wider variations are found for the neutron-capture elements (Gilroy et al. 1988), especially Sr (Ryan et al. 1991, 1996), but this is not a general property of all elements. Prochaska et al. (2000) conclude that thick-disk abundances are more in keeping with bulge (McWilliam & Rich 1994) and halo values than with those of the thin disk (e.g. Edvardsson et al. 1993), at the same [Fe/H].

Although the observational picture seems clear, the theory has one long-running deficiency. Ti is synthesised with Fe in SN models, and the transition from SN II dominating in the halo to SN Ia dominating in the disk is not expected to alter the [Ti/Fe] ratios. The observations show, however, that Ti behaves in much the same way as the α elements that are produced primarily in SN II progenitors. Models of nucleosynthesis in the cores of SN II and/or SN Ia are clearly problematic in this regard. Relatedly, the iron yields of SN II are notoriously uncertain due to the inability of models to predict where the mass-cut will lie.

Sc is also unusual. Whereas halo stars and even metal rich disk stars with $[Fe/H] > 0$ (Feltzing & Gustafsson 1998) show solar [Sc/Fe] ratios, thin disk and thick disk stars (Edvardsson et al. 1993; Prochaska et al. 2000) show Sc to be overabundant by up to 0.2 dex. This pattern is not predicted by the tabulated SN yields, so it is unclear whether they are indicating a difference between SN II and SN Ia, and/or whether there is a metallicity dependence in the massive star yields. The *latter* is proposed by Timmes et al. (1995) as the explanation of the [Mn/Fe] underabundance in metal-poor stars with $-2.5 < [Fe/H] < -1.0$. They find that SN Ia are insignificant Mn producers, and massive stars alone are responsible for the observed behaviour.

In metal-poor stars with $[Fe/H] < -2.5$, [Cr/Fe] and [Mn/Fe] underabundances are seen (McWilliam et al. 1995; Ryan et al. 1996) which are not explained by current models. Efforts to vary the yields by repositioning the mass-cut (Nakamura et al. 1999) were partially successful, but were not able simultaneously to reproduce the observed [Co/Fe] overabundance without simultaneously over-producing Ni.

6. ZINC

Nucleosynthesis descriptions (e.g. Woosley & Weaver 1995; Prantzos, Hashimoto & Nomoto 1990; Raiteri et al. 1991a,b) indicate that zinc is produced in two ways:

- via an α-rich freeze-out from explosive Si-burning in supernovae, where in the hot, low-density neutrino-heated bubble above a newly-formed neutron star, excess α particles combine with iron-peak nuclei to produce species just beyond the iron-peak; and
- in the weak s-process in massive ($M > 15\ M_\odot$) core He- and C-burning stars.

However, these clear views are not encapsulated within existing Galactic chemical evolution models. Timmes et al's (1995, Figure 35) models underproduce Zn, but with the explanation that the Woosley & Weaver (1995) yields explicitly exclude what they regard as the most probable source of the dominant isotope ^{64}Zn, an extreme α-rich freeze-out. Similarly, Nomoto et al. (1997) show that the IMF-weighted yields of 10–50 $M_\odot$ stars underproduce the solar system Zn abundance by more than 1 dex, an outcome again explainable by their (intentional) neglect of an important pathway, the weak s-process (Thielemann, Nomoto & Hashimoto 1996). The models of Goswami & Prantzos likewise fail to fit the observations. *Until numerical models are produced that do indeed fit the most up-to-date data (see Figure 3), it is premature to claim that we understand the nucleosynthesis of this element.*

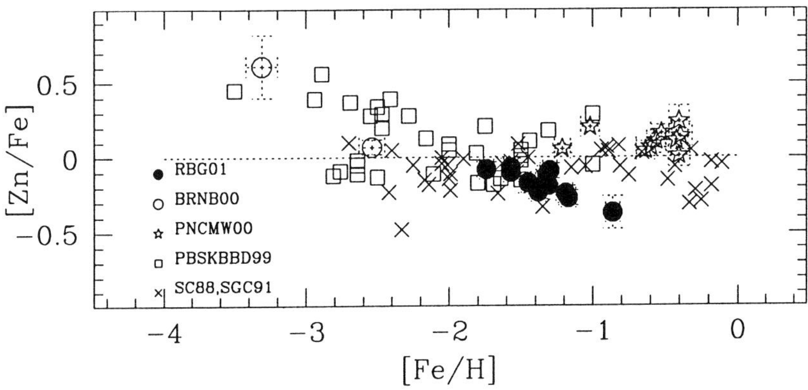

Figure 3 Zinc abundances from: crosses: Sneden & Crocker (1988) and Sneden, Gratton & Crocker (1991), adjusted to more recent solar values; squares: preliminary abundances from Primas et al. (2000); stars: Prochaska et al. (2000); open circles: Blake et al. (2001); solid circles: Ryan, Blake, & Gregory (2001b)

References

Asplund, M., Nordlund, Å, Trampedach, R., & Stein, R. F. 1999, A&A, 346, L17

Asplund, M. et al. 2001, Proc IAU GA 2000, JD8 Oxygen, in prep

Aoki, W., Norris, J. E., Ryan, S. G., Beers, T. C., & Ando, H. 2000, ApJ, 536, L97

Barbuy, B. 1988, A&A, 191, 121

Beers, T. C., Preston, G. W., & Shectman, S. A. 1992, AJ, 103, 1987

Blake, L. A. J., Ryan, S. G., Norris, J. E., & Beers, T. C. 2001, Nucl.Phys.A. in press

Boesgaard, A. M., Deliyannis, C. P., Stephens, A., & King, J. R. 1998, ApJ, 493, 206

Boesgaard, A. M., King, J. R., Deliyannis, C. P., & Vogt, S. S. 1999, AJ, 117, 492

Bonifacio, P., Centurion, M., & Molaro, P. 1999, MNRAS, 309, 533

Briley, M. M., Smith, V. V., Suntzeff, D. L., Lambert, D. L., Bell, R. A., & Hesser, J. E. 1996, Nature, 383, 604

Busso, M., Gallino, R., & Wasserburg, G. J. 1999, ARAA, 37, 239

Carbon, D. F., Barbuy, B., Kraft, R. P., Friel, E., & Suntzeff, N. B. 1987, PASP, 99, 335

Charbonnel, C. & Balachandran, S. C. 2000, A&A, preprint

Chen, Y. Q., Nissen, P. E., Zhao, G., Zhang, H.W., & Benoni, T. 2000, A&AS, 141, 491

Chiappini, C., Matteucci, F., Beers, T. C., & Nomoto, K. 1999, ApJ, 515, 226

Cottrell, P. L. & Da Costa, G. S. 1981, ApJ, 245, L79

Cowan, J. J., Burris, D. L., Sneden, C., McWilliam, A., & Preston, G. W. 1995, ApJ, 439, L51

De Donder, E. & Vanbeveren, D. 2001, this volume

Edvardsson, B., Andersen, J., Gustafsson, B., Lambert, D. L., Nissen, P. E., & Tomkin, J. 1993, A&A, 275, 101

Feltzing, S. & Gustafsson, B. 1998, A&AS, 129, 237

Feltzing, S. & Gonzales, G. 2000, A&A, in prep

Fujimoto, M. Y., Ikeda, Y., & Iben, I. Jr 2000, ApJ, 529, L25

Fulbright, J. P. & Kraft, R. P. 1999, AJ, 118, 527

Gallino, R., Arlandini, C., Busso, M., Lugaro, M., Travaglio, C., Straniero, O., Chieffi, A., & Limongi, M. 1998, ApJ, 497, 388

Gilroy, K. K., Sneden, C., Pilachowski, C. A., & Cowan, J. J. 1988, ApJ, 327, 298

Goswami, A. & Prantzos, N. 2000, A&A, 359, 191

Gratton, R. G., Carretta, E., Matteucci, F., & Sneden, C. 2000, A&A, 358, 671

Hachisu, I., Kato, M., Nomoto, K., & Umeda, H. 1999, ApJ, 519, 314

Han, Z., Eggleton, P. P., Podsiadlowski, P., & Tout, C. A. 1995, MNRAS, 277, 1443

Hartwick, F. D. A. 1976, ApJ, 209, 418

Hill, V., et al. 2000, A&A, 353, 557

Israelian, G., García López, R., & Rebolo, R. 1998, ApJ, 507, 805

Israelian, G., Garcia Lopez, & Rebolo, R. 2000, The First Stars, A. Weiss et al. (eds) (Springer, Berlin), 56

Iwamoto, K., Brachwitz, F., Nomoto, K., Kishimoto, N., Umeda, H., Hix, W. R., & Thielemann, F.-K. 1999, ApJS, 125, 439

Jehin, E., Magain, P., Neuforge, C., Noels, A., Parmentier, G., & Thoul, A. A. 1999, A&A, 341, 241

Jehin, E., Magain, P., Neuforge, C., Noels, A., & Thoul, A. A. 1998, A&A, 330, L33

Laird, J. B. 1985, ApJ, 289, 556

Lang, K. R. 1980, Astrophysical formulae, (Springer-Verlag, Berlin)
Kobayashi, C., Tsuijmoto, T., Nomoto, K. Hachisu, I., & Kato, M. 1998, ApJ, 503, L155
Kraft, R. P., Sneden, C., Langer, G. E., & Shetrone, M. D. 1993, AJ, 106, 1490
Maeder, A. 2001, this conf.
McClure, R. D. 1997, PASP, 109, 536
McWilliam, A., Preston, G. W., Sneden, C., & Searle, L. 1995, AJ, 109, 2757
McWilliam, A., & Rich, R. M. 1994, ApJS, 91, 749
Mowlavi, N. 2001, this conf.
Nakamura, F. & Umemura, 1999, ApJ, 515, 239
Nakamura, T., Umeda, H., Nomoto, K., Thielemann, F.-K. & Burrows, I. 1999, ApJ, 517, 193
Nissen, P. E., & Schuster, W. J. 1997, A&A, 326, 751
Nomoto, K., Hashimoto, M., Tsujimoto, T., Thielemann, F.-K., Kishimoto, N., Kubo, Y., & Nakasato, N. 1997, Nuc.Phys.A, 616, 79c
Norris, J. E. 1999, the Third Stromlo Symposium: The Galactic Halo, ASP Conf.Ser. 165, 213
Norris, J. E. & Da Costa, G. S. 1995, ApJ, 447, 680
Norris, J. E., Ryan, S. G., & Beers, T. C., 1997a, ApJ, 488, 350
Norris, J. E., Ryan, S. G., & Beers, T. C., 1997b, ApJ, 489, L169
Peterson, R. C. 1980, ApJ, 237, L87
Pilachowski, C., Wallerstein, G., & Leep 1980, AJ, 236, 508
Prantzos, N., Hashimoto, M., and Nomoto, K. 1990, A&A, 234, 211
Primas, F., Brugamyer, E., Sneden, C., King, J. R., Beers, T. C., Boesgaard, A. M., & Deliyannis, C. P. 2000, The First Stars, A. Weiss, T. Abel, & V. Hill (eds), (Springer-Verlag, Berlin), 51
Prochaska, J. X., Naumov, S. O., Carney, B. W., McWilliam, A., & Wolfe, A. M. 2000, AJ, in press
Raiteri, C. M., Busso, M., Gallino, R., Picchio, G., & Pulone, L., 1991, ApJ, 367, 228
Raiteri, C. M., Busso, M., Gallino, R., & Picchio, G. 1991, ApJ, 371, 665
Romano, D., Matteucci, F., Molaro, P., & Bonifacio, P. 1999, A&A, 352, 117
Rossi, S., Beers, T. C., & Sneden, C. 1999, ASP Conf.Ser. 165, 264
Ryan, S. G., 2000, in The Galactic Halo; from Globular Clusters to Field Stars, 35th Liège Int. Ap. Coll., A. Noels et al. (eds), U. Liège
Ryan, S. G., Aoki, W., Blake, L. A. J., Norris, J. E., Beers, T. C., Gallino, R., Busso, M., & Ando, H. 2001a, Mem.S.A.It., in press
Ryan, S. G., Blake, L. A. J., & Gregory, S. 2001b, in prep.

Ryan, S. G., Kajino, T., Beers, T. C., Suzuki, T., Romano, D., Matteucci, F., & Rosolankova, K. 2001c, ApJ, 548, (20 Feb)

Ryan, S. G. & Norris, J. E. 1991, AJ, 101, 1865

Ryan, S. G., Norris, J. E., & Beers, T. C. 1996, ApJ, 471, 254

Ryan, S. G., Norris, J. E., & Beers, T. C. 1999, ApJ, 523, 654

Ryan, S. G., Norris, J. E. & Bessell, M. S. 1991, AJ, 102, 303

Sneden, C., McWilliam, A., Preston, G. P., Cowan, J. J., Burris, D. L., & Armosky, B. J., 1996, ApJ, 467, 819

Sneden, C. & Crocker, D. A. 1988, ApJ, 335, 406

Sneden, C., Gratton, R. G., & Crocker, D. A. 1991, A&A, 246, 354

Steidel, C. C., Adelberger, K. L., Giavalisco, M., Dickinson, M., & Pettini, M., 1999, ApJ, 519, 1

Thielemann, K.-F., Nomoto, K., & Hashimoto, M., 1996, ApJ, 460, 408

Timmes, F. X., Woosley, S. E., & Weaver, T. A. 1995, ApJS, 98, 617

Tinsley, B. 1980, Fund.Cosmic Phys, 5, 287

Travaglio, C., Randich, S., & Galli, D. 2001, Nucl.Phys.A, in prep

Venn, K. A. 1999, ApJ, 518, 405

Woosley, S. E. & Weaver, T. A. 1995, ApJS, 101, 181

HYPERNOVA NUCLEOSYNTHESIS AND GALACTIC CHEMICAL EVOLUTION

Ken'ichi Nomoto, Keiichi Maeda, Hideyuki Umeda
Takayoshi Nakamura
Department of Astronomy and Research Center for the Early Universe
School of Science, University of Tokyo

Keywords: Supernovae, Nucleosynthesis, Chemical evolution

Abstract We study nucleosynthesis in 'hypernovae', i.e., supernovae with very large explosion energies ($\gtrsim 10^{52}$ ergs) for both spherical and aspherical explosions. The hypernova yields compared to those of ordinary core-collapse supernovae show the following characteristics: 1) Complete Si-burning takes place in more extended region, so that the mass ratio between the complete and incomplete Si burning regions is generally larger in hypernovae than normal supernovae. As a result, higher energy explosions tend to produce larger [(Zn, Co)/Fe], small [(Mn, Cr)/Fe], and larger [Fe/O], which could explain the trend observed in very metal-poor stars. 2) Si-burning takes place in lower density regions, so that the effects of α-rich freezeout is enhanced. Thus ^{44}Ca, ^{48}Ti, and ^{64}Zn are produced more abundantly than in normal supernovae. The large [(Ti, Zn)/Fe] ratios observed in very metal poor stars strongly suggest a significant contribution of hypernovae. 3) Oxygen burning also takes place in more extended regions for the larger explosion energy. Then a larger amount of Si, S, Ar, and Ca ("Si") are synthesized, which makes the "Si"/O ratio larger. The abundance pattern of the starburst galaxy M82b may be attributed to hypernova explosions. Asphericity in the explosions strengthens the nucleosynthesis properties of hypernovae as summarized above. We thus suggest that hypernovae make important contribution to the early Galactic (and cosmic) chemical evolution.

1. INTRODUCTION

Massive stars in the range of 8 to $\sim 150 M_\odot$ undergo core-collapse at the end of their evolution and become Type II and Ib/c supernovae (SNe II and SNe Ib/c) unless the entire star collapses into a black hole with no mass ejection (e.g., Arnett 1996). These SNe II and SNe Ib/c (as

D. Vanbeveren (ed.), The Influence of Binaries on Stellar Population Studies, 507–533.

well as Type Ia supernovae) release large explosion energies and eject explosive nucleosynthesis materials, thus having strong dynamical, thermal, and chemical influences on the evolution of interstellar matter and galaxies. Therefore, the explosion energies of core-collapse supernovae are fundamentally important quantities, and an estimate of $E \sim 1 \times 10^{51}$ ergs has often been used in calculating nucleosynthesis (e.g., Woosley & Weaver 1995; Thielemann et al. 1996) and the impact on the interstellar medium. (In the present paper, we use the explosion energy E for the final kinetic energy of explosion.)

SN1998bw called into question the applicability of the above energy estimate for all core-collapse supernovae. This supernova was discovered in the error box of the gamma-ray burst GRB980425, only 0.9 days after the date of the gamma-ray burst and was very possibly linked to it (Galama et al. 1998). SN1998bw is classified as a Type Ic supernova (SN Ic) but it is quite an unusual supernova. The very broad spectral features and the light curve shape have led to the conclusion that SN 1998bw was a hyper-energetic explosion of a massive ($\sim$ 14 $M_\odot$) C + O star with E = 3 - 6 $\times$ 10^{52} ergs, which is about thirty times larger than that of a normal supernova (Iwamoto et al. 1998; Woosley et al. 1999; Branch 2000; Nakamura et al. 2001a). The amount of ^{56}Ni ejected from SN1998bw is found to be $M(^{56}\mathrm{Ni}) \simeq 0.4$ - 0.7 $M_\odot$ (Nakamura et al. 2001a; Sollerman et al. 2000), which is about 10 times larger than the 0.07$M_\odot$ produced in SN1987A, a typical value for core-collapse supernovae.

In the present paper, we use the term 'Hypernova' to describe such an extremely energetic supernova with $E \gtrsim 10^{52}$ ergs without specifying without specifying the nature of the central engine (Nomoto et al. 2000).

Recently, other hypernovae candidates have been recognized. SN1997ef and SN1998ey are also classified as SNe Ic, and show very broad spectral features similar to SN1998bw (Iwamoto et al. 2000). The spectra and the light curve of SN1997ef have been well simulated by the explosion of a 10$M_\odot$ C+O star with $E = 1.0 \pm 0.2 \times 10^{52}$ ergs and $M(^{56}\mathrm{Ni}) \sim$ 0.15 $M_\odot$ (Iwamoto et al. 2000; Mazzali, Iwamoto, & Nomoto 2000). SN1997cy is classified as a SN IIn and unusually bright (Germany et al. 2000; Turatto et al. 2000). Its light curve has been simulated by a circumstellar interaction model which requires $E \sim 5 \times 10^{52}$ ergs (Turatto et al. 2000). The spectral similarity of SN1999E to SN1997cy (Cappellaro et al. 1999) would suggest that SN1999E is also a hypernova. Note that all of these estimates of E assume a spherically symmetric event.

Hypernovae may be associated with the formation of a black hole as has been discussed in the context of the GRB-SNe connection (Woosley 1993; Paczynski 1998; Iwamoto et al. 1998; MacFadyen & Woosley

1999). In these models, the gravitational energy, or the rotational energy would be released via pair-neutrino annihilation or the Blandford-Znajek mechanism (Blandford & Znajek 1977). Alternatively, large magnetic energies are released from a possible magnetar (Nakamura 1998; Wheeler et al. 2000). The explosion may also be aspherical (Höflich et al. 1999; MacFadyen & Woosley 1999; Khokhlov et al. 1999).

We investigate the characteristics of nucleosynthesis in such energetic core-collapse hypernovae, the systematic study of which has not yet been done. We examine both spherical and aspherical explosion models and discuss their contributions to the Galactic chemical evolution.

2. SPHERICAL HYPERNOVA EXPLOSIONS

We use various progenitor models with the main sequence masses of 20, 25, 30, and $40M_\odot$ (whose He core masses are 6, 8, 10, $16M_\odot$, respectively) (Nomoto & Hashimoto 1988; Umeda, Nomoto, & Nakamura 2000), metallicity of $Z = 0$ to solar, and explosion energies of $E_{51} = E/10^{51}$ ergs = 1 - 100 (Nakamura et al. 2001b; Umeda & Nomoto 2001).

In core-collapse supernovae/hypernovae, stellar material undergoes shock heating and the subsequent explosive nucleosynthesis. Iron-peak elements including Cr, Mn, Co, and Zn are produced in two distinct regions, which are characterized by the peak temperature, $T_{\rm peak}$, of the shocked material (Fig. 2). For $T_{\rm peak} > 5 \times 10^9$K, material undergoes complete Si burning whose products include Co, Zn, V, and some Cr after radioactive decays. For $4 \times 10^9\text{K} < T_{\rm peak} < 5 \times 10^9$K, incomplete Si burning takes place and its after decay products include Cr and Mn (e.g., Hashimoto, Nomoto, Shigeyama 1989; Woosley & Weaver 1995; Thielemann, Nomoto & Hashimoto 1996).

Figure 1 shows nucleosynthesis in hypernovae and normal supernovae for $E_{51} = 30$ and 1. The progenitor is $40M_\odot$ star model (Nakamura et al. 2001b). From this figure, we note the following characteristics of nucleosynthesis with very large explosion energies.

1) The region of complete Si-burning, where ^{56}Ni is dominantly produced, is extended to the outer, lower density region. The large amount of ^{56}Ni observed in hypernovae (e.g., $\sim$ 0.4 - 0.7 $M_\odot$ for SN1998bw and $\sim 0.15M_\odot$ for SN1997ef) implies that the mass cut is rather deep, so that the elements synthesized in this region such as ^{59}Cu, ^{63}Zn, and ^{64}Ge (which decay into ^{59}Co, ^{63}Cu, and ^{64}Zn, respectively) are ejected more abundantly than in normal supernovae. In the complete Si-burning region of hypernovae, elements produced by α-rich freezeout are enhanced because nucleosynthesis proceeds at lower densities than in normal su-

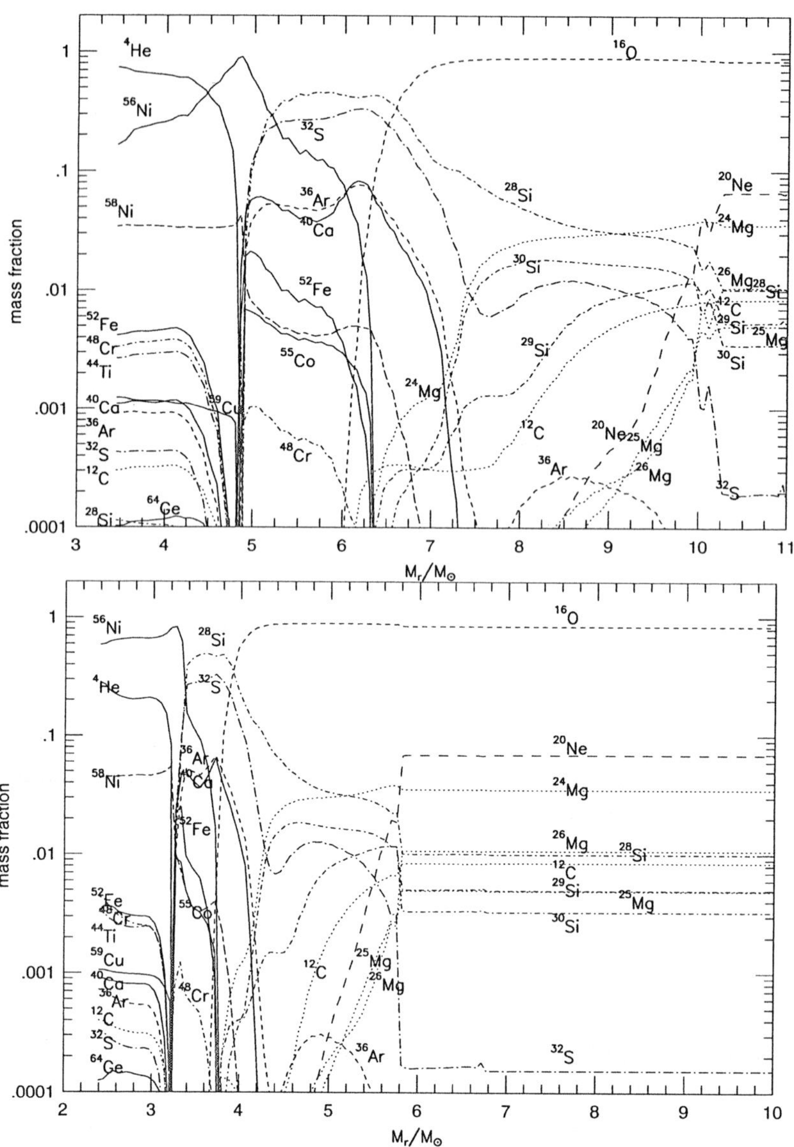

Figure 1 Nucleosynthesis in hypernovae and normal core-collapse supernovae for explosion energies of $E = 30$ (left) and 1 (right) $\times\ 10^{51}$ ergs. The progenitor model is the $40M_\odot$ star model (Nakamura et al. 2001b).

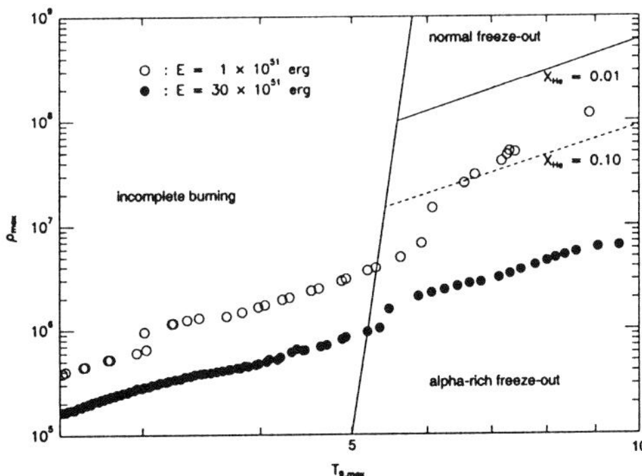

Figure 2 The ρ - T conditions of individual mass zones at their temperature maximum in hypernovae ($E = 30 \times 10^{51}$ ergs: filled circles) and normal supernovae ($E = 1 \times 10^{51}$erg: open circles) (Nakamura et al. 2001b).

pernovae (Fig. 2). Figure 1 clearly shows the trend that a larger amount of ^{4}He is left in hypernovae. Hence, elements synthesized through capturing of α-particles, such as ^{44}Ti, ^{48}Cr, and ^{64}Ge (decaying into ^{44}Ca, ^{48}Ti, and ^{64}Zn, respectively) are more abundant.

2) More energetic explosions produce a broader incomplete Si-burning region. The elements produced mainly in this region such as ^{52}Fe, ^{55}Co, and ^{51}Mn (decaying into ^{52}Cr, ^{55}Mn, and ^{51}V, respectively) are synthesized more abundantly with the larger explosion energy.

3) Oxygen and carbon burning takes place in more extended, lower density regions for the larger explosion energy. Therefore, the elements O, C, Al are burned more efficiently and the abundances of the elements in the ejecta are smaller, while a larger amount of burning products such as Si, S, and Ar is synthesized by oxygen burning.

The nucleosynthesis is characterized by large abundance ratios of intermediate mass nuclei and heavy nuclei with respect to ^{56}Fe for more energetic explosions, except for the elements O, C, Al which are consumed in oxygen and carbon burning. In particularly, the amounts of ^{44}Ca and ^{48}Ti are increased significantly with increasing explosion energy because of the lower density regions which experience complete Si-burning through α-rich freezeout. To see this more clearly, we add

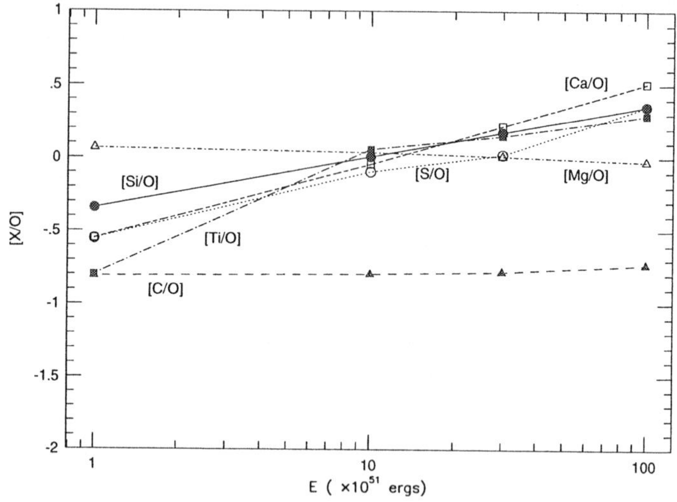

Figure 3 Abundance ratios to oxygen relative to solar values as a function of the explosion energy E. The progenitor model is the $25M_{\odot}$ star model (Nakamura et al. 2001b).

the isotopes and show in Figure 3 their ratios to oxygen relative to the solar values for various explosion energies. We note that [C/O] and [Mg/O] are not sensitive to the explosion energy because C, O, and Mg are consumed by oxygen burning. On the other hand, [Si/O], [S/O], [Ti/O], and [Ca/O] are larger for larger explosion energies because Si and S are produced in the oxygen burning region, and Ti and Ca are increased in the enhanced α-rich freezeout.

3. ASPHERICAL EXPLOSIONS

Despite the success of the hypernova model in reproducing the observed features of SN 1998bw at early times, some properties of the observed light curve and spectra at late times are difficult to explain in the context of a spherically symmetric model. (1) The decline of the observed light curve tail is slower than that of the synthetic curve, indicating that at advanced epochs γ-ray trapping is more efficient than expected (Nakamura et al. 2001a; Sollerman et al. 2000). (2) In the nebular epoch, the OI]6300Å emission is narrower than the emission centered at 5200Å. This latter feature can be identified as FeII], although it may also contain allowed FeII transitions. If the identification is correct, the difference in width may suggest that the mean expansion velocity of iron is higher than that of oxygen and that a significant amount of oxygen exists at low velocity (Danziger et al. 1999; Nomoto et al. 2000; Patat et al. 2001). Both these features are in conflict with what is expected from a spherically symmetric explosion model, where γ-ray deposition is a decreasing function of time and where iron is produced in the deepest layers and thus has a lower average velocity than oxygen. We suggest that these are signatures of asymmetry in the ejecta. Therefore in this paper we examine aspherical explosion models for hypernovae.

On the theoretical side, MacFadyen & Woosley (1999) showed that the collapse of a rotating massive core can form a black hole with an accretion disk, and that a jet emerges along the rotation axis. The jet can produce a highly asymmetric supernova explosion (Khokhlov et al. 1999). However, these studies did not calculate explosive nucleosynthesis, nor spectroscopic and photometric features of aspherical explosions. Nagataki et al. (1997) performed nucleosynthesis calculations for aspherical supernova explosions to explain some features of SN 1987A, but they only addressed the case of a normal explosion energy.

Maeda et al. (2000) have examined the effect of aspherical (jet-like) explosions on nucleosynthesis in hypernovae. They have investigated the degree of asphericity in the ejecta of SN 1998bw, which is critically important information to confirm the SN/GRB connection, by computing synthetic spectra for the various models viewed from different orientations and comparing the results with the observed late time spectra (Patat et al. 2001).

3.1. NUCLEOSYNTHESIS

Maeda et al. (2000) have constructed several asymmetric explosion models. The progenitor model is the 16 $M_\odot$ He core of the 40 $M_\odot$ star. The explosion energy is $E = 1 \times 10^{52}$.

Figure 4 shows the isotopic composition of the ejecta of asymmetric explosion model in the direction of the jet (upper panel) and perpendicular to it (lower panel). In the z-direction, where the ejecta carry more kinetic energy, the shock is stronger and post-shock temperatures are higher, so that explosive nucleosynthesis takes place in more extended, lower density regions compared with the r-direction. Therefore, larger amounts of α-rich freeze-out elements, such as ^{4}He and ^{56}Ni (which decays into ^{56}Fe via ^{56}Co) are produced in the z-direction than in the r-direction. Also, the expansion velocity of newly synthesized heavy elements is much higher in the z-direction.

Comparison with the nucleosynthesis of spherically symmetric hypernova explosions (Nakamura et al. 2001b) shows that the distribution and expansion velocity of newly synthesized elements ejected in

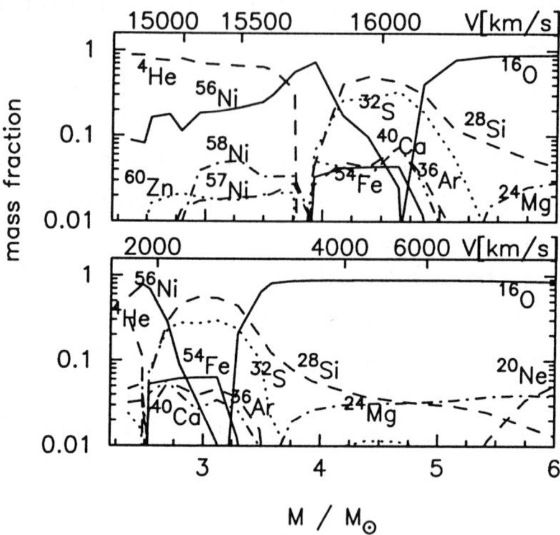

Figure 4 The isotopic composition of the ejecta in the direction of the jet (upper panel) and perpendicular to it (lower panel). The ordinate indicates the initial spherical Lagrangian coordinate (M_r) of the test particles (lower scale), and the final expansion velocities (V) of those particles (upper scale) (Maeda et al. 2000).

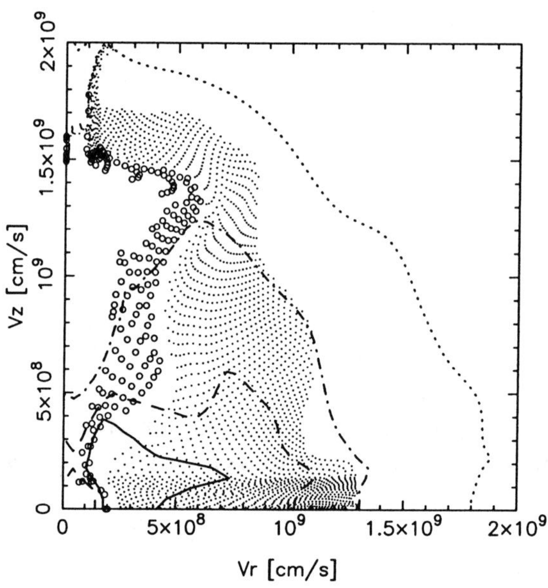

Figure 5 The distribution of ^{56}Ni (open circles) and ^{16}O (dots). The open circles and the dots denote test particles in which the mass fraction of ^{56}Ni and ^{16}O, respectively, exceeds 0.1. The lines are density contours at the level of 0.5 (solid), 0.3 (dashed), 0.1 (dash-dotted), and 0.01 (dotted) of the max density, respectively (Maeda et al. 2000).

the z-direction is similar to the result of the spherical explosion model with $E \sim 3 \times 10^{52}$ ergs, although the integrated kinetic energy is only $E = 1 \times 10^{52}$ ergs.

On the other hand, along the r-direction ^{56}Ni is produced only in the deepest layers, and the elements ejected in this direction are mostly the products of hydrostatic nuclear burning stages (O) with some explosive oxygen-burning products (Si, S, etc). The expansion velocities are much lower than in the z-direction.

Figure 5 shows the 2D distribution of ^{56}Ni and ^{16}O in the homologous expansion phase. In the direction closer to the z-axis, the shock is stronger and a low density, ^{4}He-rich region is produced. ^{56}Ni is distributed preferentially in the jet direction. The distribution of heavy elements is elongated in the z-direction, while the ^{16}O distribution is less aspherical.

3.2. NEBULA SPECTRA OF SN 1998BW

Mazzali et al. (2001) calculated synthetic nebular-phase spectra of SN 1998bw using a spherically symmetric NLTE nebular code, based on the deposition of γ-rays from ^{56}Co decay in a nebula of uniform density and composition. They showed that the nebular lines of different elements can be reproduced if different velocities are assumed for these elements, and that a significant amount of slowly-moving O is necessary to explain the zero velocity peak of the O I] line. Different amounts of ^{56}Ni ($0.35 M_{\odot}$ and $0.6 M_{\odot}$, respectively) are required to fit the narrow O I] line and the broad-line Fe spectrum (see figures in Danziger et al. 1999 and Nomoto et al. 2000). This suggests that an aspherical distribution of Fe and O may explain the observations.

In order to verify the observable consequences of an axisymmetric explosion, we calculated the profiles of the Fe-dominated blend near 5200Å, and of O I] 6300, 6363ÅṪhese are the lines that deviate most from the expectations from a spherically symmetric explosion. Line emissivities were obtained from the 1D NLTE code, and the column densities of the various elements along different lines of sight were derived from the element distribution obtained from our 2D explosion models.

The nebular line profiles of iron and oxygen with various view angles are shown in Figure 6 together with the spherical model (Maeda 2001). In this figure, the observed spectrum on day 139 (Patat et al. 2001) is also plotted for comparison.

The 'full-width-at-half-maximum' of our synthetic lines should be compared to the observational values, which are 440Å for the Fe-blend and 200Å for the O I] doublet, and were estimated from the late time

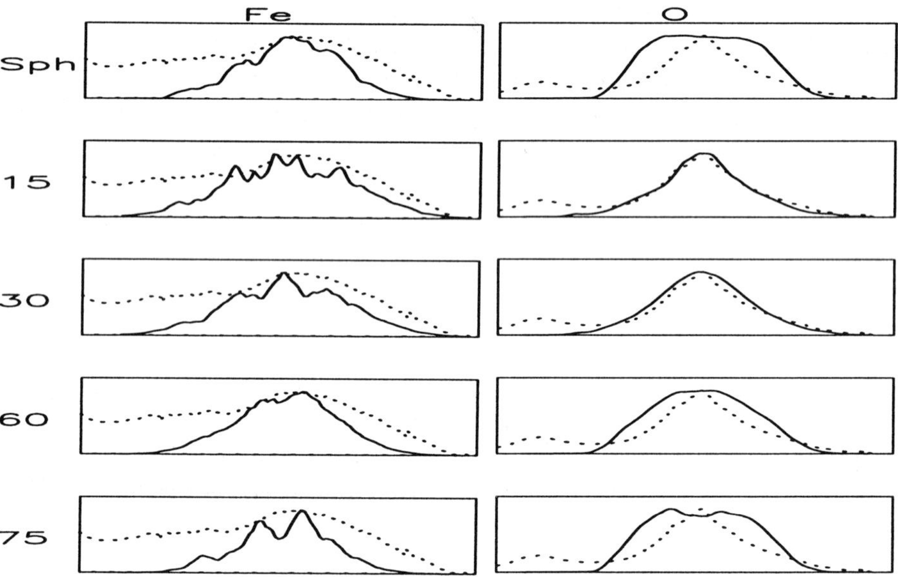

Figure 6 The profiles of the Fe-blend (left panels) and of O I] 6300, 6363 Å (right panels) viewed at 15°, 30°, 60° and 75° from the jet direction (Maeda 2001). The top panel shows the profiles of the spherically symmetric model. The observed lines at a SN rest-frame epoch of 139 days are also plotted for comparison (dotted lines, Patat et al. 2000). The intensities of the strongest lines, normalized to O I] 6300.3Å are: FeII 5158.8Å: 0.207; FeII 5220.1Å: 0.043; FeII 5261.6Å: 0.139; FeII 5273.3Å: 0.066; FeII 5333.6Å: 0.099; FeIII 5270.4Å: 0.086; and OI 6363.8Å: 0.331.

spectrum of 12 Sept 1998, 139 days after the explosion in the SN rest frame, assuming that the continuum level is the value around 5700 Å where the flux has the minimum value.

The observed line profiles are not explained with a spherical model. In a spherical explosion, oxygen is located at higher velocities than iron, and although the Fe-blend can be wider than the O line if the O and Fe regions are mixed extensively, the expected ratio of the width of the Fe-blend and the O line even in fully mixed model is $\sim 3:2$ (Mazzali et al. 2001). However, the observed ratio is even larger, $\sim 2:1$.

Because Fe is distributed preferentially along the jet direction, our aspherical explosion models can produce a larger ratio. If we view the aspherical explosion model from a near-jet direction, the ratio between the Fe and O line widths is comparable to the observed value. For a larger explosion energy, the width of the oxygen line is too large to be compatible with the observations.

The aspherical explosion models can produce a larger ratio than the spherical model because Fe is distributed preferentially along the jet direction. If the explosion energy is too large, the width of the oxygen line is too large to be compatible with the observations.

The iron and oxygen profiles viewed at an angle of 15° from the jet direction are compared to the observed spectrum on day 139 in Figure 6. When the degree of asphericity is high and the viewing angle is close to the jet direction, the component iron lines in the blend have double-peaked profiles, the blue- and red-shifted peaks corresponding to Fe-dominated matter moving towards and away from us, respectively. This is because Fe is produced mostly along the z (jet) direction. Because of the high velocity of Fe, the peaks are widely separated, making the blend wide. This is the case for the synthetic Fe-blend shown in Figure 6. In contrast, the oxygen line is narrower and has a sharper peak, because O is produced mostly in the r-direction, at lower velocities and with a less aspherical distribution.

The Fe-blend line computed with the aspherical model shows some small peaks (Fig. 6), which are not seen in the observations. However, the detailed profile of the Fe-blend is very sensitive to the matter distribution. Mixing of the ejecta may distribute the fast-moving ^{56}Ni to lower velocities, which would reduce the double-peaked profiles of the Fe lines.

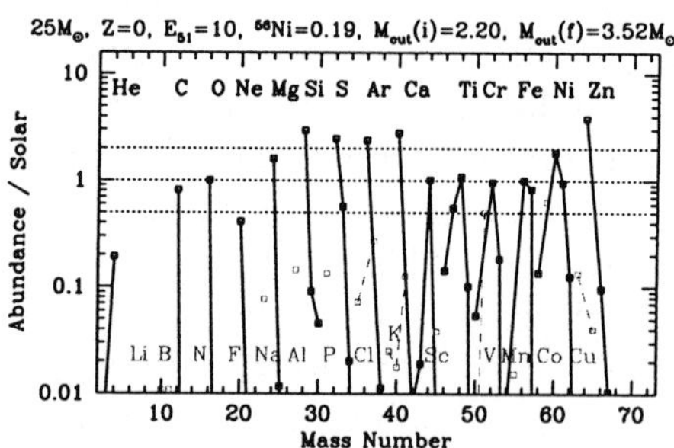

Figure 7 Abundance pattern in the ejecta normalized by the solar ^{16}O abundances for the $25M_{\odot}$ model with $E_{51} = 10$. The mass-cuts are chosen to give large [Zn/Fe] and [O/Fe]=0 (Umeda & Nomoto 2001).

4. CONTRIBUTION OF HYPERNOVAE TO THE GALACTIC CHEMICAL EVOLUTION

The abundance pattern of metal-poor stars with [Fe/H] < -2 provides us with very important information on the formation, evolution, and explosions of massive stars in the early evolution of the galaxy. Contributions of hypernova nucleosynthesis to the Galactic chemical evolution could be important because of their large amounts of heavy elements production. Figure 7 shows the abundance pattern of the explosion of the Pop III (metal-free) 25 $M_{\odot}$ star.

If hypernovae occurred in the early stage of the Galactic evolution, the abundance pattern of a hypernova may be observable in some low-mass halo stars. This results from the very metal-poor environment, where the heavy elements synthesized in a single hypernova (or a single supernova) dominate the heavy element abundance pattern (Audouze & Silk 1995). It is plausible that hypernova explosions induce star formations. The low-mass stars produced by this event should have the hypernova-like abundance pattern and still exist in the Galactic halo. The metallicity of such stars is likely to be determined by the ratio of ejected iron mass from the relevant hypernova to the mass of hydrogen gathered by the hypernova, which might be in the range range of $-4 \lesssim$ [Fe/H] $\lesssim -2.5$

(Ryan et al. 1996; Shigeyama & Tsujimoto 1998; Nakamura et al. 1999; Argast et al. 2000).

4.1. IRON, TITANIUM

One of the most significant feature of hypernova nucleosynthesis is a large amount of Fe. One hypernova can produce 2 - 10 times more Fe than normal core-collapse supernovae, which is almost the same amount of Fe as produced in a SN Ia. This large iron production leads to small ratios of α elements over iron in hypernovae (Figure 7). In this connection, the abundance pattern of the very metal-poor binary CS22873-139 ([Fe/H] $= -3.4$) is interesting. This binary has only an upper limit to [Sr/Fe] < -1.5, and therefore was suggested to be a second generation star (Spite et al. 2000). The interesting pattern is that this binary shows almost solar Mg/Fe and Ca/Fe ratios, as would be the case with hypernovae. Another feature of CS22873-139 is enhanced Ti/Fe ([Ti/Fe] $\sim +0.6$), which could be explained by an (especially, aspherical) hypernova explosion.

It has been pointed out that Ti is deficient in Galactic chemical evolution models using supernova yields currently available (e.g., Timmes et al. 1995; Thielemann et al. 1996), especially at [Fe/H] $\lesssim -1$ when SNe Ia have not contributed to the chemical evolution. However, if the contribution from hypernovae to Galactic chemical evolution is relatively large (or supernovae are more energetic than the typical value of 1×10^{51} erg), this problem could be relaxed. As we have seen in the previous section, the α-rich freezeout is enhanced in hypernovae because nucleosynthesis proceeds under the circumstance of lower densities and incomplete Si-burning occurs in more extended regions. Thus, ^{40}Ca, ^{44}Ca, and ^{48}Ti are produced and could be ejected into space more abundantly.

4.2. MANGANESE, CHROMIUM, COBALT

The observed abundances of metal-poor halo stars show quite interesting pattern. There are significant differences between the abundance patterns in the iron-peak elements below and above [Fe/H]~ -2.5. For [Fe/H]$\lesssim -2.5$, the mean values of [Cr/Fe] and [Mn/Fe] decrease toward smaller metallicity, while [Co/Fe] increases (McWilliam et al. 1995; Ryan et al. 1996). This trend cannot be explained with the conventional chemical evolution model that uses previous nucleosynthesis yields.

4.2.1 Mass Cut Dependence.

Nakamura et al. (1999) have shown that this trend of decreasing Cr and Mn with increasing Co is

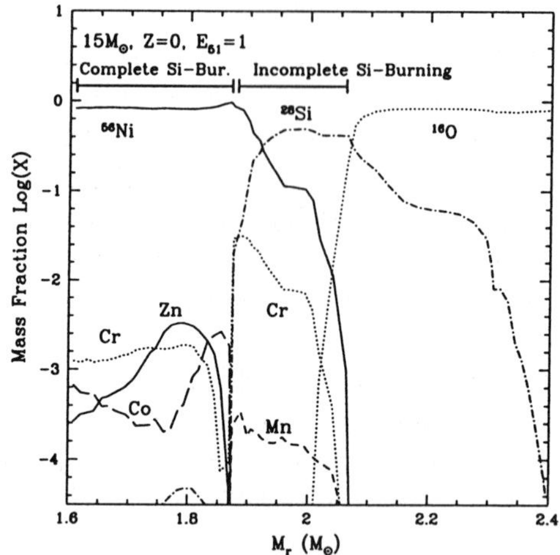

Figure 8 Abundance distribution just after supernova explosion for the Pop III model with mass $15M_\odot$. The lines labeled by Cr, Mn, Co and Zn actually indicate unstable ^{52}Fe, ^{55}Co, ^{59}Cu, and ^{64}Ge, respectively, which eventually decay to the labeled elements (Umeda & Nomoto 2001).

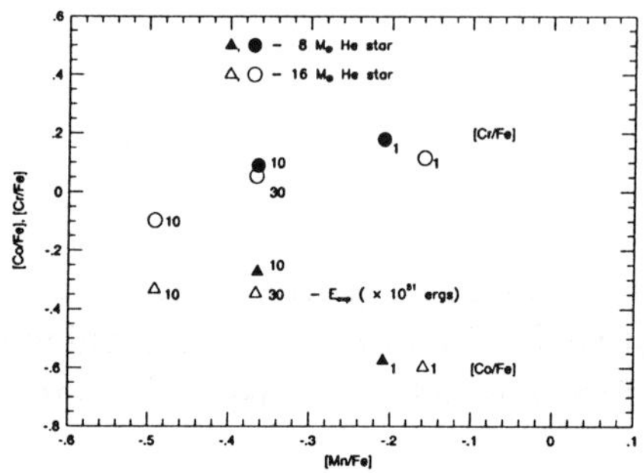

Figure 9 [Mn/Fe] vs. [Co/Fe] and [Cr/Fe] for the 40 $M_\odot$ star (16 $M_\odot$ He star) models with $E = (1.0$ - $30) \times 10^{51}$ ergs, and the 25 $M_\odot$ star (8 $M_\odot$ He star) models with $E = (1$ - $10) \times 10^{51}$ ergs (Nakamura et al. 2001b).

reproduced by decreasing the mass cut between the ejecta and the collapsed star for the same explosion model. This is because Co is mostly produced in complete Si-burning regions, while Mn and Cr are mainly produced in the outer incomplete Si-burning region (Fig. 8). If the mass cut is located at smaller M_r, the mass ratio between the complete and incomplete Si-burning region is larger. Therefore, mass cuts at smaller M_r increase the Co fraction but decrease the Mn and Cr fractions in the ejecta. Nakamura et al. (1999) have also shown that the observed trend with respect to [Fe/H] may be explained if the mass cut tends to be smaller in M_r for the larger mass progenitor.

4.2.2 Energy Dependence.

Nakamura et al. (2001b) have investigated whether the observed trend of these iron-peak elements in metal-poor stars can be explained with the the abundance pattern of hypernovae. In Figure 9, we plot [Mn/Fe] vs. [Co/Fe] and [Cr/Fe] for the 16 $M_\odot$ He star models with E = (1 - 30) $\times$ 10^{51} ergs, and 8 $M_\odot$ He star models with E = (1 - 10) $\times$ 10^{51} ergs. This figure clearly shows the correlation between [Mn/Fe] and [Cr/Fe], and anti-correlation between [Mn/Fe] and [Co/Fe], which are the same trends as observed in the metal-poor stars.

To understand the dependence on E, let us compare the models with 1×10^{51} and 10×10^{51} ergs which have almost the same mass cut. In the model with $E = 10 \times 10^{51}$ ergs, both complete and incomplete Si-burning regions shift outward in mass compared with $E = 1 \times 10^{51}$ ergs because of a larger explosion energy. Thus, the model with larger E has a larger mass ratio between the complete and incomplete Si-burning regions. (This relation between E and the mass ratio does not hold if the explosion energy is too large, because incomplete Si-burning extends so far out that Mn and Cr increase too much to fit the metal-poor star data.)

In metal-poor stars, [Mn/Fe] increases with [Fe/H]. Hypernova yields are consistent with this trend if hypernovae with larger E induce the formation of stars with smaller [Fe/H]. This supposition is reasonable because the mass of interstellar hydrogen gathered by a hypernova is roughly proportional to E (Cioffi et al. 1988; Shigeyama & Tsujimoto 1998) and the ratio of the ejected iron mass to E is smaller for hypernovae than for canonical supernovae.

4.3. ZINC

For Zn, early observations have shown that [Zn/Fe]$\sim$ 0 for [Fe/H] $\simeq -3$ to 0 (Sneden, Gratton, & Crocker 1991). Recently Primas et al. (2000) have suggested that [Zn/Fe] increases toward smaller metallicity as seen in Figure 10, and Blake et al. (2001) has one with [Zn/Fe] $\simeq 0.6$ at [Fe/H] $= -3.3$ (see Ryan 2001).

As for Zn, its main production site has not been clearly identified. If it is mainly produced by s-processes, the abundance ratio [Zn/Fe] should decrease with [Fe/H]. This is not consistent with the observations of [Zn/Fe]~ 0 for [Fe/H] $\simeq -2.5$ to 0 and the increase in [Zn/Fe] toward lower metallicity for [Fe/H]$\lesssim -2.5$. Another possible site of Zn production is SNe II. However, previous nucleosynthesis calculations in SNe II appears to predict decreasing Zn/Fe ratio with Fe/H (Woosley & Weaver 1995; Goswami & Prantzos 2000).

Understanding the origin of the variation in [Zn/Fe] is very important especially for studying the abundance of Damped Ly-α systems (DLAs), because [Zn/Fe] $= 0$ is usually assumed to determine their abundance pattern. In DLAs super-solar [Zn/Fe] ratios have often been observed, but they have been explained that assuming dust depletion for Fe is larger than it is for Zn. However, recent observations suggest that the assumption [Zn/Fe] $= 0$ may not be always correct.

4.3.1 Spherical Hypernova Models. Umeda & Nomoto (2001) have calculated nucleosynthesis in massive Pop III stars and explored the parameter ranges ($M_{\rm cut}(i)$, Y_e, M, and E) to reproduce [Zn/F] ~ 0.5 observed in extremely metal-poor stars. Their main results are summarized as follows.

1) The interesting trends of the observed ratios [(Zn, Co, Mn, Cr)/Fe] can be understood in terms of the variation of the mass ratio between the complete Si burning region and the incomplete Si burning region. The large Zn and Co abundances observed in very metal-poor stars are obtained if the mass-cut is deep enough, or equivalently if deep material from complete Si-burning region is ejected by mixing or aspherical effects. Vanadium also appears to be abundant at low [Fe/H]. Since V is also produced mainly in the complete Si-burning region, this trend can be explained in the same way as those of Zn and Co.

2) The mass of the incomplete Si burning region is sensitive to the progenitor mass M, being smaller for smaller M. Thus [Zn/Fe] tends to be larger for less massive stars for the same E.

3) The production of Zn and Co is sensitive to the value of Y_e, being larger as Y_e is closer to 0.5, especially for the case of a normal explosion energy ($E_{51} \sim 1$).

4) A large explosion energy E results in the enhancement of the local mass fractions of Zn and Co, while Cr and Mn are not enhanced. This is because larger E produces larger entropy and thus more α-rich environment for α-rich freeze-out.

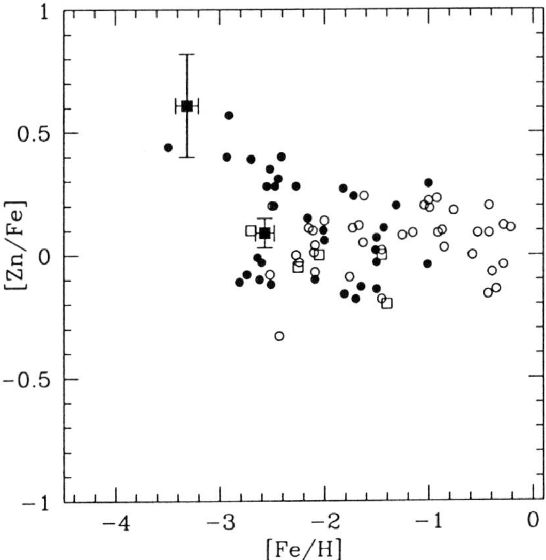

Figure 10 Observed abundance ratios of [Zn/Fe]. These data are taken from Primas et al. (2000) (filled circles), Blake et al. (2001) (filled square) and from Sneden et al. (1991) (others) (Umeda & Nomoto 2001b).

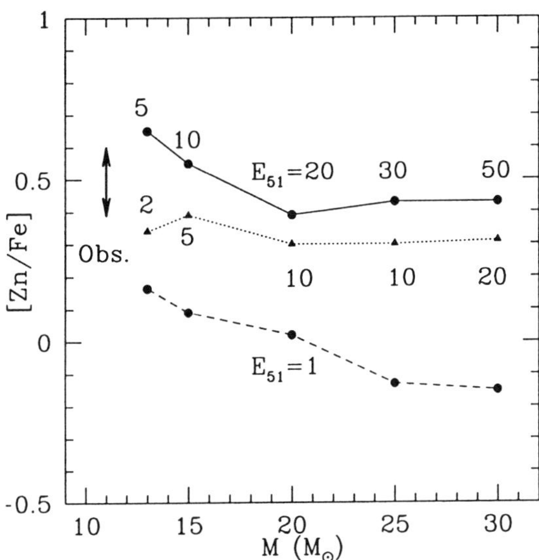

Figure 11 The maximum [Zn/Fe] ratios as a function of M and E_{51} (Umeda & Nomoto 2001). The observed large [Zn/Fe] ratio in very low-metal stars ([Fe/H] < -2.6) found in Primas et al. (2000) and Blake et al. (2001) are represented by a thick arrow.

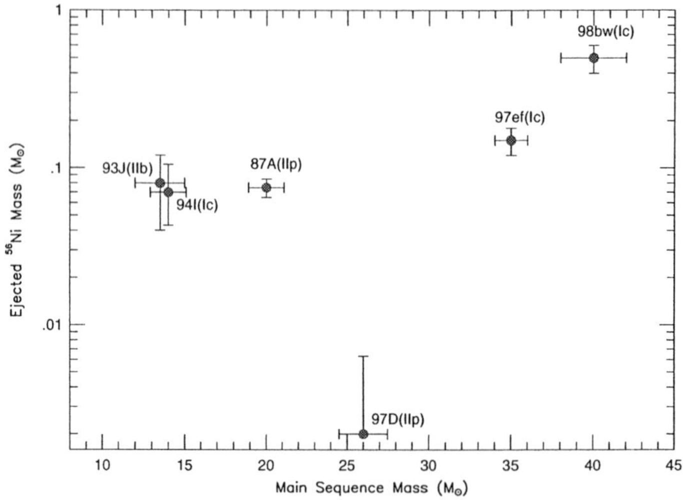

Figure 12 Ejected ^{56}Ni mass versus the main sequence mass of the progenitors of several bright supernovae obtained from light curve models (Iwamoto et al. 2000).

5) To be consistent with the observed [O/Fe] $\sim$ 0 - 0.5 as well as with [Zn/Fe] $\sim$ 0.5 in metal-poor stars, they propose that the "mixing and fall-back" process or aspherical effects are significant in the explosion of relatively massive stars.

The dependence of [Zn/Fe] on M and E is summarized in Figure 11. They have found that models with $E_{51} = 1$ do not produce sufficiently large [Zn/Fe]. To be compatible with the observations of [Zn/Fe] $\sim$ 0.5, the explosion energy must be much larger, i.e., $E_{51} \gtrsim 2$ for $M \sim 13 M_\odot$ and $E_{51} \gtrsim 20$ for $M \gtrsim 20 M_\odot$.

Observationally, the requirement of the large E might suggest that large M stars are responsible for large [Zn/Fe], because E and M can be constrained from the observed brightness and light curve shape of supernovae as follows (Fig. 12). The recent supernovae 1987A, 1993J, and 1994I indicate that the progenitors of these normal SNe are 13 - 20 $M_\odot$ stars and $E_{51} \sim$ 1 - 1.5 (Nomoto et al. 1993, 1994; Shigeyama et al. 1994; Blinnikov et al. 2000). On the other hand, the masses of the progenitors of hypernovae with $E_{51} > 10$ (SNe 1998bw, 1997ef, and 1997cy) are estimated to be $M \gtrsim 25 M_\odot$ (Nomoto et al. 2000; Iwamoto et al. 1998, 2000; Woosley et al. 1999; Turatto et al. 2000). This could be related to the stellar mass dependence of the explosion mechanisms and the formation of compact remnant, i.e., less massive stars form neutron stars, while more massive stars tend to form black holes.

To explain the observed relation between [Zn/Fe] and [Fe/H], we further need to know how M and E of supernovae and [Fe/H] of metal-poor halo stars are related. In the early galactic epoch when the galaxy is not yet chemically well-mixed, [Fe/H] may well be determined by mostly a single SN event (Audouze & Silk 1995). The formation of metal-poor stars is supposed to be driven by a supernova shock, so that [Fe/H] is determined by the ejected Fe mass and the amount of circumstellar hydrogen swept-up by the shock wave (Ryan et al. 1996).

Explosions with the following two combinations of M and E can be responsible for the formation of stars with very small [Fe/H]:

i) Energetic explosions of massive stars ($M \gtrsim 25 M_\odot$): For these massive progenitors, the supernova shock wave tends to propagate further out because of the large explosion energy and large Strömgren sphere of the progenitors (Nakamura et al. 1999). The effect of E may be important since the hydrogen mass swept up by the supernova shock is roughly proportional to E (e.g., Ryan et al 1996; Shigeyama & Tsujimoto 1998).

ii) Normal supernova explosions of less massive stars ($M \sim 13 M_\odot$): These supernovae are assumed to eject a rather small mass of Fe (Shigeyama & Tsujimoto 1998), and most SNe are assumed to explode with normal E irrespective of M.

The above relations lead to the following two possible scenarios to explain [Zn/Fe] ~ 0.5 observed in metal-poor stars.

i) Hypernova-like explosions of massive stars ($M \gtrsim 25 M_\odot$) with $E_{51} > 10$: Contribution of highly asymmetric explosions in these stars may also be important. The question is what fraction of such massive stars explode as hypernovae; the IMF-integrated yields must be consistent with [Zn/Fe] ~ 0 at [Fe/H] $\gtrsim -2.5$.

ii) Explosion of less massive stars ($M \lesssim 13 M_\odot$) with $E_{51} \gtrsim 2$ or a large asymmetry: This scenario, after integration over the IMF, can reproduce the observed abundance pattern for [Fe/H]$\gtrsim -2$. However, the Fe yield has to be very small in order to satisfy the observed [O/Fe] value ($\gtrsim 0.5$) for the metal-poor stars. For example, the ^{56}Ni mass yield of our $13 M_\odot$ model has to be less than $0.006 M_\odot$, which appears to be inconsistent with the nearby SNe II observations.

It seems that the [O/Fe] ratio of metal-poor stars and the E-M relations from supernova observations (Fig. 12) favor the massive energetic explosion scenario for enhanced [Zn/Fe]. However, we need to construct detailed galactic chemical evolution models to distinguish between the two scenarios for [Zn/Fe].

4.3.2 Aspherical Hypernova Models. Iron-peak elements are produced in the deep region near the mass cut, so that their pro-

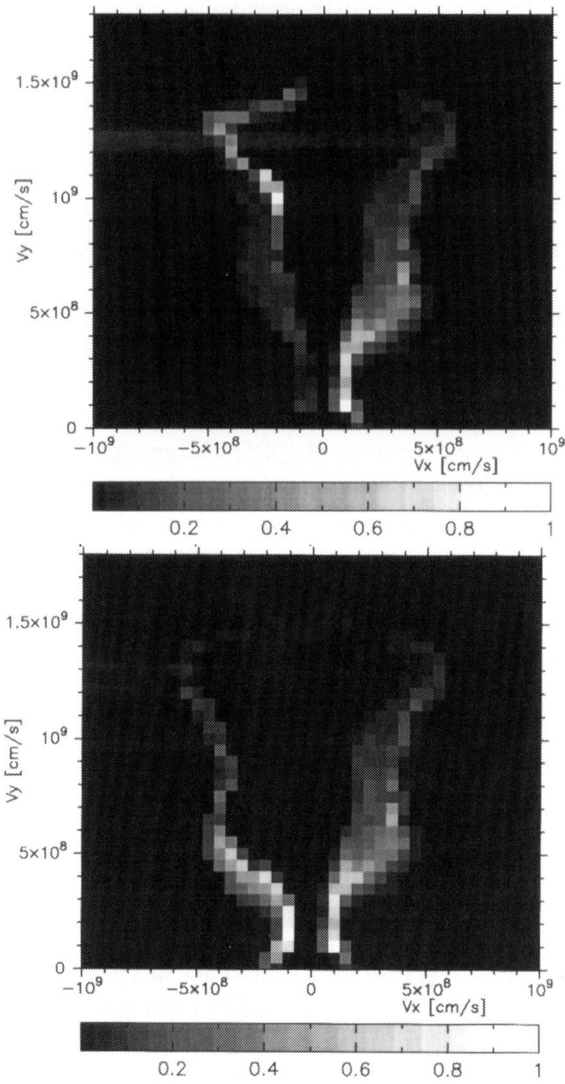

Figure 13 The density distributions of Zn (top: left half) and Fe (top: right half) and that of Mn (bottom: left half) and Fe (bottom: right half). The densities of each element are represented with linear scale, from 0 to the max density of each element (Maeda 2001).

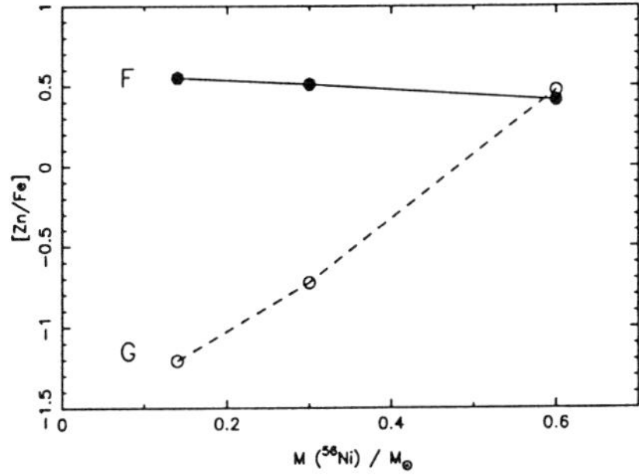

Figure 14 [Zn/Fe] as a function of the mass of ejected ^{56}Ni (Maeda 2001). Larger ^{56}Ni mass corresponds to deeper mass cut. The filled and open circles denote the aspherical and spherical models, respectively. The progenitor is the $40M_{\odot}$ star (Umeda & Nomoto 2001).

duction is strongly affected by asphericity of an explosion. Figure 13 shows the 2D density distribution of ^{64}Ge (which decays into ^{64}Zn) and that of ^{55}Co (which decays into ^{55}Mn). ^{64}Zn, which is produced by the strong α-rich freezeout, is distributed preferentially in the z-direction. Moreover, ^{64}Zn production is strongly enhanced compared with a spherical model, since the post-shock temperature along the z-direction is much higher than that of a spherical model. On the other hand, ^{55}Mn, which is produced by incomplete silicon burning, surrounds ^{56}Fe and located preferentially in the r-direction. Accordingly, materials with higher [Zn/Fe] and lower [Mn/Fe] are ejected along the Z-direction with higher velocities, and materials with lower [Zn/Fe] and higher [Mn/Fe] are ejected toward the r-direction with lower velocities.

Figure 14 shows that [Zn/Fe] is larger for deeper mass cut (i.e., larger ^{56}Ni mass) for spherical explosion, while [Zn/Fe] $\sim$ 0.5 is realized irrespective of the mass cut. This comes from the difference in the site of the Zn production. In the spherical case, Zn is produced only in the deepest layer, while in the aspherical model, the complete silicon burning region is elongated to the z (jet) direction (Figure 13), so that [Zn/Fe] is enhanced irrespective of the mass cut.

In this way, larger asphericity in the explosion leads to larger [Zn/Fe] and [Co/Fe], but to smaller [Mn/Fe] and [Cr/Fe]. Then, if the degree of the asphericity tends to be larger for lower [Fe/H], the trends of [Zn, Co, Mn, Cr/Fe] follow the observed ones.

It is likely that in the real situations, the asphericity, mixing, and the effects of the mass-cut (Nakamura et al. 1999) work to make observed trends. Then, the restrictions encountered by each of them (e.g., requirement of large Y_e, large amounts of mixing) may be relaxed.

5. OTHER OBSERVATIONAL SIGNATURES OF HYPERNOVA NUCLEOSYNTHESIS

5.1. STARBURST GALAXIES

X-ray emissions from the starburst galaxy M82 were observed with ASCA and the abundances of several heavy elements were obtained (Tsuru et al. 1997). Tsuru et al. (1997) found that the overall metallicity of M82 is quite low, i.e., O/H and Fe/H are only 0.06 - 0.05 times solar, while Si/H and S/H are $\sim$ 0.40 - 0.47 times solar. This implies that the abundance ratios are peculiar, i.e., the ratio O/Fe is about solar, while the ratios of Si and S relative to O and Fe are as high as $\sim$ 6 - 8. These ratios are very different from those ratios in SNe II. The age of M82 is estimated to be $\lesssim 10^8$ years, which is too young for Type Ia supernovae to contribute to enhance Fe relative to O. Tsuru et al. (1997) also estimated that the explosion energy required to produce the observed amount of hot plasma per oxygen mass is significantly larger than that of normal SNe II (here the oxygen mass dominates the total mass of the heavy elements). Tsuru et al. (1997) thus concluded that neither SN Ia nor SN II can reproduce the observed abundance pattern of M82.

Compared with normal SNe II, the important characteristic of hypernova nucleosynthesis is the large Si/O, S/O, and Fe/O ratios. Figure 15 shows the good agreement between the hypernova model and the observed abundances in M82 (Umeda et al. 2001). Hypernovae could also produce larger E per oxygen mass than normal SNe II. We therefore suggest that hypernova explosions may make important contributions to the metal enrichment and energy input to the interstellar matter in M82. If the IMF of the star burst is relatively flat compared with Salpeter IMF, the contribution of very massive stars and thus hypernovae could be much larger than in our Galaxy.

5.2. BLACK HOLE BINARIES

X-ray Nova Sco (GRO J1655-40), which consists of a massive black hole and a low mass companion (e.g., Brandt et al. 1995; Nelemans et al. 2000), also exhibits what could be the nucleosynthesis signature of a hypernova explosion. The companion star is enriched with Ti, S, Si, Mg, and O but not much Fe (Israelian et al. 1999). This is compatible with heavy element ejection from a black hole progenitor. In order to eject large amount of Ti, S, and Si and to have at least $\sim 4\ M_\odot$ below mass cut and thus form a massive black hole, the explosion would need to be highly energetic (Figure 1; Israelian et al. 1999; Brown et al. 2000;

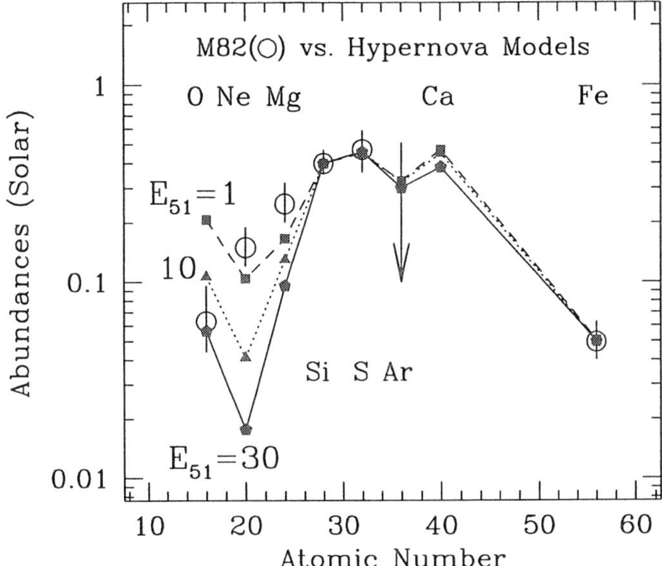

Figure 15 Abundance patterns in the ejecta of $25M_{\odot}$ metal-free SN II and hypernova models compared with abundances (relative to the solar values) of M82 observed with ASCA (Tsuru et al. 1997). Here, the open circles with error bars show the M82 data. The filled square, triangle, and pentagons represent E_{51}=1, 10, and 30 models, respectively, where E_{51} is the explosion energy in 10^{51} ergs. Theoretical abundances are normalized to the observed Si data, and the mass cuts are chosen to eject 0.07, 0.095, and 0.12 ($M_{\odot}$) Fe for E_{51} = 1, 10, and 30, respectively (Umeda et al. 2001).

Podsiadlowski et al. 2001). A hypernova explosion with the mass cut at large M_r ejects a relatively small mass Fe and would be consistent with these observed abundance features.

Alternatively, if the explosion which resulted from the formation of the black hole in Nova Sco was asymmetric, then it is likely that the companion star captured material ejected in the direction away from the strong shock which contained relatively little Fe compared with the ejecta in the strong shock (Maeda et al. 2000).

6. SUMMARY AND DISCUSSION

We investigated explosive nucleosynthesis in hypernovae. Detailed nucleosynthesis calculations were performed for both spherical and aspherical explosions. We also studied implications to Galactic chemical evolution and the abundances in metal-poor stars. The characteristics of hypernova yields compared to those of ordinary core-collapse supernovae are summarized as follows:

1) Complete Si-burning takes place in more extended region, so that the mass ratio between the complete and incomplete Si burning regions is generally larger in hypernovae than normal SNe II. As a result, higher energy explosions tend to produce larger [(Zn, Co)/Fe], small [(Mn, Cr)/Fe], and larger [Fe/O]. If hypernovae made significant contributions to the early Galactic chemical evolution, it could explain the large Zn and Co abundances and the small Mn and Cr abundances observed in very metal-poor stars. The large [Fe/O] observed in some metal-poor stars and galaxies might possibly be the indication of hypernovae rather than Type Ia supernovae.

2) In hypernovae, Si-burning takes place in lower density regions, so that the effects of α-rich freezeout is enhanced. Thus ^{44}Ca, ^{48}Ti, and ^{64}Zn are produced more abundantly than in normal supernovae. The large [(Ti, Zn)/Fe] ratios observed in very metal poor stars strongly suggest a significant contribution of hypernovae.

3) Oxygen burning also takes place in more extended regions for the larger explosion energy. Then a larger amount of Si, S, Ar, and Ca ("Si") are synthesized, which makes the Si/O ratio larger. The abundance pattern of the starburst galaxy M82, characterized by abundant Si and S relative to O and Fe, may be attributed to hypernova explosions if the IMF is relatively flat, and thus the contribution of massive stars to the galactic chemical evolution is large.

More direct signature of hypernovae can be seen in the black hole binary X-ray Nova Sco (GRO J1655-40), which indicates the production of a large amount of Ti, S, and Si when a massive black hole is formed as a compact remnant.

We also examine aspherical explosion models. The explosive shock is stronger in the jet-direction. The high explosion energies concentrated along the jet result in hypernova-like nucleosynthesis. On the other hand, the ejecta orthogonal to the jet is more like ordinary supernovae. Therefore, asphericity in the explosions strengthens the nucleosynthesis properties of hypernovae as summarized above in 1) - 3).

It is shown that such a large Zn abundance as [Zn/Fe] $\sim$ 0.5 observed in metal-poor stars can be realized in certain supernova models (Umeda & Nomoto 2001). This implies that the assumption of [Zn/Fe] $\sim$ 0 usually adopted in the DLA abundance analyses may not be well justified. Rather [Zn/Fe] may provide important information on the IMF and/or the age of the DLA systems.

Properties of hypernova nucleosynthesis suggest that hypernovae of massive stars may make important contributions to the Galactic (and cosmic) chemical evolution, especially in the early low metallicity phase. This may be consistent with the suggestion that the IMF of Pop III stars is different from that of Pop I and II stars, and that more massive stars are abundant for Pop III (e.g., Nakamura & Umemura 1999; Omukai & Nishi 1999; Bromm, Coppi & Larson 1999).

Acknowledgments

We would like to thank P. Mazzali, N. Patat, S. Ryan, C. Kobayashi, M. Shirouzu, F. Thielemann for useful discussion. This work has been supported in part by the grant-in-Aid for Scientific Research (07CE2002, 12640233) of the Ministry of Education, Science, Culture, and Sports in Japan.

References

Argast, D., Salmand, M., Gerhard, O.E., & Thielemann, F.-K. 2000, A&A, 356, 873

Arnett, W. D., Supernovae and Nucleosynthesis (Princeton Univ. Press)

Audouze, J., & Silk, J. 1995, ApJ, 451, L49

Blake, L.A.J., Ryan, S.G., Norris, J.E., & Beers, T.C. 2001, Nucl.Phys.A.

Blandford, R.D., & Znajek, R.L. 1977, MNRAS, 179, 433.

Blinnikov, S., Lundqvist, P, Bartunov, O., Nomoto, K., & Iwamoto, K. 2000, Apj, 532, 1132

Branch, D. 2000, in "Supernovae and Gamma Ray Bursts" eds. M. Livio, et al. (Cambridge: Cambridge University Press), in press

Brandt, W.N., Podsiadlowski, P., Sigurdssen, S. 1995, MNRAS, 277, L35

Brown, G.E., Lee, C.-H., Wijers, R.A.M.J., Lee, H.K., Israelian, G., & Bethe, H.A. 2000, New Astronomy, 5, 191

Bromm, V., Coppi, P.S., & Larson, R. B. 1999, ApJ, 527, L5

Cappellaro, E., Turatto, M., & Mazzali, P. 1999, IAU Circ. 7091

Cioffi, D.F., McKee, C.F. & Bertschinger, E. 1988, ApJ, 334, 252

Danziger, I.J., et al. 1999, in The Largest Explosions Since the Big Bang: Supernovae and Gamma Ray Bursts, eds. M. Livio, et al. (STScI), 9

Galama, T.J., et al. 1998, Nature, 395, 670

Germany, L.M., Reiss, D.J., Schmidt, B.P., Stubbs, C.W., & Sadler, E.M. 2000, ApJ, 533, 320

Hachisu, I., Matsuda, T., Nomoto, K., & Shigeyama, T. 1992, ApJ, 390, 230

Hachisu, I., Matsuda, T., Nomoto, K., & Shigeyama, T. 1994, A&AS, 104, 341

Hashimoto, M., Nomoto, K., & Shigeyama, T. 1989, A&A, 210, 5

Höflich, P., Wheeler, J. C., & Wang, L., 1999, ApJ, 521, 179

Israelian, G., Rebolo, R., Basri, G., Casares, J., & Martin E.L. 1999, Nature, 401, 142

Iwamoto, K., et al. 1998, Nature, 395, 672

Iwamoto, K., et al. 2000, ApJ, 534, 660

Iwamoto, K., Nomoto, K., Höflich, P., Yamaoka, H., Kumagai, S., & Shigeyama, T., 1994, ApJ, 437, L115
Khokhlov, A.M., Höflich, P.A., Oran, E.S., Wheeler, J.C., Wang, L., & Chtchelkanova, A.Yu. 1999, ApJ, 524, L107
MacFadyen, A.I. & Woosley, S.E. 1999, ApJ 524, 262
Maeda, K. 2001, Master Thesis, University of Tokyo
Maeda, K., Nakamura, T., Nomoto, K., Mazzali, P.A., & Hachisu, I. 2000, ApJ, submitted (astro-ph/0011003)
Mazzali, P.A., Iwamoto, K., & Nomoto, K. 2000, ApJ, 545, 407
Mazzali, P.A., Nomoto, K., & Patat, F. 2001, ApJ, submitted
McWilliam, A., Preston, G.W., Sneden, C., & Searle, L. 1995, AJ, 109, 2757
Nagataki, S., Hashimoto, M., Sato, K., Yamada, S. 1997, ApJ, 486, 1026
Nakamura, F., & Umemura, M. 1999, ApJ, 515, 239
Nakamura, T. 1998, Prog. Theor. Phys., 100, 921
Nakamura, T., Mazzali, P. A., Nomoto, K., & Iwamoto, K. 2001a, ApJ, 550, in press (astro-ph/0007010)
Nakamura, T., Umeda, H., Iwamoto, K., Nomoto, K., Hashimoto, M., Hix, R.W., Thielemann, F.-K. 2001b, ApJ, in press (astro-ph/0011184)
Nakamura, T., Umeda, H., Nomoto, K., Thielemann, F.-K., & Burrows, A. 1999, ApJ, 517, 193
Nelemans, G., Tauris, T.M., van den Heuvel, E.P.J. 2000, A&A, 352, 87
Nomoto, K., Suzuki, T., Shigeyama, T., Kumagai, S., Yamaoka, H., Saio, H. 1993, Nature, 364, 507
Nomoto, K., Yamaoka, H., Pols, O. R., van den Heuvel, E. P. J., Iwamoto, K., Kumagai, S., & Shigeyama, T. 1994, Nature, 371, 227
Nomoto, K., & Hashimoto, M. 1988, Phys. Rep., 256, 173
Nomoto, K., et al. 2000, in "Supernovae and Gamma Ray Bursts" eds. M. Livio, et al. (Cambridge Univ. Press) (astro-ph/0003077)
Omukai, K., & Nishi, R. 1999, ApJ, 518, 64
Paczynski, B. 1998, ApJ, 494, L45
Patat, F., et al. 2001, ApJ, in press (astro-ph/0103111)
Podsiadlowski, Ph., Nomoto, K., Maeda, K., Nakamura, T., Mazzali, P.A., & Schmidt, B. 2001, in preparation
Primas, F.,Reimers, D., Wisotzki, L., Reetz, J., Gehren, T., & Beers, T.C. 2000, in The First Stars, ed. A. Weiss, et al. (Springer), 51
Ryan, S.G. 2001, this volume
Ryan, S.G., Norris, J.E. & Beers, T.C. 1996, ApJ, 471, 254
Shigeyama. T., & Tsujimoto, T. 1998, ApJ, 507, L135
Sollerman, J., Kozma, C., Fransson, C., Leibundgut, B., Lundqvist, P., Ryde, F., & Woudt, P. 2000, ApJ 537, L127
Sneden, C., Gratton, R.G., & Crocker, D.A. 1991, A&A, 246, 354

Spite, M., Depagne, E., Nordström, B., Hill, V., Cayrel, R., Spite, F., & Beers, T.C. 2000, A&A, 360, 1077
Thielemann, F.-K., Nomoto, K., & Hashimoto, M. 1996, ApJ, 460, 408
Tsujimoto, T., Nomoto, K., Yoshii, Y., Hashimoto, M., Yanagida, Y., & Thielemann, F.-K. 1995, MNRAS, 277, 945
Tsuru, T. G., Awaki, H., Koyama K., Ptak, A. 1997, PASJ, 49, 619
Turatto, et. al. 2000, ApJ, 534, L57
Umeda, H., Nomoto, K. 2001, ApJ, submitted (astro-ph/0103241)
Umeda, H., Nomoto, K. & Nakamura, T., 2000, in The First Stars, eds. A. Weiss, T. Abel & V. Hill (Berlin: Springer), 121
Umeda, H., Nomoto, K. Tsuru, T., et al. 2001, in preparation
Wheeler, J. C., Yi, I., Höflich, P. A., & Wang, L. 2000, ApJ, 537, 810
Woosley, S.E. 1993, ApJ, 405, 273
Woosley, S.E., Eastman, R.G., & Schmidt, B.P. 1999, ApJ, 516, 788
Woosley, S. E., & Weaver, T. A. 1995, ApJS, 101, 181

THE EFFECT OF BINARIES ON THE CHEMICAL EVOLUTION OF THE SOLAR NEIGHBOURHOOD.

Erwin De Donder and Dany Vanbeveren
Astrophysical Institute, Vrije Universiteit Brussel, Pleinlaan 2, 1050 Brussels, Belgium
ededonde@vub.ac.be; dvbevere@vub.ac.be

Keywords: binaries, population synthesis, supernovae, chemical evolution

Abstract We follow the chemical evolution of the solar neighbourhood (the elements He, C, N, O, Ne, Mg, Si, S, Ca and Fe) using a galactic chemical evolutionary code that includes the most recent results of single and binary star evolution. First we discuss the differences in the overall net yields produced seperately by binary and single star evolution for different initial metallicities. Then we implement binary evolution into a galactic code that simulates the chemical evolution of the solar neighbourhood as function of time. Our galactic code makes it possible to study the evolution in time of the rate of the different types of supernovae.

1. INTRODUCTION

The chemical evolution of the Milky Way has been extensively studied during the last years. New observational data on chemical abundances and abundance ratios in stars and gaseous objects of different type, provide an important set of constraints that have to be fulfilled by a chemical evolutionary model (CEM) (see the contribution of S. Ryan in this volume). On theoretical side, much effort has been done in computing nucleosynthesis yields for stars of all masses and of different initial metallicity Z (e.g. Woosley & Weaver, 1995 hereinafter WW95, Marigo et al. 1998). Yields for an extensive set of isotopes, masses and metallicities are fundamental to make a detailed comparison between theory and observation. Other main ingredients in CEMs are the initial mass function (IMF), the star formation rate (SFR) and the infall rate of (extra) galactic gas which are treated as free parameters.

D. Vanbeveren (ed.), The Influence of Binaries on Stellar Population Studies, 535–555.

Although no unique scenario exists yet for the evolution of the Milky Way, present CEMs that use Z-dependent yields and consider the formation of the Milky Way through infall of primordial and extragalactic gas, are able to reproduce at least qualitatively (within a factor of two to three) the major observational constraints i.e. the age-metallicity relation, the star metallicity distribution, abundance ratios of many chemical elements in halo and disk stars and the radial abundance gradients. This may look quite surprising since these CEMs usually account in detail for single star evolution only whereas the effect of Type Ia supernovae (SN Ia) is studied in most cases by introducing a free parameter (without modelling in detail the evolution of the underlying progenitor population) which is choosen so that the presently observed frequency of SN Ia is reproduced. Remark that SN Ia are major producers of Fe and are thought to be responsible for at least one third of the observed Fe abundance.

Because of the success of these simple CEM, one could be tempted to conclude that except for the iron production through SN Ia, there is no general need for binary evolution to explain the observed average trends in the chemical evolution (for the most abundant isotopes) of our Galaxy. This is obviously ad hoc and to justify or reject this conclusion, detailed binary evolution has to be included into chemical evolutionary calculations. This has been the scope of a project in Brussels. Results are presented here and in De Donder and Vanbeveren, 2001.

First we make a comparative study on the net yields generated by a population of single stars only and by a population of binary stars only. We use the Brussels population number synthesis code that treats the evolution of both binary stars and single stars in detail. Following the stars through all their evolutionary phases, we compute the total net production in helium and metals (C, N, O, Mg, Si, S, Ca and Fe) for one generation of binaries and one generation of single stars.

The next step is to include binary evolution into a galactic code that simulates the chemical evolution of our Galaxy. We use the two infall galactic model of Chiappini et al. (1997) and perform our computations for the solar neighbourhood. For the SN Ia we consider different progenitor models (Sect.2.1).

We also investigate in detail the influence of black hole (BH) formation on our chemical evolutionary computations. It will be demonstrated that theoretical predictions are significantly affected by the BH-formation model.

2. POPULATION NUMBER SYNTHESIS

A description of our population number synthesis code (PNS) is given in the contribution by D. Vanbeveren in this volume. We summarize the main ingredients that are important for this paper .

Initial distributions.

- The single star and binary primary initial mass functions are equal and constant in space and time; they obey the following law: φ(M)$\propto M^{-\gamma}$ with γ=2.3 (Salpeter, 1955) for M $\leq 2M_{\odot}$ and γ=2.7 (Scalo, 1986) for M $> 2M_{\odot}$.

- We use different binary mass ratio distributions ϕ(q), i.e. a Hogeveen type distribution (Hogeveen, 1991), a flat distribution or one that peaks at q=1 (Garmany et al. 1980).

- The initial binary period distribution Π(p) is flat in the log (Popova et al., 1982; Abt, 1983) between 1 day and 10 years.

- We account for asymmetric SN explosions with kicks that are arbitrarily directed in space and have magnitudes that are χ^2-distributed. We perform our calculations with an average kick velocity $\bar{v}_k$ = 450 km/s and 150 km/s.

Evolutionary related PNS input.

- PNS results of stars depend on their evolution. The latter depends on the effect of convective core overshooting (we use the model of Schaller et al., 1992) and for the massive ones also on SW mass loss during the different evolutionary phases (we use the formalisms described by D. Vanbeveren this volume).

- We have an extended library of explicitely calculated evolutionary sequences of single stars and of close binaries till the end of core helium burning (CHeB). Using this library our PNS code can handle the evolution of both single and binary stars for 0 < Z $\leq$ 0.02, and allows to explore a population of binaries with initial parameters $3 \leq M_1/M_{\odot} \leq 120$ (M_1 = primary mass), 0 < q $\leq$ 1 and ($1 \leq P(days) \leq 3650$ and a population of single stars with $0.1 \leq M/M_{\odot} \leq 120$.

- The orbital period evolution of a binary is governed by SW mass loss in the case of massive binaries and by the different binary interaction processes i.e. conservative and non-conservative Roche lobe overflow (RLOF), common envelope (CE), spiral-in and asymmetric SN explosion; in our model the period evolution of each binary is followed in detail. Depending on the process type the following parameters are important:

- β, the fraction of matter, lost by the primary star during RLOF, that is accreted by the secondary star. In case A/B_r systems with q≥0.2 the RLOF occurs completely or partly conservative. In the latter case it is assumed that the matter leaves the system via the second Langrangian point and forms a ring around the binary. In our calculations we assume that $\beta = 1$ when q≥0.4 whereas between q=0.2 and 0.4 β increases from 0 to 1 (we made simulations using different functions)
- α, the efficiency of the conversion of orbital energy into potential energy during CE and spiral-in. We distinguish three different CE phases: non evolved binaries with an initial mass ratio q ≤0.2 evolve through a CE phase, the CE phase of case B_c/C binaries and the spiral-in of a compact component (WD, NS or BH). We use respectively α_1, α_2 and α_3.
- $\bar{\mathbf{v}}_{\mathbf{k}}$, the average kick velocity imparted to the NS during the SN explosion.

BH formation.

- At the end of their life all stars with initial mass larger than some minimum value Min(BH) collapse directly into a BH without a SN outburst. We use Min(BH) = 40 $M_\odot$.
  The final CO cores of single stars with initial mass smaller than 40 $M_\odot$ are larger than those of their primary counterparts of case B binaries. We therefore also made some test calculations where Min(BH) = 25 $M_\odot$ for single stars

- Stars with initial mass smaller than Min(BH) may form low(er) mass BHs. We relate the final CO-core masses of our evolutionary calculations with the computations of WW95 and we use their case B supernova model. To investigate the effect of the SN asymmetry, we use the same kick velocity distribution as for neutron stars.

The standard set of PNS parameters. Our calculations are always performed, unless mentioned, for a standard set of PNS parameters: Min(BH) = 40 $M_\odot$ for binaries and single stars, the case B supernova model of WW95 for stars with initial mass smaller than

Min(BH), α_1=α_2=α_3=1, $\bar{v}_k$=450 km/s, the initial mass ratio distribution of Hogeveen type.

2.1. SUPERNOVA IA PROGENITOR MODEL

It is generally accepted that type Ia SNe are exploding CO WDs. The nature of the binary progenitors is still a matter of debate. We consider here the presently most favoured scenarios i.e. the "single degenerate" (SD) scenario in which the CO WD grows to the Chandrasekhar mass by accreting matter from a main-sequence (MS) or red giant companion (RG), and the "double degenerate" (DD) scenario in which two CO WDs with a combined mass larger than or equal to the Chandrasekhar mass, merge as a consequence of orbital decay by gravitational wave radiation (GWR).

In the SD scenario, for the WD to reach the Chandrasekhar mass hydrogen and helium burning must proceed in a quasi-stationary non-degenerate way on top of the CO core which requires more or less stable mass transfer. Using the strong wind model of Hachisu (1996), Hachisu et al. (1999) propose the following WD binary progenitors of SN Ia:

- the mass of the WD ranges between 0.75 and 1.1$M_\odot$
- the binary has a period P $\sim$ (0.5 – 10) days and the WD companion is a low mass MS star with mass $\sim (2-3.5)M_\odot$ for Z=0.02 or with a mass $\sim (1.8-2.6)M_\odot$ for Z=0.004
- the binary has a period P $\sim$ (70 – 750) days with a RG companion of solar mass.

If the strong wind depends on the Fe opacity, it can be expected that the wind will be less efficient the lower the metallicity (Kobayashi et al., 1998) and fewer SN Ia's will be formed. We performed our calculations with the SD model where the stellar wind is Z-dependent but we also explored the case where the stellar wind is not Z-dependent and we used the SD model holding for Z=0.02 for all Z-values.

The formation of double WD systems mainly depends on the energy conversion efficiency during CE and spiral-in. Once a double WD binary is formed, its further evolution is governed by the GWR model (Landau and Lifshitz, 1958) and it is straightforward to calculate the characteristic timescale when both WDs will merge and eventually produce a SN Ia event.

To illustrate the difference between both scenario's, let us assume a constant star formation rate for the solar neighbourhood ($\sim 1M_\odot$/yr) and an initial binary fraction of 100%. Our PNS model then calculates the present SN Ia rate for both progenitor scenarios separately.

Using the SD scenario with the above given criteria on the WD binary precursors, we find a typical rate of ~0.0016/yr which is the expected observational minimum rate (Cappellaro, 1999; see also these proceedings). The majority of our type Ia's comes from WD+MS systems and form between 0.3 and 1.5 Gyr with a maximum rate at ~0.8 Gyr. Remind that these results where obtained for a standard set of PNS parameters. Other combinations give lower rates, especially when lower values for α are taken and/or when a Garmany q-distribution applies.

With the DD scenario, we produce a rate of ~0.002/yr and ~0.005/yr when using respectively a Hogeveen and flat q distribution with $\alpha_2 = \alpha_3 = 1$. Remark that our flat-q results are very similar to the results of Yungelson et al., present proceedings. The DD model predicts quite a different lifetime of the SN Ia's compared to the SD model, because the majority of the double WD binaries merge by GWR decay on a time scale much longer than 10^8 yr. Only 20% merge within 1.5 Gyrs and 60% within Hubble time.

Both models have pro and contra. The SD model produces a rather low SN Ia rate while the merging of two WDs could result in the formation of a NS (e.g. Saio & Nomoto, 1998). The latter would imply that the amount of ejected matter is smaller than the Chandrasekhar mass making the interpretation of SN Ia observations very difficult.

If the accretion rate is too low or too high for the WD to perform stable burning on its surface and to grow in mass, the WD undergoes either nova outbursts or a CE is formed with an ensuing spiral-in phase. In the latter case, the system may merge before the secondary star turns into a WD. What happens during merging and what kind of object is finally formed is not known. It was suggested by Sparks & Stecher (1974) that if the combined mass (WD+mass of the helium core of secondary) exceeds the Chandrasekhar mass, the merging would lead to a SN explosion with the formation of a NS. However, their calculations do not include hydrodynamical effects during the movement of the WD inside the envelope and may be speculative. Nevertheless, we estimated their number with our PNS code and it is interesting to notice that if they indeed produce SNe that are probably of type II, they can make up a significant fraction (~19%) of the total type II SN population produced by binary evolution. Their zero-age main sequence progenitors are mainly initial short period systems with an intermediate initial mass ratio where the RLOF was partly conservative.

2.2. NET YIELDS PER GENERATION OF STARS.

We first look at the differences between the net yields produced by one generation of binaries only and one generation of single stars only. This means, given a closed system of gas that forms one generation of stars what is the net production of the different considered element isotopes? The net yield $\mathbf{y_i}$ is the mass of the new element i ejected by a generation of stars per unit mass of matter that remains locked into the stars.

For single stars (Maeder, 1992):

$$y_{s,i} = \frac{\int M \cdot Q_i(M) \cdot \varphi(M) dM}{\int w(M) \cdot \varphi(M) dM} \tag{37.1}$$

with $Q_i(M)$ = the mass fraction of a star with mass M that is converted to the element i and is finally ejected and w(M) = the remnant mass of the star with initial mass M.

For binary stars:

$$y_{b,i} = \frac{\int\int\int M \cdot Q_i(M_1, q, P) \cdot \varphi(M_1) \cdot \phi(q) \cdot \Pi(P) dP dq dM_1}{\int\int\int w(M, q, P) \cdot \varphi(M_1) \cdot \Phi(q) \cdot \Pi(P) dP dq dM_1} \tag{37.2}$$

with $Q_i(M, q, P)$ = the mass fraction of a binary system of initial total mass M (including the contribution of both primary and secondary) that is converted to the element i and is finally ejected; $w(M, q, P)$ = the remnant mass of the binary system.

The integration is performed over the total range of single and/or binary systems.

Nucleosynthesis prescriptions. For the individual stellar yields we adopt the following prescriptions:

a. single stars:

Our evolutionary tracks are computed till the end of CHeB. We therefore have to link them to calculations that go beyond CHeB in order to obtain realistic yields. We proceed as follows:

- $1 \leq M/M_\odot < 8$: we use the yields of Renzini & Voli (1981), case A, α=1.5, η=0.33. At first glance this may look inconsistent. We use overshooting tracks till the end of CHeB whereas Renzini and Voli do not account for overshooting. However, the amount of overshooting in our calculations is very moderate and test computations allow us to conclude that the Renzini and Voli yields are more than sufficient for our purposes.

- For the high mass stars ($> 8M_{\odot}$), to be consistent with our PNS model, we first account for the SW yields and then we relate our final CO-core masses with the nucleosynthesis computations of WW95 to estimate the corresponding SN yields. To account for the formation of low mass BHs, we use the WW95 Case B supernova model.

b. interacting binary stars:

For interacting binary stars and more specific accretion and merger stars no detailed nucleosynthesis computations have been done yet and so we have to estimate them. We do this by combining single star yields with binary properties according the following recipe:

- all primaries in interacting binaries lose most of their hydrogen rich layers by RLOF and/or the common envelope process. Part of this matter has been nuclear processed during core hydrogen burning and is enriched in ^{4}He en ^{14}N. Some of these layers are accreted by the secondary, some may leave the binary. In the latter case we account explicitely for the interstellar enrichment
- for the massive CHeB remnant star after RLOF we account for further mass loss by a SW (D. Vanbeveren, present proceedings). Similarly as for single stars we link the primary CO core to the nucleosynthesis results of WW95 to calculate the SN yields and we also use the WW95 case B results
- the remnants after RLOF and/or common envelope of intermediate mass primaries form WDs with a mass $\approx$ the final CO core mass. We assume that the mass layers outside the CO core (mainly helium) are lost by stellar wind
- the element yields ejected by the secondary star depend on its evolution in the binary system. If mass accretion occurred we have to account for the rejuvenation process which depends on the adopted accretion model. We apply the standard accretion model (Neo et al., 1977) when the gas stream flowing through the inner Langrangian point L_1 directly hits the secondary and the full mixing model (Vanbeveren & de Loore, 1994) when mass accretion proceeds via an accretion disk
- the evolution of merged objects has not yet been studied in detail. We treat them as follows:

- the merging of two main sequence stars is simulated by a single main sequence star that has been rejuvenated through accretion and has a mass equal to the total mass of the binary at the moment of contact
- for Thorne-Zytkow objects (TZOs) with a neutron star in the core we use the model of Cannon et al. (1992). In case of a BH in the centre we assume that the secondary star is completely swallowed without ejecting any material. The chemical contribution by WDs that merge with a MS/RG companion is estimated by taking the SN yields of a single star that has a final He core of mass equal to the sum of the mass of the WD and the core mass of the companion star at the moment of merging. This may differ from reality. Therefore, the computed yields should only be taken as indicative

- for the SN Ia we adopt the yields from Thielemann et al.(1986)
- we do not account for the nucleosynthesis of nova outbursts and for low mass binaries ($M_{1,0} < 3\ M_{\odot}$). Their contribution to the chemically enrichment of the considered isotopes is small but may be important for other isotopes like ^{15}N, ^{17}O, etc.

Results. The overall net yields are computed for the most abundant isotopes of the elements He, C, N, O, Ne, Mg, Si, S, Ca and Fe. Figure 1 illustrates typical differences between the overall single star and binary net yield relative to the overall single star net yield for each isotope i $(= (y_{s,i}\text{-}y_{b,i})\ /y_{s,i})$. They are computed for the metallicities Z=0.002 (open circles) and Z=0.02 (filled circles) and for different q-distributions combined with different SN Ia progenitor models. Because of the accumulation of many effects it is not straightforward to disentangle the processes responsible for the differences in the yields of the individual isotopes. Though from evolutionary theory the following arguments are derived:

- primaries of case B type binaries form smaller He and CO cores than corresponding single stars. These primaries will therefore produce less metals than single stars of initially the same mass.
- in case of non-conservative RLOF, part of the material that leaves the system may enrich the ISM in He and N if CNO processed layers were peeled off from the mass losing star during the second phase of the RLOF.
- secondary stars that have been accreting matter from the primary star during RLOF, have increased He and CO core masses resulting

in a larger production of metals. After thermohaline mixing of the accreted matter in the secondary's envelope, the envelope is enhanced in N and a SW during CHB may enrich the ISM with N.

- intermediate mass secondaries may become massive after mass accretion in a Case B type binary
- intermediate mass stars that have evolved into a CO WD may explode after accreting matter from their companion or may merge with the secondary star.

It is not the scope of the present paper to discuss the effect of all processes and parameters in the PNS code. This has been done elsewhere (De Donder and Vanbeveren, 2001). Notice however that the differences between single stars and binaries are always within a factor 2-3.

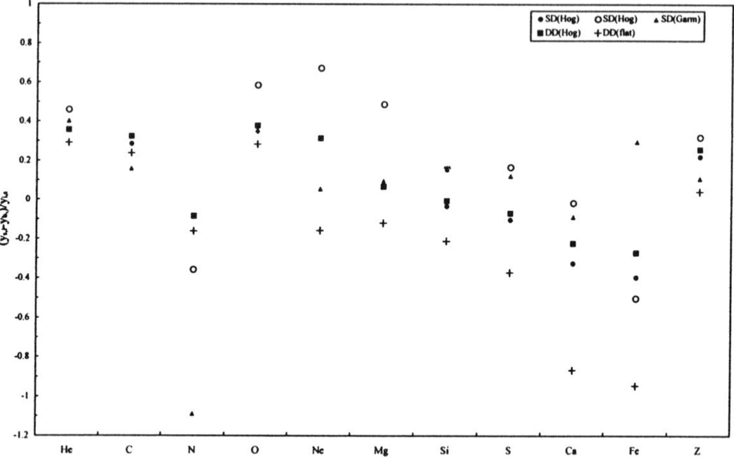

Figure 1 The difference between the overall single and binary star yields, normalized to the overall single star yields for the different considered isotopes. Results are given for the standard PNS model with Z=0.002 (open circles) and Z=0.02 (filled circles), the SD SN Ia model with a Garmany q-distribution (triangles), the DD SN Ia model combined with a flat q-distribution ('+' symbol) and Hogeveen q-distribution (squares).

3. CHEMICAL EVOLUTION

3.1. ADOPTED MODEL

In order to follow the chemical evolution in the solar neighbourhood we use the galactic two-infall model of Chiappini et al. (1997). In this model the Galaxy forms in two phases of major infall, the first corresponding with the formation of the halo-thick disk on a timescale of ~1 Gyr and the second one during which the thin disk is built in ~8 Gyr. The infalling gas has a primordial chemical composition and proceeds at a rate given by :

$$(\frac{dG_i(r,t)}{dt})_{inf} = \frac{A(r)}{\sigma(r,t_{now})} \cdot (X_i)_{inf} \cdot e^{-t/t_I} + \frac{B(r)}{\sigma(r,t_{now})} \cdot (X_i)_{inf} \cdot e^{-(t-t_{max})/t_{II}} \tag{37.3}$$

with $G(r,t)_{inf}$ the normalised surface gas density of the infalling gas, σ(r,t) the surface mass density, t_{now} the present age of the Galaxy (~15 Gyr), t_I and t_{II} the time-scale for mass accretion in the halo-thick disk and thin disk respectively and t_{max} the maximum period of gas accretion onto the disk which is taken to be 2 Gyr, marking the end of the halo-thick disk phase. The index i refers to the identity of the considered chemical element. For the solar neighbourhood (r $= r_\odot \approx 10$ kpc) $t_I = 1$ Gyr and $t_{II} = 8$ Gyr. The quantities A(r) and B(r) are determined from the current surface mass density distribution in the solar neighbourhood as given by Rana (1991). For the star formation rate we use the following relation that is based on energy and dynamic considerations during the formation of the Galaxy (Talbot & Arnett, 1975; Chiosi, 1980):

$$\psi(r,t) = \tilde{\nu} \cdot [\frac{\sigma(r,t)}{\tilde{\sigma}(\tilde{r},t)}]^{2(k-1)} \cdot [\frac{\sigma(r,t_{now})}{\tilde{\sigma}(\tilde{r},t)}]^{(k-1)} \cdot G^k(r,t) \tag{37.4}$$

with $\tilde{\nu}$ a parameter for the star formation efficiency and $\tilde{\sigma}$(r,t) the total surface mass density at given galactocentric distance $\tilde{r}$ (here = 10 kpc) which is used as normalisation factor. If the surface gas density is below 7 $M_\odot/\mathrm{pc}^2$ (Gratton et al., 1996) a star formation stop is assumed.

The evolution of the chemical elements is followed in time in the solar neighbourhood by solving the following set of equations:

$$\frac{dG_i(t)}{dt} = -X_i \cdot \psi(t) + (1-f_b) \cdot \int \psi(t-t_M) \cdot R_{Mi}(t-t_M) \cdot dM + f_b \cdot \int\int\int \varphi(M_{1,0}) \cdot \phi(q) \cdot \Pi(P) \cdot (\psi(t-t_{M_1}) \cdot R_{M1,i}(t-t_{M_1}) + \psi(t-t_{M_2}) \cdot R_{M2,i}(t-t_{M_2})) \cdot dP \cdot dq \cdot dM + (\frac{dG_i(t)}{dt})_{inf} \tag{37.5}$$

with f_b the binary frequency and R_{Mi} the mass fraction of a star of mass M that is ejected into the ISM in the form of element i. In case of a binary system the mass fraction is separated in the primary $R_{M1,i}$ and the secondary $R_{M2,i}$. The ejection time depends on the evolutionary channel of the binary system and is different for the two components unless the binary has merged before the first RLOF. So delayed recycling is considered for both binary and single stars. We also use metallicity dependent stellar lifetimes.

3.2. RESULTS

Our chemical simulations are performed for the standard set of PNS parameters; for the type Ia we separately consider the SD and the DD model. There are presently no clear indications from theory or observations that the binary formation rate f_b may depend on metallicity, star formation rate or other factors. Yet, we consider two cases i.e. f_b increases linear with f_b=0% at Z=0 and f_b=Present Value at Z=0.02 (bin-rate 1), and a binary formation rate that is constant in time f_b=Present Value (bin-rate 2). Accounting for the observed characteristics of the OB type close binary population (B. Mason and W. van Rensbergen, these proceedings) and for statistical biases the present total OB type binary formation rate in the solar neighbourhood may be very large. Therefore we made our calculations for Present Value=70%.

Star Formation Rate. The local Galactic star formation rate is given in Figure 2 which has a shape typical for the two infall model with a star formation treshold. A first short phase of high activity and a much longer second phase of moderately continuous star formation. It predicts a present local star formation rate of ~2.54 $M_\odot/pc^2$/Gyr. Although interacting binaries return less matter to the ISM due to a higher formation rate of NSs and BHs, our computations reveal that this effect is very small in the overall SFR even for a constant binary frequency of 70%.

Age Metallicity Relation. The age metallicity relation (AMR) gives the temporal evolution of the ratio [Fe/H] which is generally considered as a measure of the metallicity of the Galaxy. Figure 3 shows typical predicted AMRs in the solar neighbourhood. During the first few Gyrs, Fe increases due to massive stars. Two processes are important here: the variation as function of Z of the binary formation rate (increasing or constant) and the BH formation. After a few Gyrs the SN Ia appear and become the major Fe-producers. Obviously the results depend on the adopted SN Ia model, but also on the binary mass ratio distribution and on the parameters α which describes the CE phase of binary evolution.

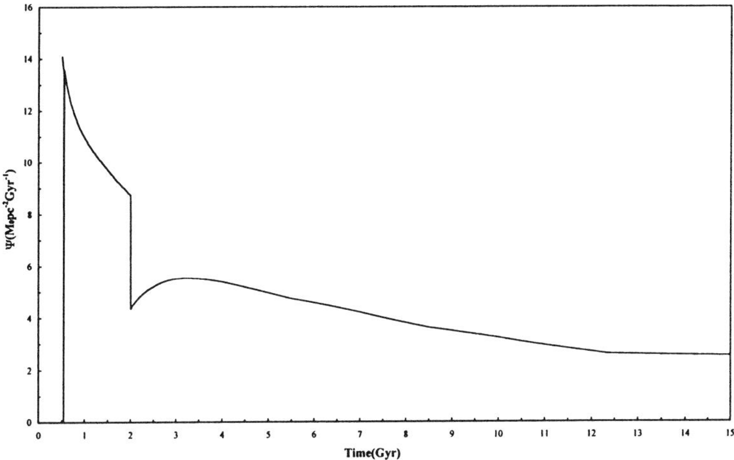

Figure 2 The theoretical predicted star formation rate in the solar neighbourhood as function of time. We skipped in our drawings the effects of the threshold in the formation rate, which mainly appear during the thick disk formation and at the end (last 5 Gyrs) of the thin disk evolution.

In Figure 3 we also plot the available observations. As can be noticed most of the theoretical AMRs predict too much Fe. The worst cases correspond to those where the binary formation rate is constant in time, SN Ia are the result from DD binaries, the mass ratio distribution of unevolved binaries is flat or (even worsts) peaks around unity.

The [O/Fe] - [Fe/H] Relation. We have followed the evolution of the abundance ratios [element/Fe] (=log((X_i/Fe)/(X_i/Fe)$_\odot$) as function of [Fe/H]. Here we restrict ourselves to the [O/Fe] vs [Fe/H] evolution since general conclusions resulting from a comparison with observations are very similar when other elements are considered. Figure 4 shows the [O/Fe] ratio as function of metallicity computed with the standard PNS parameters, the different type Ia progenitor models i.e. SD model without metallicity effect, SD model with metallicity effect and the DD model and this for both binary formation rates. All our abundance ratios (predicted and observed) are relative to the OBSERVED Solar ratio. For the latter we used the value of Anders and Grevesse (1989). In the evolutionary behaviour of [O/Fe] type Ia SNe play a major role as iron producers. They are responsible for a downturn of [O/Fe] at [Fe/H] $\sim$ -0.3 (SD model) (-0.2 for the DD model) after which the ratio strongly declines. Since the time of appearance of type Ia SNe is completely determined by the stellar lifetime of the progenitor binary system (and

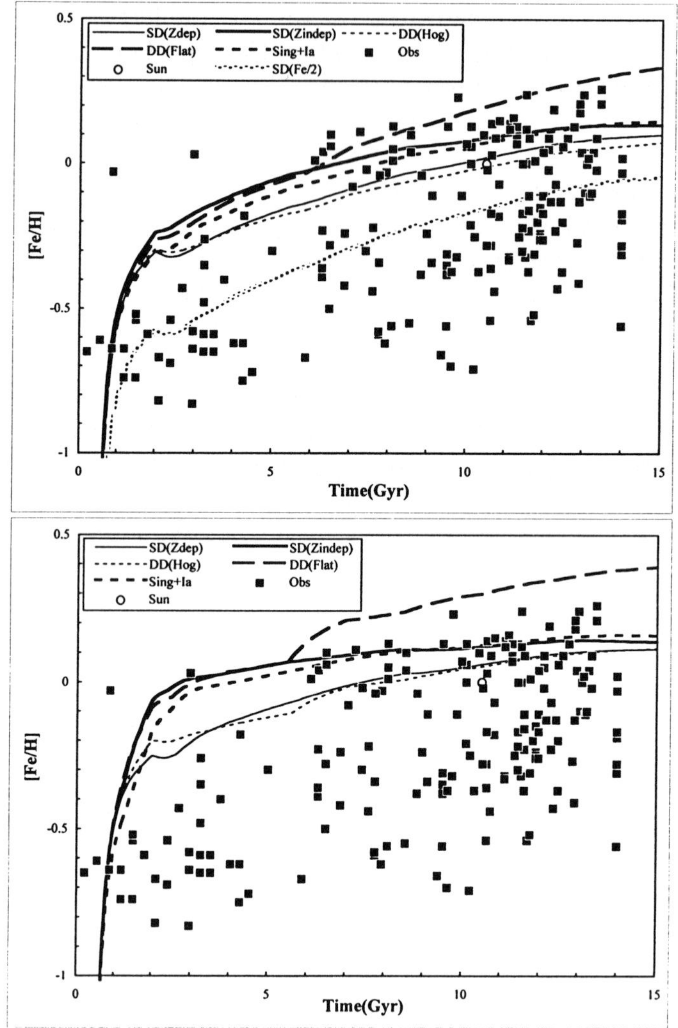

Figure 3 The theoretically predicted age-metallicity relation in the solar neighbourhood. In the top figure we assume a varying binary frequency, in the bottom figure a constant binary frequency = 70% is adopted. We use the standard PNS model with the SD SN Ia model with a Z-dependent stellar wind and with a Z independent stellar wind, with the DD model and with the DD model but with a flat q-distribution. We also show the calculations with the Z-dependent SD model but we lowered the Fe production in massive stars by a factor 2. And finally, the figure shows a model where we account for single stars only but the SN Ia progenitor population is followed in detail according to the Z-independent SD model. The bold points on the graph are observed values of Edvardsson et al. (1993).

also the orbital decay time in case of the DD model) the downturn (if real) can be considered as a constraint on the type Ia progenitor model. Obviously, before the downturn, the [O/Fe]-[Fe/H] relation is governed primarily by the element production in massive stars.

In Figure 4 we also plot the observations listed by Gratton et al. (1996) and we add the recent observations of Boesgaard et al. (1999). First let us remark that when considering all observations, the evidence for a downturn is much less pronounced than is the case in the observations of Gratton et al. (1996) only. Furthermore, the majority of the observations ly above the theoretical predictions (also the solar value at [0,0]). It looks as if the Fe increases too fast relative to O. The difference between theory and observation is also very pronounced before the downturn, i.e. for [Fe/H]$\leq$-0.3. A variable binary formation rate gives slightly better results but the correspondence is still far from good.

We also calculated the [C/Fe] - [Fe/H] relation. Similarly as for O, the C production is too low during the early phases of galaxy evolution and similarly as for O, C is produced primarily by the massive stars during these phases.

Possible reasons for both discrepancies. It is clear from Figure 3 that if we can reduce the Fe ejecta in massive stars the correspondence between the predicted and observated AMR will increase significantly. As noted by Woosley et al. (1995) the explosion and ejecta mechanism itself contains a number of uncertainties. As an example, if we increase slightly the mass of the low mass BHs in the mass range 20-30 $M_\odot$, the amount of iron that is ejected by massive stars could be much smaller. Similarly as was done by Timmes et al. (1995), we calculated the age-metallicity relation assuming that the Fe-production in massive stars is two times smaller. Figure 3 shows the much better agreement between observations and predictions.

Figure 4 indicates that smaller Fe-ejecta may not be the only reason for the [O/Fe]-[Fe/H] relation discrepancy. A second major uncertainty is the $^{12}C(\alpha,\gamma)^{16}O$ reaction rate during CHeB which may significantly affect the amount of oxygen ejected by massive stars. Furthermore, we assumed that all stars with initial mass larger than 40 $M_\odot$ end their life as a BH without SN explosion. This may not be correct. It was concluded by Nelemans et al. (1999) that some matter ejection $\sim (3-4)M_\odot$ is necessary to explain the runaway status of the BH X ray binaries Cyg X-1 and Nova Sco 1994. Both BHs are probably the end product of stars with initial mass larger than $40M_\odot$ so that it can not be excluded that also in the mass range $\geq 40\ M_\odot$ the O-ejecta is significant. We simulated the effect of the latter by assuming that during their collapse into a BH, all stars with mass $\geq 40\ M_{\odot}$ eject 4 $M_{\odot}$. Our CHeB evolutionary calculations of massive stars reveal that ~50% is oxygen and the other 50% is maily carbon. Figure 4 shows the results of the simulation. We conclude that correspondence with observations has improved considerably whereas we also reproduce the Solar values. Interestingly, also the C discrepancy during the early phases of galaxy evolution almost disappears when we account for the processes discussed in this paragraph.

Supernova Rates. The evolution of the local SN rates of different type is given in Figure 5 for the standard PNS parameters, a varying binary formation rate and the SD model for SN Ia's. It is obvious that the history of the SN rates and especially the evolution of the ratios II/Ibc and Ia/total depends on PNS parameters, the binary formation rate and on whether the SD or DD model is used as type Ia progenitors. This is illustrated in Figure 5 as well but for the SN Ia's only. We conclude:

- in the beginning of Galaxy evolution when Z is small, type II SNe dominate. They result from massive stars that have retained their hydrogen envelope after evolution because of the absence of a strong SW. With the formation of binaries and increasing metallicity type Ibc's and Ia's start to form. After 15 Gyrs of evolution we find a total rate of ~1.4 per century of which ~ 14% SNIa, ~ 20% SNIbc and ~ 66% SNII in fair agreement with the observed rates (Cappellaro, present volume1). We remark that we haven't included possible type II SNe coming from WDs that merge with a MS or RG companion. Including them may increase the SN II rate by 20%

- the variation as function of Z of the different types of SN depends crutially on two processes: the effect of Z on the charateristics of the population of binaries (frequency, period and mass ratio distribution) and on the effect of Z on stellar wind in general (type Ia's depend on the stellar wind in accreting WDs whereas type II's and type Ib/c's depend on the stellar wind in massive stars). Supernova studies at high red shift can give more hints on SN rates at low metallicities

- to recover the present SNIa rate with the standard set of PNS parameters, an interacting binary frequency larger than ~(40-50)% is needed among intermediate mass stars

- the standard set of PNS parameters includes a binary mass ratio distribution of Hogeveen type. We made some test computations

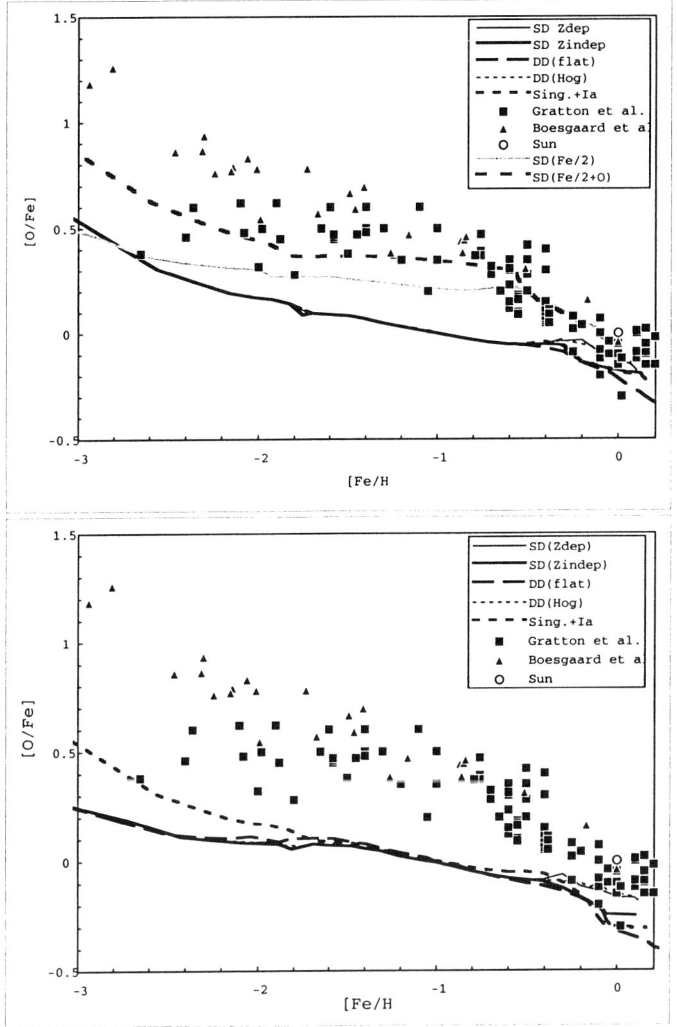

Figure 4 The theoretically predicted [O/Fe] vs. [Fe/H] relation for the different models defined in Figure 3. We also show the simulation where all stars with mass larger than 40 $M_{\odot}$ eject 4 $M_{\odot}$ during the collapse into a massive BH (label SD(Fe/2+O)). Remark the improved correspondence with observations. Similarly as in Figure 3, we also computed a model with single stars only + the population of SN Ia progenitors but the difference with the other models is barely visible.

with a distribution that favors the formation of binaries with equal component masses. The SD model then predicts a SN Ia rate that is a factor 5 at least too small. The DD model works when we lower the present value of the intermediate mass interacting binary frequency to ~(20-30)%.

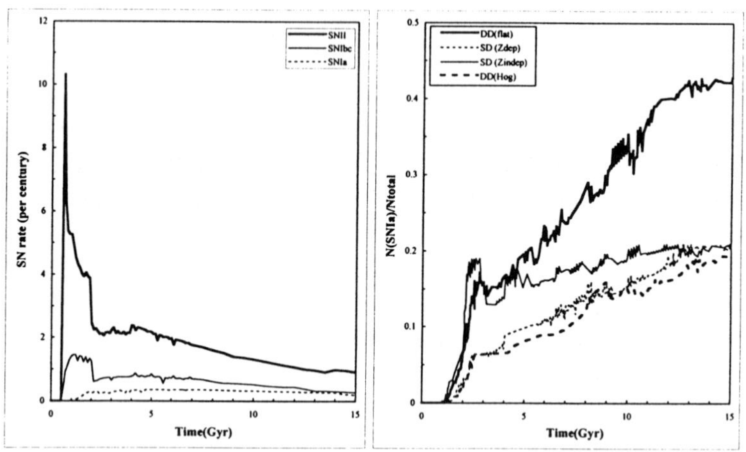

Figure 5 Left: the local supernova rate as function of time for SNII (thick full line), SNIbc (thin full line) and SNIa (thin dashed line). Right: the time variation of the SN Ia rate for different SN Ia progenitor models: the DD model, the SD model with and without the Z-effect on the stellar wind, the standard PNS parameters but we separately consider a flat and a Hogeveen q-distribution.

4. THE EFFECT OF STELLAR ROTATION ON CEM RESULTS

The effect of rotation on the evolution of stars has been discussed by A. Maeder in the present volume. Rotation is essential to explain some observed properties of individual stars but it probably does not change significantly the overall results of present PNS studies (D. Vanbeveren, this volume).

Rotational mixing may be the mechanism that was responsible for the production of primary ^{14}N early in the evolution of galaxies (Maeder, 2001). However, for C, O, Ne, Mg, Si, S, Ca and Fe we are tempted to conclude that it is unlikely that stellar rotation will alter considerably the results of present CEMs. Indeed, CEMs rely on stellar chemical yields and the latter depend on the extend of the convective cores during hydrogen and helium burning. This extend depends on the stars' rotation. As a matter of fact, the cores in rotating stars are on the AVERAGE similar as the cores of non-rotating stars but with a moderate amount of convective core overshooting. Therefore, the yields of a population of rotating stars should be similar to the yields of a population of non-rotating stars with moderate convective core overshooting. The latter are used in most of the present CEMs, i.e. CEMs already accounted for stellar rotation *avant la lettre.*

5. GENERAL CONCLUSION

We have studied the chemical evolution of the most abundant isotopes in the solar neighbourhood with a galactic model that includes the formation and detailed evolution of binaries of intermediate and high mass and we discuss the effects of black hole formation. The latter affects significantly the production of iron, oxygen and carbon in massive stars and controls the timevariation of the chemical elements during the early phases of the evolution of the Galaxy. We propose a realistic model where the predictions and the observations correspond fairly well.

To predict the evolution of the rates of the different types of SN explosions, it is indispensable to include PNS models which follow the evolution of the binary population in detail. The results are entirely ad hoc when binaries are omitted.

Net yields ejected by a stellar population obviously depend on the binary frequency. However, we also performed all our calculations summarized above with a model where all stars are treated as single stars and where we only accounted in detail for those binaries that produce SN Ia's (SD and/or DD model, which obviously still requires a binary PNS model but for intermediate mass binaries only). The resulting evolution is shown in Figure 3 and in Figure 4. We conclude that the differences with the detailed models are rarely larger than a factor 2. Therefore, accounting for

- uncertainties in chemical evolutionary codes that contain a considerable number of parameters
- the uncertainties in overall binary statistics
- the uncertainties in the physics of supernova processes and the resulting yields
- the uncertainties in stellar evolution in general
- the uncertainties in observations

except for the effect of SN Ia's, including in detail the effect of binaries in CEMs is not the only priority. There are still too many uncertainties. To conclude, before we will have really reliable *galactic models*, much more research is necessary in *single star and binary* astrophysics.

References

Abt, H.A., 1983, ARA&A., 21, 343.

Anders, E., Grevesse, N., 1989, Geochim. Cosmochim. Acta, 53, 197.

Boesgaard, A.M., King, F.R., Deliyannis, C.P., Vogt, S.S., 1999, Astron. J. 117, 492.

Brown, G.E., Weingartner, J.C., Wijers, R.A.M.J., 1996, ApJ., 463.

Cannon, R., Eggleton, P.P., Zytkow, A.N., Podsiadlowski, P., 1992, ApJ., 386, 206.

Cappellaro, E., Turatto, M., Tsvetkov, D. Yu., Bartunov, O.S., Pollas, C., Evans, R., Hamuy, M., 1997, A&A., 322, 431.

Cappellaro, E., Evans, R., Turatto, M., 1999, A&A., 351.

Carlberg, R. G., Dawson, P. C., Hsu, T., Vandenberg, D. A., 1985, ApJ., 294, 674.

Chiappini, C., Matteucci, F., Gratton, R., 1997, ApJ., 477.

Chiosi, C., 1980, A&A., 83, 206.

De Donder, E., Vanbeveren, D., 2001, A&A. (in press)

Edvardsson, B., Andersen, J., Gustafsson, B., Lambert, D. L., Nissen, P. E., Tomkin, J., 1993, A&A, 275, 101.

Fryer, C., 1999, ApJ., 522, 413.

Garmany, C.D., Conti, P.S., Massey, P., 1980, ApJ, 242, 1063.

Gratton, R., Carretta, E., Matteucci, F., Sneden, C., 1996, in The Formation of the Galactic Halo, ASP.

Greggio, L., Renzini, A., 1983, A&A, 118, 217.

Hachisu, I., Kato, M., Nomoto, K., 1996, ApJ., 470, L97.

Hachisu, I., Kato, M., Nomoto, K., Umeda, H., 1999, ApJ., 519.

Hogeveen, S.J., 1991, Ph.D. Thesis, University of Amsterdam.

Israelian, G., Rebolo, R., et al., 1999, Nature, 401, 142.

Kobayashi, C., Tsujimoto, T., Nomoto, K., Hachisu, I., Kato, M., 1998, ApJ, 503, L155.

Kippenhahn, R., Ruschenplatt, G., Thomas, H.C., 1980, A&A., 91, 175.

Lorimer, D.R., Bailes, M., Harrison, P.A., 1997, MNRAS, 289, 592.

Maeder, A., 1992, A&A. 264, 105.

Maeder, A., 2001, in *The evolution of the Milky Way*, eds. F. Matteucci and F. Giovannelli, Kluwer Academic Publ.: Dordrecht, p. 403.

Marigo, P., Bressan, A., Chiosi, C., 1998, A&A., 331, 564.

Meusinger, H., Reimann, H. G., Stecklum, B., 1991, A&A., 245, 57.

Nelemans, G., Tauris, T.M., van den Heuvel, E.P.J., 1999, A&A, 352, L87-L90.

Neo, S., Miyaji, S., Nomoto, K., Sugimoto, D.: 1977, Publ. Astron. Soc. Japan 29, 249.

Popova, E.I., Tutukov, A.V., Yungelson, L.R., 1982, Astron. Space Sci., 88, 55.
Rana, N., 1991, ARA&A., 29, 129.
Renzini, A., Voli, M., 1981, A&A., 94, 175.
Saio, H., Nomoto, K., 1998, ApJ., 500, 388.
Salpeter, E.E., 1955, ApJ., 121, 161.
Scalo, J.M., 1986, Fund. of Cosm. Phys., 11, 1.
Schaller, G., Shaerer, D., Meynet, G., Maeder, A., 1992, A&AS., 96, 269.
Sparks, W.M., Stecher, T.P., 1974, ApJ., 188, 149.
Talbot, R.J., Arnett, D.W., 1975, ApJ., 197, 551.
Thielemann, F.K., Nomoto, K., Yokoi, Y.: 1986, A&A., 158, 17.
Vanbeveren, D., De Loore, C.: 1994, A&A., 290, 129.
Woosley, S. E., Weaver, T. A., 1995, ApJ.S.S., 101.

VII

POSTER PAPERS

MULTI-COLOUR CCD PHOTOMETRY OF INTERMEDIATE VISUAL DOUBLE STARS

A. Strigachev[1,2], P. Lampens[1] and D. Duval[1]
[1] *Koninklijke Sterrenwacht van België, Ringlaan 3, B-1180 Brussels, Belgium*
[2] *Institute of Astronomy, Bulgarian Academy of Sciences, Sofia 1784, Bulgaria*
Anton.Strigachev@oma.be, Patricia.Lampens@oma.be

Keywords: binaries: visual - stars: techniques: photometric - astrometry

Abstract We are presently obtaining CCD V, R and I magnitude and colour differences for the individual components of a sample of selected "intermediate" visual double stars (with angular separations between 1 and 15"). Observations were performed with the CCD camera attached to the Ritchey- Chrétien focus of the 2m telescope of the Rozhen National Astronomical Observatory in Bulgaria. We present the photometric and astrometric results for two cases: the 9th mag standard star HD 218587, hereby reported as new visual double, and HIP 104210, a probable optical system with a higher proper motion foreground star.

Observations and results

Accurate magnitude and colour differences of double stars allow to characterize the evolutionary stage of each of the components of the system but such data are seldomly available, even for the brightest and nearest systems. We hereby present multi-colour differential photometric as well as new relative astrometric data of high precision for some selected "intermediate" visual double stars.

We used the V (Johnson) and R, I (Cousins) broad-band filters. Typical exposures were 1, 1.5 and 2 seconds. We have astrometric and photometric results for 15 visual double stars as well as for a field centered on the open cluster M15 for the purpose of astrometric calibration (Le Campion et al. 1996). The primary reductions were done in the usual way using ESO-MIDAS standard routines. Further reduction was performed using the simultaneous profile fitting algorithm developed within the ESO-MIDAS package (Cuypers 1997).

Accurate photometric and astrometric results for two double stars are

D. Vanbeveren (ed.), The Influence of Binaries on Stellar Population Studies, 559–560.

presented in Table 1. We list the epoch, position angle, angular separation and the instrumental differential magnitude in the filters V, R and I. The quoted errors on the mean values are typical for this range of separation (e.g. Cuypers & Seggewiss 2000). We report the detection of a new visual double star, the 9th mag standard star HD 218587. In the case of HIP 104210 we have very good agreement with the last measurement (Mason 1999). There is however a difference in relative position of about 0.7" with the Hipparcos one (ESA 1997). Analysis of all available data in the WDS catalogue suggests that it is an optical pair whose components are moving apart at the relative speed of 0.1"/yr. Using the colour difference between the components we find that the distance of the secondary is probably within 100 pc (A type-primary and F type-secondary). Since the Hipparcos distance of the primary indicates 150 ± 30 pc, we conclude that the secondary is the probable foreground star.

Table 1 Photometric and astrometric results for two visual double stars (Nr is the number of frames).

Identifier	Epoch	Nr	P.A.(°)	Sep.(")	ΔV	ΔR	ΔI
HD 218587	1998.79	8	146.21	3.073	3.08	2.71	2.50
			±0.03	±0.002	±0.02	±0.01	±0.05
HIP 104210	1998.79	6	30.61	3.512	0.72	0.63	0.48
			±0.05	±0.010	±0.02	±0.01	±0.01

Acknowledgments

A. Strigachev gratefully acknowledges a grant received from the Belgian Federal Office for Scientific, Federal and Cultural Affairs (OSTC) in the frame of the project "Multicolour photometry and astrometry of double and multiple stars".

References

Cuypers, J., 1997, Proc. of the Workshop "Visual Double Stars: Formation, Dynamics and Evolutionary Tracks", Docobo, J.A. et al. (eds.), Kluwer Academic, 35

Cuypers, J., Seggewiss, W., 2000, A&AS 139, 425

ESA, 1997, The Hipparcos and Tycho Catalogues, ESA SP-1200

Le Campion, J.-F., Colin, J., Geffert, M., 1996, A&AS 119, 307

Mason, B.D., 1999, The Washington Double Star Catalog of Observations, USNO, Washington, DC (WDS)

SIMULATIONS OF THE X-RAY OUTPUT OF EVOLVING STELLAR POPULATIONS

L. Norci and E.J.A. Meurs
Dunsink Observatory, Castleknock, Dublin 15, Ireland
ln@dunsink.dias.ie, ejam@dunsink.dias.ie

Abstract We describe the design of a population synthesis programme that monitors the X-ray emission from a stellar population as it evolves in time. The emphasis is on strong starformation activity, with an obvious role for massive stars, SNRs and X-ray binaries. The programme is applicable for the studies of extragalactic starburst regions, such as the 30 Dor complex, IRAS galaxies and indeed proper Starburst galaxies.

Progress report

Extensive X-ray studies with recent space missions -notably *ROSAT*- have established the high energy characteristics of various categories of object. We have set out to make use of such extensive databases in order to simulate the X-ray emission from evolving stellar populations. These simulations are meant as a tool for interpreting the X-ray emission of starburst regions, both galactic and extragalactic.

Several studies have shown that the observed X-ray emission from starburst regions is a combination of several contributions. Stars are one such component with X-ray luminosity depending on spectral type, produced by stellar winds or coronal emission. Close binary stars produce X-rays through an accretion process from the primary to the secondary star and supernova remnants are producing X-rays from shocked material. The contribution of a diffuse hot interstellar medium has also to be taken into account.

The evolving stellar populations are simulated by generating a population of stars with a particular metallicity and total mass. These stars are randomly assigned a mass in accordance with the overall mass distribution specified by the Initial Mass Function (IMF). We have chosen a Salpeter IMF as is observed for many starburst regions. A steeper slope might have to be used for the more massive stars. The stars are born either in an instantaneous burst or continuously, which represents two extreme situations. We are using for the stellar evolution input

D. Vanbeveren (ed.), The Influence of Binaries on Stellar Population Studies, 561–562.

the evolutionary tracks of the Geneva group (see Schaller et al. 1992). The theoretical parameters obtained in this simulation are converted to observable quantities by using the bolometric corrections from Lejeune et al. (1997) and the LogT/$(B-V)_o$ relation from Landolt-Börnstein (1992). An X-ray luminosity is assigned to each simulated star, assuming for the late type stars the X-ray luminosity function of the Pleiades cluster (Cassinelli 1994) and for the early type stars the well known relation between X-ray and bolometric luminosity (see Berghöfer et al. 1997). The normal stellar component accounts for a few percents of the total X-ray luminosity.

The major contribution to the X-ray luminosity is expected to come from supernova remnants and evolved binary stars. The number of supernovae produced by a certain region follows naturally from the stellar evolution adopted. Their subsequent contribution to the X-ray luminosity is evaluated using the age and L_X distribution of SNRs in the Large Magellanic Cloud as an example. The contribution of the binary stars is accounted for by including a population of binaries in the simulations. A binary fraction is assumed and the periods of the binary systems are randomly generated. The components of the binary system are left to evolve as for single stars when mass transfer is not taking place. The X-ray emission during mass transfer is computed along with the changing system configuration, using simple analytical recipies. The ISM contribution to the total X-ray luminosity of the region will also be included.

The programme is applicable for the studies of extragalactic starburst regions, such as the 30 Dor complex, IRAS galaxies and indeed proper Starburst galaxies.

References

Berghöfer T. W., Schmitt J.H.M.M., Danner R., Cassinelli, J.P.(1997). *Astronomy and Astrophysics*, 322,167.

Cassinelli J.P.(1994).*Astrophysics & Space Science*, 221, 277.

Landolt-Börnstein (1982). *Numerical data and Functional Relationships in Science and Technology - New series*, Astronomy and Astrophysics, vol2b.

Lejeune Th., Cuisinier F., Buser R. (1997).*Astron. Astrophys. Suppl. Ser.*, 125, 229.

Schaller G., Schaerer D., Meynet G., Maeder A. (1992).*Astron. Astrophys. Suppl. Ser.*, 96, 269.

THE CV MINIMUM PERIOD PROBLEM

John Barker and Ulrich Kolb
Dept. of Physics and Astronomy, The Open University, Milton Keynes, UK.
john.barker@open.ac.uk, u.c.kolb@open.ac.uk

Abstract We investigate the period minimum and missing period spike for CVs.

Minimum period too long ?

In the observed period distribution of cataclysmic variables (CVs) with hydrogen rich donor stars there is a sharp short-period cut-off at 78 minutes. This is thought to arise from a change in the secular $\dot{P}$ from negative to positive when degeneracy begins to dominate the donor star's structure. If mass transfer is driven by gravitational radiation alone, current stellar structure codes place this minimum period typically at $P_{min} \simeq 65$ min.

One way to increase the theoretical minimum period is to increase the mass transfer rate from the secondary, e.g. by increasing the orbital angular momentum (AM) loss rate of the system. We investigate a form of consequential AM loss (CAML), where the AM is lost as a "consequence" of the mass transferred from the secondary. We define the "CAML efficiency", η, as the fraction of the AM the transfered material had on leaving the L_1 point that is removed from the system. Assuming a $0.6 M_{\odot}$ white dwarf we find that P_{min} rises to 70 min if $\eta = 0.95$, close to its maximum value 1 (see Figure 2).

Table 1 Flat parent distribution (respectively parent distribution as in Figure 2, spike model). (bin width is in minutes)

bin width	χ^2_{obs}	*rejection probability*
1	0.987 (0.230)	0.57 ($> 1 - 10^{-4}$)
2	0.972 (0.016)	0.43 ($> 1 - 10^{-4}$)
4	0.791 (0.002)	0.68 ($> 1 - 10^{-4}$)

D. Vanbeveren (ed.), The Influence of Binaries on Stellar Population Studies, 563–564.

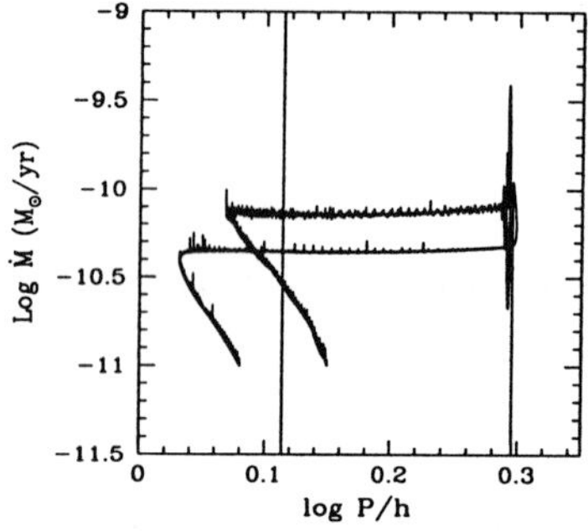

Figure 1 Evolutionary tracks for $\eta = 0$ (lower curve) and for $\eta = 0.95$, generated using Mazzitelli's stellar code (Mazzitelli 1989)

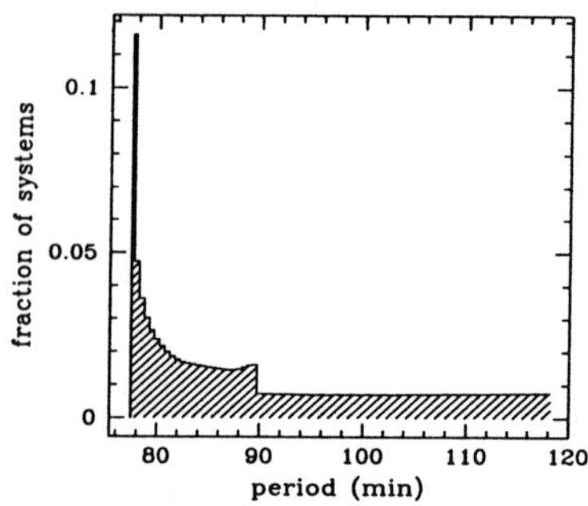

Figure 2 The period distribution generated from a single evolutionary track (the step is an artifact of age restrictions imposed on the model).

No period spike ?

As the likelihood of detecting a system in a given period range is proportional to the time taken for the system to evolve through the range, the probability of observing a system close to P_{min} (where $\dot{P} = 0$) increases, causing a spike in the theoretical period distribution (see fig 2). As shown below such a feature is not found in the observed distribution. We used a χ^2 test to determine the quality of fit for model parent distribution's (PDs) with the observed distribution (Ritter & Kolb 1998 updated) as follows. A given PD, such as the one in Figure 2, is used to generate model samples containing 135 systems (the observed number of CVs with $77.5 \leq P/\text{min} \leq 118$). For each PD we generated 10000 model samples. Each model sample as well as the observed distribution was tested against the PD using a χ^2 test. Table 1 shows, the fraction of model samples with a greater χ^2 value than χ^2_{obs}, the χ^2 value for the observed sample. This fraction is a measure of the significance level of rejecting the hypothesis that the observed distribution is drawn from this PD. We show results for a distribution that is flat in P, and the model distribution shown in Figure 2.

The tables show that the spike model can be rejected as a model for the PD. The observed sample is fully consistent with a flat PD.

References

Mazzitelli, I. 1989, ApJ, 340, 249

Ritter, H., & Kolb, U. 1998. A&AS, 129, 83 (updated)

RESONANT TIDES IN CLOSE BINARIES

B. Willems
Physics & Astronomy, Open University, Walton Hall, Milton Keynes, MK7 6AA, UK
B.Willems@open.ac.uk

Keywords: Binaries, tides, resonances, nonadiabatic effects, orbital evolution

Abstract In this paper, we present the results of a theoretical treatment of the combined effects of a resonance of a dynamic tide with a free oscillation mode and the nonadiabaticity of the tidal motions on the orbital evolution in close binaries.

Basic assumptions

Consider a close binary system of stars that are moving around each other in an orbit with semi-major axis a and eccentricity e. The first star is assumed to rotate uniformly around an axis perpendicular to the orbital plane and the companion is treated as a point mass. We neglect the effects of the Coriolis force and the centrifugal force, so that the star can be considered to be spherically symmetric, with a total radius R_1.

Following Polfliet & Smeyers (1990), we expand the tide-generating potential in Fourier series in terms of multiples of the companion's mean motion. As a result of this expansion, the tidal action of the companion induces an infinite number of forcing frequencies in the star. When one of the forcing frequencies is close to the eigenfrequency of a free oscillation mode, a resonance occurs and the tidal action exerted by the companion is enhanced. Here, we consider the influence of the perturbations of the *energy generation* and the *radiative energy flux* on a resonant dynamic tide.

Orbital evolution

The nonadiabaticity of the tidal motions generates a phase shift between the tides and the tide-generating potential, so that a *torque* is applied to the tidally distorted star. This torque is normal to the orbital plane and causes time-dependent variations of the orbital elements and the

D. Vanbeveren (ed.), The Influence of Binaries on Stellar Population Studies, 565–566.

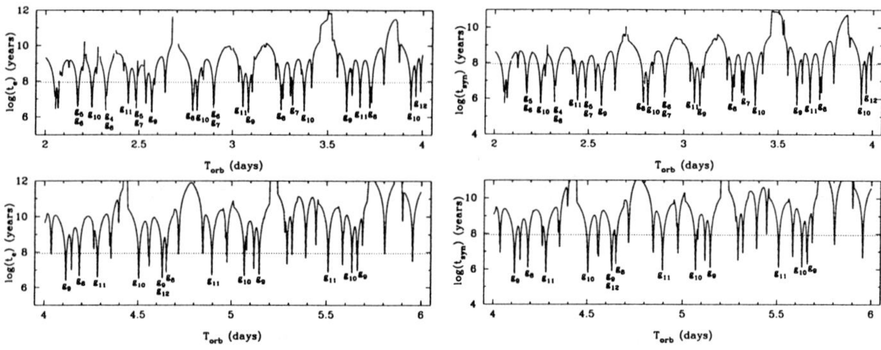

Figure 1 The logarithm of the circularisation (left) and the synchronisation (right) time scale. The dotted horizontal line represents the star's nuclear time scale.

star's angular velocity which may contribute to the circularisation of the orbit and the synchronisation of the star's rotation. The time scales of circularisation and synchronisation associated with a resonant dynamic tide are of the order of $(R_1/a)^{-7/2}$ and $(R_1/a)^{-3/2}$, respectively. These time scales are considerably shorter than those found by Zahn (1977) for the limiting case of small forcing frequencies.

The variations of the circularisation and the synchronisation time scale as a function of the orbital period are displayed in Fig. 1 for a $5M_\odot$ ZAMS star and an orbital eccentricity $e = 0.5$. Numerous resonances occur and lead to time scales shorter than the star's nuclear life time. Several orbital periods even give rise to resonances with two or more oscillation modes. The influence of the nonadiabatic effects during the resonances decreases with increasing orbital periods.

As a possible application, we suggest that, through their nonadiabatic character, resonant dynamic tides may be an efficient mechanism for the circularisation of the orbit in X-ray binaries with short orbital periods.

References

Polfliet, R., Smeyers, P. (1990), *Astronomy and Astrophysics* **237**, 110

Zahn, J.-P. (1977), *Astronomy and Astrophysics* **57**, 383

Acknowledgments

The author thanks Em. Prof. Dr. P. Smeyers for fruitful discussions and Dr. A. Claret for providing the $5M_\odot$ ZAMS model. He also acknowledges the support of the Fund for Scientific Research Flanders and PPARC grant PPA/G/S/1999/00127.

IMPROVED BASIC PHYSICAL PROPERTIES OF THE OE-STAR BINARY V1007 SCO

P. Mayer[1], P. Harmanec[1,2], R. Lorenz[3], H. Drechsel[3], P. Eenens[4], L.J. Corral[5], N. Morrell[6]

[1] *Institute of Astronomy of the Charles University, V Holešovičkách 2, CZ-180 00 Praha 8, Czech Republic*
[2] *Astronomical Institute, Academy of Sciences, CZ-251 65 Ondřejov, Bohemia - Czech Republic*
[3] *Dr. Remeis Observatory, University Erlangen-Nürnberg, Sternwartstrasse 7, D-96049 Bamberg, Germany*
[4] *Departamento de Astronomia, Universidad de Guanajuato, Apartado 144, 36000 Guanajuato, GTO, Mexico*
[5] *Instituto de Astrofísica de Canarias, c/Vía Lactea s/n, E-38200, La Laguna, España*
[6] *Observatorio Astronomico, Paseo del Bosque S/n, 1900 La Plata, Argentina*

mayer@mbox.cesnet.cz; hec@sunstel.asu.cas.cz; drechsel@sternwarte.uni-erlangen.de; eenens@astro.ugto.mx; lcorral@ll.iac.es; nidia@fcaglp.edu.ar

Abstract An analysis of new and published sets of spectroscopic and photometric observations of the astrophysically important eclipsing O+O binary system V1007 Sco in NGC 6231, leads to significantly improved basic physical properties of the system. We arrived at the following preliminary values for masses and radii: $M_1 = 28.4$ M$_\odot$, $M_2 = 29.5$ M$_\odot$, $R_1 = 15.6$ R$_\odot$, $R_2 = 17.1$ R$_\odot$. These values clearly indicate log g of about 3.5 [CGS], therefore the stars are Oe giants, not Of supergiants. Similarity to Be stars follows also from the fact that the Balmer and He I emission in the spectra from the year 2000 is much stronger than in the spectra taken several years ago. We also report the discovery of apsidal motion with a period of 132 years. Our solution implies a constant of internal structure essentially in agreement with its theoretical value and with the age of NGC 6231. We suggest that the masses, which still differ from the evolutionary ones, and the observed variable Balmer and He I emission may be related to the fact that both stars come very close to their respective limits of dynamical stability near periastron, and large mass loss from the system is likely.

D. Vanbeveren (ed.), The Influence of Binaries on Stellar Population Studies, 567.

THE RP-PROCESS ON COMPACT BINARIES: IMPORTANCE OF EXACT PARAMETERS.

J.L. Fisker[a], F. Rembges[a],V. Barnard[b], M.C. Wiescher[b]
[a] *Department of Physics and Astronomy, University of Basel, Switzerland.*
[b] *Department of Physics, University of Notre Dame, Notre Dame, Indiana 46556, USA*

Abstract We investigate the specific influence of exact (p, γ) reaction rates on the rp-process in the intermediate mass range by comparing the rates of the current REACLIB with the rates of new shell model based calculations in the mass range A=44–63, which are believed to be more accurate. Using the case of a type I x-ray burst model we demonstrate that just a few rates reached early in the rp-process may significantly alter the burst profile.

Preliminary studies

The rp-process consisting of a series of successive (p, γ) reactions and β^+-decays powers the explosive hydrogen burning above temperatures of $T_9 > 0.3$ and is believed to occur at several different sites e.g. accretion onto compact objects. Specific knowledge of the reaction path and the dynamics of the system allows one to determine the energy spectrum evolution of the event, which can then be compared to observations. However, in order to determine the spectrum, the rp-process network equations, whose behavior is given by the reaction rates, the corresponding photodisintegration rates, and the beta decay rates along the reaction path, must be solved.

To study the influence of the new shell model (SM) based rates of Fisker *et al.* on the nucleosynthesis large scale network calculations were performed for $\rho = 10^4$ g/cm^3 and T=0.8 GK, T=1.0 GK, and T=1.5 GK with an evolution time of 1000 s to allow comparison with previous calculations of van Wormer *et al.* We comment on the most significant changes in terms of abundances. The SM-rates for lower temperatures is found to increase the ^{48}Cr abundances by $\approx$ two orders of magnitude, and for high temperatures by up to four orders of magnitude, compared to the abundaces calculated with Hauser Feshbach (HF) rates. The abundances of ^{52}Fe and ^{57}Ni on the other hand decrease significantly. This is not due to changes in single rates but is caused by tempera-

D. Vanbeveren (ed.), The Influence of Binaries on Stellar Population Studies, 569–571.

tures T<2 GK the HF predictions for (p,γ) rates on nuclei below ^{56}Ni along the rp-process path being typically several orders of magnitude higher than the SM-rates. For nuclei along the rp-process path above ^{56}Ni, the HF predictions at T<2 GK are generally about one order of magnitude smaller than the SM-rates. Comparison of HF-, SM-, and experimental proton capture rates on stable nuclei however show excellent agreement within a factor of two. This seems to indicate that level densities for proton rich nuclei A<56 are overestimated in the HF calculations. Therefore, the use of SM-rates causes globally slower processing of sd-shell isotopes along the rp-process path material which results in an enrichment of these isotopes correspondingly reducing the abundances in the iron nickel range.

Type I X-ray bursts

Type I x-ray bursts have been associated with the unstable burning on H/He-accreting neutron stars in LMXB systems resulting in regular outbursts of periodic timescales from a few minutes to days. Precise knowledge of the burst spectrum as determined by the dynamics and the rp-process powering the burst would therefore allow these sources to be used as standard candles.

We use an implicit one-dimensional code, which solves the GR hydrodynamical equations suitable for a neutron star. The accretion layer is divided into 150 zones covering the range $\rho = 10^4$–10^7 g/cm^3 with a high grid point resolution in the hot burning zone ($\rho \sim 10^6$g/cm^3). The hydrodynamics is coupled explicitely to a full network of 208 select isotopes between ^{1}H and ^{94}Pd. We have investigated the significance of the new rates by comparing them to the rates of the present REACLIB (see Schatz *et al*) by a series of model runs using various configurations of shell model rates and REACLIB-rates.

Inserting all the new rates an earlier burst is observed. Though the SM-rates above A=50 are generally faster by up to a few factors in about 80% of the cases at the temperatures and densities considered in our simulations, they alone produce relatively small and insignificant changes in the burst evolution, whence they are not sufficiently strong to trigger the burst, though they may reinforce an already triggered burst.

Therefore the reactions mainly responsible for the altered burst behavior are ^{45}V(p,γ), ^{46}V(p,γ), ^{47}V(p,γ), and ^{49}Mn(p,γ). These rates compromise all the new reactions below A=50 in our model network, and they are about a factor two stronger than the corresponding HF-rates, except the ^{49}Mn(p,γ) which is weaker by some 3-4 orders of magnitude in the temperature window of the burst.

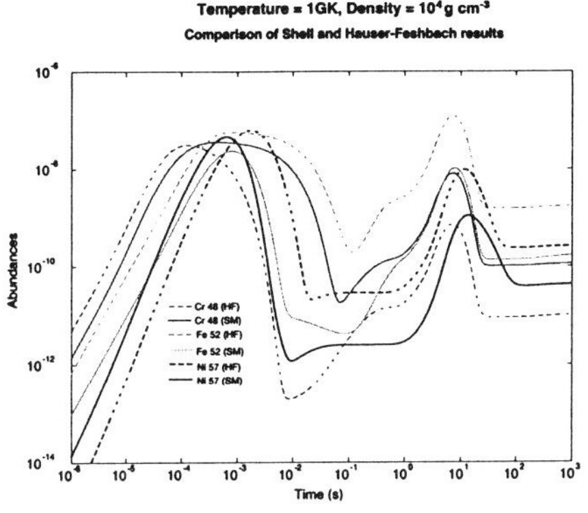

Figure 1 The graph shows the time evolution of the abundances at a temperature of T=1.0GK for the most significant abundance changes due to the shell model rates.

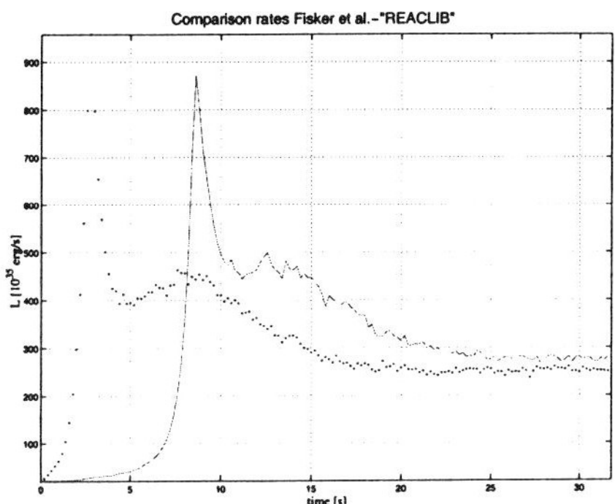

Figure 2 A simulation of an x-ray burst comparing the effect of the REACLIB rates and the shell model rates. An an earlier and more luminous burst is observed in the latter case.

The comparably stronger $^{45}V(p,\gamma)$ reaction makes the burst earlier, but it is not sufficient to carry the burst, whence the $^{46}V(p,\gamma)$ and the $^{47}V(p,\gamma)$ reactions are also important to carry on the burst.

However, the most significant new rate to be introduced is the relatively slow $^{49}Mn(p,\gamma)$ rate – if this is not included, the early burst does not appear! Since the β^+ decay rate of ^{50}Fe is correspondingly faster than the SM-rate by some six orders of magnitude at T_9=1 this effectively blocks the reaction path at this point. This scenario should be compared to the REACLIB, where the corresponding rates have about the same strength at the peak temperature of the burst. Therefore using the new rates, the reaction path must progress through an alternate path through the isotopes above A=50, which tend to react faster, ultimately resulting in a stronger and earlier burst.

In conclusion: As it is indeed possible that some of the Hauser-Feshbach based reaction rates in the intermediate mass segment are overestimated at XRB temperatures and densities (see Fisker *et al.*), we have shown that a more precise determination of specific reaction rates are important, when the network is coupled to the hydrodynamics of the scenario, if precise knowledge of the burst evolution is desired.

References

J.L. Fisker *et al.*, submitted to ADNDT.

L. van Wormer *et al.*, Astrophys.J. **432**, 326 (1994)

H. Schatz *et al.* Phys. Rep. **294**, 167 (1998)

MAGNETIC ACTIVITIES OF NEAR CONTACT BINARIES

Young Woon Kang
Dept. of Earth Sciences, Sejong University, Seoul, 143-747, Korea
kangyw@sejong.ac.kr

Ho-il Kim and Woo-Baik Lee
Korea Astronomy Observatory, Taejon, 305-348, Korea

Kyu-Dong Oh
Dept. of Earth Sciences, Chonnam National University, Kwangju, 500-757, Korea

Keywords: Binary stars, magnetic activity

Abstract The light curves of 19 NCBs, whose X-ray luminosities are greater than those of normal late type main-sequence stars, have been analyzed by the spot model for their magnetic activities. 14 systems have a cool spot while one system has a hot spot on the secondary system. Their chromospheric activities have been investigated by measureing Mg II emissionline of the IUE high resolution spectra.

Introduction

The research of the near-contact binaries (NCBs) is one of the long term projects among Korea Astronomy Observatory, Sejong University Observatory and Chonnam National University. The research includes observations, period variations, light curve analyses, UV and X-ray data analyses, and theoretical works for the NCBs. In this paper we present the magnetic activities of 19 NCBs whose X-ray luminosities are greater than those of normal late type main-sequence stars. First we have analyzed the light curves by the method of Wilson and Devinney Differential Corrections. Then the IUE spectra and X-ray luminosities have been investigated for the selected NCBs.

D. Vanbeveren (ed.), The Influence of Binaries on Stellar Population Studies, 573–574.

Photospheric spot

The light curve of the NCBs belongs to the typical beta Lyr type. The light curves of the most NCBs show a high degree of asymmetry, namely an O'Connell effect is presented. It is generally accepted that a cool star with convective envelope and with rapid rotation has magnetic activity. The magnetic activity causes photospheric spot, chromospheric activity, coronal X-ray, and flare activity. If a star with a significant convective atmosphere is a member of a close binary system, tidal forces can cause the star to spin rapidly and maintain a high level of magnetic activity through out most of its life time.

RS CVn, BY Dra, Algol, and W UMa type binaries are known as chromospherically active binaries that emit large amounts of X-ray radiation from their hot, magnetically heated corona. This is why close binaries with cool, synchronously rotating components are so active and fairly numerous. The majority of the near-contact binaries is one of such binaries. We tested two possible spot models (Kang and Wilson, 1989), a cool spot and a hot spot on the cooler star. The cool spot model is better fitted to the most NCBs light curves.

Conclusion

15 of 19 binary systems required the spot model to fit to their observed light curves. 14 systems have a cool spot while one system has a hot spot on the secondary system. More than half of the systems have X-ray luminosity greater than that of normal late type main-sequence stars. We confirmed the UV excess for only two systems using the chromospheric emission lines in the IUE spectra because most NCBs are too faint to get the IUE spectra. But we can infer the NCBs have quite strong magnetic activity using the asymmetrical light curves and X-ray luminosity.

Acknowledgments

This work was supported by Korea Reseach Foundation Grant (KRF-00-DP-0443).

References

Kang, Y. W. and Wilson, R. E. 1989, AJ, 97, 848.

METAL ABUNDANCE OF THE ECLIPSING BINARY YZ CAS

E. Lastennet
Astronomy Unit, QMW, Mile End Road, London E1 4NS, UK

D. Valls-Gabaud
Lab. d'Astrophysique, UMR CNRS 5572, OMP, 14 av. Belin, F-31400 Toulouse, France

C. Jordi
Dep. d'Astronomia i Meteorologia, Av. Diagonal 647, E-08028 Barcelona, Spain

Abstract We review current and new estimates of the eclipsing binary YZ Cassiopeiae metallicity (Z). Since the individual components cover a quite large range of mass (1.35-2.31$M_\odot$), YZ Cas is potentially one of the best stellar laboratories to understand the structure and evolution of 1 to 2 $M_\odot$ stars. The derivation of Z from IUE spectra, as well as from photometric indices, provides the chemical composition of the atmosphere ($Z_{atmospheric}$), while the fit of evolutionary tracks provides the initial chemical composition ($Z_{initial}$). While a disagreement is expected between $Z_{atmospheric}$ and $Z_{initial}$ because the primary component is an Am star (one expects $Z_{atmospheric}$ to be larger), we find some unexpected discrepancy between atmospheric determinations of Z for this star.

$Z_{atmospheric}$ from IUE spectra

From IUE spectra of the YZ Cas primary, De Landtsheer and Mulder (1983) found that the metals add up to a Z value of over 0.03 (Z=0.036±0.005). We try to assess to what extent this high Z-value has to be revised. It is well known that the abundance of iron is enhanced in Am stars with respect to normal A-F stars. [LM83] found for YZ Cas A that Fe is ~6 times overabundant, which implies that Fe contributes to ~43% of Z. In order to update the De Landtsheer and Mulde study, we applied two corrections to this value: 1) They assumed an iron solar abundance A_{Fe}=7.60. More detailed studies give A_{Fe} = 7.50±0.05 (Grevesse and Sauval, 1999) implying that Fe contributes to ~57% of Z. 2) Their analysis assumed T_{eff}=10,300K. Recent determi-

D. Vanbeveren (ed.), *The Influence of Binaries on Stellar Population Studies*, 575–576.

nations give 9100±300K (Ribas et al. 2000). This would reduce the De Landtsheer and Mulder iron abundance by about 0.15-0.20 dex, and then imply that Fe contributes to ~33% of Z. Both corrections are opposite and imply altogether that Fe contributes to ~40% of Z. Hence, our attempt to update some of the De Landtsheer and Mulder assumptions does not change their high-Z conclusion.

$Z_{\rm atmospheric}$ from photometric calibrations

Since *uvbyβ* Strömgren colours are available for both components of YZ Cas (Lacy, 1981; Jordi et al., 1997), it is possible to derive [Fe/H] from existing calibrations. The [J97] photometric indices do not suggest a metallic behaviour of the Am star atmosphere: [Fe/H] is −0.02±0.22, i.e $Z=0.017^{+0.012}_{-0.007}$ (assuming $Z_\odot=0.018$). The upper limit is only marginally compatible with the Z derived from the IUE spectra. Assuming the individual photometric indices from Lacy (1981), the derived atmospheric [Fe/H] is 0.12±0.05, i.e. Z=0.024±0.003, again only marginally compatible with the IUE determination. Using Strömgren photometry from Jordi et al. (1997) and accurate gravities, Lastennet et al. (1999) derived simultaneous $T_{\rm eff}$-[Fe/H] estimates from the BaSeL photometric calibrations (Lejeune et al., 1998). They obtain [Fe/H] strictly lower than 0.1 for YZ Cas A, i.e Z<0.022. This result assumes no reddening which is justified by experiments we performed with the BaSeL models (see Lastennet et al. (1999) for details on the method) for E(b−y) ranging from −0.034 (*negative* value of [L81]) to 0.09 (maximum value quoted by De Landtsheer and Mulder). In summary, even if the results of the present paper and Lastennet et al. (1999) are not completely independent because based on the same colour indices, they suggest that the atmospheric metallicity of YZ Cas A is solar or sub-solar, and exclude Z-values greater than 0.03.

References

De Landtsheer A.C., Mulder P.S. 1983, A&A 127, 297
Grevesse N., Sauval A.J. 1999, A&A 347, 348
Jordi C., Ribas I., Torra J., Giménez A. 1997, A&A 326, 1044
Lacy C.H. 1981, ApJ 251, 591
Lastennet E., Lejeune Th., Westera P., Buser R. 1999, A&A 341, 857
Lejeune Th., Cuisinier F., Buser R. 1998, A&AS 130, 65
Ribas I., Jordi C., Torra J., Giménez A. 2000, MNRAS 313, 99

RELATIVISTIC EFFECTS TO NEUTRON STAR MERGER EJECTA

R.Oechslin and F.K.Thielemann
Department of Physics and Astronomy, Klingelbergstr. 82, CH-4056 Basel, Switzerland
Roland.Oechslin@unibas.ch

Abstract We perform 3D smooth particle hydrodynamics (SPH) simulations to investigate the effects of general relativity to the amount of ejected matter in a neutron star merger event. General Relativity (GR) is approximated by the conformal flat approximation which imposes a simplified metric. As Equation of State (EoS), we use a polytrope with adiabatic index $\Gamma = 2.6$.

Formulation

To evolve the matter and field distribution in time, we approximate GR with a conformal flat approximation: While we still solve the fully relativistic hydrodynamic equations, the equations for the field distribution are simplified. For the details of the formulation, see Shibata et al. (1998) and Baumgarte et al. (1998).
On top of our conformal flat formalism, we account for gravitational radiation reaction. We add a radiation reaction force which is accurate to 1PN order, see Blanchet (1997). The mass quadrupole is calculated according to Wilson et al. (1996), the mass-current quadrupole is calculated to Newtonian order.

Merger Calculations

The starting point of our merger calculation is a corotational equilibrium configuration. We choose Γ=2.6 for the adiabatic index, M_0=1.6$M_\odot$ for the rest-mass of one star and d_0=30 km for the initial binary separation. (By eliminating the polytropic constant K out of the system of equations and nondimensionalizing these quantities, we get $\bar{M}_0 = K^{-n/2}M_0$=0.0490 and $\bar{d}_0 = K^{-n/2}d_0$=0.612, where $n = 1/(\Gamma - 1)$ is the polytropic index.). The final state is a dense, rapidly rotating central object with elongated spiral arms that contain potentially unbound matter.

D. Vanbeveren (ed.), The Influence of Binaries on Stellar Population Studies, 577–578.

Ejecta

To obtain a relativistic criterion to distinguish bound from unbound matter we assume that the spiral arms are surrounded by a stationary metric generated by the dense central object. We also neglect the pressure gradient force compared to the gravity force in this region. In this case, we can find a conserved energy density

$$\tau_{stationary} = \rho^* \left(v_i \tilde{u}_i + \frac{\epsilon}{u^0} + \frac{1}{u^0} - 1 \right). \tag{46.1}$$

(For the signification of the notation, see Shibata 1998).
Similar to the Newtonian case, $\tau_{stationary}$ is zero for a piece of matter that just manages to escape the system. Therefore imposing $\tau_{stationary} > 0$ gives us a criterion for matter being unbound. The Newtonian limit of this criterion is the usual expression $e_{kin} + e_{int} + e_{pot} > 0$.

We apply criterion (46.1) to our simulation and register the total amount of unbound matter. It turns out however, that no matter at all gets ejected within the simulation timescale that reaches to about 2 ms beyond the gravitational radiation luminosity peak. Therefore, we expect that the case considered will eject no matter at all. This stays in contrast to the analogue Newtonian merger case where about $M_{ej} = 1.5 \cdot 10^{-2} M_\odot$ are ejected (see Rosswog 1999).

References

Shibata, M., Baumgarte, T., Shapiro, S.L., 1998, Phys.Rev.D, 58, 023002

Baumgarte, T., Cook, G.B., Scheel, M.A., Shapiro, S.L., Teukolsky, S.A., 1998, Phys.Rev.D, 57, 7299

Blanchet, L., 1997, Phys.Rev.D, 55, 714

Wilson, J.R., Mathews, G.R., Marronetti, P., 1996, Phys.Rev.D, 54, 1317

Rosswog, S., Liebendörfer, M., et al., 1999, A&A, 341, 499

ON THE FORMATION OF OXYGEN-NEON WHITE DWARFS IN BINARY SYSTEMS

Pilar Gil–Pons[1] & Enrique García–Berro[1,2]
[1] *Departament de Física Aplicada, Universitat Politècnica de Catalunya*
[2] *Institute for Space Studies of Catalonia*
pilar,garcia@fa.upc.es

Abstract The evolution of a star of initial mass 10 $M_\odot$ and solar metallicity in a close binary system (CBS) is followed from its main sequence until an ONe white dwarf forms.

Introduction

The evolution of isolated intermediate mass stars has been carefully studied in a series of recent papers (Ritossa et al., 1999; and references therein). Close attention was paid to all the relevant phenomena such as the inward advance of the thermonuclear flame, the composition of the remnant core, the thermal pulses and the dredge-up processes. Here we follow the evolution of a 10 $M_\odot$ star in a CBS, aimed to understand the formation of ONe white dwarfs in cataclysmic variables.

Results and discussion

The presence of a close companion induces a series of Roche lobe (RL) overflows and has a considerable influence in the evolution of our model when compared to the evolution of an isolated star of the same mass. The first mass loss process begins when the primary star becomes a giant and its radius reaches the Roche lobe (RL) radius. We have assumed that the RL of the primary remains constant during this phase ($\sim 210 R_\odot$), and let $\dot{M}$ change accordingly. As a convective envelope forms and moves deeper, the mass loss rates reach high values ($\dot{M} \sim 10^{-3} M_\odot$ yr^{-1}), and a common envelope forms. This, in turn, has two important consequences: first, most of the hydrogen rich envelope is lost during the first RL overflow, and second, a large amount of the orbital angular momentum is lost as well. The values of $\dot{M}$ combined with the requirement of stable accretion onto the secondary determine the ratio between the

D. Vanbeveren (ed.), The Influence of Binaries on Stellar Population Studies, 579–580.

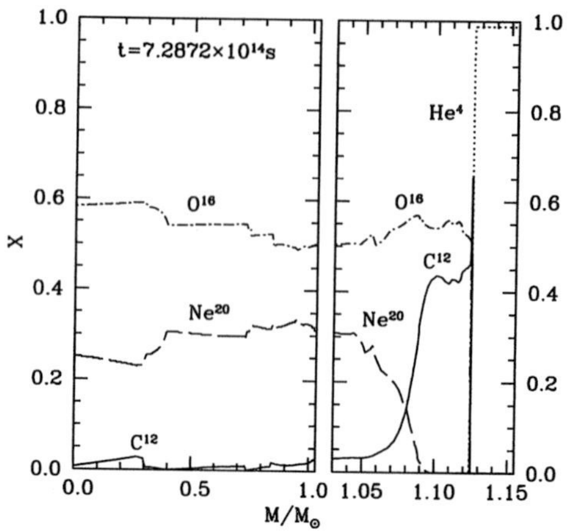

Figure 1 Abundance profiles of the remnant core.

mass accreted by the secondary and the mass lost by the primary ($\beta \sim 0.1$), which means that the episode is highly non-conservative.

The second mass loss episode starts after core He-burning ceases and the He-burning shell is established. During this process, the rest of the hydrogen rich envelope and part of the helium envelope are lost. The values of $\dot{M}$ are $\sim 10^{-6}\ M_\odot\ \mathrm{yr}^{-1}$ except during a initial phase where the small remaining hydrogen envelope is lost and just after the very beginning of carbon burning, where $\dot{M} \sim 10^{-5}\ M_\odot\ \mathrm{yr}^{-1}$. Thus, it is reasonable to assume that $\beta \sim 1$. This is furthermore supported by the fact that luminosity associated to accretion is considerably smaller than the Eddington luminosity. Mass loss stops when carbon burning starts.

The resulting carbon rich nucleus is smaller than in the case of an isolated star ($M_c = 0.86 M_\odot$). Thus, carbon ignition starts off-center at a larger mass coordinate ($0.38 M_\odot$). As far as the nucleus is concerned, during this phase the evolutionary differences with the isolated model are minor. It is, however, important to mention that, even though a convective envelope forms, it does not go deep enough to produce a third dredge-up. After the bulk of carbon burning takes place, steady carbon burning starts in a shell at approximately $1.08 M_\odot$ from the center. The He-burning shell continues active at $1.12 M_\odot$, very close to the base of the envelope. There are not thermal pulses due to the alternate burning of these shells, and carbon luminosity merely decreases. Figure 1 shows the final abundance profiles of the remnant core, which are of importance for understanding the nucleosynthesis in novae.

References

Ritossa, C., García–Berro, E., & Iben, I., Jr., 1999, ApJ, **515**, 381

BINARY STATISTICS FROM HIPPARCOS

Staffan Söderhjelm
Lund Observatrory
staffan@astro.lu.se

Introduction

An obvious use of the Hipparcos data (ESA, 1997) is for making improved binary statistics for the stars close to the sun. The most straightforward way of using the data is to use only reasonably complete samples, but with a completeness-limit at around magnitude 8, this gives small spatial volumes and star numbers. Not many such studies have so far been published, but there are strong indications (Quist and Lindegren 2000, Söderhjelm 2000) that the 'standard' binary frequency given by Duquennoy and Mayor (1991) should be increased significantly.

Principles

A more efficient use of the data consists of extensive modelling of both the Galaxy and its binaries and of the Hipparcos observations. In principle, we simulate the Hipparcos observations of a model Galaxy, compare it with the actual HIP data, and then adjust the model until the fit is reasonable. For consistency, we model the local Galaxy in both time and space, synthesizing the present stellar content (including the binaries) by following the creation and subsequent evolution of all the stellar generations over the whole ($\sim$ 12 Gyr) history of the galactic disk. In this way, a thing like the fraction of white dwarf binary companions is defined by the model, but only indirectly through (probably variable) Initial Mass Functions and Star Formation Rates.

Present Status

Such population synthesis modelling is in progress, and we have verified that the (uncertain) choice of background (single-star) Galaxy model does not greatly affect the differential results of changes in the binary distribution functions. For the present, the observational constraints come mainly from observations of resolved (HIP 'C') and astrometric

D. Vanbeveren (ed.), The Influence of Binaries on Stellar Population Studies, 581–582.

(HIP 'O' and 'G') binaries, and e.g. the binary frequency is obtained for lg P(yr) in the intervals 0-1 and 2-4. The mass-ratio distribution for the wider pairs follows mainly from the Δm-distribution of the resolved binaries, and there are indications of an interesting peak around q=1 (Söderhjelm, 2000). Ideally, we should include a reasonably realistic evolution also for close (mass-exchange) binaries, getting a handle on their initial distributions from the observed eclipsing binaries. As shown at this meeting (e.g. P. Eggleton present proceedings), the number of qualitatively different evolutionary paths is however uncomfortably large, and the Hipparcos-data may in the end be insufficient to disentangle them.

Prospects for the future

Even if the Hipparcos data are interesting in their own right, the methods have been developed with more extensive future space astrometry surveys in mind. The model Galaxy has to be extended to fainter magnitudes/larger distances, and the instrument properties and observing strategy have to be specified in detail. The use of CCD-detectors gives a vast improvement in star numbers and overall accuracy as compared with Hipparcos even for the 'miniature' DIVA project (Röser, 1999). The FAME satellite (Horner et al., 1999) will observe approximately the same stellar sample at about the same epoch (2004-), but with several times higher precision. The GAIA project (ESA, 2000) is still an order of magnitude more ambitious, as regards both the accuracy and the number of objects observed, but it may not fly before 2012. The present work was started in order to clarify the astounding binary star observing capabilities of GAIA, but realistic double star modelling is an important part of the preparations for all these projects. The post-mission results are yet distant, but they are bound to confirm the important role of binaries and multiple stars in the history of our Galaxy.

Acknowledgment

This work is supported by the Swedish National Space Board.

References

Duquennoy A., Mayor M. 1991, Astron.& Astrophys. 248, 485
Horner S.D. et al. 1999, ASP Conference Series vol 194, p. 114
Quist, F., Lindegren, L. 2000, Astron.& Astrophys. 361, 770
Röser, S. 1999, ASP Conference Series vol 194, p. 121
Söderhjelm S. 2000, Astron. Nachr. 321, 165